ALLE ZEIT WACH 1842

V. S. Kirpichnikov

Genetic Bases of Fish Selection

Translated by G. G. Gause

With 81 Figures

Springer-Verlag
Berlin Heidelberg New York 1981

Prof. VALENTIN S. KIRPICHNIKOV
Institute of Cytology
Academy of Sciences
Leningrad F-121/USSR

Translation of: Geneticheskiye osnovy selektsii ryb

ISBN-13:978-3-642-68162-2 e-ISBN-13:978-3-642-68160-8
DOI: 10.1007/978-3-642-68160-8

Library of Congress Cataloging in Publication Data. Kirpichnikov, Valentin Sergeevich. Genetic bases of fish selection. Translation of: Geneticheskie osnovy selektsii ryb. Bibliography: p. Includes index. 1. Fish breeding. 2. Fishes – Genetics. I. Title. SH155.5.K5713 639.3 81-14335 AACR2

Softcover reprint of the hardcover 1st edition 1981

2131/3130-543210

Preface

Fish resources in natural water bodies are tending to decrease due to intensified fishing, the extensive construction of hydropower plants on rivers, and the pollution of seas and freshwater basins by industrial and agricultural wastes. Nowadays only artificial fish rearing can meet man's requirements in fish products.

Fish breeding is still very young as compared to plant breeding and animal husbandry. Although fishes have been reared artificially since ancient times in certain Asian countries, this usually included the cultivation of embryos and larvae caught in rivers and lakes. Among the exceptions, only the common carp *Cyprinus carpio* and the domesticated variety of the crucian carp, the goldfish *Carassius auratus*, which were cultivated in the East, may be mentioned. Common carp breeding began in China about 2000 years ago but was later banned by one of the emperors and started again only relatively recently. The goldfish has been cultivated for decorative purposes for about 1000 years. Many remarkable varieties of the goldfish have been developed in China and later in Japan.

The first improved breeds (German "races") of the common carp known in Europe appeared after the domestication of the Danube wild carp in the seventeenth and eighteenth centuries. Local breeds of the carp were probably established somewhat later in China, Japan and Indonesia; even now these breeds have only minor differences as compared to their ancestor, the Asian wild carp.

Several other species of freshwater fishes were domesticated at the beginning and in the middle of the twentieth century. These include the tench (*Tinca tinca*), the crucian carp (*Carassius carassius*, Cyprinidae); the rainbow and the brown trout (*Salmo gairdneri* and *S. trutta fario*), the brook and the lake trout (*Salvelinus fontinalis, S. namaycush*, Salmonidae), the gouramy (*Osphronemus gourami*, Osphronemidae) and a few others. Many strains and breeds of aquarium fishes such as the platy, the swordtail, the guppy, the fighting and paradise fishes and some others were created by aquarium breeders for decorative purposes.

The intensive work aimed at the domestication of new fish species for pond fish farms began only during the second half of this century. Successful breeding of the three fishes belonging to the so-called Chinese flatland complex: the grass carp (*Ctenopharyngodon idella*), the silver carp and bighead (*Hypophthalmichtys molitrix, Aristichthys nobilis*) have been developed in many countries. Phytophagous grass and silver carp are of particular interest because their cultivation markedly shortens the food chains in water bodies and decreases the expenditure on food. Unisexual strains of the silver crucian carp (*Carassius auratus gibelio*) have been domesticated. The cultivation of several perch species belonging to the genus *Micropterus* such as *M. salmoides, M. dolomieu,* as well as *Lepomis gibbosus,* has been mastered. *Ictalurus* species and the buffalo (*Ictiobus cyprinellus* and others, Catostomidae) have now also become important in fish pond breeding. The lake whitefishes such as *Coregonus peled* and *C. nasus* have been domesticated in the U.S.S.R. The methods of rearing the perch pike *Lucioperca lucioperca,* the sheat-fish *Silurus glanis* and several tilapia species (*Tilapia mossambica* etc.) have been improved. Attempts have been made to domesticate large Indian carps belonging to the genera *Catla, Labeo* and *Cirrhina* as well as the sturgeon hybrids (*Huso huso* × *Acipenser ruthenus*).

The domestication of a fish species is inevitably followed by selection involving the change of heredity of the cultivated species and the development of breeds adapted to life under new environmental conditions.

Selection is also necessary in the work aimed at replenishing fish resources in natural water bodies. This is true primarily of anadromous salmonids and sturgeon growing in seas and oceans, but spawning in freshwater basins. Partial or even complete control of the reproduction of such fishes is feasible and frequently necessary. Generally, however, only some populations of a given species are used for reproduction; this results in a change in its genetic structure and not infrequently leads to the genetic deterioration of the species as a whole. Therefore selection measures become particularly important in the work with anadromous migrating species. The involvement of selection and genetics is also essential for freshwater and marine fish species, the reproduction of which is not controlled by man. The establishment of marine fish farms, restricted areas of the sea suitable for the cultivation of fishes, invertebrates and algae will make selection work with several marine species possible in future.

Finally, geneticists are playing a highly important part in the work aimed at planning the catch, prevention of the harmful consequences of selection caused by fishing (catch of the best part of the population) and of overfishing.

This book mainly deals with fish genetics, selection problems are considered only in the last chapter. Such distribution of the material appears to be justified, since in many countries the work on fish selection is predominantly based on the individual experience and intuition of the selectionist rather than on precise genetic knowledge.

Meanwhile, modern selection should have a solid genetic foundation in order to be effective. The selective breeding of fishes should take into account both the genetic regularities common to all organisms and the results of the special genetic study of the species subject to breeding. As far as I know, no single sufficiently complete review dealing with the genetics and selection of fishes has been published, only a few popular brochures have appeared. This book is the first attempt to summarize the information in the field of fish genetics and the genetic bases of fish selection, the information that has been accumulated in world literature over a relatively short period of time. This literature is very vast and I do not expect to cover all the papers published. Nevertheless, I have done my best to bring into the picture all the most significant studies and to summarize their results.

As well as the data derived from the literature, I have widely used the materials collected by myself, my associates and students over more than 40 years. I take this opportunity to acknowledge my colleagues who have participated in these studies, working both at scientific institutions and on state fishfarms. It is a particular pleasure for me to acknowledge the contribution made by my oldest associates E. I. Balkhashina and K. A. Golovinskaya, in co-operation with whom I started a genetic and selection investigation of fishes. The development of the Ropsha and the Krasnodar carp breeds would have been impossible without the participation of a large team of scientists and fish breeders among whom I wish specifically to mention A. G. Konradt, M. A. Andriyasheva-Nikitina, R. M. Tzoy, K. V. Kryazheva-Ponomarenko, A. S. Zonova, A. M. Sakharov, M. K. Tchapskaya, E. S. Slutzky, G. I. Djakova, L. A. Shart, Yu. I. Iljasov and many others. The writing of this book has been greatly aided by N. B. Cherfas who prepared, at my request, the chapter dealing with gynogenesis and hybridogenesis of fishes.

I am deeply and sincerely grateful to G. V. Sabinin, T. I. Faleeva, N. A. Butzkaya, Yu. L. Gorotschenko, Ya. V. Barchiene, T. I. Kaydanova and V. Ya. Katasonov for submitting original cytological preparations and photographs. Some of the illustration were prepared by my daughter O. V. Kirpichnikova.

The book *The Genetic Bases of Fish Selection* was published in the U.S.S.R. in 1979 in the Russian language and was therefore not accessible to most Western readers. This extensively revised and greatly expanded edition contains much new information on the genetics and selection of fishes. I hope that the book will be useful not only for specialists, but for all those who are interested in the general problems of genetics, selection and the evolution of animals.

Leningrad, Summer 1981 V. S. KIRPICHNIKOV

Contents

Chapter 1

The Material Bases of Heredity in Fishes. The Structure of Chromosomes

The chromosomes, small bodies located in the cell nucleus, represent the most important cellular structures responsible for the hereditary transmission of different traits. After specific staining they can clearly be seen during mitotic cell division, particularly at the metaphase. Both the size and the shape of chromosomes differs in different fish species. One can distinguish three or four types of chromosomes (Fig. 1):

1. the acro(telo)centric chromosomes: the centromere (region attached to the spindle filaments during mitosis) is located very close to one of the ends of the chromosome; [1]
2. the subtelocentric chromosomes: the centromere is located close to the end of the chromosome, but the short arm may be seen distinctly;
3. the submetacentric chromosomes: the centromere is located close to the middle of the chromosome; the two chromosomal arms are, however, of unequal length;
4. the metacentric chromosomes: the chromosomes with arms of equal length and strictly median location of the centromere.

In certain species of fishes the chromosomal complement (karyotype) visible during the metaphase is only represented by rod-like chromosomes, moderate in

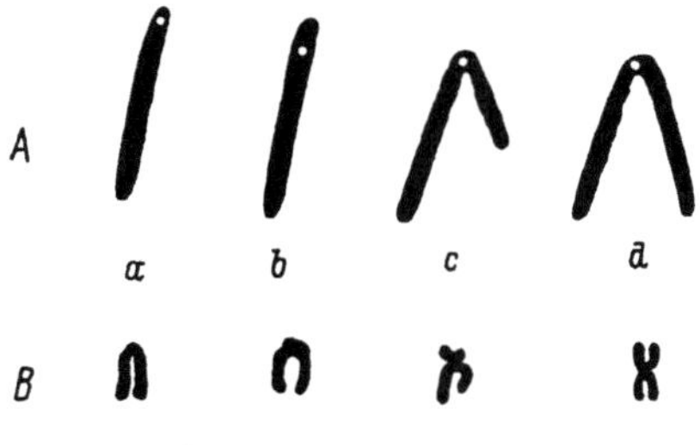

Fig. 1 A, B. Most common types of chromosomes in fishes. **A** the scheme, **B** actual chromosomes of *Megupsilon aporus* (Cyprinodontidae); chromatids are connected in the centromeric region (Uyeno and Miller 1971). *a* acro(telo)centric (t); *b* subtelocentric (st); *c* submetacentric (sm or sM); *d* metacentric (m or M)

---

1 According to some authors, telocentric chromosomes are defined as chromosomes that are completely devoid of a second arm following the terminal centromere, in contrast to acrocentric ones, which have a very small arm. The classification of chromosomes into four morphological groups is made nowadays in most cases according to Levan et al. (1964), that is on the basis of the length ratio of chromosomal arms

size and having an acrocentric or subtelocentric structure; in other species the karyotypes consist exclusively of larger metacentric or submetacentric elements. Most frequently, however, fish chromosomal complements contain chromosomes of two, three or even all four types. In a number of teleostean fishes, but particularly frequently in the cartilaginous ones (sharks and skates) as well as in cartilaginous and bone ganoids (Acipenceridae, Amiidae), additional very small "microchromosomes" have been found alongside the large ones; generally it is very difficult to quantitate such chromosomes (Ohno et al. 1969 b; Nygren and Jahnke 1972 b; Fontana and Colombo 1974). Prokofieva (1935) has found chromosomes with "satellites" among other chromosomes in salmonids. In such chromosomes small regions are separated from the main body by narrow strangulations.

It can be demonstrated that each chromosome contains two identical parallel structures or chromatids. Each chromatid in its turn consists of one or several thin filaments, chromonemata or genonemata containing characteristic condensed stainable regions called chromomers. The chromonema represents a very long double filament helically wound in a dividing cell and despiralled to a large extent in the nuclei of non-dividing "resting" cells during the interphase. In such resting cells long thin chromonemata fill the nucleus completely, forming a complex network, which is poorly visible under the optical microscope; one or two nucleoli can be distinguished in this network. The network is not, however, random, the location of different chromosomes within the nucleus being strictly fixed. Association with the nucleolus is accomplished through a specialized region in one of the chromosomes, this region is called the nucleolar organizer.

Genes, elementary units of heredity, are located throughout the whole length of the chromosome. In fishes and other vertebrate animals each chromosome is known to contain hundreds and perhaps even thousands of genes. The basis of the chromonema is represented by DNA or deoxyribonucleic acid. The structure of this substance playing the crucial role in biological processes has been elucidated by Watson and Crick (1953).

The DNA molecule represents a double helix. In each of the two strands forming the double helix residues of phosphoric acid and deoxyribose are repeated in a linear order many thousand times. To each deoxyribose residue one of the following four nitrogenous bases is covalently linked: adenine or guanine (purines), cytosine or thymine (pyrimidines). Bases located opposite each other in two DNA strands are associated by hydrogen bonds. The configuration of bases is such that adenine may be bonded only with thymine and guanine only with cytosine (Fig. 2). Thus, the two strands within the DNA double helix are complementary to each other.

Regions of the DNA molecule containing typically from 500 to 1500 base pairs but sometimes more and sometimes less correspond to individual genes. Pairs of nitrogenous bases may have a quite different sequence in different genes; this base sequence determines the structural and functional specificity of a given gene.

Building blocks of a DNA molecule, consisting of phosphate, sugar and one of the bases, are termed nucleotides. Depending on the type of the base, four main nucleotides can be distinguished: deoxyadenosine-5′-phosphate, deoxyguanosine-5′-phosphate, deoxythymidine-5′-phosphate and deoxycytidine-5′-phosphate. Other nucleotides occurring in lower quantities and therefore called "minor" nucleotides are also present in DNA. In chromosomes DNA is complexed with a number of

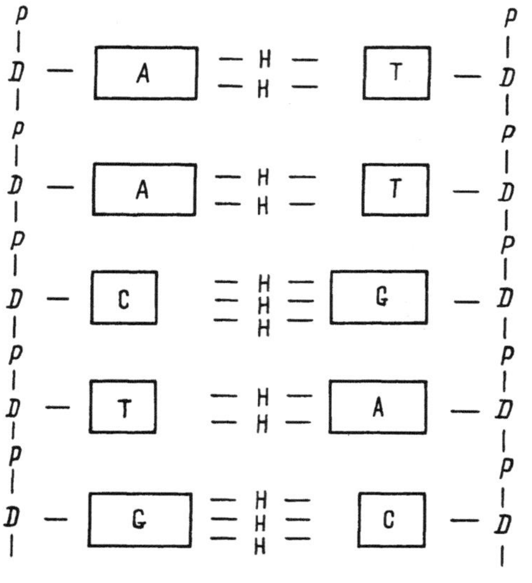

Fig. 2. The structure of DNA. *A, G* adenine and guanine bases (purines); *T, C* thymine and cytosine bases (pyrimidines); *P* phosphoric acid residue; *D* sugar deoxyribose; *H* hydrogen bonds between the bases

chemical substances, primarily with proteins such as histones and protamines; these DNA-protein complexes are called nucleoproteins. Chromosomes also contain ribonucleic acids (RNA) differing from DNA in the type of sugar (ribose instead of deoxyribose) and the base (uracil instead of thymine).

The DNA present in chromosomes and in some other structural elements of the cell such as mitochondria and plastids fulfils two highly important biological functions. One of these involves support of the hereditary continuity of living beings; another involves the template function for the synthesis in the organism of a strictly specific set of high molecular weight compounds, ribonucleic acids, and in particular the proteins necessary for its metabolism and development.

The surprising ability of DNA to replicate its structure precisely in each cell division provides a basis for the transmission of hereditary information from one generation to another, from parents to children. Replication starts from the sequential rupture of hydrogen bonds between consecutive complementary nitrogenous bases that is from the separation of two strands of the DNA duplex. New daughter chains are then synthesized on the templates of the parental DNA molecule (Fig. 3). Two double helices are formed in this way and these helices are identical along their whole length; they have identical nucleotide sequences. It is difficult to imagine any other mechanism so ideally suited to the conservation of the specific structure of a complex substance during its replication.

Another vital function of DNA, coding of protein synthesis, is accomplished in several stages. At the first stage (transcription), a single-stranded molecule of messenger or informational RNA is synthesized on one of the two strands of the DNA double helix. After its synthesis is completed, this molecule is detached from the chromosome and transferred to the cytoplasm; the cell generally contains a certain pool of mRNA. At the second stage (translation), mRNA serves as a template for the synthesis of protein chain-polypeptides. Here we can present only an extremely simplified view of the complex process of translation which is still incompletely deciphered at present. Three adjacent nucleotides in mRNA represent a codon carry-

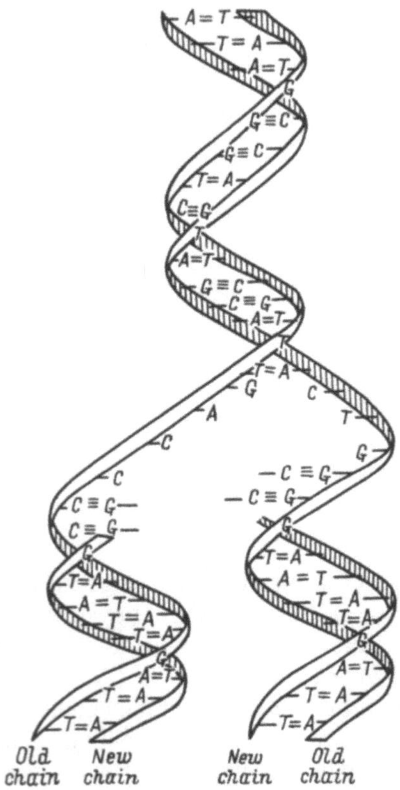

Fig. 3. DNA replication (Watson 1976). For abbreviations see Fig. 2

ing the information about the selection of one from 20 common amino acids which is to be incorporated into the synthesized polypeptide. The order of codons is of decisive importance. The number of codons is equal to 64.

The rule of correspondence between the structure of the codons and the amino acids incorporated into proteins was called the genetic or biological code (Table 1). The code is degenerate; certain amino acids are coded by two, three, four or even six different codons. The third letter in the codon is the least significant. Three codons are called nonsense codons – they do not contain any information about the amino acids but their presence in messenger RNA results in the termination of polypeptide synthesis. The code was completely deciphered in the mid-sixties and was found to be common for all animals, plants and microorganisms; its significance in the process of life cannot be overexaggerated. In most cases one gene codes for one final polypeptide products, but nowadays there are many exceptions to this concept, which was generally acceptable until recently: the relationships between genes, the mRNA transcribed from them and proteins may sometimes be very complex and unequivocal. However, even now we may accept the functional criterion as the main criterion of the gene.

The general scheme of protein synthesis is this: mRNA comes into contact with ribosomes – small intracellular structures consisting of specific molecules of

Table 1. The genetic code

First letter of mRNA triplet	Base triplets (codons) and corresponding amino acids								Third letter
	Second letter								
	U		C		A		G		
U	UUU	Phe	UCU	Ser	UAU	Tyr	UGU	Cys	U
	UUC		UCC		UAC		UGC		C
	UUA	Leu	UCA		UAA	–	UGA	–	A
	UUG		UCG		UAG	–	UGG	Trp	G
C	CUU	Leu	CCU	Pro	CAU	His	CGU	Arg	U
	CUC		CCC		CAC		CGC		C
	CUA		CCA		CAA	Gln	CGA		A
	CUG		CCG		CAG		CGG		G
A	AUU	Ile	ACU	Thr	AAU	Asn	AGU	Ser	U
	AUC		ACC		AAC		AGC		C
	AUA		ACA		AAA	Lys	AGA	Arg	A
	AUG	Met	ACG		AAG		AGG		G
G	GUU	Val	GCU	Ala	GAU	Asp	GGU	Gly	U
	GUC		GCC		GAC		GGC		C
	GUA		GCA		GAA	Glu	GGA		A
	GUG		GCG		GAG		GGG		G

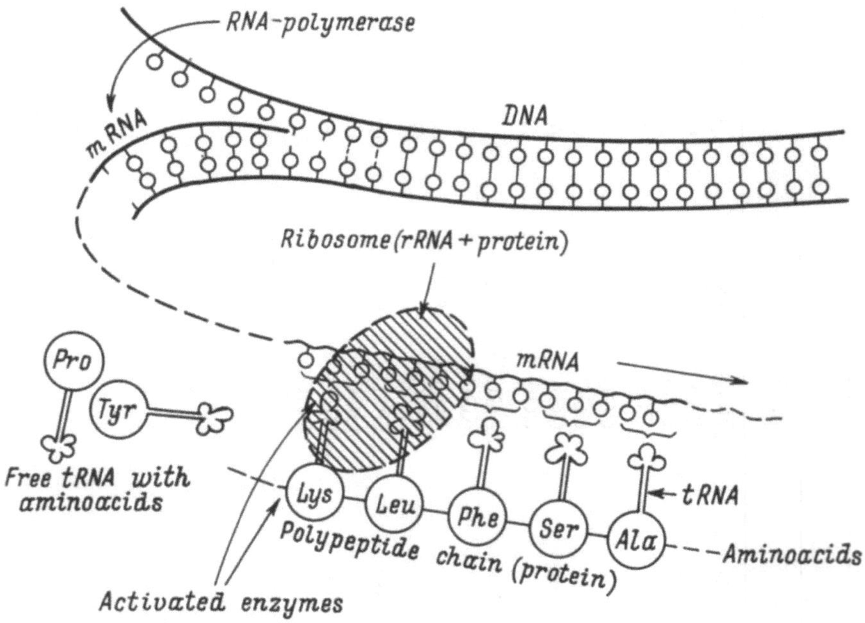

Fig. 4. The scheme of protein synthesis

ribosomal ribonucleic acid (rRNA) and proteins. When ribosomes are attached to mRNA they sequentially bind relatively small molecules of the transfer RNA. Each molecule of the transfer RNA carries one of 20 amino acids bound to one of its termini. As this process continues, the amino acids are bound into a continuous polypeptide chain and the residues of deacylated tRNA leave the protein synthesizing complex (Fig. 4). Specialized genes present in the genome in many copies code for the synthesis of rRNA and tRNA species.

The high fidelity of synthesis of polypeptide chains depends on the fact that each codon in messenger RNA corresponds to an anti-codon in the tRNA. The complementary base sequences of the codon and anticodon transiently interact with each other. Cell cytoplasm contains molecules of tRNA able to interact with all the sense codons present in mRNA. Transfer RNA having a certain anti-codon serves as the carrier of the corresponding amino acid.

Not all genes in chromosomes code for the synthesis of proteins. For certain genes only transcription products are known.

The Behaviour of Chromosomes: The Main Patterns

Chromosome Pairing and the Behaviour of Chromosomes During Mitosis and Meiosis. In fish karyotypes all the chromosomes, with the exception of the sex chromosomes, are represented by pairs of elements similar in shape and size.

Such pairs of homologous chromosomes have been observed in studies of karyotypes from the cells in various tissues and organs. The presence of pairs of chromosomes is conserved in successive generations because of the precise segregation of all the chromosomes during mitosis, the twofold reduction of the chromosomal number during the maturation of gametes (meiosis) and the recovery of the initial diploid number during fertilization.

In fishes mitosis does not differ from that in other animal species. After the four stages of mitosis, prophase, metaphase, anaphase and telophase, one parental cell divides into two daughter cells having two nuclei, each daughter cell containing in its nucleus the complement of chromosomes identical to that present in the parental nucleus. The telophase is followed by the interphase (interkinesis); during this period the chromosomes are despiralled and are not visible in the nucleus; they are too thin to be seen under the light microscope.

The interphase is generally divided into three periods: postmitotic (G_1), synthetic (S) and premitotic (G_2). Replication of the chromosomes in the cell nucleus takes place during the S period; most active RNA synthesis also takes place during this time.

All the chromosomes contain the so-called heterochromatic regions where the chromonemata (and DNA as well) are tightly coiled and RNA synthesis is therefore suppressed. The regions adjacent to the centromere and to the nucleolar organizer (Kiknadze 1972) represent permanent or "genuine" heterochromatic regions. Different genes are active in different tissues and at different stages of development, the activity of these genes being regulated by specialized mechanisms.

Ribosomal RNA (rRNA) is synthesized in large amounts in the nucleoli on special genes located in the nucleolar organizer, these genes being present in many copies. The nucleolar organizer also apparently represents the site of the formation of ribosomes.

Meiosis involves two sequential divisions of the germinal cells resulting in the transfer of just one chromosome of each pair of homologues into the mature gamete (Fig. 5). The gonia or primary germinal cells undergo several divisions and then start to grow; this is accompanied by complex rearrangements of the chromosomes in the nuclei of such cells. In the male testes the gonia are gradually transformed into the primary spermatocytes. Soon afterwards they enter the first meiotic maturational division. The prophase and metaphase of this division may be divided into a number of characteristic stages: the leptotene (the stage of thin chromosomal filaments), the zygotene (the stage of conjugation of homologous chromosomes), the diplotene (the beginning of the disjunction of the conjugated chromosomes) and diakinesis (the gradual coiling of the chromosomes). At the diplotene stage, the chromosomes within each pair are still connected with each other, forming peculiar structures called chiasmata. During the rupture of the chiasmata homologous chromosomes may exchange regions, resulting in crossing-over. Crossing-over has been found in fishes, particularly in genetic studies of several aquarium species belonging to the family Poeciliidae (Winge 1923; Dzwillo 1959; Morizot et al. 1977; Leslie 1979; and others). Recently crossing-over has been demonstrated in the carp (Cherfas 1977; Nagy et al. 1978) and in the plaice (Purdom 1976).

In the course of spermatogenesis (Figs. 5 and 6) the diploid primary spermatocytes undergo reductional division and form secondary spermatocytes which are smaller in size and contain the haploid set of chromosomes. The second (equational) division results in the formation of two spermatids, which are then gradually

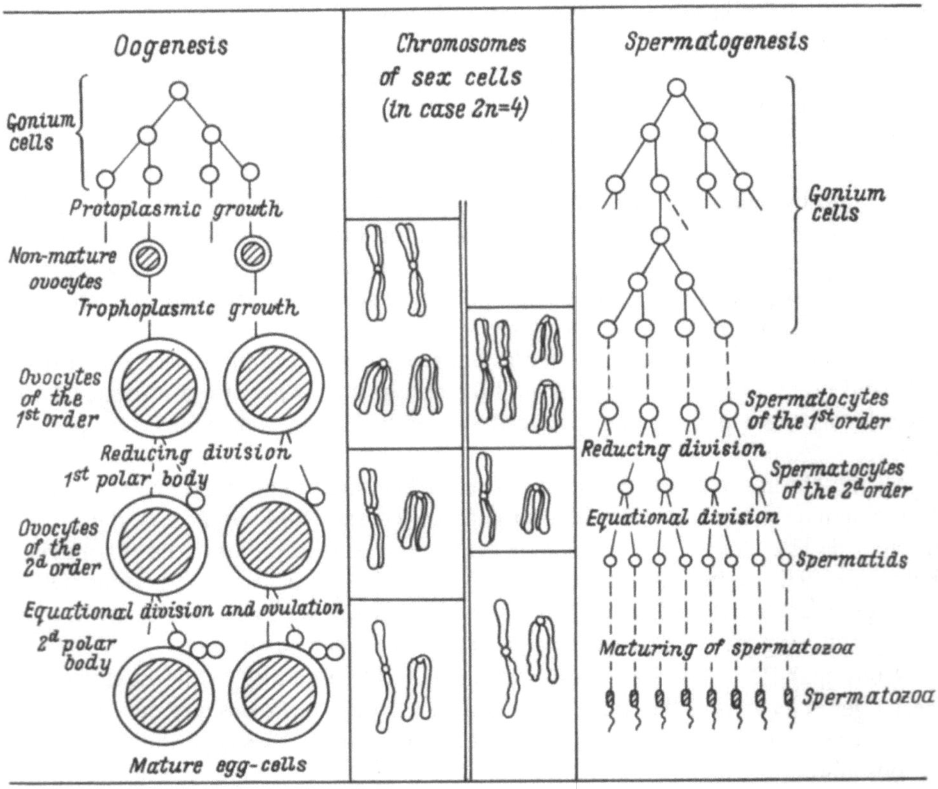

Fig. 5. The scheme of oogenesis and spermatogenesis in fishes

transformed into mobile spermia. The replication of chromosomes only occurs before the first maturational division and mature spermatozoa contain the haploid set of chromosomes.

By spawning time the bulk of the testes in teleostean fishes is filled with spermatozoa, but at the same time regions may be found containing ampules with spermatids, primary and secondary spermatocytes, and with primary gonia continuing to divide (Fig. 6c, d).

The main difference between oogenesis and spermatogenesis lies in the fact that by the beginning of the meiotic divisions the egg attains a very large size due to the accumulation of the yolk. Both divisions take place at the end of the prolonged period of germinal cell growth. During the protoplasmic or slow growth (Fig. 6a) the oogonial cell is transformed into the primary oocyte and increases severalfold in size due to the enlargement of the cytoplasmic volume. During the trophoplasmic or rapid growth which continues for one or two years or even more in certain fish species the yolk is accummulated in the oocytes in large quantities and the diameter of the oocyte increases several dozen times (Fig. 6b). The first maturational division generally coincides in time with ovulation, that is the liberation of oocytes from the surrounding follicular cells followed by the transfer of the eggs into the body cavity or into the cavity of the ovary. The eggs become free-flowing and able to undergo

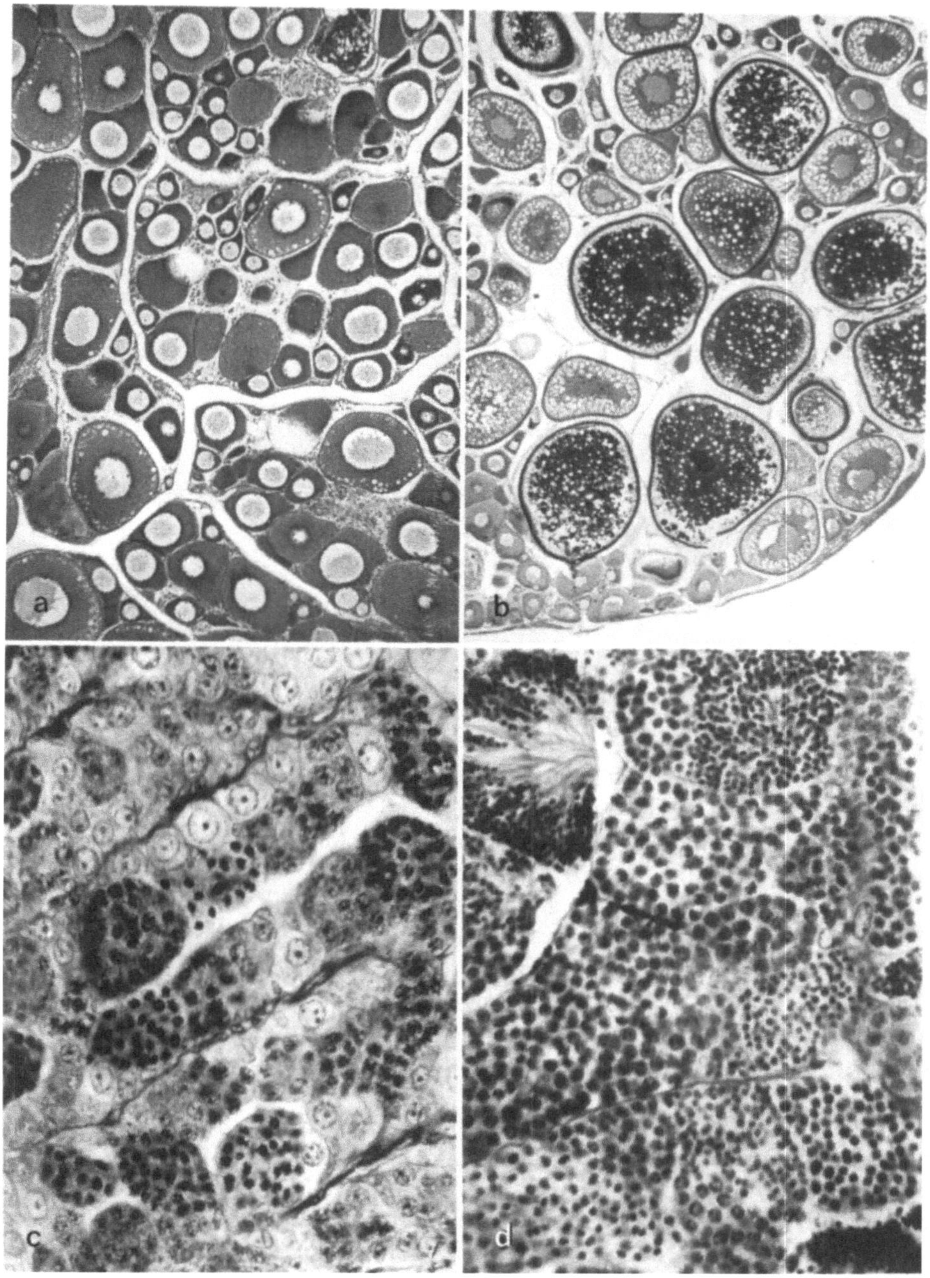

Fig. 6a–d. Oogenesis and spermatogenesis in fishes. Preparations were kindly provided by T. I. Faleeva – oogenesis, and N. A. Butskaya – spermatogenesis; photographed by G. M. Sabinin. **a** ovary of the ruff (*Acerina cernua*), oocytes during protoplasmic growth; **b** the same but during trophoplasmic growth (yolk accumulation); **c** testes of the ruff with spermatogonia at different stages; **d** the same but primary and secondary spermatocytes, spermatids and mature spermatozoa can be seen

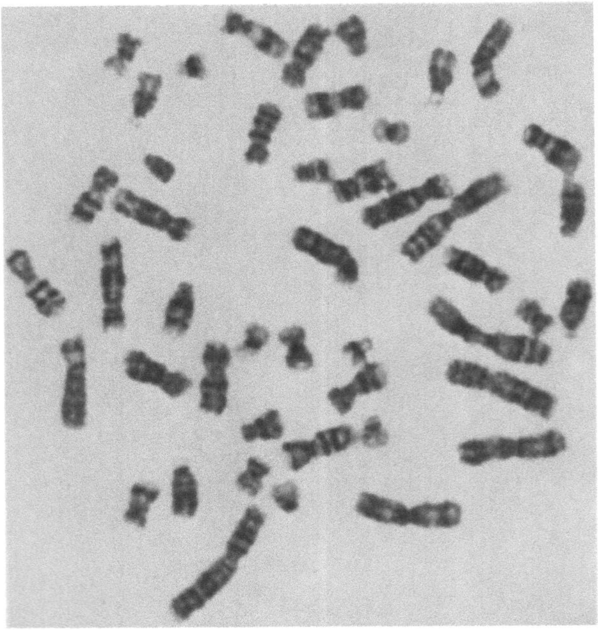

Fig. 7. Individuality of chromosomes. Human chromosomes at the stage of metaphase, differential G-banding. (Preparation and photography by Yu. L. Goroshchenko)

fertilization and subsequent development. The equational division in fishes occurs simultaneously with the penetration of the spermium into the egg or immediately thereafter. In the course of the first (reductional) division half of the chromosomal complement remains in the egg cytoplasma, while the second half passes into a small polar body visible only under the microscope. Another such body of similar size separates during the second (equational) division (sometimes the first polar body divides into two single ones at this time). After the first division the primary oocyte with the diploid set of chromosomes undergoes transformation into a haploid secondary oocyte and then into the mature egg. The female haploid nucleus (pronucleus) fuses with the male one and forms the diploid nucleus of the fertilized egg. In the course of maturation each oocyte gives rise to only one egg, retaining all the yolk necessary for the subsequent nutrition of the embryo. Polar bodies undergo degeneration.

A reduction in the number of chromosomes during meiosis is characteristic of most fish species in which the sexual process is normal. As a result of meiosis, one half of the mature germinal cells receives one of the two chromosomes of each pair and the other half receives the second chromosome of the pair. A random combination of these chromosomes after the fusion of the pronuclei forms the basis of the first law of Mendel, according to which the segregation in the F_2 of parents differing in a single genetic trait occurs in the ratio 3:1 or 1:2:1.

Sometimes reductional division does not occur during the parthenogenetic or gynogenetic reproduction of fishes, when the male nucleus does not participate in the development of the embryo. In such cases the number of chromosomes in the nuclei of gametes (or eggs in this case) is equal to the number present in the somatic cells of the maternal organism. The occasional absence of the reduction in fishes whose sexual process is normal results in the formation of diploid male and female gametes. Subsequent fertilization may lead to the appearance of triploid (3n) or even tetraploid (4n) embryos.

The Individuality of Chromosomes. The chromosomes of each pair differ in their size and structure from the chromosomes of other pairs. These differences may involve the position of the centromeric region (at the centre or at the end of the chromosome), they may involve the length ratio of the arms, the presence or absence of strangulations, satellites etc. The administration of low doses of colchicine to fishes prior to the fixation of tissues and the use of the squash preparations (without sectioning) have extended the possibilities of identifying different chromosomes (McPhail and Jones 1966). Nowadays cytologists widely use various differential staining techniques for chromosomes or banding procedures; these techniques enable one to elucidate the more subtle differences in the structure of different pairs (Fig. 7). Hopeful results were obtained, when chromosomal preparations were treated with quinacrine, inducing the fluorescence of certain regions of the chromosomes (Q-banding). In experiments using this technique clear-cut differences between the chromosomes of the different pairs have been found in salmonids (*Salvelinus leucomaenis, S. malma*) (Abe and Muramoto 1974). G-banding, involving the treatment of chromosomes with trypsin and subsequent Giemsa staining (Fig. 8), as well as C-banding have successfully been used in studies of fishes (Zenzes and Voiculescu 1975; Barshiene 1978; Ojima and Takai 1979; and others).

The Independent Assortment of Chromosomes in the Course of Meiosis. In most animal and plant species, as well as in man, chromosomes of different pairs assort independently between the daughter cells in the course of germ cell maturation.

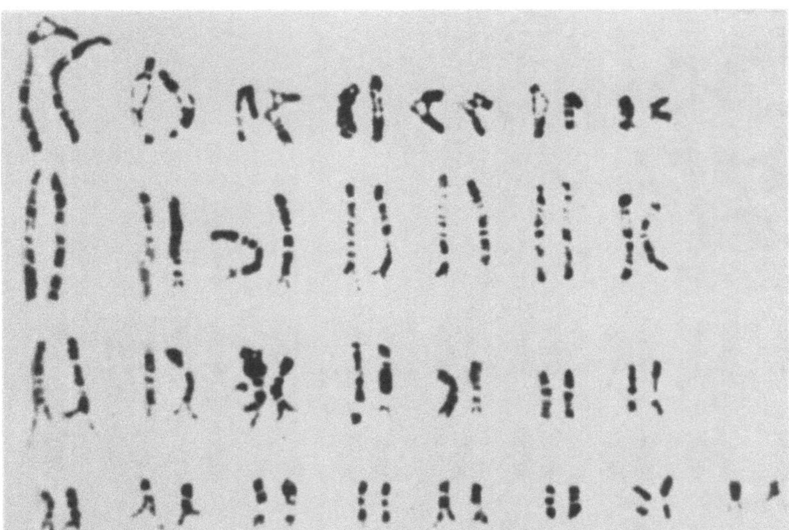

Fig. 8. Individuality of fish chromosomes. Chromosomes of the Atlantic salmon (*Salmo salar*), differential G-banding. Preparation and photography of Ya. V. Barshiene

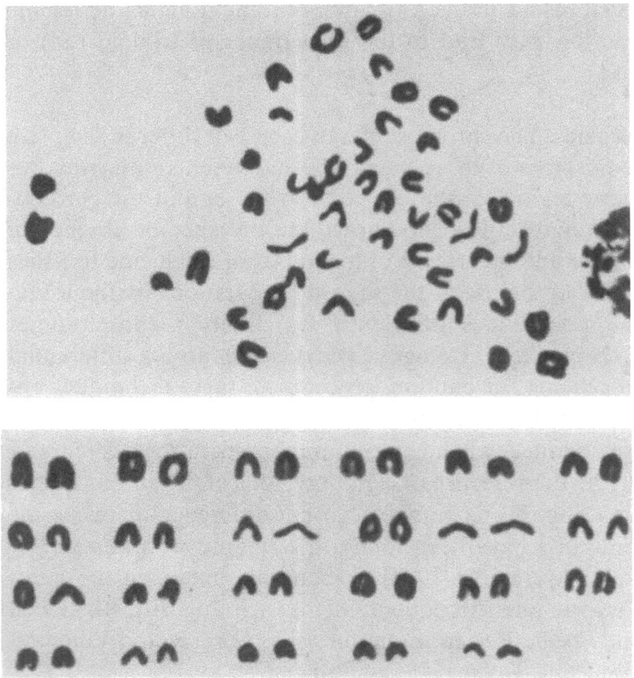

a

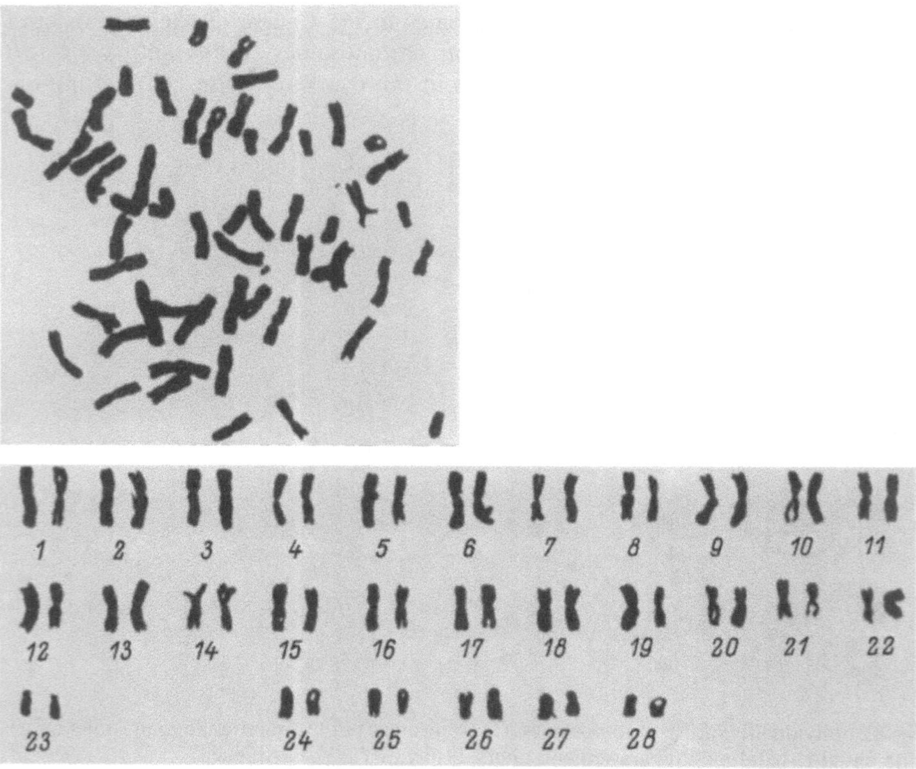

b

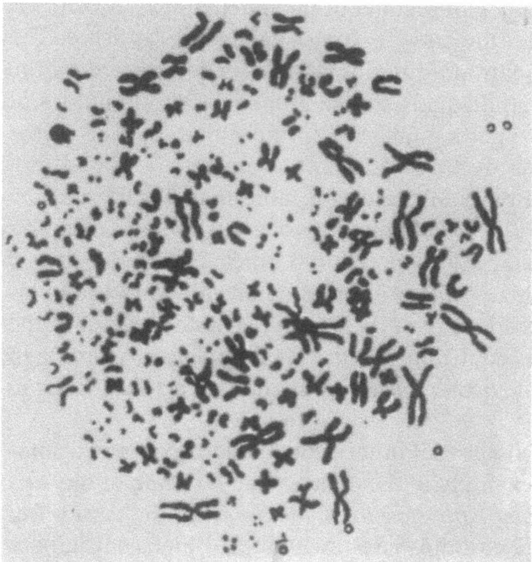

c

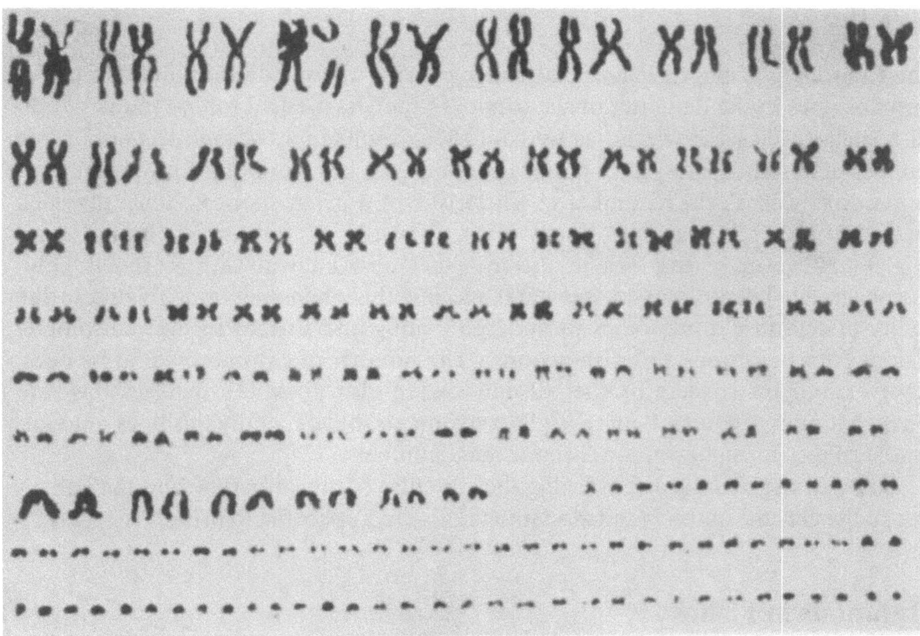

d

Fig. 9a–d. Fish karyotypes. **a** the goby *Gobius melanostomus,* 2 n = 46, metaphase plate and karyogram (Ivanov, 1975); **b** the sockeye salmon *Oncorhynchus nerka,* 2 n = 56, metaphase plate and karyogram (photographed by E. V. Chernenko); **c, d** the sturgeon *Acipenser naccarii,* 2 n = 239 ± 7, metaphase plate and karyogram (Fontana and Colombo 1974)

This principle, underlying the second law of Mendel of the independent assortment of traits in the second generation after the cross, is completely valid for fishes. The number of different combinations of chromosomes in gametes is determined by the number of pairs of chromosomes (n) and equals 2^n; the number of combinations of these chromosomes in the nuclei of zygotes equals 4^n. In more than half of all the fishes studied, the karyotype consists of 48 or 50 chromosomes (n=24 and 25). When n=24, the number of possible types of gametes is more than $16 \cdot 10^6$ and the number of possible zygotes attains 10^{14}. In polyploid species such as the common and the crucian carp, catostomids, and other species for which 2n is close to 100 (n=50), the number of different gametes equals approximately 10^{30} and the number of zygotes exceeds 10^{60}. The wide diversity of the possible types of zygotes in the course of sexual reproduction of fishes contributes to the maintenance of a very high level of genetic variability of morphological, physiological and biochemical traits in natural fish populations.

Cases are known, however, when the law of independent assortment of chromosomes is not fulfilled. A remarkable example of this kind is found in one of the Asian fishes, the chocolate gourami *Schaerichthys osphromonoides* from the family Belontiidae (Perciformes). Only 16 large chromosomes including 14 metacentric ones are found in this species. These chromosomes form five structures in spermatocytes: three normal bivalents, one tetravalent (having a circular structure) and one hexavalent; in the latter case, six chromosomes are connected by the ends and form a large ring (Calton and Denton 1974). Apparently, the non-random distribution of the chromosomes limiting genetic variability is advantageous for this highly specialized species.

The Constancy of the Chromosomal Number. After mitotic division the number of chromosomes in the daughter nuclei (diploid number) is equal to that in the parental nucleus. Meiosis, however, as we could see, results in a twofold decrease in the number of chromosomes; after fertilization the even number of the chromosomes is restored, as well as their number characteristic of a given species. Generally each species of fishes contains a strictly defined constant number of chromosomes (Fig. 9). However, exceptions to this rule are also known. Chromosomal polymorphism has been observed in several species; this phenomenon is related to the ability of certain chromosomes to undergo joining and fissions in the centromeric regions (centric fusions and disjunctions). The number of chromosomal arms does, however, remain constant. Other deviations are also possible. In fishes they are somewhat more frequent than in birds or mammals, but generally this does not violate the rule of constancy of the chromosomal number.

Thus, all the main laws governing the structure of the karyotype and the behaviour of the chromosomes in mitosis and meiosis are applicable to fishes.

Mutations in Fishes

In spite of all its perfection, the chromosomal apparatus of fishes does not remain unchanged; from time to time genes and chromosomes are affected by mutations – changes in the structure, transmittable to subsequent generations.

Several types of mutations are known. Gene mutations also known as point mutations affect the structure of genes. They may result from a nucleotide substitution in a certain region of the gene as a consequence of a replication error (for example, when thymine is substituted for guanine); mutations may be a consequence of the deletion of one or several nucleotides from a long DNA chain of the nucleotide insertion or vice versa, possibly accompanying random chain scissions. If such changes result in the formation of a triplet belonging to nonsense codons in the middle of a gene, such a mutation results in the premature termination of the synthesized protein molecule which becomes short and unable to function normally. In other cases, only the order of the amino acids in a protein changes. The substitution of just one or two amino acids in a protein molecule may not be essential for the viability of the organism, and the mutation may be captured by natural selection.

No data are as yet available on the frequency of gene mutations in fishes, but this frequency appears not to be high: for example, after the examination of 260 thousand individuals of the common carp, no single mutation affecting the S and N genes responsible for the pattern of scales had been found (Tzoy et al. 1974b). Even if we assume that there was one unnoticed mutation in this population, the mutation frequency would not exceed $4 \cdot 10^{-6}$. As regards mutability in nature, we can only judge of it from indirect observations. The occurrence of gene mutation follows first of all from the high biochemical (genetic) variation characteristic of all fishes. The populations of many fish species appear to be saturated by mutant forms of genes coding for the synthesis of many different proteins (Kirpichnikov 1973a; Lewontin 1974). A similar conclusion follows from observations showing the high genetic variability manifested in morphological and physiological traits, as well as from the finding of large numbers of aberrant forms in natural fish populations (see Dawson 1964).

The second large group of mutations is chromosomal rearrangements. They involve translocations (exchanges of regions within one chromosome or between different chromosomes), inversions (changes in the position of the chromosomal regions by 180 °), duplications (the doubling of genes or of small parts of chromosomes) and deletions of certain regions.

Translocations (Fig. 10a–e) appear to occur quite frequently since even close species of fishes (and sometimes races inside the species) differ in the structure of the karyotype, these differences in many cases resulting from the evolutionary fixation of translocations. Individuals carrying reciprocal translocations, i.e., the exchanges, region by region, of chromosomes without any loss or addition of the genetic material are generally quite viable. Even individuals with additional genetic material can sometimes survive. Intrachromosomal translocations may result in changes in the relative length of the chromosomal arms or in the change in the number of arms, but the probability of the survival of such rearrangements during meiosis is very low. Reciprocal interchromosomal exchanges are of great evolutionary significance and may be accompanied (but not always) by a decrease or an increase in the numer of chromosomal arms. The so-called Robertsonian translocations or centric fusions are very important and are apparently fairly frequent. The breakage of one acrocentric chromosome occurs near the centromere and another acrocentric chromosome is joined to the site of the breakage. One or two small regions adjacent to the centromere (with one of the centromeres) are lost and two acrocentric chromo-

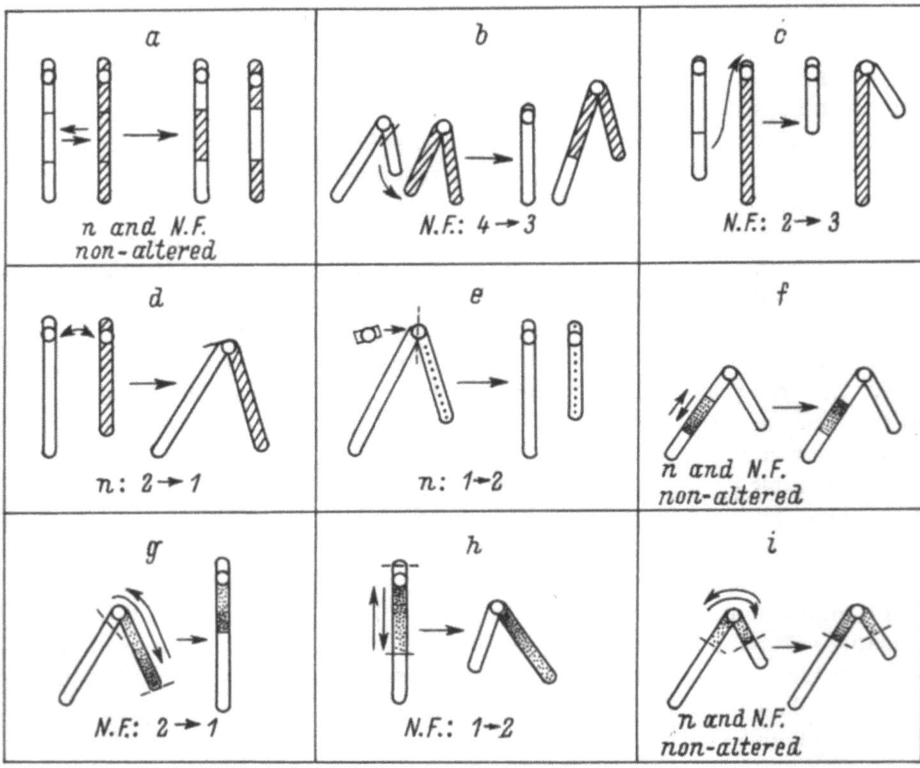

Fig. 10a–i. Chromosomal rearrangements in fishes (schematic). **a** non-apparent translocation within one pair of chromosomes; **b** interchromosomal translocation resulting in the decrease of arm number; **c** the same as in **b**, but with the increase in the number of arms; **d** centric fusion of Robertsonian type leading to a decrease in the number of chromosomes; **e** centric fission resulting in the increased number of chromosomes; **f** non-apparent paracentric inversion; **g** pericentric inversion with the decrease in the number of arms; **h** the same as in **g**, but with increase in the number of arms; **i** non-apparent pericentric inversion

somes fuse into a single metacentric one. The number of chromosomal arms remains unchanged. The reverse process of centric fissions is rarer because an additional centromere is required. According to recent data, however, the direct division of one centromere into two may be possible (Imai 1978).

Inversions (Fig. 10f–i) can be classified into two main types. Paracentric inversions, which do not involve the centromeric regions, are difficult to detect. They do apparently occur in fishes quite frequently, but their presence can only be established by analyzing the inheritance of the linked genes. Pericentric inversions involving the centromere are quite frequent. If the two breakage events take place at equal distances from the centromere, one cannot detect the inversion without the analysis of marker genes. When the sites of breakage are located asymmetrically, the relative length, or even the absolute number of chromosomal arms, will be changed.

Duplication of chromosomal regions is known to occur in fishes, although these events are probably infrequent. The presence of duplicated genes has been established by purely genetic techniques, since duplications cannot be detected under the microscope; the most probable duplication mechanism involves unequal crossing-over, i.e., the exchange by portions of imperfectly conjugated chromosomes. According to many authors, duplications play a particularly important role in the evolution of fishes (Ohno 1970a, b). Deletions undoubtedly occur among fishes as well and are perhaps even more frequent than duplications, but most of the deletions lead to a drastic loss of viability, and the organisms carrying these deletions are rapidly eliminated from the populations.

The spontaneous frequency of chromosomal rearrangements in fishes remains unknown.

The third group of mutations is represented by ploidy changes: the appearance in diploid fish species (2n) of individuals with a haploid karyotype (n), with a modified chromosomal number – triploid (3n) and tetraploid (4n) sets, or with the change in the number of individual chromosomes – aneuploid.

Haploid individuals among fishes are non-viable. When the egg development is stimulated by spermatozoa with destroyed nuclei, almost all the developing embryos become haploid, the development proceeding with malformations (haploid syndrome) and resulting in embryonic death at later stages of embryogenesis.

Triploids in fishes appear quite frequently. Apparently, in most fish species the reductional division may be omitted during meiosis (predominantly in females), and the resulting gametes are diploid. The fertilization of such an egg by a normal spermatozoon (as well as the fertilization of the normal egg by a diploid spermatozoon) leads to the appearance of triploid organisms. They can be quite viable: for example, certain varieties of *Carassius auratus gibelio* (Fig. 11) and of the viviparous fishes *Poeciliopsis* and *Poecilia* are triploid (Rasch et al. 1965; Cherfas 1966b; Schultz 1969; Kobayashi et al. 1970). Triploid organisms have recently been found in the rainbow trout as well as in the Californian minnow *Hesperoleucus symmetricus* (Cuellar and Uyeno 1972; Gold and Avise 1976; Thorgaard and Gall 1979). The decreased fertility of the triploids in bisexual fishes (associated with the irregular segregation of chromosomes in meiosis) apparently explains the absence of such forms in the natural populations of most species. Triploids appear in fishes from time to time as a consequence of distant hybridization (Vasilyev et al. 1975).

Probably tetraploids also appear from time to time as a result of fusion of the diploid gametes, but there are no data on the frequency of these chromosomal mutations in nature.

The existence of aneuploidy in fishes was until recently regarded as a controversial issue. The finding of an individual with 85 chromosomes, with three chromosomes carrying the Ldh-B2 gene in the brook trout (2n = 84) (Davisson et al. 1972, 1973) enables one to assume that trisomic organisms may survive under certain circumstances, at least in fishes which have undergone polyploidization (Salmonidae, Catostomidae, etc.).

The frequency of mutations in fishes can be markedly increased by X-rays and chemical treatment. A number of the most important facts will be discussed.

X-irradiation of fish gametes results in the appearance of various genic and chromosomal mutations (Samokhvalova 1938; Penners 1959; Schröder 1969a, e; 1973, 1976; Purdom 1972; Egami and Hyodo-Taguchi 1973; Purdom and Woodhead 1973). In particular, I should like to mention the paper by Anders and co-

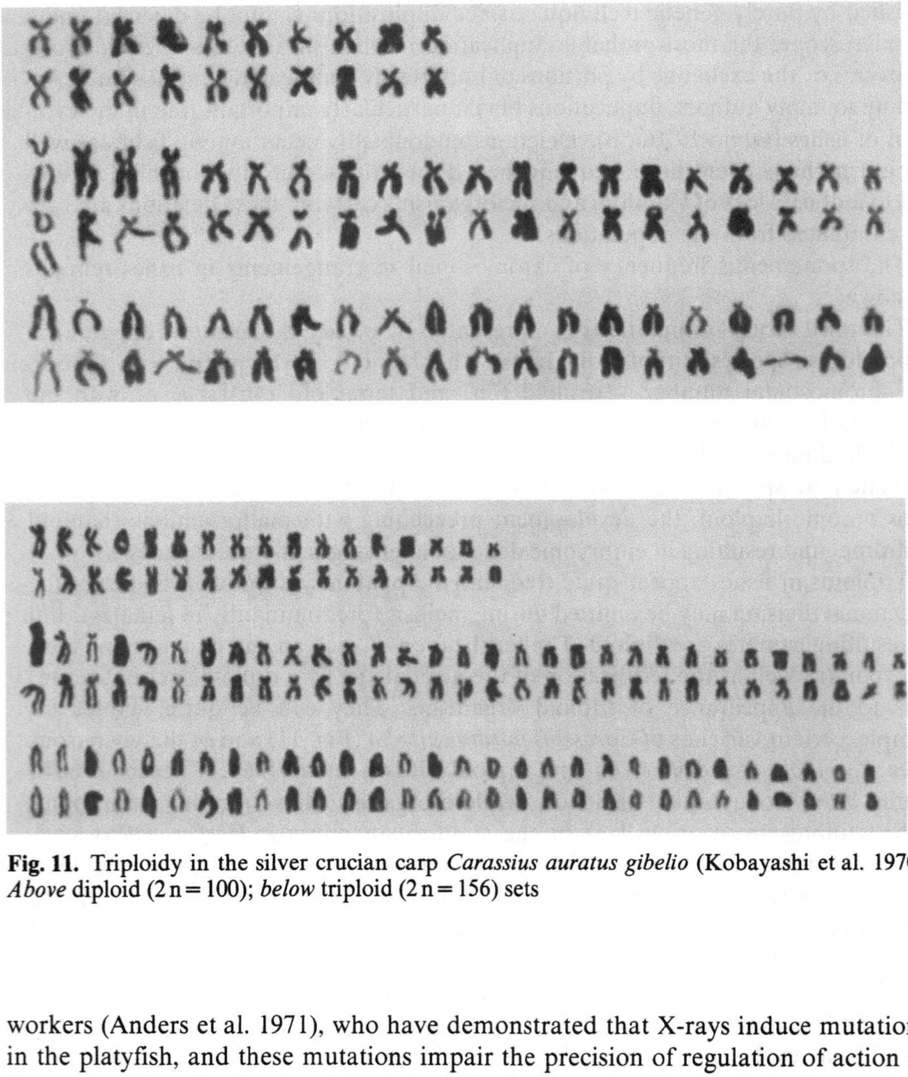

Fig. 11. Triploidy in the silver crucian carp *Carassius auratus gibelio* (Kobayashi et al. 1970). *Above* diploid (2n = 100); *below* triploid (2n = 156) sets

workers (Anders et al. 1971), who have demonstrated that X-rays induce mutations in the platyfish, and these mutations impair the precision of regulation of action of the gene coding for the development of black pigment cells – melanophores. As a result, these fishes develop a tumour, a premelanoma, resembling that which emerges after the hybridization of the platyfish with the swordtail.

Chemical mutagens, particularly nitrosoethylurea (NEU) are highly effective in inducing mutations in fishes. If spermatozoa or eggs are treated with NEU, embryos with many chromosomal defects are usually found; these defects may be easily observed when mitotic figures are inspected at the blastula stage. These observations lead to the conclusion that the frequency of chromosomal mutations induced by such treatment is very high (Tzoy 1969a, b). The frequency of genic mutation was measured in the carp using the S and n scale genes as a model; it was equal to 0.02%–0.04% after dimethyl sulphate treatment. In one experiment with NEU 40 mutations of the n gene were observed among the 11 500 fishes examined, the frequency being equal to 0.36% (Tzoy 1971b; Tzoy et al. 1974a, b). It is quite probable, however, that the mutation induced in this gene (n) is a chromosomal aberration (deletion); then the surprisingly high mutation frequency in the latter case is due to the induction of deletions in the region of the gene n and is not a consequence of point (genic) mutations. No matter which interpretation of Tzoy's experiments is correct, they have shown that the rate of mutation can be increased dozens or even hundreds of times by chemical mutagenes.

X-irradiation of fish spermatozoa in large doses and their treatment by high doses of chemical mutagenes result in gynogenesis, that is the development in the absence of the male parent's chromosomes. In such conditions embryos are haploid, but up to 1% of the embryos turn out to be spontaneous diploids (Cherfas 1975; Tzoy 1976). The use of a temperature shock (incubation at lowered temperatures) has made it possible to markedly increase the yield of diploid larvae due to gynogenesis in the loach (Romashov and Belyaeva 1965b) and in the common carp (Cherfas 1975). Cooling of eggs also resulted in an increase in the number of triploids after normal fertilization (Swarup 1959; Vasetzky 1967; Valenti 1975). It can be mentioned that cytochalazine increases the number of polyploid embryos in salmonids (Refstie et al. 1977b).

To sum up, using X-rays, chemical mutagens and temperature treatment, one can drastically increase the mutational variability in fishes.

The Evolution of Karyotypes Among Cyclostomes and Fishes

The chromosomal complements of cyclostomes and fishes have been studied by many authors. A number of reviews are available containing data on the number of chromosomes and the number of chromosomal arms. Unfortunately, not one of these reviews is sufficiently complete. At present, chromosomal numbers have been determined for more than 1300 species. The information regarding the variation of karyotypes is reviewed in Table 2 showing the distribution of chromosomal numbers among representatives of the various taxons. The chromosomal complements turn out to be quite variable, the diploid numbers varying between 12 and 250. The total DNA content in the cell nucleus is even more variable (Table 3); the haploid DNA content increases from 0.4 pg ($0.4 \cdot 10^{-12}$ g) in a species of Tetraodontidae to 140 pg in Dipnoi, that is, by a factor of 350. Such surprising variability can perhaps be explained by the assumption that fishes and cyclostomes represent a very ancient, highly heterogeneous group of animals which has been undergoing divergent evolution in different directions throughout the hundreds of millions years of its existence.

We shall now consider the main trends in karyotype evolution in several taxons for which sufficient data are available.

The data available for *Cyclostomata* are too scarce to provide a basis for any generalizations. It can only be pointed out that the two groups, hagfishes and lampreys, have greatly diverged in the course of evolution. Complements with small chromosomal numbers and a marked DNA haploid content are characteristic of the hagfishes *Myxinoidea;* in lampreys, however, the number of chromosomes is high,

Table 2. Chromosome number in cyclostomes and fishes

Taxon	Diploid number of chromosomes[a]											
	12–20	22	24	26	28	30	32	34	36	38	40	42
Cyclostomata												
Myxiniformes					1				2			
Petromyzoniformes												
Fishes												
Chondrichthyes					1							
Dipnoi								1		1		
Ganoidomorpha:												
Acipenseriformes												
Polypteri-, Amii-, Lepidosteiformes									6			
Teleostei												
Gonorhynchiformes, Clupeiformes					1		1					1
Salmoniformes:												
Diploids (Osmeroidei a. oth.)	1	3				1			2	1		
Polyploids (Salmonoidei)												
Myctophiformes												
Osteoglossiformes, Mormyriformes								1				3
Anguilliformes										7	2	1
Cypriniformes:												
1. Lebiasinidae		1	2	1		1			3	1	1	3
2. Characidae					1		1		1	1	1	2
3. Cobitidae												
4. Cyprinidae												
5. Catostomidae												
6. Other families			1									
Siluriformes					1			1		2	2	7
Cyprinodontiformes	6	2	1	3	3	5	4	7	12	9	23	7
Atheriniformes, Beloniformes, Gadiformes, Beryciformes				1	1				1	1	1	1
Gasterosteiformes												6
Mugiliformes, Synbranchiformes			1		1							
Perciformes												
1. All families, except (2, 3)	1			1			1	1		2	4	5
2. Cichlidae										2	2	2
3. Gobioidei						1				2	1	3
Scorpaeniformes								1	1		2	1
Pleuronectiformes					1					2	2	2
Tetraodontiformes, Gobiesociformes, Lophiiformes								4	1	1	3	4
Total	8	6	5	6	11	8	7	16	29	32	44	48

[a] Sources used for the construction of the Table 2 are marked with an asterisk in the list of references

[b] Sturgeons are operationally divided into two groups. The numbers obtained by Serebryakova are corrected by addition of 60 or 120 microchromosomes as described by Fontana and Colombo (1974)

44	46	48	50	52	54	56	58	60	62 – 70	72 – 80	82 – 90	92 – 100	102 – 150	152 – 200	> 200	Total
	1	2														6
										2				11		13
							1		7	2	5	3	1			20
				1	1											4
													7[b]		5[b]	12
	1								1							8
1	2	3		2						1(?)						12
1		3	7		2	2		1	1	1						26
				1		3	4	2	5	30	8	3				56
1	27	2														30
	3	4				2										13
		2														12
2	1															16
	1	13	38	29	3											91
		3	7						1			4				15
5	3	32	151	14	1					1		6	1[c]	1[c]		215
												14				14
1		1	1	3	16	2	1	1	5							32
3	9	5	4	4	10	7	10	4	7	4	1	5	1			87
8	24	124	4		1				1[d]	1[d]						245
1	5	15	1	1									1			30
	2															8
	1	10														13
5	19	125	3			1				1						169
17	8	46	1	1				1								80
23	18	8	2	1					1							60
1	2	21	1													30
2	3	13	1													26
3	2	3	1													22
74	132	435	222	57	34	17	16	9	29	43	14	35	11	12	5	1365

[c] The karyotypes of the triploid (2 n = 150) and tetraploid (2 n = 200) crucian carps *Carassius auratus langsdorfii* and *C. a. gibelio*

[d] The karyotypes of the triploid gynogenetic tooth carps from the genera *Poecilia* and *Poecilopsis*

and this is particularly true of the species living in the Northern hemisphere (Zanandrea and Capanna 1964; Howell and Denton 1969; Robinson and Potter 1969; Potter and Rothwell 1970; Howell and Duckett 1971; Potter and Robinson 1971; Nygren and Jahnke 1972a; Robinson et al. 1974, 1975), but the DNA content in the nucleus is moderate (Hinegardner 1976b). Any mechanisms underlying this divergent evolution of chromosomes still remain unclear at present.

Among fishes per se (*Pisces*) the most primitive groups, in particular chondrosteans such as sharks, skates or chimeras, have been evolving towards an increase in the number of chromosomes, and the DNA content has been increasing in parallel. Thus, in certain skate species the karyotype consists of almost 100 chromosomes (Nygren et al. 1971d); only the electric skate from the family Torpedinidae – *Narcine brasiliensis,* which have 28 chromosomes, is an exception (Donahue 1974). In all cartilaginous fishes with the exception of the chimeras the DNA content in the cell nuclei is greatly increased (Table 3).

The evolution of karyotypes in Dipnoi occurred in the direction of a very pronounced increase in the DNA content of the nucleus: in the modern species it is equal to 80–140 pg (Pedersen 1971). Meanwhile, the number of chromosomes in certain species has decreased (see Table 2). The DNA content in the species related in evolution to Dipnoi such as *Latimeria* or *Polypterus* is somewhat increased (3.3 and 5.0 pg respectively), but these values are within the limits characteristic of chondrosteans or cartilaginous fishes (Bachmann 1972). The number of chromosomes in these species is also moderate (Capanna and Cataudella 1973; Denton and Howell 1973).

Studies of the cell size in the fossil ancestors of the present-day Dipnoi have shown that this characteristic increased gradually but permanently (Thomson 1972). The DNA content in the nuclei appeared to increase in parallel. Certain assumptions about the mechanisms underlying this process can be made. Dipnoi differ from all other fishes, first of all in the presence of a kind of "lungs", an organ existing in the place of the air bladder and allowing the atmospheric air to be breathed. The two coexisting respiration mechanisms required greater plasticity of the physiological processes. It might well be that a large increase in the DNA con-

References: 1. Bachman (1972); 2. Bachman et al. (1974); 3. Beamish et al. (1971); 4. Booke (1968); 5. Booke (1974); 6. Cimino (1974); 7. Ebeling et al. (1971); 8. Fontana (1976); 9. Hinegardner (1968); 10. Hinegardner (1976a); 11. Hinegardner (1976b); 12. Hinegardner and Rosen (1972); 13. Kang and Park (1973a); 14. Kornfield et al. (1979); 15. Mauro and Micheli (1979); 16. Mirsky and Ris (1951); 17. Muramoto et al. (1968); 18. Ohno (1970b); 19. Ohno and Atkin (1966); 20. Ohno et al. (1967a); 21. Ohno et al. (1969a); 22. Ojima et al. (1972); 23. Park and Kang (1976); 24. Pedersen (1971); 25. Rasch et al. (1965); 26. Robinson et al. (1975); 27. Scheel (pers. commun); 28. Stingo et al. (1980); 29. Szarski (1970); 30. Thomson (1972); 31. Uyeno and Smith (1972); 32. Vervoort (1979); 33. Vialli (1957); 34. Wolf et al. (1969)

Table 3. DNA quantity in the genomes of various fish taxons

Taxon	DNA, pg/n	Number of species	Reference
1	2	3	4
Cyclostomata			
Myxini	2.8	1	10
Petromyzoni	1.3 – 1.6	2	10, 19, 26
Pisces			
Selachomorpha	2.8 – 32.8	14	10, 11, 24, 28, 33
Batomorpha	2.8 – 16.2	25	10, 11, 28
Chimeridae	1.5 – 1.6	1	10, 27
Dipnoi	80 – 142	3	24, 29, 30
Coelacanthidae	3.0 – 7,0	1	24
Polypteridae	4.7 – 4.9	1	1, 10, 30
Acipenseridae, 2 n	1.7 – 3.6	3	2, 8, 10, 21
Acipenseridae, 4 n	5.1 – 5.7	2	8, 10
Amiidae	1.2 – 1.4	2	9, 21
Clupeidae, Esocidae, Osmeridae and others	0.8 – 2.7	9	3, 7, 9, 21, 27
Bathylagidae	1.7 – 6.4	4	7
Salmonidae	2.8 – 3.5	3	4, 5, 9, 12 17 – 20, 22, 24
Anguillidae	1.5 – 2.0	2	23
Cypriniformes:			
Gyrinocheilidae, Anostomidae, Characidae	0.7 – 1.7	9	12, 27
Serrasalmidae	1.7 – 2.0	2	27
Cobitidae, Cyprinidae (2 n)	0.7 – 1.1	3	13, 17, 19, 20, 21 22, 24, 35
Cobitidae, Cyprinidae (4 n)	1.7 – 2.2	3	20, 22, 24, 27, 35
Catostomidae (4 n)	2.0	1	31
Siluriformes:			
Bagridae, Ariidae	1.1 – 2.4	2	12
Callichthyidae (Coridorus)	4.1 – 4.4	3	27
Cyprinodontidae	1.4 – 1.5	2	6, 25, 27
Poeciliidae	0.6 – 1.0	17	6, 25, 27
Atherinidae	1.3	1	27
Hemirhamphidae	0.7	1	27
Serranidae, Percidae, Cichlidae, Gobiidae	1.0 – 1.4	10	9, 12, 14, 27
Echeneidae, Pomatomidae, Anabantidae, Mastacembelidae	0.6 – 0.9	6	12, 27
Pleuronectidae, Bothidae	0,7 – 0.8	6	20, 24
Tetraodontidae	0.4	1	27

tent was necessary in order to supply the organism with the necessary amounts of different enzymes. The progressing specialization of the Dipnoi and the limited variety of habitats suitable for their existence (shallow sludge-containing freshwater basins well heated by the sun) contributed to the decrease in the number of chromosomes and to the formation of more stable gene complexes.

The question regarding the evolution of the nuclei in the Dipnoi has still not been settled. This evolution could perhaps proceed by sequential, so-called tandem, duplications of certain chromosomal regions (Ohno 1970a, b), but it is not to be excluded that the polytenization or the increase in the number of chromonemata in chromosomes was involved.

Certain authors (Matthey 1949; Pedersen 1971) believe that the similarity in the DNA content between the Dipnoi and certain amphibia (Urodela) is associated with their philogenetic relatedness. It is also possible that the migration of water-living vertebrates to the land was initially accompanied by a marked increase in the DNA content in their nuclei (Bachman 1972) just as in the Dipnoi which had adapted to life in two different environments.

The chondrostean ganoids (family Acipenseridae) have a large number of chromosomes and a marked DNA content in the genome. In this respect, they resemble selachia. Perhaps such a trend in the evolution of the karyotype is related to the mode of life and the size of these fishes, in particular to their high mobility and ability to grow rapidly. These two groups of fishes have another common feature, that is, the presence of small microchromosomes in many species. The role of microchromosomes is not fully understood, although some authors suggest that they contain the "redundant" genetic material necessary for the cells where there is increased protein synthesis. The number of such chromosomes in the complement may probably vary; their presence can be considered as a peculiar mechanism for an increase in the cromosomal variability without impairing the integrity of the main genome.

The family Acipenseridae may be divided into two groups according to the criteria of the number of chromosomes and DNA content in the nucleus. One of these groups includes the sturgeon with a large number of chromosomes, such as *Acipenser güldenstädti, A. baeri, A. schrencki* and *A. naccarii.* The other group includes the white sturgeon (beluga) and the East Siberian sturgeon (kaluga), *Huso huso* and *H. dauricus,* the shovel-nosed sturgeon *Scaphirhynchus platorhynchus,* sterlet *Acipenser ruthenus, A. stellatus* (sevrjuga), *A. nudiventris* (ship), as well as the West European sturgeon *A. sturio,* which has half as many chromosomes in the genome (Ohno et al. 1969b; Serebriakova 1969, 1970; Fontana and Colombo 1974; Fontana 1976; Vasilyev 1980). Different authors hold different opinions on techniques of quantitating the number of sturgeon's chromosomes; Soviet authors do not appear to have taken microchromosomes into account. The existence of two groups of acipenserid with different karyological characteristics has, however, been proved beyond doubt. The differences in the content of DNA per genome and in the size of the erythrocytes between these two groups provide evidence in favour of the hypothesis on the polyploid origin of several sturgeon species (Fontana 1976).

Evidence of the tetraploidy of the karyotype in the North American paddlefish *Polyodon spathula* (Polyodontidae) has been also reported (Dingerkus and Howell 1976).

It is interesting to discuss the pattern of karyotype evolution in the orders Clupeiformes and Salmoniformes. Herrings, anchovi and related families have generally conserved chromosomal numbers (2n equals from 48 to 52) characteristic of the ancestors of modern teleost fishes (Ohno 1970 b; Scheel 1974). Some species are an exception to this. About 80 chromosomes (80 ± 4) have been found in *Sardina pilchardis* (Tsitsugina 1970), but these data are not very reliable. In *Gonostoma bathyphilum* from the family of Gonostomatidae only 12 large chromosomes have been found (Post 1974). So far, this is the smallest number of chromosomes that has ever been found in fishes. In another species (*G. elongatum*) of the same genus the karyotype consists of 48 chromosomes. Since six circular tetrades are observed in *G. bathyphilum* during meiosis, its large chromosomes appear to have been formed as a result of several centric fusions. The chromosomal complex of this species resembles the chromosomal complement of *Oenothera.*

A moderate reduction in the number of chromosomes has been observed in the *Notopterus* species (Srivastava and Kaur 1964; Uyeno 1973) and in several other families.

The salmon and the grayling have characteristically peculiar karyotypes. It is generally proved that all families of fishes belonging to this group have passed through the duplication of the chromosomal complement at the middle of the Tertiary period (Ohno et al. 1969 a; Ohno 1970 a). The number of chromosomes in the original tetraploid (which was apparently a common ancestor of all the families in the group) must be to 96–100. Later, in the course of the migration of fish species, extensive divergence of the karyotype occurred, which was accompanied by the secondary diploidization of the genome. The number of chromosomes in all the species with the exception of the grayling have decreased to a different extent. Polyploidy of Salmoninae and Coregoninae not only follows from karyological and cytochemical evidence, but also from genetic data, specifically from the presence of a number of duplicated loci in the genome (Ohno 1970 a; Engel et al. 1971 b; Kirpichnikov 1973 a; Ferris and Whitt 1977 a, b, c).

The divergence of Salmonidae has resulted in the prominent variation of the karyotypes even within a single genus (Table 4). For example, the number of chromosomes in the Pacific salmon *Oncorhynchus* varies from 52 in *Oncorhynchus gorbusha* to 74 in *Oncorhynchus keta;* the number of arms or the fundamental number (N.F.) does, however, remain almost constant (Simon 1963; Chernenko 1968, 1971; Chernenko and Viktorovsky 1971; Fukuoka 1972 a; Viktorovsky 1978 b). Viktorovsky (1975 b) has assumed that almost all the chromosomal rearrangements that have occurred in the divergent evolution of *Oncorhynchus* belong to the type of Robertsonian translocations (centric fusions). In his opinion, *O. keta* represents the species most closely related to the original form that had about a hundred acrocentric chromosomes. According to Victorovsky's definition, in having only metacentric chromosomes, *O. gorbusha* represents a species where the karyotype is most highly advanced as compared to the ancestral type.

The number of chromosomes in the diploid complement of the various representatives of the genus *Salmo* varies from 80–82 to 54; the number of arms, however, is only decreased in *S. trutta, S. ischchan* and *S. salar.* The karyotype of *S. salar* is par-

Table 4. Number of chromosomes and chromosomal arms in salmonids (the whitefishes omitted)

Species	2n	N.F.	References	Presumed number of chromosome mutations relative to a hypothetical polyploid ancestor [a]		
				Centric fusions	Pericentric inversions	Total
1	2	3	4	5	6	7
Oncorhynchus keta	74	106 – 108	23, 25	15	1	16
O. tschawytscha	68	106	17, 25	18	1	19
O. masu (=*rodurus*)	64 – 66	104	6, 17	19 – 20	0	19 – 20
O. kisutch	58 – 60	104 – 106	25	22 – 23	0	22 – 23
O. nerka	56 – 58	102 – 104	4, 5, 11, 12, 23, 25	23 – 24	0	23 – 24
O. gorbuscha	52	104	16, 25	26	0	26
Salmo trutta	78 – 82	98 – 100	8, 19, 20, 27	12	2	14
S. ischchan	80	96	9	12	4	16
S. letnica	80	104	7	12	0	12
S. carpio	80	98	15	–	–	–
S. salar	54 – 60 (58)	72 – 74	1, 18, 20 – 22, 24, 27	22 – 24	15	37 – 39
S. (Parasalmo) gairdneri (=*irideus*)	58 – 65	104	13, 16, 20, 26, 27, 29	22	0	22
S. (P.) mykiss	60 – 62	104 – 108	29	–	–	–
S. (P.) clarkii clarkii	70	106	16, 26	17	1	18
S. (P.) clarkii henshavi	64	106	16, 26	20	1	21
S. (P.) clarkii levisi	64	106	16, 26	20	1	21
S. (P.) aguabonita	58	104	16	23	1	24
S. (P.) apache	56	106	16	24	0	24
S. (P.) gilae	56	105 (106?)	2	–	–	–
Salmothymus obtusirostris	82	94	3	10	5	15

Salvelinus fontinalis	84	100	27	10	2	12
S. namaycush	84	100	32	10	2	12
S. leucomaenis	84 – 86	100	6	9 – 10	2	11 – 12
S. alpinus	80 – 84	96 – 100	19, 28	12	2	14
S. (alpinus) cronocius	78 – 82	100	31	11 – 13	2	13 – 15
S. malma malma	76 – 78	96	6, 31	13 – 14	4	17 – 18
S. m. Krascheninnikovi	82 – 84	98	6, 31	10 – 11	3	13 – 14
S. m. curilus	84 – 86	100	31	9 – 10	2	11 – 12
Hucho taimen	84	102	10	10	1	11
Brachymystax lenok	92	102	10	6	1	7
Brachymystax lenok	90	116	14	–	–	–

References: 1. Barshiene (1977 a); 2. Beamish and Miller (1977); 3. Berberovič et al. (1970); 4. Chernenko (1971); 5. Chernenko (1977); 6. Chernenko and Viktorovsky (1971); 7. Dimovska (1959); 8. Dorofeeva (1965); 9. Dorofeeva (1967); 10. Dorofeeva (1977); 11. Fukuoka (1972 a); 12. Gorshkova and Gorshkov (1978); 13. Kaidanova (1974); 14. Kang and Park (1973 b); 15. Merlo (1957); 16. Miller (1972); 17. Muramoto et al. (1974); 18. Nygren et al. (1968 b); 19. Nygren et al. (1971 b); 20. Prokofieva (1935); 21. Rees (1967); 22. Roberts (1970); 23. Sasaki et al. (1968); 24. Sepovaara (1962); 25. Simon (1963); 26. Simon and Dollar (1963); 27. Svärdson (1945 a); 28. Vasilyev (1975 a); 29. Vasilyev (1975 b); 30. Viktorovsky (1975 b); 31. Viktorovsky (1978 b); 32. Wahl (1960)

[a] Viktorovsky (1978 b)

ticularly interesting. At present, most authors agree that the European Atlantic salmon *S. salar* has 58 chromosomes (Sepovaara 1962; Nygren et al. 1968 b, 1972; Barshiene 1977 a, b). Discrepancies do, however, remain with respect to the salmon species of the American coast.

Differences in the number of chromosomes between different populations of the American Atlantic salmon have been reported (Boothroyd 1959; Roberts 1970). The modal number in the different populations is 54, 55 or 56 chromosomes. Roberts explains these differences, as well as those between different individuals, by the Robertsonian translocations, because the number of chromosomal arms remains almost constant (72). Rees (1967) and Nygren et al. (1972) insist that the chromosomal complements in salmon of the European and Canadian populations are identical. This viewpoint is based on a thorough comparison of the karyotypes and studies of the hybrids between these two subspecies, as well as between the salmon and the lake trout. Deviations from the diploid number equal to 58 chromosomes are explained by experimental errors. It should be pointed out, in conclusion, that, according to Viktorovsky (1975 b), the Atlantic salmon has undergone the largest number of chromosomal rearrangements, no less than 37–40.

The genus *Salvelinus* has undergone much less divergent evolution. It is assumed that its divergence started later (prior to or after the first glaciation) and has not yet been completed (Viktorovsky 1975 a, b); Viktorovsky and Glubokovsky 1977). Morphological studies of the *Salvelinus* species enable one to suggest that they originated from a form similar to *Salmo trutta.* Comparatively recent divergence is also characteristic of the various species of the white lakefish (the *Coregonus*): in almost all of them the diploid number is 80 and the number of arms around one hundred. Evolution was more pronounced in the species of the genus *Prosopium,* the divergence appeared to be accelerated by the diversity of free ecological niches in their range of distribution. Graylings (*Thymallus*) retained the "ancestral" karyotype characteristics to the greatest extent, but the number of chromosomal arms has increased in the course of evolution. Families of Salmoniformes related to salmonids and whitefishes such as Osmeridae, Argentinidae and also the deep water fishes – Bathylagidae, did apparently remain at the initial diploid level. This conclusion has been made on the basis of measurements of the number of chromosomes and of the DNA content per genome (Ohno 1970 b; Ebeling et al. 1971). Species of Bathylagidae living at the lowest depth have the highest DNA content per genome, the increase in the number of chromosomes and in the DNA content not appearing to be due to polyploidization, but to an increase in the genome size caused by duplications.

It is difficult to speculate about the factors responsible for the decrease in the number of chromosomes in the course of the secondary diploidization of the genomes of Salmonidae. The following explanations may, however, be discussed:

1. the decrease in the number of chromosomes in diploid complements occurred automatically as a result of the greater probability of chromosomal fusion as compared to that of disjunction; such rearrangements were then fixed in a random manner. Polyploidy stimulated the fusion of the chromosomes related by common origin (Viktorovsky 1975 c, 1978 b);
2. the decrease in the number of chromosomes has been associated with specialization of the species in the course of which the formation of the complexes of

linked genes represented a definite adaptive advantage (Hinegardner and Rosen 1972; Kirpichnikov 1973 b);

3. centric fusions of related chromosomes were advantageous because they contributed to the speed-up of diploidization in polyploids (Roberts 1970).

The probability of the purely random fixation of chromosomal rearrangements without any involvement of selection is quite low, particularly if one takes into account that such rearrangements are quite rare. It is more probable that evolutionary transformations lead to the conservation of those rearrangements that were of adaptive value. On the basis of these considerations, the models laying emphasis on the part played by selection appear to be more justified.

It should be pointed out that the evidence in favour of the relatively recent polyploidization of Salmonidae is detectable in many species of the salmon, since circular chromosomes, quadrivalents and other chromosomal figures deviating from the normal state are formed in the course of meiosis (Svärdson 1945 a; Ohno et al. 1965, 1969 a; Nygren et al. 1968 b, 1971 a, 1972, 1975 b; Roberts 1970; Barshiene 1977 a).

Evolutionary transformations among Cypriniformes are quite interesting. The family Cyprinidae is known to be surprisingly conservative with respect to the number of chromosomes; disregarding several rare polyploids, the number of chromosomes in this group varies insignificantly, a small decrease in the number of chromosomes is found only in the specialized group of Rhodeinae. Chromosomal complements with 50 chromosomes are most common (Fig. 12). The low variation of the chromosomal complement in cyprinids is associated with their biological features such as high fertility, the diversity of ecological niches and plasticity requiring a free recombination of genes. In several cyprinid groups specifically in American *Notropis* species the constancy of the chromosomal complement is accompanied by a marked morpho-ecological divergence (Gold 1980).

Having acquired advantages in the competition with the usual diploid forms, the polyploid mutants are well able to survive precisely in such a rapidly evolving group. The polyploids have emerged in three families of the order Cypriniformes, Cyprinidae, Cobitidae and Catostomidae. Among Cyprinidae we find four polyploid species in the subfamily Barbinae and three polyploid species in the subfamily Cyprininae. The carp and the bisexual forms of the two species of the genus *Carassius* have doubled chromosomal numbers (2n = 100–104), and unisexual populations of the ginbuna *Carassius auratus langsdorfii* have triploid or even tetraploid karyotypes with 150 or 200 chromosomes respectively (Makino 1939; Cherfas 1966 a; Kobayashi et al. 1970, 1973; Ueda and Ojima 1978). Species with a large body size in the genus *Barbus* such as *B. barbus, B. plebeius, B. meridionalis,* are tetraploid, while a large number of smaller forms such as *B. tetrazona, B. oligolepis, B. fasciatus* and some others are diploid (Ohno et al. 1967 a; Fontana et al. 1970; Berberovič et al. 1973; Sofradzija and Berberovič 1973). This provides some evidence in favour of the independent emergence of polyploids in the two subfamilies of Cyprinidae. In the loaches five polyploid species have been found so far, while other species are apparently diploid (Hitotsumachi et al. 1969; Kobayashi 1976). In the loach *Misgurnus fossilis* the number of chromosomes is 100 and the number of arms 136, at the same time in *M. anguillicaudatus* the diploid number varies from 48–50 up to 100 (Raicu and Taisescu 1972; Ojima and Takai 1979). Similar charac-

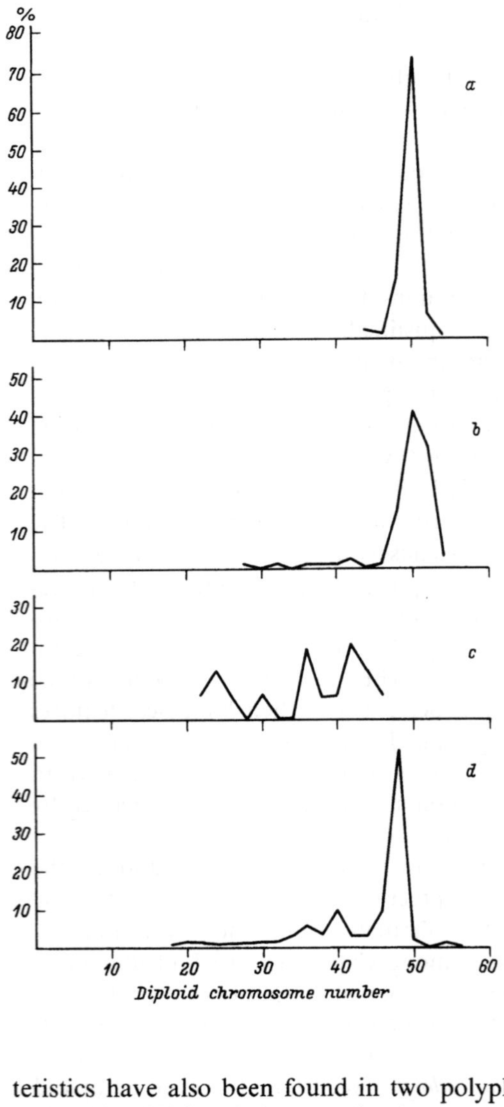

Fig. 12a–d. The distribution of species with respect to diploid number of chromosomes in different taxons. **a** Cyprinidae; **b** Characidae; **c** Lebiasinidae (Cypriniformes); **d** Cyprinodontiformes. Polyploid species are not included; maximal chromosomal numbers are shown for polymorphic species

teristics have also been found in two polyploid species of the genus *Botia* (Ferris and Whitt 1977b). Unfortunately, the amount of DNA per genome has not been determined in these species. The polyploidy of the common and the crucian carp has been confirmed by genetic observations (Ohno et al. 1967a, 1969a; Wolf et al. 1969; Engel et al. 1971a, 1975). Karyotypes of the fourteen species of catostomid fishes examined were found to contain about 100 chromosomes on average, and an increased DNA content. Genome polyploidization in Catostomidae and Cyprinidae appear to have taken place no later than the middle of the Tertiary period, that is more than 50 million years ago on the territory of South-East or South Asia (Uyeno and Smith 1972).

The very important trend in the evolution of Cypriniformes involved the reduction of a chromosomal complement. This tendency is clearly traceable among the relatively specialized families of Characidae and Lebiasinidae (Fig. 12), particularly in the latter family, where the number of chromosomes in several species has decreased to 22–24 (Scheel 1972a). In this case, the relationship between the number of chromosomes and the specialization is more distinct than among Salmonidae.

Finally, it should be pointed out that the number of chromosomes has increased to 54–64 in a number of families of this order, such as Heteropneustidae, Anostomidae, Alestidae, Serrasalmidae; most chromosomes are metacentrics (Scheel 1972a). This provides an additional indication that the reduction in the number of chromosomes per genome cannot be explained by the increased probability of centric fusions. The increase in the number of chromosomes in several families, for example in Serrasalmidae, was accompanied by a rise in the DNA content per genome, apparently due to duplications.

Thus, the evolution of the karyotype among Cypriniformes occurred simultaneously in four different directions, resulting in the prominent divergence of karyotypes in this group.

Few data are available so far on Siluriformes which are characterized by the high heterogeneity of the karyotypes. It should only be pointed out that the species belonging to the families Clariidae and Callichthyidae containing many chromosomes apparently represent polyploids. For example, in the genus *Corydorus* (family Callichthyidae) species can be found that are characterized by a proportional increase in the number of chromosomes and in the number of chromosomal arms (see Scheel et al. 1972).

Species	Number of	
	chromosomes	arms
C. arcuatus, C. axelrodi and others	46	92
C. metae	92	180
C. julii	92	184
C. aeneus	132	222

The DNA content par genome turned out to be rather high (4.1–4.3 pg) in all the species studied including one diploid, but in this case the error of species identification has been assumed (Scheel, pers. comm.).

The large, well-studied group of Cyprinodontiformes may provide an impressive illustration of the relationship between the number of chromosomes and the specialization. The hypothetical ancestral number of chromosomes, that is 48, only remained in 45% of the species among Cyprinodontidae; in other species of this group it is lowered down to 17–20 chromosomes (see Table 2 and Fig. 12). A similar picture may be observed in the family Goodeidae. In contrast, the karyotype with 48 chromosomes is most frequent among viviparus Poeciliidae.

In the family Cyprinodontidae we find varying degrees of reduction of the chromosomal number in different genera, as well as different patterns of the interspecific and of the intraspecific variation of the karyotypes. A common tendency, the presence of certain greater or lesser number of species with a reduced number of chro-

Diploid numbers in different species of the genus [a]	
Fundulus	– 48 (15), 46 (4), 44 (1), 40 (2), 34 (1), 32 (1)
Rivulus	– 48 (5), 46 (3), 44 (2), 40 (1)
Aplocheilus	– 50 (4), 48 (7), 42 (1), 40 (2), 38 (1), 34 (2)
Notobranchius	– 44 (1), 38 (3), 36 (3), 18 (1)
Aphyosemion	– 46 (2), 42 (4), 40 (19), 38 (4), 36 (4), 34 (3), 32 (1), 30 (3), 28 (2), 22 (2), 20 (3), 18 (1)

[a] The number of species in parentheses

mosomes, has been observed for many polytypic genera. A few examples of such variation will be given (see Scheel 1972 a).

The decrease in the total number of chromosomes is accompanied by a rise in the number of large metacentric chromosomes (Fig. 13); it follows that centric fusion plays a highly important part in the reduction mechanism. Another important mechanism involves pericentric inversions (Scheel 1972 a, b).

In several species the intraspecific variation of the karyotypes is no less in its magnitude than the interspecific one. Thus, Scheel (1972 a) reported the following series of karyotypes for three species of *Aphyosemion:*

A. bivittatum – 40, 38, 36, 34, 30, 26
A. calliurum – 40, 38, 36, 34, 32, 30, 26, 22, 20
A. cameronense – 34, 32, 30, 28, 26, 24.

This variation is of a systematic nature, the varieties from different geographical locations having a different number of chromosomes. In his analysis of *A. bivittatum* Scheel points out that the karyotypes consisting of 38 or 40 chromosomes are typical of the most "generalized" varieties. In the course of the population of vast areas of tropical forests which occurred millions of years ago, the original forms had to be

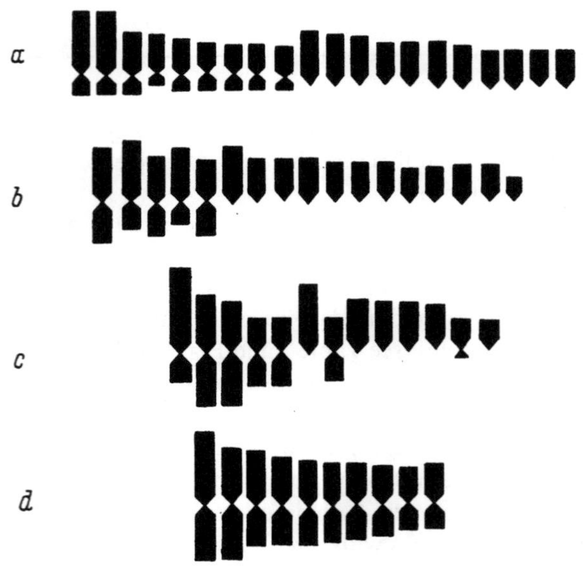

Fig. 13 a–d. Haploid chromosomal complements of different races of *Aphyosemion calliurum* (schematic, from Scheel, 1972 a). *From top to bottom* n = 20 (**a**), 17 (**b**), 13 (**c**) and 10 (**d**)

sufficiently flexible and therefore had to possess genotypes with a maximal degree of variation, in other words genotypes with a weak gene linkage. Later, after the penetration of other *Aphyosemion* species into these small isolated habitats, the maximal specialization and limitation of variability became advantageous. Consequently, individuals with low numbers of chromosomes or with more stable gene combinations acquired a selective advantage. A similar specialization process and a parallel decrease in the number of chromosomes occurred in the course of the evolutionary migration of *A. calliurum.* The survival of each of the species depended on the presence of varieties with more variable genotypes, that is with a lower degree of the linkage and a larger number of chromosomes. When more highly specialized forms were dying under the impact of various factors the free ecological niches could be occupied by these sufficiently "flexible" varieties (Scheel, pers. comm.).

This remarkable example of the peculiar "adaptive strategy" of a species demonstrates that the karyotype evolution among Cyprinodontidae in each specific case was determined by the equilibrium of several forces, in the main the advantage of specialization and the need to retain a certain plasticity. Excessively specialized varieties, species and perhaps even genera were dying out in cases of unfavourable changes in environmental conditions.

In *Poeciliopsis,* one of the genera of Poeciliidae, triploid species and varieties having 72 chromosomes have been found; their origin is apparently due to hybridization or gynogenesis (see Chap. 7). Triploids may also result from certain crosses of different *Poecilia* species.

Among other groups of fishes, several families of Perciformes belong to those better studied. This heterogeneous group contains families with markedly different levels of plasticity and specialization. Extensive adaptability is characteristic in particular of the families Percidae and Centrarchidae, and we can see that almost all the species in these families possess karyotypes with 48 chromosomes. Such families as Gobiidae, Cichlidae and Anabantidae include quite a number of specialized forms and it is among these that we find many cases of a reduction in the number of chromosomes in the karyotype. The tendency for the number of chromosomes to be reduced in the karyotype is distinctly traceable in the order Perciformes.

It is difficult to say anything definite about the orders Mugiliformes, Scorpaeniformes and Pleuronectiformes because few observations have as yet been made; it should be pointed out, however, that the karyotypes of Tetraodontiformes and related taxons (Table 2) are apparently characterized by a greatly reduced number of chromosomes.

Thus, a comparison of all the fishes and cyclostomas studied leads to the conclusion that there is a definite decrease in both the number of chromosomes and the amount of DNA per genome during the evolution from primitive to more highly organized groups. This tendency has been noted by many authors (Hinegardner 1968, 1976a; Hinegardner and Rosen 1972; Kirpichnikov 1973b; Ohno 1974). Among the more primitive groups the karyotypes with a small number of chromosomes are characteristic of the hagfishes (Myxinae) and Dipnoi, but the content of DNA per genome in these groups is quite high.

Both an increase and a decrease in the number of chromosomes may be observed in different taxons. The reduction in the chromosomal number cannot apparently be explained away by the random fixation of Robertsonian translocations. It is more probable that natural selection determined by the conditions of life of one group or another and by the pattern of adaptation of the fish to its environment is involved. The decrease in the number of chromosomes in the karyotype in a number of cases is definitely related to the specialization of the species, requiring some limitation in the assortment of the genetic material. This relationship, is not, however, absolute, because the change in the number of chromosomes and in the DNA content per nucleus may result from adaptation to specific environmental conditions and from the need for the metabolic level to be adjusted.

Centric fusions and apparently fissions, as well as pericentric inversions, played the main part in the process of evolution of fish karyotypes. It can be assumed by analogy with other organisms that paracentric inversions, duplications and small deletions also played an important part in this process. At present, however, we do not possess adequate methods of detecting these chromosomal changes in fishes.

Polyploidization has occurred many times during fish evolution. It has been proved that polyploids appeared independently in the family Acipenseridae, among Salmoniformes and Cypriniformes and to a lesser extent among Siluriformes (the genus *Corydorus*). It cannot be excluded at present that polyploidization has occurred in the course of the evolution of several other taxons as well. The survival of polyploid forms among fishes is due to a relatively simple type of genetic sex determination and may even by related to the complete absence of sex chromosomes in several groups (Viktorovsky 1969).

The evolution of the chromosomal apparatus in the dipnoan represents an exceptional example of the tremendous increase in the content of genetic material and a parallel increase in the cell size. This example may be useful in studying ways of penetrating aquatic organisms in dry habitats.

Claims that there are differences in the number of chromosomes between freshwater and marine fishes as well as between deep-water and coastal forms (Nikolsky 1973; Nikolsky and Vasilyev 1973) appear to be very doubtful. Thus, the increase in the number of chromosomes in freshwater fishes established by an analysis of the data for Clupeiformes and Perciformes hardly exists at all, the differences in the mean values result from insufficient data on several groups and the exclusion of purely freshwater and purely marine groups from calculations. Among the deep-water fishes there are many species with an increased number of chromosomes and a high DNA content per genome. Claims that there are chromosomal differences related to the geographical latitude manifested as a larger number of chromosomes in Arctic forms are perhaps more justified but in this case as well, one should not forget the main factor responsible for the divergence of fish karyotypes. I refer to the degree of specialization which is undoubtedly greater in species living in tropical and subtropical areas.

Chromosomal Polymorphism

We have already seen that the number of chromosomes in different species varies to a large extent and that one of the most important mechanisms responsible for such variation involves centric fusions of the chromosomes of different pairs, in other words, Robertsonian translocations. The first consequence of such a rearrangement, if it survives in the population, is polymorphism with respect to the number of chromosomes. One can discuss three levels of chromosomal polymorphism:

1. polymorphism as manifested between different cells of one and the same animal (the intra-individual level);
2. differences between individuals within one family or one population (intra-populational level);
3. differences between populations within one species (inter-populational level).

Variation of the Number of Chromosomes in Different Cells. Such variation has been found in many fishes. In rainbow trout embryos at the eye stage, the variation turned out to be quite considerable; the following chromosomal numbers have been found in embryos with a high mitotic level (Ohno et al. 1965):

Number of chromosomes						
Total number of chromosomes	59	60	61	62	63	64
Including telocentrics	14	16	18	20	22	24
Number of metaphases						
First embryo	2	15	6	1	–	3
Second embryo	3	4	–	8	1	2
Third embryo	1	21	2	1	1	–

Examination of mitoses in different organs such as the liver, the kidney, the spleen, and the gonads at later developmental stages from one to 18 months has shown that the variation in the chromosomal number is also characteristic of growing and adult trout; different numbers of chromosomes are found in different organs.

Similar results have been obtained in the familial analysis of the rainbow trout embryos (population Ropscha) (Table 5).

In this case, great differences have been found both between different families and between different embryos within one family. The scope of variation in each embryo is not very large. The somewhat greater relative abundance of karyotypes containing an even number of chromosomes has been observed just as in the experiments of Ohno.

The good quality of the photographs made by Kaidanova (Fig. 14) allows us to suggest that the errors made during the counting were minimal. The number of arms in almost all cases was identical and equal to 104.

The variation of the chromosomal number in the rainbow trout has been described in other papers as well (Fukuoka 1972b; Thorgaard 1976).

The marked variability "within the individual" has been found by Roberts (1970) in the Atlantic salmon (*Salmo salar*). Unfortunately, Roberts' figures appear

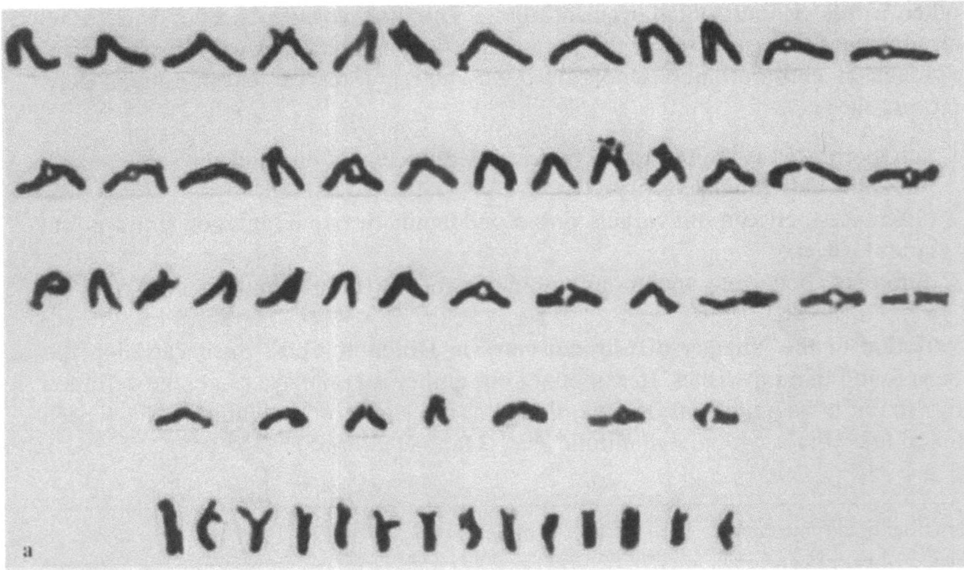

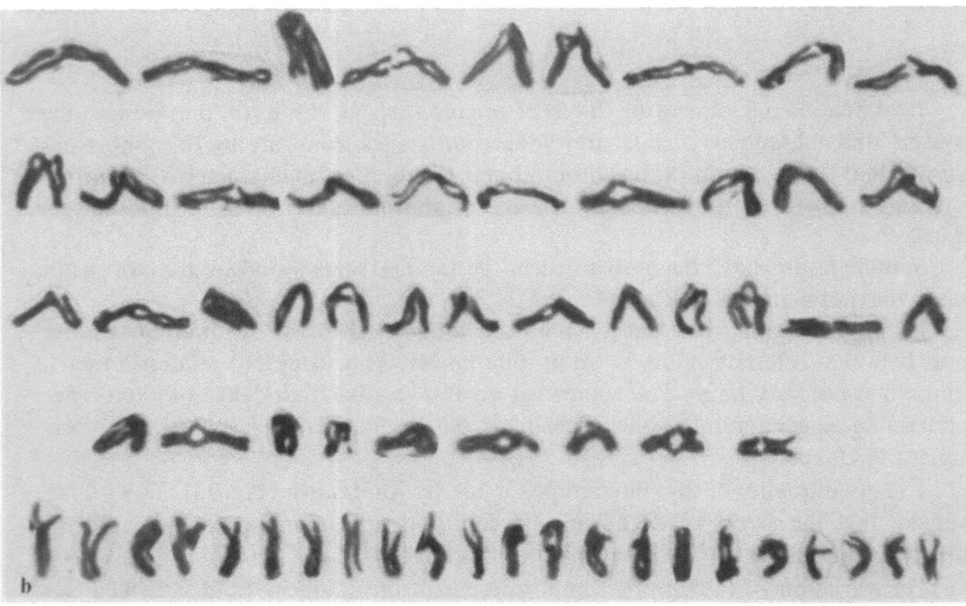

Fig. 14 a, b. Chromosomal polymorphism in the rainbow trout *Salmo gairdneri*. (Preparations and photography of T. I. Kaidanova). **a** 2n = 59; **b** 2n = 63. *First four lines in each picture* show metacentric and submetacentric chromosomes; the *lowest lines* show acrocentric chromosomes

Table 5. Karyotype variation in the rainbow trout (Kaidanova 1974)

Family No.	Embryo No.	Karyotypes (2 n)							The total number of counted metaphases
		58	59	60	61	62	63	64	
1	1	2	2	1	1				6
	2		1	4	1				6
	3			1	1	6			8
2	1				2	3	1		6
	2					6	1		7
3	1					6	4	1	11
	2					2	4	4	10

doubtful since mitoses were observed in cell cultures, in this specific case in the cells of ovaries. Occasionally, chromosomes might be lost during the cell division in the process of cultivation and in addition the metacentric chromosomes having two arms (7–8 pairs in the case of the Atlantic salmon) could undergo disjunction under cultivation.

Cellular polymorphism in the Atlantic salmon is quite prominent (Barshiene 1977a, 1978, 1980), but it appears to be only weakly related to Robertsonian translocations; this polymorphism apparently depends on the incorrect disjunction of homologous chromosomes and the elimination of some of them during mitosis. Two types of Atlantic salmon karyotypes have a different number of chromosomal arms.

Considerable data on the chromosomal variation "within the individual" at different stages of development of the sockeye salmon *Oncorhynchus nerka* have been presented by Chernenko (1968, 1971). Each embryo is characterized by a specific pattern of chromosomal variability, the karyotypes with 56, 57 or 58 chromosomes being found most frequently. The number of acrocentric chromosomes increases as the total number of chromosomes in the karyotype grows; this provides evidence that Robertsonian translocations and also other types of chromosomal rearrangements may be involved. Changes in ploidy have sometimes been found. In some cells haploid, triploid and tetraploid karyotypes have been detected. In the coho salmon *Oncorhynchus kisutch* polymorphism has been found as well, but the number of chromosomes varied over a narrower range (58 or 60) (Ohno et al. 1969a). Robertsonian translocations in these species only involve two pairs of chromosomes, the number of arms remaining strictly constant and equal to 104. The regular variation of the number of chromosomes in the karyotype has been found in one domesticated population of the green sunfish, *Lepomis cyanellus* (Beçak et al. 1966). Normal individuals had 46 acrocentric chromosomes, and individuals heterozygous with respect to the Robertsonian translocation had 46, 47 and 48 chromosomes. Somatic segregation with respect to the number of chromosomes occurred in heterozygotes, resulting in the formation of cells of three types (Fig. 15). In *Acheilognathus rhombea,* an Asian fish from the family Cyprinidae, somatic segregation had a somewhat different character – cells with three different but always an even number of chromosomes have been found in one and the same individual. In cells with 48 chromosomes all of them were acrocentrics, when 46 chromosomes were

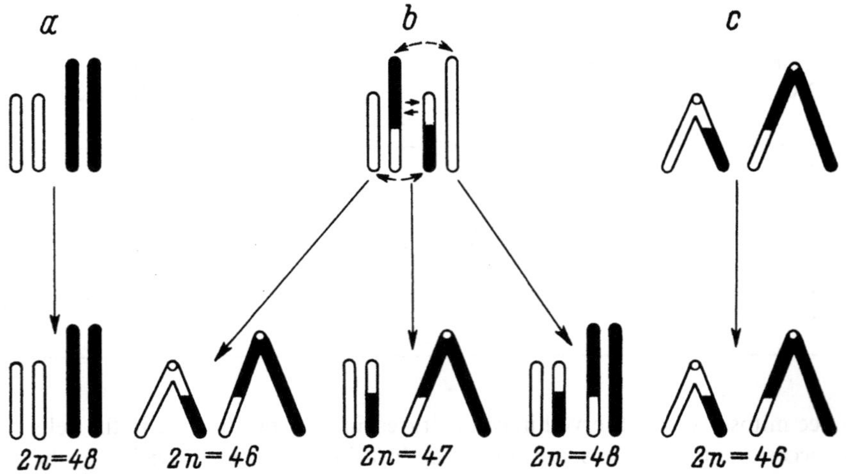

Fig. 15 a–c. The scheme of somatic segregation with respect to the number of chromosomes in the green sunfish *Lepomis cyanellus* (Beçak et al. 1966). Only two pairs of chromosomes involved in the translocation are shown. **a** normal homozygotes with 48 acrocentric chromosomes; **b** heterozygotes with respect to the Robertsonian translocation; **c** homozygotes with respect to the translocation having 46 chromosomes including two metacentrics

present, two metacentric chromosomes were seen in the karyotype and when the karyotype had 44 chromosomes four metacentric chromosomes were found (Nogusa 1955 b). The two translocations appeared to occur simultaneously in this case: four acrocentric chromosomes constituting two pairs underwent conversion into two metacentric ones, and, when eight acrocentric chromosomes were involved, four metacentrics resulted.

The chromosomal intra-individual polymorphism has been most frequently observed in salmonids. Ohno (1970 a) explains this polymorphism by the polyploid origin of this group and by the remaining affinity between the pairs of duplicated chromosomes. The chromosomal variation in the rainbow trout may be a direct consequence of its hybrid origin. The other mechanisms responsible for polymorphism in the course of development in addition to Robertsonian translocation may also operate among salmonids. They may involve non-disjunction, the elimination of smaller chromosomes and the complex interchromosomal rearrangements. Such transformations may affect the number of chromosomal arms as well (Barshiene 1977; Chernenko 1977; Gorshkova 1979).

Intrapopulational Chromosomal Variability. Differences in the chromosomal number in different individuals within a population have been found in several fish species. First of all such a variation was observed in most species with ontogenetic chromosomal variations due to centric fusions. Such species include *Acheilognathus rhombea, Lepomis cyanellus, Salmo irideus, Salvelinus malma* and *S. leucomaenis.* In one and the same stock of the rainbow trout individuals with different chromosomal modal numbers, 58, 60, 62 and even 64, have been found (Kaidanova 1974, 1976). The coincidence between the intraindividual and intrapopulational variation of the Robertsonian type is not accidental: the translocation affecting the centromeres, if present in heterozygous state, destabilizes the karyotype in the course of development.

Intrapopulational polymorphism has also been observed in a Black sea fish *Spicara flexuosa* (family Centracanthidae); the number of chromosomes in this species varies in the range 44–46 (Vasilyev et al. 1980). Individuals having 52 or 54 chromosomes have been found in the White Sea herring (Krysanov, 1978). The variability of the karyotype in certain populations of African toothcarps such as *Aphyosemion cognatum, A. bivittatum, A. calliurum* and *A. cameronense* has been observed (Scheel 1966, 1972 a). In all these cases, the Robertsonian translocation played an important part in the evolution of the karyotypes.

Intrapopulational chromosomal polymorphism unrelated to Robertsonian translocation definitely exists among fishes as well. The pattern of variation described by Barshiene (1977 a) in the Atlantic salmon and by Gorshkova (1978) in the sockeye salmon appears to depend on rearrangements resembling paracentric and pericentric inversions, but such rearrangements are difficult to demonstrate among fishes. The presence of trisomic karyotypes in populations of *Salvelinus* (Davisson et al. 1972) suggests that in the species containing many chromosomes, aneuploidy may exist as well. The variation in the number of microchromosomes in selachia and sturgeons has been proved beyond doubt. Finally, polymorphism with respect to ploidy has been observed. Populations of *Carassius auratus gibelio* with karyotypes containing about 100 and 150 chromosomes have been described (Cherfas 1966 b) and in the subspecies *Carassius auratus langsdorfii* three karyotypes have been reported containing 100, 150 and 200 chromosomes (Kobayashi et al. 1970, 1973).

The Chromosomal Variability Between Different Populations. The differences in the number of chromosomes between populations of one and the same species are particularly pronounced in the family Cyprinodontidae (Scheel 1972 a). A similar difference has been earlier established for *Aphanius chantrei,* belonging to the same family: for one of the populations 2n was equal to 46 and for another to 48 (Öztan 1954). Races with 46 and 48 chromosomes have been described in *Lepomis cyanellus* (family Centrarchidae) (Roberts 1964). In the loach *Cobitis biwae* (Cobitidae) two races with 48 and 96 chromosomes have been detected; the fishes of these races are easily distinguished by the size of their erythrocytes (Sezaki and Kobayashi 1978). Among Salmonidae the karyotypes of *Salmo gairdneri* (Kaidanova 1974), *Salmo clarkii* and *Salmo aguabonita* (Simon and Dollar 1963; Gold and Gall 1975), *Salvelinus malma* and *Salvelinus leucomaenis* (Viktorovsky 1975 a) also showed some interpopulation variation. The "spring" and "summer" races of *Oncorhynchus nerka* (Kamchatka variety) differ in the number of chromosomes, the modal number in the first case is 58 and in the second 56 (Gorshkova and Gorshkov 1978; Chernenko 1980; Gorshkova 1980). In two varieties of the whitefish, *Coregonus lavaretus maraenoides* and *C. l. ludoga,* living in Lakes Chudskoye and Ladoga the number of chromosomes was identical. However, the number of metacentrics in the ludoga karyotype was four less than in the other subspecies (Ruchkyan and Arakelyan 1980); the hybrid population acclimated in Lake Sevan was polymorphic.

Chromosomal differences between populations are generally observed in those taxons for which intrapopulational variations of the chromosomal number or variations associated with development can be found. This type of variation is not very widespread among fishes. The rearrangement of karyotypes during the formation of species appears to occur relatively quickly and at present it can only be detected in a few "hot spots" of fish evolution.

Sex Chromosomes

The authors studying fish chromosomes were unable to find heterochromosomes in them for quite a long time. This is explained by the fact that many fishes possess a primitive mechanism of sex determination as compared to higher vertebrates. Genuine "synchronous" hermaphrodites can occasionally be found among modern teleost fishes; in particular, these include species of the family Serranidae and a few species of other groups, for example, *Rivulus marmoratus* from the family Cyprinodontidae (Harrington 1963; Atz 1964; Ohno 1967). Synchronous hermaphroditism in a number of cases gives way to a sequential one, more frequently of protogynic type. Initially the gonad acts as an ovary, then the oocytes degenerate and regions with male germinal cells appear (Clark 1959; Fishelson 1970; Smith 1975). The genetic mechanism of sex determination is apparently lacking in all hermaphrodite fish species, both synchronous and gonochoric ones.

Occasional hermaphrodite individuals have been normally described in many fish species normally having two distinct sexes. In particular, hermaphrodites have been described among the whitefishes (Porter and Corey 1974), in the common carp (Kossmann 1971) and in the guppy (Spurway 1957). It should be pointed out that self-fertilization is possible in the two latter species.

Genuine genetic mechanisms of sex determination are, however, present in most fishes, including those with occasional hermaphrodites. Different species and sometimes even whole taxons are at different levels in the sequence of evolution of these mechanisms. The polygenic determination of sex belongs to the most primitive type: male and female genes are located in many chromosomes and the determination of sex depends on the balance of these genes. Examples of such mechanisms of sex determination are provided by the swordtail *Xiphophorus helleri,* as well as by *Limia vittata* and *L. caudofasciata* from the family of *Poeciliidae* (Kosswig 1965). Each individual gene affecting gonad development is relatively weak in action and therefore a change in environmental conditions and the variation in the genotype may easily be accompanied by changes in the sex ratio.

Sex determination mediated by sex chromosomes represents a more advanced type of sex regulation. Within the framework of this mechanism, the sex chromosomes, X and Y or W and Z are different, but the difference merely involves the existence of one or several specific male and female sex genes. In other respects, sex chromosomes are similar to one another and have one and the same size and shape. In certain viviparous fishes of the family Poeciliidae such as *Poecilia reticulata, P. variatus* and others as well in *Oryzias latipes* from the family Cyprinodontidae, the sex is determined via this mechanism; sex chromosomes have been marked using the genes responsible for pigmentation. The crossing-over between the X and Y chromosomes or between the W and Z chromosomes already appears to be difficult, but both the Y chromosome and the W chromosome are not destroyed; they are, however, indistinguishable under the microscope. Several populations of *P. maculatus* contain different X chromosomes, with female genes having different strength of action (Kallman 1973).

The determination of sex by chromosomes differing in one or a few of the genes is apparently characteristic of very many fish species, in particular of most species of Salmonidae and Cyprinidae. The genetic factors affecting the sex (acting apparently via the hormones determining the sex) are, however, present in autosomes as well. Nevertheless, the main part is played by the sex genes located in the sex chromosomes. Transdetermination of sex by environmental treatment such as temperature etc. is possible among fishes with sex chromosomes as well. Feeding the brood fishes with the male hormone (testosterone) or the female hormone (estrone) or the addition of these hormones to water sometimes result in the complete transformation of sex and in the appearance of the so-called "sex reversal" (Yamamoto 1955, 1958; Johnston et al. 1978); however, in the offspring of the transformed individuals the sex is again determined by sex chromosomes.

The hormonal transdetermination of sex allows us to estimate the type of heterogamety relatively easily without making a microscopic examination of the chromosomes and without any detailed genetic analysis. When females are homogametic, their transformation into males by testosterone and crosses of such sex reversal males with normal females results in offspring consisting solely of females. When females are heterogametic, a similar cross results in progeny consisting of both females and males:

1. ♂XX (transformed) × ♀XX = 100% ♀♀XX,
2. ♂WZ (transformed) × ♀WZ = 25% ♂♂ZZ + 75% ♀♀WZ + WW.

Experiments conducted with the goldfish have shown that in this species males are heterogametic (Yamamoto and Kajishima 1968; Yamamoto 1975a). The homogamety of females in the grass carp has been established with the aid of gynogenesis (Stanley 1976b).

The evolution of sex chromosomes has proceeded much further in many other fish families. True heterochromosomes, often markedly different in their size and structure, have been found recently in a number of species from different groups of taxons of fishes (Table 6). Male heterogamety of the type XX (♀♀), XY (♂♂) was most common. Less frequent was the complete absence of the Y-chromosome in males (XX-XO type) or the presence of multiple sex chromosomes (♀♀XXXX, ♂♂XXY).

By the end of 1979 male heterogamety had been established in 35 fish species. Multiple X-chromosome have been found in three representatives of Cyprinodontiformes (Uyeno and Milleer 1971, 1972; Levin and Foster 1972). In *Megupsilon aporus* (Cyprinodontidae) a large metacentric Y-chromosome formed apparently by the fusion of two acrocentric chromosomes is characteristic of the male karyotype; the X-chromosomes of this species are acrocentric (Fig. 16).

Clearly distinguishable sex chromosomes have been described in males of several species of deep-water fishes.

The female heterogamety has so far been found only in 14 fish species, including five species of Poeciliidae.

Table 6. Heterochromosomes in fishes

Order, family	Genera, species	Heterochromosomes	References
1	2	3	4
Selachiiformes			
Dasyatidae	*Dasyatis sabina*	XY	10
Salmoniformes			
Salmonidae	*Salmo gairdneri*	XY	34
	Oncorhynchus nerka	XO (?)	35
Galaxiidae	*Galaxias platei*	XO	3
Sternoptychidae	*Sternoptyx diaphana*	XO	11
Bathylagidae	*Bathylagus milleri*	XY	12
	B. ochotensis	XY	11
	B. stibbins	XY	11
	B. wesethi	XY	11
Myctophiformes			
Neoscopelidae	*Scopelengys tristis*	XY	11
	Two species (non-ident.)	XY	11
Myctophidae	*Lampanyctus ritteri*	XO	7
	Symbolophorus californiensis	XY	7
	Two species (non-ident.)	XY	5
Anguilliformes			
Anguillidae	*Anguilla anguilla*	XY (?)	17, 25
	A. rostrata	ZW	22
	A. japonica	ZW	24
Congridae	*Astroconger myriaster*	ZW	24
Cypriniformes			
Cyprinidae	*Vimba vimba*	XY	32
	Carassius auratus	XY	23
Erythrinidae	*Hoplias lacerdae*	XY	2
Siluriformes			
Bagridae	*Mystus tengara*	ZW	26
Loricariidae	*Plectostomus anastroides*	XY	18
	Pl. macrops	XY	18
Cyprinodontiformes			
Cyprinodontidae	*Fundulus diaphanus*	XO	9
	F. parvipinnis	XO	9
	Germanella pulchra	XXY, XXXX	16
	Megupsilon aporus	XXY, XXXX	36
Poecilidae	*Gambusia affinis*	ZW	4,6
	G. nobilis	ZW	4,6
	G. gaigei	ZW	4,6
	G. hurtadoi	ZW	4,6
	Mollienesia sphenops	ZW	31
	Xiphophorus maculatus	XY	13
	X. xiphidium	XY	13
	Sp. nova	XXY, XXXX	37
Bericiformes			
Melamphaeidae	*Melamphaeus parvus*	XY	11
Anoplogasteridae	*Scopelogadus mizolepis*	XY	11
	Scopeloberyx robustus	XY	11

Table 6 (continued)

Order, family	Genera, species	Heterochro-mosomes	Refer-ences
1	2	3	4
Gasterosteiformes			
Gasterosteidae	*Gasterosteus wheatlandi*	XY	8
	Apeltes quadracus	ZW	8
Perciformes			
Percidae	*Perca fluviatilis*	XY (?)	17
	Acerina cernua	XY (?)	17
Periophthalmidae	*Boleophthalmus boddaerti*	ZW	33
Gobiidae	*Gobiodon citrinus*	XO	1
	Mogrunda obscura	XY	20
Belontiidae	*Colisa lalius*	ZO	28
	C. fasciatus	ZO	30
Scatophagidae	*Scatophagus argus*	XY	14
Osphronemidae	*Trichogaster fasciatus*	ZW	27
Cichlidae	*Callichromus bimaculatus*	XY	29
	Geophagus brasilliensis	XY	19
Scorpaeniformes			
Cottidae	*Cottus pollyx*	XY	20
Pleuronectiformes			
Gynoglossidae	*Symphurus plagiusa*	XO	15

References: 1. Arai and Sawada (1974); 2. Bertollo et al. (1978); 3. Campos (1972); 4. Campos and Hubbs (1971); 5. Chen (1969); 6. Chen and Ebeling (1968); 7. Chen and Ebeling (1974); 8. Chen and Reisman (1970); 9. Chen and Ruddle (1970); 10. Donahue (1974); 11. Ebeling and Chen (1970); 12. Ebeling and Setzer (1971); 13. Foerster and Anders (1977); 14. Khuda-Bukhsh and Manna (1974); 15. Le Grande (1975); 16. Levin and Foster (1972); 17. Lieder (1963); 18. Michele et al. (1977); 19. Michele and Takahashi (1977); 20. Nogusa (1955a); 21. Nogusa (1957); 22. Ohno et al. (1973); 23. Ojima and Takai (1979); 24. Park and Kang (1976); 25. Passakas and Klekowski (1972); 26. Rishi (1973); 27. Rishi (1975); 28. Rishi (1976a); 29. Rishi (1976b); 30. Rishi (1979); 31. Rishi and Gaur (1976); 32. Rudek (1974); 33. Subrahmanyam (1969); 34. Thorgaard (1977); 35. Thorgaard (1978); 36. Uyeno and Miller (1971); 37. Uyeno and Miller (1972)

It can be assumed that Y-chromosomes (or W-chromosomes in the case of female heterogamety), in all cases when they differ greatly from their partners (X- or Z-chromosomes), are destroyed to a greater or a lesser extent. There is virtually no crossing-over between the X and Y or between the W and Z in such cases.

Taken together, as regards mechanisms of sex determination fishes represent an extremely heterogeneous group.

Nonchromosomal Heredity

No single example of nonchromosomal inheritance has so far been found in fishes. The only exception is the rather widespread phenomenon of matroclinal inheritance or the preferential transmission of maternal traits in certain crosses. Matroclinal in-

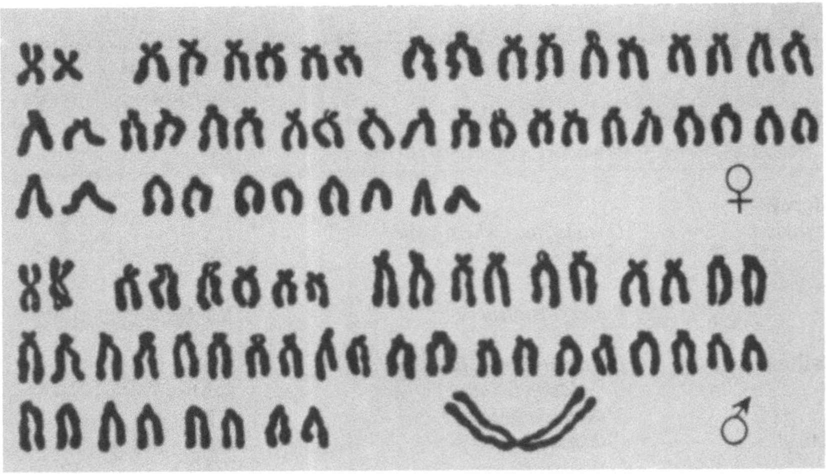

Fig. 16. The female and male karyotypes of *Megupsilon aporus*. The female has 44 autosomes and four X-chromosomes; the male has 44 autosomes, two X-chromosomes and a gigantic Y-chromosome (Uyeno and Miller 1971)

heritance can be very pronounced in crosses of the domesticated carp with the wild-living carp variety (Kirpichnikov 1949), as well as in crosses between the common and crucian carps, *Cyprinus carpio* and *Carassius carassius* (Nikoljukin 1952). The matroclinal inheritance can be based on one or more of the following mechanisms:

a) The accumulation of large quantities of products of the maternal chromosomes (mRNA and proteins) in the cytoplasm and in the yolk of the mature egg. This results in a certain dominance of maternal traits at the early stages of embryonic development; it should also be recalled that the male parent's genes start to work only at the stage of the late blastula and sometimes even later.

b) The presence in the cytoplasm of the structures containing DNA and directing the synthesis of several specific proteins. The primary candidates are mitochondria. The transmission of such non-chromosomal genes is maternal, that is through the egg cytoplasm, and can result in stable maternal inheritance.

Most cases of matroclinal inheritance in fishes appears to be due to the after-effects of the maternal genotype; this follows from the transmission of maternal traits only in the first generation and their disappearance in subsequent generations.

Evidence of the genuine non-chromosomal inheritance will apparently be found in fishes; such inheritance has been now reported in a number of other animals, in *Drosophila* in particular, and in many plants.

Chapter 2

The Genetics of Fish Grown in Fish Ponds and Living in Natural Water Bodies

The Main Principles of Mendelian Inheritance

Long before the advent of the chromosomal theory of inheritance, Mendel in his classical paper *Versuche über Pflanzenhybriden* (Mendel 1865) formulated the main principles of the transmission of hereditary traits to the offspring. These principles include the rule of uniformity of the first generation of hybrids, the rule of segregation of traits in the second generation of hybrids, the rule of gamete purity and the rule of the independent assortment of different traits in the offspring. These laws were rediscovered and confirmed at the beginning of the twentieth century simultaneously by Correns, Chermak and De-Freese.

Mendelian rules are based on the specificity of chromosomal behaviour in the course of gamete maturation (meiosis) and fertilization. The most important part is played by the rule of gamete purity. Chromosomally, the purity of the gametes is based on the presence of just one of the two homologous chromosomes in the nucleus of each gamete. Each gene (the anlage according to Mendel) comes to the gamete together with the chromosome in which it is located. It is not affected by the genes located in another homologue. Different forms of one and the same gene are called alleles. If two different alleles of one gene (locus) are present in the hybrid in two homologous chromosomes, only one of them can come to each gamete after reductional division. As both types of gametes are formed in equal numbers, this provides a basis for the rules of assortment of traits (and genes) in the second generation. After the fusion of gametes during fertilization, chromosomes with different alleles are combined in accordance with the laws of statistics. In the absence of complicating circumstances four types of zygotes are formed in approximately equal numbers (Fig. 17).

The inheritance of many traits in plants and animals follows the law of dominance: one of the alleles in the hybrid zygote is expressed "more strongly" than the other and suppresses the weaker allele in the course of development. Such stronger alleles are called the dominant ones; correspondingly, the weaker alleles are called recessive. When the dominance of the allele A over the allele a is complete, all the descendants having allele A in one of the chromosomes will resemble the parent which has donated the chromosome with the gene A. In the second generation of hybrids we observe the classical Mendelian ratio 3:1 (75% AA and Aa, 25% aa). When the dominance is complete, the AA homozygotes are indistinguishable from the heterozygotes Aa.

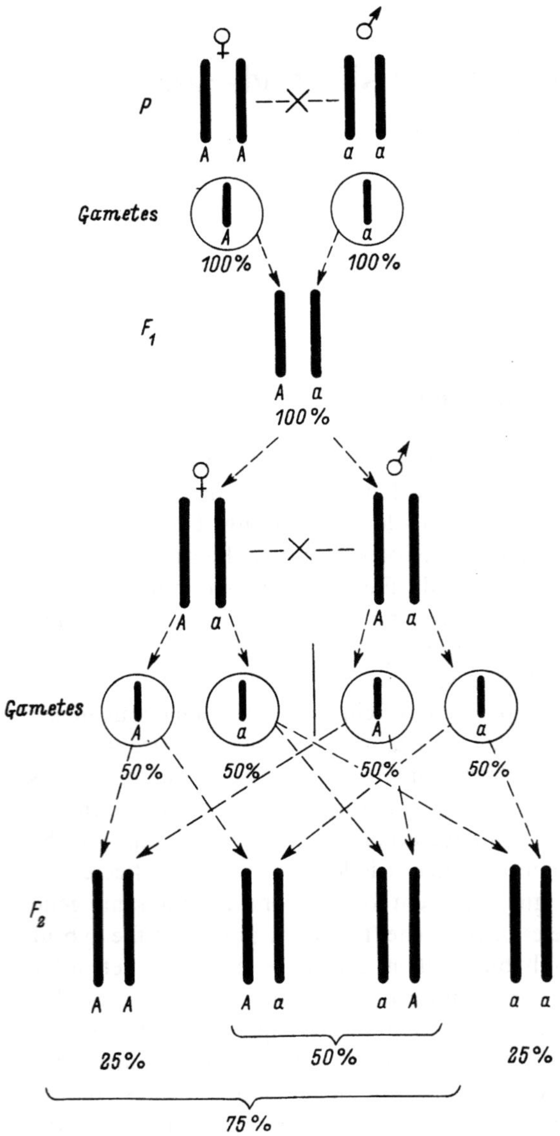

Fig. 17. The scheme of assortment of chromosomes and genes in meiosis and fertilization illustrating the basis of Mendelian rules of homogeneity of the first hybrid generation and segregation in the second hybrid generation

If AA homozygotes differ from Aa heterozygotes, the second generation of hybrids in assorted into three groups, AA, Aa and aa in the proportion 1:2:1 (25% AA, 50% Aa and 25% aa). This proportion represents a more common universal consequence of the principle of gamete purity, because it directly reflects the main feature of chromosome assortment in meiosis: the random segregation of the chromosomes of each pair in the course of the reduction and random fusion of gametes possessing different sets of allelic genes.

Dominance is, certainly, rarely complete. In a number of criteria, most frequently external ones, a gene may be dominant, however, in other criteria, sometimes even very important ones, the same gene turns out to be semidominant. The end products of most genes with the exception of those responsible for the synthesis of ribonucleic acids are polypeptides. In the homozygote AA only one such polypeptide coded by the allele A is formed. A different polypeptide coded by the a allele is synthesized in the aa homozygote; in the heterozygote Aa both polypeptides can be synthesized and, in addition the third hybrid form can be synthesized. Therefore in most cases all three genotypes, AA, Aa and aa differ distinctly in the products of protein synthesis.

After the so-called backcrosses of the Aa × AA type, the ratio 1:1 is observed in the offspring because two types of gametes, A and a (each representing 50% of the total) are only formed in heterozygous individuals.

Perhaps the most important manifestation of the rule of gamete purity is the division of gametes of any individual heterozygous with respect to any one gene into two groups of identical size. This division basically represents the foundation of all, even the most complex principles of inheritance.

The independent segregation of traits is completely explained by the independent assortment of chromosomes of different pairs into gametes during meiosis. Having two pairs of genes and traits (A and a, B and b) and two pairs of chromosomes, we find that 16 combinations of genes are possible in the second hybrid generation (F_2). These combinations can be most illustratively presented using a Punnett square giving the formulae of gametes and zygotes (see p. 48).

When the dominance is complete, the offspring can, according to Mendel, be divided into four phenotypic groups: AB, Ab, aB and ab, their proportion being equal to 9:3:3:1. If dominance is incomplete, this proportion becomes more complex (4:2:2:2:2:1:1:1:1). Nine classes of descendants appear, the most abundant of which are the double heterozygotes AaBb.

When segregation involves three independent anlages, i.e., three genes located in different pairs of chromosomes, the number of different gamete types in each of the parents is equal to eight (2^3), and the number of different zygotes calculated using the Punnett square is equal to 64 (4^3).

Segregation, resulting from the rules of gamete purity and the random assortment of chromosomes during meiosis may be changed if one of the genes suppresses the manifestation of the other non-allelic gene. This phenomenon is called epistasis. For example, in cases of albinism, the manifestation of the other genes responsible for pigmentation is frequently suppressed in albino individuals homozygous with respect to the recessive genes and the ratio 9:3:3:1 changes to the ratio 9:3:4. If in the presence of the gene A, differences between the genotypes Bb and bb are indistinguishable, we obtain the ratio 12:3:1 when dominance is complete. Many other numerical relations are also possible.

The genes causing death in a homozygous state or lethal genes are found very frequently in the analysis of the inheritance of different traits. If the aa homozygotes are non-viable then only individuals with the genotypes AA and Aa survive in the second hybrid generation. This generation will therefore be either uniform or, in the presence of differences between AA and Aa, can be divided into two groups in the proportion two heterozygotes to one homozygote. Early embryonic mortality equal to 25% of the offspring has been observed in many animal and plant species for certain combinations of parents. The lethal genes are found in

	♂ \ ♀	Gametes: A B	A b	a B	a b
Gametes	A B	$\frac{A\ B}{A\ B}$	$\frac{A\ B}{A\ b}$	$\frac{A\ B}{a\ B}$	$\frac{A\ B}{a\ b}$
	A b	$\frac{A\ B}{A\ b}$	$\frac{A\ b}{A\ b}$	$\frac{A\ B}{a\ b}$	$\frac{A\ b}{a\ b}$
	a B	$\frac{A\ B}{a\ B}$	$\frac{A\ B}{a\ b}$	$\frac{a\ B}{a\ B}$	$\frac{a\ B}{a\ b}$
	a b	$\frac{A\ B}{a\ b}$	$\frac{A\ b}{a\ b}$	$\frac{a\ B}{a\ b}$	$\frac{a\ b}{a\ b}$

Zygotes

natural popoulations to no lesser extent than among cultivated plants and domestic animals.

Using the fruit fly *Drosophila* as a very convenient laboratory model with a short generation time, Morgan and co-workers found the so-called gene linkage resulting from the presence of different genes in one and the same chromosome. Such genes are inherited together but during meiosis, when homologous chromosomes transiently form conjugates, these chromosomes frequently exchange parts in the process of crossing-over. Crossing-over leads to the appearance of a certain number of new combinations of genes. The frequency of crossing-over is proportional to the distance between genes in the chromosomes, and this rule makes it possible to construct genetic maps of chromosomes with a rather high resolution.

When, for instance, the frequency of crossing-over is equal to 10%, part of the gametes (10%) will carry cross-over chromosomes. Consequently, the probability of the appearance of zygotes carrying cross-over chromosomes will be equal to 10%. The unit of distance between the genes is defined as a distance giving a crossing-over frequency equal to 1%, this unit is called the morganid.

It should be pointed out that genes located in sex chromosomes are inherited in a somewhat unusual fashion. The females of *Drosophila* carry the XX sex chromosomes and the males the XY chromosomes. The X-chromosome contains many genes while the Y-chromosome of *Drosophila* is almost empty and has no genes allelic with respect to those located in the X. The gene w (a recessive gene coding for the white pigmentation of the eyes in the flies) is located in the X chromosome. When homozygous females having white eyes are crossed with males having red-eyes, a peculiar crosswise inheritance is observed:

$$♀X^wX^w \times ♂X^WY = ♀♀X^WX^w + ♂♂X^wY$$

white eyes | normal eyes | normal eyes | white eyes.

An unusual pattern of inheritance is also found when heterozygous females are crossed with a normal male having red eyes:

♀X^WX^w ×	♂X^WY =	♀♀X^WX^W +	♀♀X^wX^W +	♂♂X^WY +	♂♂X^wY
normal eyes	normal eyes	normal eyes	normal eyes	normal eyes	white eyes.

The "crosswise" inheritance is characteristic of all the traits and genes linked with sex.

In the above presentation we have outlined the main features of the inheritance of genes located in autosomes and in sex chromosomes of all the diploid bisexual organisms. These principles are completely applicable to fish. The hereditary variation in fishes is as diverse as in other animals. The following four groups of hereditary differences can be distinguished.

1. Qualitative morpho-anatomical traits of alternative type inherited according to Mendel and yielding distinct and clear segregation in the offspring after crosses of individuals differing in these traits.

2. Quantitative differences with respect to various morphological and physiological traits having a polygenic inheritance. The expression of such traits not only depends on genetic factors, but also on many variable environmental factors. The interactions of genes affecting a specific trait are frequently additive in nature, the effects of different genes are simply summed up for such situations. More complex non-additive interactions have also been described; these involve the dominance, i.e., the suppression of one gene by another "allelic" gene; epistasis, the suppressing or modifying action on the manifestation of a non-allelic gene, and finally the so-called overdominance or increased manifestation of a character in a heterozygote as compared to both homozygotes.

Typical quantitative traits showing variation in most fish species include the number of vertebrae and of fin rays, the number of scales and gill rakers, exterior features such as the length and weight of the body, oxygen consumption and resistance to high and low temperatures, etc.

3. The biochemical differences expressed as a variation with respect to blood groups or the presence of several forms of one and the same protein synthesized under the control of different genes or of different alleles of one gene. The presence of multiple forms of proteins, isozymes and isoforms has recently been established for practically all genes coding for enzymes and other proteins (Markert 1975). Allelic differences in proteins are inherited similarly to differences in the qualitative traits in accordance with Mendelian laws.

The great advances achieved recently in protein chemistry and in particular the development of methods of achieving the high resolution of proteins using electrophoresis in starch and polyacrilamide gels have resulted in the appearance of a new, rapidly developing branch of genetics – biochemical genetics. A large number of studies in this field has been made with fishes.

4. Phenodeviants (Lerner 1954); this concept refers to malformations and other aberrations, which are generally poorly manifested and inherited in a complex fashion. Phenodeviants are frequently found in natural fish populations, they are even more frequent in populations of domesticated species. The frequency of their occurrence and the degree of change in a trait do, to a large extent, depend on

Fig. 18 a–d. Types of scaling in the common carp *Cyprinus carpio.* **a** scaled (SSnn and Ssnn); **b** scattered (ssnn); **c** linear (SSNn and SsNn); **d** nude or leather (ssNn)

1. ♀ssNn nude	+ ♂ssnn scattered	= ssnn scattered	+ ssNn nude	
2. ♀ssNn nude, the same	× ♂ssNn nude	= ssnn scattered	+ 2ssNn nude	+ ssNN perished

Fertilized eggs were glued to a glass and incubated in running water at a temperature of 14°–18 °C. The quantitation of embryo mortality on specially selected squares has provided the following results:

	Cross No. 1	Cross No. 2
Death during the first days of incubation (%)	18.8	20.0
Death on the last day of incubation (%)	1.5	19.5
The total number of eggs in the experiment	6072	7156

We expected that the mortality in the cross No. 2 would be equal to about 25% of the total number of fertilized eggs, but the actual difference in the survival between crosses No. 2 and No. 1 was 18%. Presumably about 7% of the embryos hatched successfully and perished at a later stage.

Similar experiments were performed in the late 1940's. at higher temperatures, 20°–25° (Probst 1950). Prior to hatching mortality was similar in the experiment and in the control, but the offspring from the cross of a nude female with nude and linear males contained about 25% of the larvae with characteristically distorted "curved" appearance. These larvae did not as a rule survive later stages of development. The total number of non-viable larvae having the very characteristic shape of "the comma" was as follows: lethal crosses (Nn × Nn) – 25.9% (461 from 1778 embryos); control (Nn × nn) – 0.8% (14 from 1805 embryos). As we can see, 25% of all the embryos died. The viable progeny in lethal crosses could be divided into two groups according to scale cover and the proportion of these two groups was equal to 2:1. Segregation typical of recessive lethals has been observed in our experiments just as in those of Probst (1953).

The elevation of environmental temperature during the embryonic development of the carp results in the hatching occurring at earlier stages of development; as a result the death of lethal homozygotes takes place after hatching. When temperatures are low the embryos die while still in the membranes. Unfortunately, no detailed morphological and histological examination of non-viable embryos and larvae has been conducted so far.

The gene S is dominant with respect to the gene s; in crosses of two heterozygous scaled carps we obtained the classical Mendelian ratio 3:1 in the offspring; the ratio in the backcross was 1:1. The results of three crosses are presented as an example:

1. Ssnn × Ssnn = 15,690 scaled (75.9%) + 4980 scattered (24.8%) (Kirpichnikov 1948),
2. Ssnn × Ssnn = 3526 scaled (76.4%) + 1089 scattered (23.6%) (Probst 1953),
3. Ssnn × ssnn = 3616 scaled (50.5%) + 3544 scattered (49.5%) (Probst 1953).

The scattered carps with the ss genotype are quite viable, but when the environmental conditions are unfavourable the number of scaled individuals after backcrosses increases to 52%–55%, this being due to their higher viability (Kirpichnikov 1945).

The dominance of the gene S is apparently incomplete. Heterozygous scaled carps grow somewhat faster than homozygous ones (Kirpichnikow 1966 b); irregularities in the position of the scales on the body are more frequently found in heterozygous individuals (Steffens 1966).

Differences in the viability between carp with the genotypes Nn and nn are much greater; the survival of heterozygotes is greatly decreased. Even under comparatively favourable conditions of growth marked deviations from the expected ratios take place. For example, in crosses of the wild scaled carp with the nude carp an equal proportion of linear and scaled descendants might be expected:

♀ wild carp × ♂ nude carp

SSnn ssNn

↓ ↓

SsNn (50%) linear — Ssnn (50%) scaled

In practice, however, about 60% of the hybrid fry are scaled and about 40% are linear at the age of three to four months.

The hypothesis of inheritance of the main types of carp scale cover advanced by us has been confirmed by subsequent studies conducted in many countries. Apparently, the genes N and s emerged (or were detected) in the European carp *Cyprinus carpio carpio* soon after its domestication as two independent mutations. The mutation S → s has resulted in the appearance of mirror carp of scattered type, while the mutation n → N has resulted in the emergence of linear carp. Both scattered and linear carp show a partial reduction in the number of scales and an increase in their size. A subsequent combination of the two mutant genes in one individual has resulted in the appearance of nude carp (ssNn) which are almost completely and sometimes absolutely devoid of scales.

Individuals with a scattered scaling pattern, and linear and nude specimens have also been found in Japan among domesticated carps belonging to another subspecies isolated from the European one long ago, *C. carpio haematopterus*. The inheritance of the patterns of scales in European and Japanese carp is similar (Katasonov 1971). Scattered and linear carp have also been described among the Vietnamese subspecies *C. carpio viridiviolaceus* (fossicola) (Kirpichnikov 1967 b; Trâň dinh-Trong 1967) although no analysis of the type of inheritance has been made.

Thus, the homologous genes responsible for the pattern of scaling have appeared among carp belonging to three subspecies with different distribution ranges and in addition differing greatly both morphologically and physiologically. This provides an impressive example of the manifestation in fishes of the law of homologous series of hereditary variation discovered by N. I. Vavilov in 1920.

We shall now present the theoretical results of all the possible crosses of carp with different types of scale cover (Table 7). In all cases, segregation in the experiments corresponded to the expected one. The results of crosses between linear carp can provide a specific example (theoretical frequencies are given in parentheses):

	Scaled	Scattered	Linear	Nude	
1. ♀SsNn × ♂ssNn	= 758	+ 758	+ 1406	+ 1426	
linear nude	(725)	(725)	(1450)	(1450)	(Golovinskaya 1946)
2. ♀SsNn × ♂SsNn	= 2263	+ 721	+ 4454	+ 1455	
linear linear	(2223)	(741)	(4447)	(1488)	(Probst, 1949 b)
3. ♀SsNn × ♂SsNn	= 343	+ 109	+ 568	+ 184	
linear linear	(301)	(100)	(602)	(201)	(Wohlfarth et al. 1963)

The differences in the survival of carp carrying the gene N and devoid of this gene markedly increase when fishes are kept under unfavourable conditions. In the experiment conducted in 1940 the offspring resulting from the cross of the scaled carp with the nude carp (♀Ssnn×♂ssNn) were placed into three ponds with very different levels of productivity (Kirpichnikov 1945). The number of one-year-old fishes caught was as follows (the data are given as a percentage of the total):

	Scaled	Scattered	Linear	Nude
Under favourable conditions	27.8	25.4	24.4	22.4
Under intermediate conditions	34.8	29.4	18.4	17.4
Under poor conditions	38.4	36.5	14.0	11.1

The decreased survival of carp carrying the gene N is a result of the unfavourable action of this gene upon a large number of traits (Kirpichnikov et al. 1937). The alleles of the other gene S and s also have a pleiotropic effect, but it is much less pronounced. Many organs of the carp show changes depending on the genes responsible for the scaling patterns. This is accompanied by changes in the morphological and physiological characteristics as well (Table 8). The weight of scaled carp is generally somewhat greater than that of the scattered individuals, particularly if the fishes do not obtain additional food (Zelenin 1974). Linear and nude

Table 7. Inheritance of the pattern of scaling in the carp

Parents (independently of sex)	Number of offspring (%)			
	Scal.	Scat.	Lin.	Nude
Scaled × scaled	100	–	–	–
	75	25	–	–
Scaled × scattered	100	–	–	–
	50	50	–	–
Scaled × linear	50	–	50	–
	37.5	12.5	37.5	12.5
Scaled × nude	50	–	50	–
	25	25	25	25
Scattered × scattered	–	100	–	–
Scattered × linear	50	–	50	–
	25	25	25	25
Scattered × nude	–	50	–	50
Linear × linear	33.3	–	66.7	–
	25	8.3	50	16.7
Linear × nude	33.3	–	66.7	–
	16.7	16.7	33.3	33.3
Nude × nude	–	33.3	–	66.7

Here and below: *Scal.* scaled, *Scat.* scattered, *Lin.* linear

carp grow more slowly than the others, this delay becoming more prominent where there is insufficient food (Kirpichnikov 1948). Although linear and nude carp grow somewhat more slowly, the coefficients of food conversion determined for them are greater (Tscherbina and Tsvetkova 1974). Another peculiar feature of these latter carp is a more intensive fat metabolism. Fat accumulates faster in linear and nude carp during the summer and is consumed in greater quantities during the winter as compared to scaled and scattered carp (Tsvetkova 1969). Perhaps the decreased resistance of the fingerlings carrying the gene N to wintering is related to these particular characteristics.

The reduction of the gill organs and the decrease in the number of pharingeal teeth are perhaps the main factors responsible for the delayed growth of linear and nude carps. Instead of the three rows of teeth characteristic of the common carp living in nature (the formula 1.1.3–3.1.1) the linear and nude carp possess two rows of teeth and sometimes even just one row; the tooth formula in these animals shows a high diversity (1.3–3.1.1, 1.3–3.1, 1.3–3 and sometimes even 3–3).

The action of the genes s and N upon the fins of the carp is very characteristic. In the scattered carp the number of soft rays in the dorsal fin is generally somewhat decreased if they are compared with the carp with normal scaling pattern. The number of rays in the ventral and pectoral fins is decreased as well. In the presence of the gene N the action of the gene S is stronger. The reduction of all the fins (dorsal, anal, ventral and pectoral) in the nude carp (s, N) is expressed to a markedly greater extent than in linear ones (S, N). The reducing effect of the gene N is, however, many times greater. The fins of the scaled and scattered carp have a normal structure. In linear and nude carp certain soft rays in the middle and

Table 8. The pleiotropic action of "scale" genes in the carp

Indices	Carp phenotypes and genotypes				Reference
	Scaled SSnn, Ssnn	Scattered ssnn	Linear SSNn, SsNn	Nude ssNn	
Weight of one-year-old fishes, favourable conditions [a]	100	93–96	85–88	79–80	4, 8, 11
Same, but unfavourable conditions [a]	100	83–94	42–70	37–72	4, 11
Weight of two-year-old fishes [a]	100	94–96	86–91	83–84	11, 12
Mean number of the soft rays in the dorsal fin (D)	18.8 (17–22)	18.7 (17–22)	16.4 (12–19)	15.4 (5–18)	1, 4
Same but in the anal fin (A)	4.96	5.00	3.82	3.56	4
Mean ray number in the ventral fin (V)	8.91	8.68	8.76	8.47	4
Mean number of the soft rays in the pelvic fin (P)	14.7	14.3	14.3	13.1	12
Number of the gill rakers (variation of the mean)	24.6–25.1	24.3–24.8	19.4–21.6	18.5–20.5	1, 4, 12
Mean number of the gill lamellae	88.6	83.5	82.3	83.2	12
Mean number of the pharyngeal teeth	9.22	9.58	7.63	7.44	4
l/H index (the ratio of body length to the maximal body height), variation of the mean, German carps	2.33–2.77	2.26–2.74	2.35–2.86	2.35–2.82	12
Ability to fin regeneration [a]	100	76	39	19	11
Length ratio of the posterior and anterior chambers of the air bladder	> 1	< 1	–	–	1, 2, 4, 6
Erythrocyte count (10^6/ml)	1.93	1.99	1.76	1.69	9
Haemoglobin (g/%)	9.02	8.87	8.18	8.28	9
Critical temperature (°C)	37.6	37.5	36.8	36.6	9
Survival time (min) under oxygen deficit	210	210	132	132	9
Immunologic reactivity	Fast	Fast	Slow	Slow	5
Resistance to dropsy	–	Increased	–	Decreased	10
Intensity of fat metabolism	Low	Low	High	Very high	3, 7
Total survival of one-year-old fishes, optimal conditions	100	91–98	87–93	80–92	4, 11
Same, unfavourable conditions [a]	100	93–95	36–37	28–60	4, 11

[a] Expressed as percentage of the value in scaled carps taken for 100

References: 1, 2. Golovinskaya (1940, 1965); 3. Golovinskaya et al. (1974b); 4. Kirpichnikov (1945, 1948); 5. Lukyanenko and Sukacheva (1975); 6. Popova (1969); 7, 8. Tsvetkova (1969, 1974); 9. Chan May-Tchien (1969); 10. Merla (1959); 11. Probst (1953); 12. Steffens (1966)

Fig. 19. Different degrees of reduction of the dorsal fin in the linear and nude (leather) common carp

sometimes in the posterior part of the dorsal fin do not develop at all and the total number of rays decreases markedly. As a consequence the dorsal fin acquires a peculiar form (Fig. 19). The great reduction generally involves the anal fin as well, the ventral and pectoral fins are affected to a lesser extent. The number of hard rays in the dorsal and anal fins is decreased as well.

We have seen thus that the gene N leads to the disorder of developmental homeostasis in the common carp. It is particularly clear in the case of the interaction of the N and s genes. In the presence of the gene N (in Nn heterozygotes) the action of the gene s is generally increased while this latter gene is relatively weakly expressed in the homozygotes nn. This stimulation of the gene action involves many traits including the number of gill rakers, the content of erythrocytes in the blood, the intensity of fat metabolism, etc.

It is interesting to consider the differences between the two groups of carp with Nn and nn genotypes observable in the resistance to higher temperatures and oxygen deficiency, in the level of erythrocytes and haemoglobin content in the blood and in the ability of the fins to regeneration. By all these indices linear and nude carp are inferior as compared to scaled and scattered varieties, and the differences are statistically significant (Probst 1953; Chan May-Tchien 1969). The surprisingly wide range of action of the gene N can be explained solely by the assumption that this gene is expressed at very early stages of development and affects some very essential morphogenetic processes. According to Probst (1953) the action of the gene N is associated with a defect in the development of the mesenchyme. The gene N apparently represents a result of the large mutation, most probably, a chromosomal rearrangement involving deletion affecting a small region of the chromosome. In NN homozygotes the synthesis of one or several vitally

other genes present in the genotype as well as on the environment. Inbreeding and unfavourable environmental conditions contribute to the appearance of phenodeviants and stimulate the manifestation of such traits. The genetic analysis of traits of this type is generally very difficult.

In this chapter we shall consider the genetics of qualitative traits in common carp and in other pond, lake and sea fish species. Chapter 3 will be devoted to the genetics of aquarium fishes. Genetics of the quantitative characters and biochemical genetics will be examined in the following chapters.

The Inheritance of Qualitative Traits in the Common Carp (Cyprinus carpio L.)

Until recently the common carp represented the only model for genetic studies among the commercially valuable fish species. More recently communications have appeared describing the pattern of inheritance of certain morphological and physiological traits in other commercially valuable fishes, mainly among cyprinids, salmonids, catostomids, percids and silurids. Extensive studies have been conducted with aquarium fishes, predominantly with several species from the families Cyprinodontidae and Poeciliidae (oviparous and viviparous tooth carp).

The Genetics of the Differences in the Scale Cover. Rudzinski (1928) was the first author to publish data on crosses of carp with different scale cover. He showed that the continuous pattern of scaling in the carp dominates over the "mirror" type of scale distribution. Subsequent studies (Kirpichnikov and Balkhashina 1935, 1936; Kirpichnikov 1937, 1945, 1948; Golovinskaya 1940, 1946; Probst 1949 b, 1950, 1953) made it possible to identify in the carp two pairs of autosomal genes not linked with one another, that is located in different pairs of chromosomes, and determining the pattern of scaling. The following genotypes and phenotypes of the common carp are possible (Fig. 18):

SSnn, Ssnn	– scaled
ssnn	– "scattered" mirror
SSNn, SsNn	– linear mirror
ssNn	– "nude" or "leather"

Carps with the genotypes SSNN, SsNN and ssNN are non-viable embryos which have received two N genes perish at the stage of hatching or soon after the emergence of the larvae from the membranes. This phenomenon was established by Kirpichnikov and Golovinskaya in 1937 (Golovinskaya 1946). The mature eggs of a nude female were divided into two portions, each of which was mixed with the sperm of one of two males differing in the phenotype and the genotype. The presumed formulae of the crosses were as follows:

Fig. 20. Hereditary variations in the location and number of scales in the scattered common carp

important proteins is apparently impaired and therefore the homozygous individuals are unable to survive. The protein synthesis is apparently greatly impaired in the heterozygotes as well. Possible differences in the protein spectra in these two groups of carp are inferred also from the differences in the erythrocyte antigens detected in them (Altukhov et al. 1966; Pokhiel 1969).

The commercial significance of the different forms of carp varies. In their growth rate the scaled carp are generally somewhat better than the scattered ones, and the linear ones are better than the nude ones, however, inverse relationships can sometimes be found. The linear and nude carp always grow more slowly than the scaly ones and scattered ones. The delayed rate of growth and the lower viability of carp carrying the gene N become more pronounced under unfavourable rearing conditions (Kirpichnikov 1945; Lieder 1957; Schäperclaus 1961).

Scaled and scattered carp can be found among many varieties of this fish species. Nude carp are also found among some strains, but in recent years they have been discarded and not used for industrial carp breeding in the USSR, the GDR, the FRG and a number of other countries. Linear carp possess particularly low resistance to diseases and to wintering. Recently fish breeders have abandoned the breeding of linear carp in some countries.

The impressive variability of the mutant forms of the carp involves the structure of the scales themselves. The number and location of scales in the mirror and nude carp depends on the number of modifier genes, the inheritance of which has not as yet been traced. Extreme variants are represented by the so-called big-scaled scattered carp whose body is completely covered with scales, and the "framed" carp with a broad frame of scales along the periphery (Fig. 20). A comparable variation is found in the linear carp (Fig. 21).

The segregation of genes coding for the type of scale cover in crosses of the common carp with the crucian carp *Carassius carassius* and the tench *Tinca tinca* follows the same rules as in crosses between different common carp; apparently, these species have homologous genes located in homologous chromosomes. In the offspring from the crosses of the crucian carp (*Carassius carassius*) with the nude common carp, scaled and linear hybrids are present in the proportion equal to almost 1:1. Linear hybrids of the common and crucian carp greatly resemble linear carp (Fig. 22), but their fins are normal just as in the case of Vietnamese linear carps. As communicated to the author by A. I. Kuzema, the progeny from crosses of the nude carp with the tench can also be divided into two groups, but scattered hybrids are present instead of the linear ones.

The Genetics of the Pigmentation Types. The large diversity of pigmentation types is characteristic of all the subspecies of the carp living in nature and of different domesticated strains of the carp. The following forms have been extensively studied from a genetic point of view.

1. "Blue" carp, frequently occurring among various domesticated varieties. The "blue" colour is inherited as a simple recessive trait. German blue carp (gene bl_D) virtually do not differ in their viability and the rate of growth from ordinary carp (Probst 1949 a). Probst obtained the following proportions of phenotypes in the second generation:

	Non-blue		Blue	
	Number of fish	%	Number of fish	%
F_2	4925	76.0	1553	24.0
F_b	757	50.0	756	50.0

In this specific case the blue colour results from the underdevelopment of guanine crystals associated with the reduction of guanophores in the skin of the carp. This condition is known as alampia.

Polish blue carp found in a pond fishery at Ochaby apparently result from another mutation (bl_P). This is also a recessive gene but having a strong pleiotropic action. During the first year of life these carp grow faster than their normal counterparts. The average weight of the fingerlings in grams has been given by Wlødek (1963):

	Blue	Non-blue
1958	109.2	82.1
1959	164.7	77.8
	156.1	83.8
1961	55.1	52.7
	95.8	47.9
	62.7	58.4

The great differences in the average weight of the young fishes in 1959, and in one case in 1961 as well, are apparently explained by the increased mortality of the blue carp and by the low density of their population in ponds resulting from high mortality. Weight ratios obtained in 1958 and in two experiments conducted in 1961 are to be considered as more precise. The relative excess in the weight of one-year-old blue carp could be taken to be equal to 10%–20%. In the second and particularly in the third or fourth year of life the growth of the blue carp is drastically retarded. The weight of three-year-old fishes in 1960 was around 1812 g (blue) and 2687 g (non-blue) (Wlødek 1963).

The blue Israeli carp (gene bl_I) also represents a recessive mutant with retarded growth and lower viability of the homozygotes (Moav and Wohlfarth 1968). This mutation is at present used to mark one of the breeding stocks, which is used for the production of commercially usable hybrids.

2. "Gold" individuals, sometimes red or orange with black eyes, can be found in many countries both among cultivated strains of the carp and among populations of its wild-living ancestor. Unequivocal recessive inheritance is characteristic of the Israeli "gold" carp (the gene g); they are only slightly inferior to non-pigmented carp when compared on the basis of the growth rate and viability (Moav and Wohlfarth 1968). The gene g is also used to mark the stocks for commercial crosses.

Fig. 21. Hereditary variation in the location and number of scales in linear common carp

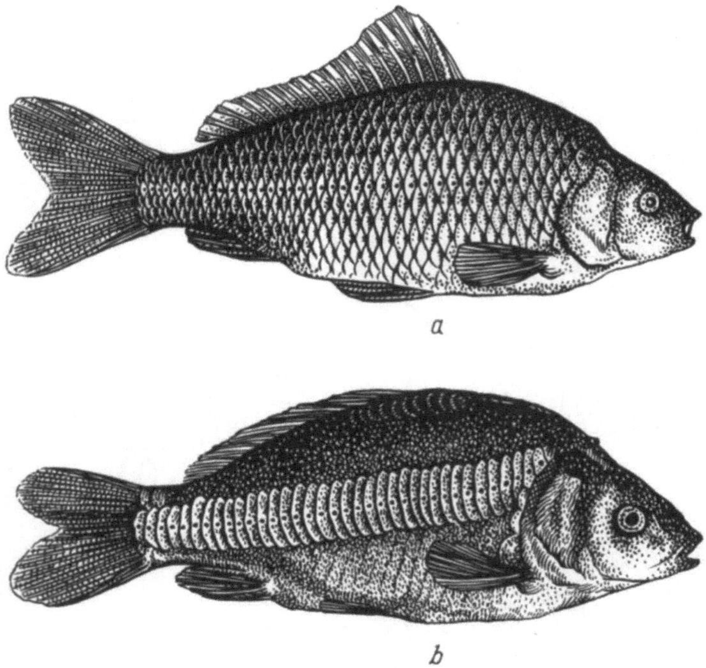

Fig. 22. a, b. Hybrids between the common and crucian carp, *Cyprinus carpio* and *Carassius carassius* (Lieder 1957). **a** scaled; **b** linear

The orange Japanese carp is the outcome of the interaction of the recessive genes b_1 and b_2; the genetic formula of these carp is $b_1b_1b_2b_2$ (Katasonov 1974 b, 1978).[2] When heterozygotes in the genes b_1 and b_2 ($B_1b_1B_2b_2$) are crossed with each other only 1/16 of the progeny receives the orange pigmentation. At the larval stage the double homozygotes are transparent; they do not contain black pigment cells, melanophores in the skin and only the eyes are black (Fig. 23). Later, isolated pigmented regions appear on their bodies, the carp become spotted, black and orange ("spotty"). Because of early expression and easy identification, the genes b_1 and b_2 can become very useful in a number of selection experiments. Specialized stocks of red and orange carp exist in Indonesia on the Island of Java (Ikan-mas and Look-mas) and in Japan (Chigoi) (Buschkiel 1933, 1938; Steffens 1975). Red carp living in nature have been described in the mountain regions of Vietnam, in certain waterpools their numbers attain 40%–50% (Thrâň dinh Trong 1967). Isolated red or dark-red individuals of the carp are occasionally found in German fish farms (Steffens 1975), they have also been described in the lakes of North America (Shoemaker 1943). The pattern of inheritance of all these "gold" mutations has not been established. It can be assumed that in most cases the mutation is associated with inhibition (although to a greater or lesser extent) of melanine formation and melanophore development.

2 Probably the same genes but called p and r have been described recently by Hungarian authors who also worked with the Japanese coloured carp (Nagy et al. 1979)

3. The "steel" pigmentation results from a recessive mutation (the gene r) found in the progeny of Japanese decorative carp (Katasonov 1974 b, 1978). The steel carp are characterized by a decreased amount of red and yellow pigment cells – xantho- and erythrophores. The inheritance of the r gene is very clear, it hardly affects carp viability and is used as a marker of certain breeding stocks. The combination of the genes r, b_1 and b_2 results in white carp devoid of both melanophores and xanthophores.

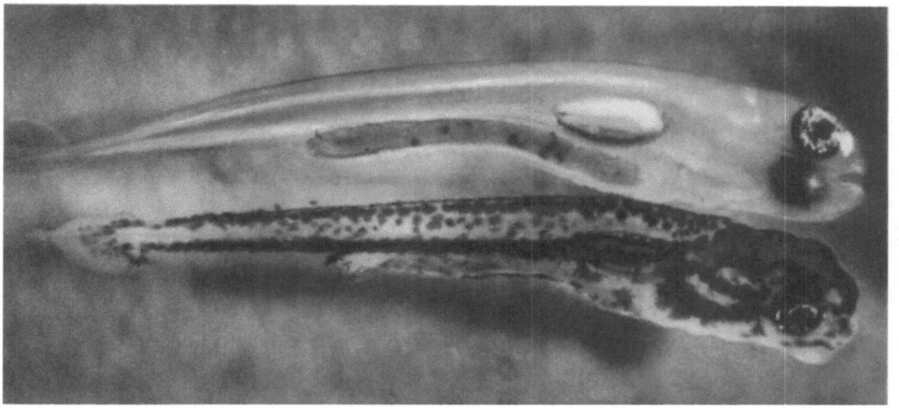

Fig. 23. Transparent (genotype $b_1b_1b_2b_2$) and normal (B_1B_2) larvae of the carp. (Photography by V. Ya. Katasonov)

4. "Grey" and "yellow" carp represent ordinary pigmentation variants among European stocks; the differences between them are easily detected in the discoloration of the abdomen. A genetic analysis remains to be conducted, but there are reasons to believe that the differences are determined by no more than two or three genes. The grey pigmentation of the Israeli carp depends on the presence of one recessive gene (Moav and Wohlfarth 1968), which, like the gene g (gold) and the gene b l (blue), serves as a marker of parental stocks in industrial crosses.

5. "Light-coloured" carp result from a dominant mutation, frequent among Japanese decorative carp (Fig. 24). Carp, homozygous with respect to the mutant gene (LL) die at the stages of the larvae or fry, no living fishes with this genotype are found among the fingerlings (Katasonov 1976); the heterozygote (Ll) survive but possess a lowered viability. The gene L in a heterozygous state has a pronounced pleiotropic action: the pectoral fins undergo elongation, the posterior chamber of the air bladder becomes shorter, and the head dimensions increase. The light-coloured carp have longer intestines, the protein content in the blood serum is decreased. Their growth rate is characteristically accelerated to about 20% during the first year of life and their behaviour is quieter. The lighter pattern of the pigmentation throughout the trunk is due to the stable contraction of melanophores (Katasonov 1978).

6. Another dominant mutation found in the Japanese carp is associated with the particular light-yellow pattern (the stripe on the back and the ornamental pattern on the head). Carp with such a pattern, both homozygous and heterozygous ones,

(DD and Dd) are reasonably viable. In the second generation (after the cross between the heterozygotes) segregation has been observed in a proportion of 3:1 with respect to the trait: presence of the pattern/absence of the pattern. For example, after one of the crosses, 1678 carp were found to possess the pattern in question (75.2%) and 554 carp were without the pattern (24.8%) (Katasonov 1973).

Fig. 24. The mutation "Light" in the common carp (*Cyprinus carpio*). *Left* normal carp (ll); *right* heterozygous "Light" carp (Ll). (Photography by V. Ya. Katasonov)

The gene D is pleiotropic just like the gene L; in individuals with this gene, the head size is increased, the posterior chamber of the air bladder is elongated, the number of vertebrae is increased and the outer appearance is changed towards the type characteristic of the wild common carp. As regards the rate of growth, carp with the gene D have only minor differences from the control carp devoid of this gene (Katasonov 1974 a).

Other variants of pigmentation can be found both among domesticated carp and wild carp living in nature. These include green carp not infrequently appearing among Amur wild carp and strains created on the basis thereof. Green carp have not been studied so far. Many other "coloured" varieties of carp living in tropical areas have also been poorly investigated; this is true of the white, yellow, lemon-yellow, violet, brown and genuine albino varieties. It can be assumed that most of these pigmentation variants depend on one, sometimes on two or three genes affecting the development of pigment cells.

Until now, no single case has been described, when the genes responsible for pigmentation would be linked to each other or to the genes S and N coding for the scale pattern. This absence of linkage provides ample opportunities to use many of these genes to mark different breeding stocks and varieties of the carp.

The Inheritance of Other Traits. There is only limited evidence in the literature regarding the inheritance of traits, other than the pattern of the scales or the pigmentation.

The dwarf common carp "Pizartsovichi" was found on one of the Polish fish farms (Rudzinski and Miaczynski 1961). Morphologically the dwarf carp resembles normal fish, but has a somewhat undersized mouth; malformations of the vertebral column due to the fusion of vertebral bodies are another frequently occurring feature in this carp strain.

Dwarf varieties and breeds of the common carp adapted for growth in shallow ponds and on rice paddies are common in China, Japan, Vietnam and Indonesia. Varieties with a very small size can also be frequently found in the populations of wild common carp in the Soviet Union. In the Aral Sea, for example, a dwarf carp has been found which is fully mature when weighing 100–300 g. The yearlings of this variety kept in ponds grew severalfold more slowly than the fishes of other varieties of the common carp (Kirpichnikov 1958 b, 1967 b). The small size was hereditary in this case, but unfortunately, no detailed analysis was conducted. Such dwarf carp also inhabit a number of lakes in the Khoresm region (Abdullaev and Khakberdiev 1972).

The absence of ventral fins is sometimes inherited as a recessive mutation (Kirpichnikov and Balkhashina 1936). An isolated population of the wild common carp lives in one of the lakes in the system of the river Illinois (USA); more than 40% of the fishes in this population are devoid of one or both ventral fins (Thompson and Adams 1936). It can be assumed that this loss is hereditary. It should, however, be pointed out that the fins in the carp are not infrequently diminished or lost when embryos or larvae are subjected to unfavourable environmental conditions (Wunder 1932; Tatarko 1963, 1966). In such cases, one can speak only of the hereditary predisposition to developmental abnormalities.

One of the common carp breeds in Indonesia ("Kumpai") is characterized by its elongated fins as well as its elongated body shape (Steffens 1975). Apparently the phenotype "elongated fins" originates from a simple gene mutation.

Several dozen fish with an additional preanal fin were found among second-generation hybrids of the common domesticated carp and the Amur wild carp. This mutation, definitely atavistic in nature, was also found in the fourth hybrid generation. The additional fin consisted of two to four large hard rays (Fig. 25). The carp with such fins were completely viable and did not differ in the rate of growth from their normal siblings. Subsequently, these highly interesting mutants were unfortunately lost.

A dolphin-like head has been found in one common carp strain in France. The crossings with non-related fishes show that this trait is dominant but has an incomplete manifestation from 62% to 76% in F_1 (Pojoga 1969).

Finally, one can propose a relatively simple genetic explanation of the body shortening in the Aischgrund "saucer-like" breed of the carp, it appears to be due to the fusion of a number of vertebrae (Hofmann 1927; Wunder 1949 a). Unfortunately, no genetic analysis has been conducted in this case as well. The numerous aberrations of a different character frequently occurring in natural populations of the common carp have not been studied either.

Thus, we possess some information about the inheritance of around 15–20 genes of the common carp. Most interesting are the lethal genes N (reduction of scales) and L (lighter pigmentation) which kill the carriers in the homozygous state. These genes have a very wide pleiotropic action and were selected by carp breeders

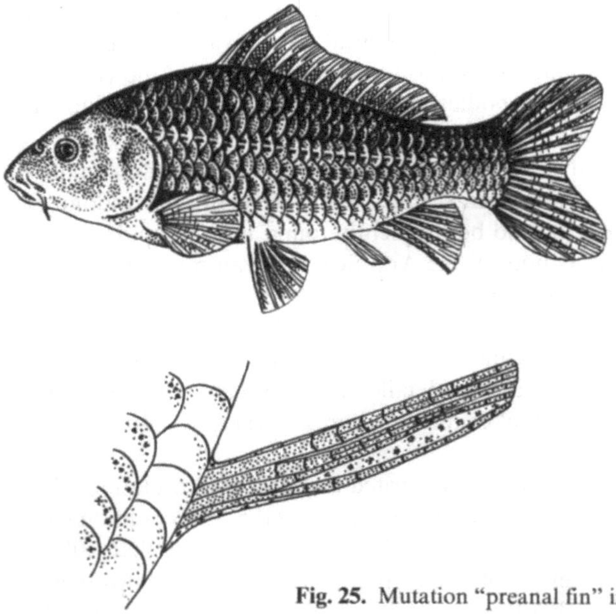

Fig. 25. Mutation "preanal fin" in the common carp

for applied purposes. The fertility decrease due to the segregation of the homozygotes carrying lethals amounting to 25% is of no danger to the carp which is able to yield up to a million or more eggs simultaneously during one spawning. The persistence of lethal mutation in selected stocks can be easily explained away by certain advantages present in the fish, heterozygous with respect to these genes. Some success has been attained among European fish breeders with nude carp (Nn) due to the virtually complete absence of scales. The light-coloured (genotype Ll) Japanese decorative carp apparently attracted breeders by their unusual external appearance, particularly in combination with other recessive pigmentation genes.

The Inheritance of Qualitative Traits in Other Pond Fishes

There is only limited information on the inheritance of qualitative traits for other fishes grown in ponds (Kirpichnikov 1969 a; Kirpichnikov 1971 a). Three genes affecting the pigment cells have been studied in the rainbow trout (*Salmo gardneri* = *S. irideus*). The recessive autosomal gene a produces a complete albino phenotype, segregation in the F_2 and F_b hardly differs from the expected one: 16,856 normally pigmented and 5679 albino fishes or 74.8% and 25.2% respectively have been obtained in the F_2; 3253 and 3145 fishes were observed in the F_b, this is equal to 50.8% and 49.2% respectively (Bridges and Limbach 1972).

The dominant gene G leads to the appearance of the gold (red) pigmentation; the eyes, however, in contrast to the genuine albino animals remain pigmented

(Beall 1963). Trout carrying two copies of the gene G have an increased sensitivity to light, somewhat lower activity and a slower rate of growth. In the dark-yellow heterozygous fishes (the "palomino" pigmentation) the rate of growth is increased by about 20% (Clark 1970; Wright 1972). The gene G is inherited as an autosomal gene.

The "metal blue" pigmentation (initially light-blue pastel, followed by some darkening of the fishes) appears in the trout after 200 days of life and is discernible only when the fishes are kept in basins. It is assumed that the inheritance is recessive and the manifestation of the trait is incomplete. Fishes possessing this trait have lower activity and grow faster than their normal counterparts (Kincaid 1975).

Many papers deal with the inheritance of different traits in the goldfish *Carassius auratus*. The albino phenotype in this case is apparently determined by two recessive genes, m and s (Yamamoto 1973; Kajishima 1977). The following combinations of genes are possible: MS and Ms – dark fishes, mS – light fishes, ms (mmss genotype) – albino fishes.

In this case, the gene M is epistatic with respect to the gene S. Below we shall present a Punnett square for the relevant cross conducted by Yamamoto:

♀MmSs × ♂MmSs
Gametes

Gametes		MS	Ms	mS	ms	
	MS	MMSS	MMSs	MmSS	MmSs	Dark
	Ms	MMSs	MMss	MmSs	Mmss	
	mS	MmSS	MmSs	mmSS	mmSs	Light
	ms	MmSs	Mmss	mmSs	mmss	Albino

Segregation should occur in the proportion 12:3:1. In actual fact, 273 dark fishes (266 expected), 62 light fishes (66 expected) and 19 albino ones (22 expected) were observed.

At the early stages of development specifically in the course of retinal melanocyte differentiation, light and albino fish (mmSS, mmSs and mmss) are in- →

Fig. 26. Genealogy of basic varieties of the goldfish (*Carassius auratus*) created by breeders by breeders of China and Japan (Matsui 1956).
1 Wild crucian carp (*Carassius auratus?*); *2* Hibuna (common goldfish); *3* Wakin; *4* Ryûkin; *5* Maruco (ovoid fishes), *5a* Rantyû, *5b* Maruko, *5c* Nankin; *6* Demekin (telescope-eyed fishes), *6a* Aka (red) demekin, *6b* Kuro (black) demekin, *6c* Sansuoku (calico) demekin; *7* Zikin; *8* Tosa-kin; *9* Tetuonaga (iron-coloured long-tail fish); *10* Oranda-sisigasira; *11* Watônai; *12* Syûkin; *13* Syubunkin; *14* Calico (hybrid of Ryûkin and Calico-demekin); *15* Azuma-nisiki (hybrid of Oranda and Calico-demekin); *16* Tetugyo (iron-fish); *17* Kinransi (hybrid of Wakin and Rantyû)

Fig. 26
(legend see p. 67)

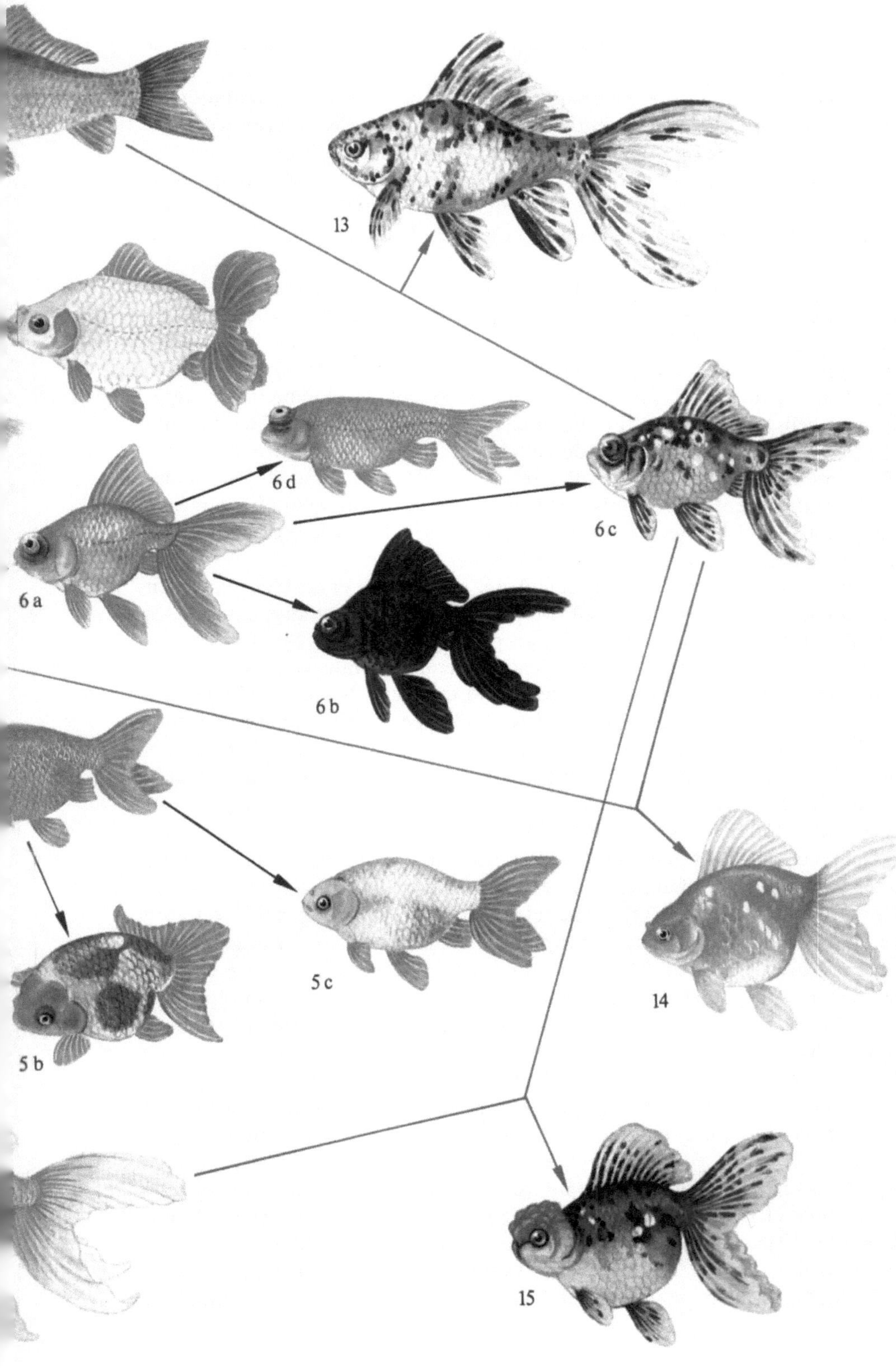
13
6d
6c
6a
6b
5c
14
5b
15

discernible and as a result segregation in the proportion 3:1 is observed (Kajishima 1977).

Chen (1934) has studied the inheritance of two other types of pigmentation in the goldfish: blue and brown pigmentation. The blue colour is determined by a simple recessive gene. The inheritance of the brown pigmentation is more complex; its expression, just like that of the albino phenotype, requires two and perhaps even four interacting recessive genes. Two thousand eight hundred and sixty-seven normally pigmented fishes and 189 brown ones were obtained in the F_2, in some crosses, however, this ratio was close to the classic 3:1.

The red pigmentation of the goldfish is associated with the presence in the genome of the genes responsible for depigmentation (Kajishima 1965, 1975; Takeuchi and Kajishima 1973; Kajishima and Takeuchi 1977). The bulk of the embryonic and larval melanophores (both skin and choroidal ones) in normally pigmented fishes undergoes degradation in the course of development; the remnants of these cells are destroyed by specialized cells, melanophages. The destruction is genetically programmed and depends on the presence of two dominant genes Dp_1 and Dp_2. Substitution of the recessive alleles dp_1 and dp_2 for these genes results in the retention of embryonic melanophores throughout the lifetime of the animals and the fishes remain dark (almost black); the variety of gold fishes with such discoloration has been called "black mavers".

Skin transparency (Chen 1928; Matsui 1934c; Matsumoto et al. 1960) has been shown to depend on the combination of two pairs of genes: the genotype ttnn leads to the peculiar network-like character of transparency; the genotype Tt in combination with either the N or the n gene leads to uniform transparency. Thus, it appears that the gene T is epistatic with respect to the genes N and n. The gene T acts to decrease the number of pigment cells of all types (melanophores, xanthophores and iridophores) and its action is manifested after embryogenesis is completed. The gene n (apparently identical to the gene g of Kajishima) only affects iridophores, thereby impairing guanine synthesis at the earliest stages of pigment cell differentiation (Kajishima 1977).

The changed shape of the fins ("veiled fins" trait), characteristic of many varieties of the goldfish, also appears to be determined by a combination of two or three pairs of genes (Matsui 1934d). The telescopic shape of the eyes is due to the recessive mutation of a single gene. When the telescopes are crossed with goldfishes having normal eyes, the usual Mendelian ratios are observed. If they are crossed with the crucian carp (the evolutionary ancestor of the goldfish) then in the second generation almost all the offspring (99.9%) are normal. If the hybrids of the first generation, however, are again crossed with the telescopic goldfish, half of the offspring have telescopic eyes (Matsui 1934b). According to Matsui telescopic eyes in goldfish depend on the presence of a special recessive gene (the gene d, according to the author) and require modifier genes stabilizing the manifestation of the gene d.

Many traits characteristic of different varieties of the goldfish await genetic studies. Varieties of goldfish have been created in the course of millenia (Berndt 1924; Schmidt 1935; Chen 1956). Complete domestication of the species was achieved in China by the middle of the twelfth century. Subsequently, many variants with respect to pigmentation and in the body structure were found, and these were used by breeders. In approximately 1500 the goldfish was introduced into Japan, and many new varieties were created, predominantly in the eighteenth and nineteenth centuries. An interesting attempt to construct a general scheme of the evolution of goldfish varieties has been undertaken (Matsui 1956). The evolution was apparently based on the selection of visible mutations fundamentally changing pigmentation, the structure of the fins and eyes, the structure of the integument, the shape of the body and other characteristics (Fig. 26). Subsequently, the manifestation of the newly selected mutations was stabilized by the selection of modifying genes; the viability of the mutant forms also increased. Most of the traits turned out to depend on two, three or an even greater number of interacting genes.

According to the recent karyological observations the ancestor of goldfishes was one of the Chinese subspecies of the crucian carp, *Carassius auratus auratus* (Ojima and Ueda 1978; Ojima et al. 1979).

In another subspecies of the crucian carp, *C. a. langsdorfii* a characteristic pearl colour has been found which was inherited as a single-gene trait (Yamamoto, 1977). The golden variety of the domesticated ide, *Leuciscus idus* ("orfa") is a recessive mutation which greatly affects many features of the fish. In particular the orfa specimens show an increased sensitivity to a number of environment factors including electric current.

Other pond fishes have not been studied at all from a genetic point of view. Some mutants have, however, been found in a few species; these include mirror phenotypes in the rudd *Scardinius erythrophthalmus* (Rünger 1934), albinotic phenotypes in the American channel catfish *Ictalurus punctatus* (Nelson 1958; Menzel 1959; Prather 1961); forms with reddish pigmentation described in *Tilapia zilii* (Chervinski 1967), four different colour patterns in the cichlid *Pseudotropheus zebra* (Holzberg 1978; Schröder 1980), "handpaint" variety (black spots on the sides of the body) in the blue-gill sunfish *Lepomis macrochirus* (Felley and Smith 1978), variants devoid of anal or ventral fins found among the Indian major carp *Catla* and *Cirrhina* (Kaushik 1960) and many other cases. The absence of any genetic information with respect to qualitative morphological traits for such economically breeding species as American trout *Salvelinus,* buffalo (*Ictiobus* spp.) and other catostomids, herbivorous fishes the grass and silver carp (*Ctenopharhyngodon, Hypophthalmichthys*), the bighead (*Aristichthys*) and finally the tilapias impedes breeding work with those species already domesticated by man.

The Genetics of the Wild Fish Species

There is only limited genetic information on the species living in nature. The stickleback and several cavernicolous species belong to the better-studied ones.

The three-spined stickleback *Gasterosteus aculeatus* living in Europe shows polymorphism with respect to the number of lateral plates on the body. Three main phenotypes have been distinguished (Fig. 27) differing in the degree of the trunk protection by these plates (Münzing 1959, 1962, 1963). These phenotypes are apparently controlled by a single pair of genes, T and t (Münzing 1959). In Northern Europe and along the Black sea coast (in the salt water) the phenotype

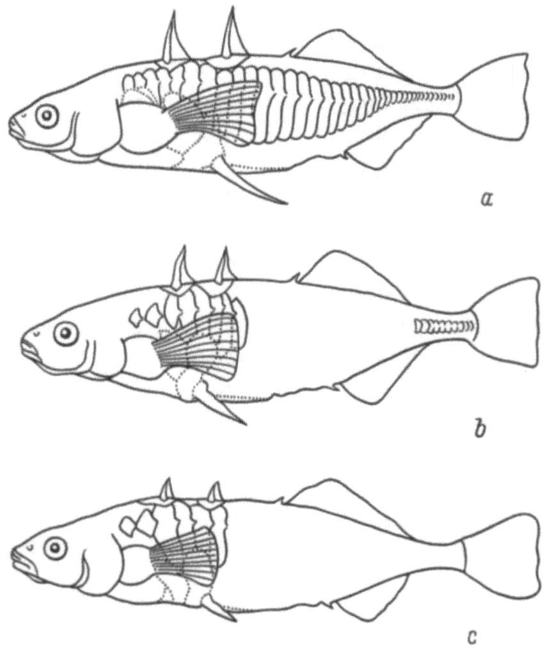

Fig. 27a–c. Phenotypes and genotypes of the three-spine stickleback *Gasterosteus aculeatus* (Münzing 1959, 1963). **a** *trachurus* (TT); **b** *semiarmatus* (Tt); **c** *leiurus* (tt)

trachurus is most numerous, in the freshwater West European habitats the variety *leiurus* is present instead. In Eastern Europe the *trachurus* variety is not limited to marine biotopes, it is found in large numbers in certain freshwater populations in Poland and the USSR; the varieties *leiurus* and *semiarmatus* generally become rare in the regions to the east of the river Oder (Münzing 1972).

The frequency of occurrence of the *trachurus* variety increases in anadromous populations of the three-spined stickleback living in the English channel and the North sea from 20% to 70%–90% in the direction from west to east, in other words a distinct geographical variation is observed. The variation in the number of lateral plates is perhaps a consequence of the postglaciation contact of fishes from the two main "shelters" where stickleback survived during glaciation: the southwestern Atlantic populated during glaciation by fishes similar to *trachurus,* and the southeast freshwater one, populated by the *leiurus* variety. Mixed populations showing geographical variation formed in the area of postglaciation intergradation (Münzing 1962, 1963, 1972; Kosswig 1973).

The constancy of the proportions of the three genetic forms in each of the local mixed populations and some prevalence of Tt heterozygotes (*semiarmatus*) in some of the populations living in the estuaries of the northwestern valley rivers provide convincing evidence of the adaptive nature of polymorphism of sticklebacks with respect to this trait (Münzing 1972). The hypothesis about the presence in sticklebacks of a single gene with two alleles controlling the number of lateral plates has not yet been conclusively proven. The inheritance of this trait appears to be more complex; this can be deduced in particular from the existence in Turkey and Canada of populations consisting completely of the intermediate variety (*semiarmatus*) (Münzing 1963; Hagen and Gilbertson 1973b). The involvement of at least two pairs of genes is also inferred from the data of American authors (see below). Apparently, populations of sticklebacks contain one or two "strong" loci, controlling the formation of the lateral plates and many weaker modifying genes. Each of the principal loci may be represented by more than two alleles.

The diversity of the stickleback varieties in America is greater than in Europe (Miller and Hubbs 1969; Hagen and Gilbertson 1973a, b; Bell 1976). The populations living in the freshwater Lake Vapato can be divided into three groups: strongly armed (C), intermediate (P) and "weak" (L). Fishes of the C type resemble the marine variety *trachurus,* but are not identical to them. The inheritance of the three forms is rather complex; for example, different results were obtained in different crosses of the intermediate fishes with the weak ones:

1. P×L = L (100%);
2. P×L = P + L (50% each);
3. P×L = L + P + C (39:23:17; 39:28:8, etc.).

A hypothesis postulating the presence of two pairs of genes A and a, B and b, with additive action has been advanced to explain this pattern of inheritance (Hagen and Gilbertson 1973b). According to this hypothesis the three phenotypes of the sticklebacks are controlled in the following way:

Phenotypes	Genotypes
Strong (C)	– AABB, AaBB, AABb
Intermediate (P)	– aaBB, AaBb, AAbb
Weak (L)	– aaBb, Aabb, aabb

There can be no doubt that this hypothesis is somewhat artificial; hopefully subsequent genetic analysis will allow a more natural explanation to be found.

Lake Matamek and the river Matamek in Canada are populated predominantly by the "intermediate" form similar to *semiarmatus;* nearby Lake Bill is populated both by *semiarmatus* and *trachurus;* the sea bay Amori Cove is populated predominantly by the *trachurus* variety (Coad and Power 1974).

Marked variation in the number of plates has also been reported for sticklebacks of the Pacific coast of North America (Bell 1976). The intermediate type disappears in California; when fishes weakly and extensively covered by plates are crossed with each other the offspring segregates into just two groups. The crosses between Californian sticklebacks and those of other regions do, however, yield all three phenotypic groups (Avise 1976b).

The genetics underlying the inheritance of lateral plates in sticklebacks is not therefore completely clear. Polymorphism in the degree of protection appears to have some adaptive value. Fishes with poorly developed plates apparently survive better in fresh waters, fishes with many lateral plates do better in the sea. The upper limits of water salinity for the guaranty of spermatozoa mobility in *trachurus* and *leiurus* types are correspondingly equal to 52% and 16% (Ziuganov 1978; Ziuganov and Khlebovich 1979).

Belonging to one of the three main types and the total number of plates correlates with the pressure of the predators: the greater the number of predators in a given basin the better is the protection of the fishes by plates and spines (Kynard 1979). The presence of a variety of predators in the basin stimulates polymorphism (Maskell et al. 1978). A relationship between the trait "lateral plates" and the temperature and other environmental factors has not been established, such a relationship is doubtless, however, with regard to such quantitative traits as the number of vertebrae, the number of gill rakers and the shape of the body (Hagen and Gilbertson 1972).

In several three-spined stickleback populations there are males with a red stripe on the belly. The presence or absence of this trait is determined by two codominant sex-limited autosome alleles. In the presence of a large number of predators (*Novumbra* spp.) red-striped fishes disappear as a result of intensive negative natural selection (Hagen and Moodie 1979).

The number and size of lateral plates in another species of sticklebacks *Pungitius* is also variable, but no genetic analysis of this variation has yet been conducted. The variation is characteristic of two widespread species *P. pungitius* and *P. platygaster,* the subspecies of each of them having characteristic patterns of the plate positioning (Münzing 1969). Among the Canadian sticklebacks *P. pungitius* and *Culaea inconstans* many fishes have no ventral fins and even no pelvis (Nelson 1971, 1977). All the fish in Lake Fox Holls are of this type, whereas in nearby Lake Rig their proportion is less than 4%. I assume that this trait is controlled by a simple recessive mutation just as in other fishes.

A small Mexican freshwater fish belonging to the family Characidae – *Astyanax mexicanus* – has close relatives living in caves (*Anoptichthys antrobius, A. jordani, A. hubbsi*). All these troglobionts can easily be crossed with *A. mexicanus.* Fishes of all three cavernicolous species are blind and their melanine and guanine pigmentation is either reduced or lacking. Blindness is determined by many additive genes (Kosswig 1963; Pfeiffer 1967; Peters and Peters 1973). The reduction of pigment in these species depends on the presence of one or two genes. The albino mutation in *Anoptichthys antrobius* is inherited as a simple recessive trait. In the second generation after the cross *Anoptichthys* × *Astyanax* 787 normally pigmented fishes and 278 albino ones have been obtained (Sadoglu 1955, 1957). Later another mutation was found in this species, due to the gene bw responsible for the brown pigmentation. The three types of pigmentation: normal (black), brown and white (albino) are determined by the interaction of two unlinked genes a and bw (Table 9).

The gene a, just like many other albino genes, is epistatic with respect to another gene of pigmentation, bw, and therefore segregation in the second generation (F_2) occurs in the proportion 9:3:4. In this case, the homozygotes aa are viable, the gene bw also exists among other cavernicolous species belonging to the genus *Anoptichthys.*

The establishment of fish populations in caves was accompanied by the gradual accumulation of alleles which weakened the vision and prevented pigment formation. As can be judged on the basis of the number of genes interacting in the process of eye degeneration (no less than 6 – Wilkens 1970) the possibility of preadaptation, that is of the random transfer of blind depigmented fishes to caves, is improbable. The accumulation of "degenerative" alleles did apparently occur in the course of selection, facilitating the destruction of the unnecessary organs and lessening the pigmentation. It has also been assumed by several authors that the establishment of populations living in caves was greatly affected by the "founder effect": a few original animals had genes (including the degenerative ones) which thereafter became automatically multiplied in the course of the increase in population size (Peters and Peters 1973). As I see it, this hypothesis is insufficiently founded, just like the hypothesis postulating preadaptation.

Table 9. Inheritance of pigmentation in cave-dwelling blind fishes and their normal river relatives (Sadoglu and McKee 1969)

Phenotype	Taxonomic position	Genotype
Normal silvery pigmentation	Astyanax mexicanus	$\frac{+\ \ +}{+\ \ +}$
Normal silvery pigmentation	Hybrid populations	$\frac{+\ \ +}{+\ \ +};\ \frac{+\ \ +}{+\ \ bw};\ \frac{+\ \ +}{a\ \ bw}$
Brown	Hybrid populations	$\frac{+\ \ bw}{+\ \ bw};\ \frac{+\ \ bw}{a\ \ bw}$
Light (albino)	Anoptichthys spp. (cavernicolous forms)	$\frac{a\ \ bw}{a\ \ bw};\ \frac{a\ \ bw}{a\ \ +}$
Light (albino)	Hybrid populations	$\frac{a\ \ +}{a\ \ +};\ \frac{a\ \ +}{a\ \ bw};\ \frac{a\ \ bw}{a\ \ bw}$

The evolution of cavernicolous fishes has also led to the loss of the defense reaction characteristic of all fishes living outside caves.

In the fishes belonging to the genus *Anoptichthys* the absence of the defense reaction is determined by the interaction of two recessive unlinked genes. The segregation in F_2 is close to 15:1 (Pfeiffer 1966). It should be pointed out for the sake of comparison that in *Poeciliopsis viriosa* (Poeciliidae) a single recessive mutation was sufficient to abolish another important behavioural reaction – the ability of males to repel enemies and competitors by a rapid, very strong intensification of colour. The mutant gene inhibits the formation of xantophores in the scale epithelium and any display coloration leading to a frightening effect becomes impossible (Vrijenhoek 1976).

In the poeciliid fish *Aphanius anatoliae* there is variation in scale cover. The frequency of four main types of scale cover – scaled, scaled with naked abdomen, slightly scaled and linear – was different in different small populations of this species and has greatly changed in the course of 30 years of investigations (Grimm 1979). We suppose that this variation is determined by two or three genes.

In the cobitid fish *Nemachilus barbatulus* living in water bodies in the Leningrad and Novgorod regions (USSR) albino individuals of two types have been repeatedly found. These were bright-red individuals with red eyes and orange fishes with black eyes. Crosses carried out by myself have shown that complete albinism in this case is associated with the recessive mutation. Unfortunately, the genotype of semialbino specimens has not been determined. Albino fishes have also been found in another species of the same family, *Cobitis taenia.*

Albino individuals have also been found in many commercially valuable fish species. For example, albino fish have been described in the shark belonging to the genus *Stegostoma* (Nakaya 1973), in the common carp (Johnson 1968), in the lake chub belonging to the genus *Kyphosus* (Sgano and Abe 1973) and in other cyprinids; albinism has also been described in channel catfish (Aitken 1937), in flatfish *Limanda yokohamae* (Abe 1972) and in many other freshwater and marine fish species.

Certain other aberrant forms found in natural fish populations also belong to genetic variants. These aberrations include the elongation or duplication of the fins, distortions of the vertebral column, alterations in the pattern of the scales, and the underdevelopment of the eyes, etc. The number of different types of aberrations is extremely large (Dawson 1964), but there is no information on the percentage of these aberrations that represent the consequences of mutations.

The data on the inheritance of qualitative traits in fishes are very limited at present; the common carp has been better studied than other species, but the material available is too scarce even for this species. Knowledge of the special genetics of fish, primarily of those used for breeding, is critically important for the correct planning of selective breeding work and for the protection of the commercially valuable species. The accumulation of such data has now become a pressing matter.

Chapter 3

The Genetics of Aquarium Fish Species

Many genetic studies have been conducted using aquarium fish species. Since there are several good reviews dealing with this chapter of fish genetics (Gordon 1957; Dzwillo 1959; Kosswig 1965; Kallman and Atz 1966; Schröder 1974, 1976; Kallman 1975; Yamamoto 1975 b), we shall consider only some selected results of these studies.

The Guppy (Poecilia reticulata)

Poecilia (Lebistes) reticulata is a small common aquarium fish from the family Poeciliidae (viviparous tooth carps).

The males of the guppy generally have a bright, diverse colour pattern, whereas the females are usually monotonously grey. Polymorphism of the male coloration has been described in natural populations of this species (Haskins a and Haskins 1951, 1954; Haskins et al. 1961; Kosswig 1964 a). Aquarists have created many breeds of the guppy differing in the pattern of the body colour as well as in the shape and colour of the fins (Fig. 28).

It is a specific feature of the genetics of the guppy that most of the genes responsible for the colour pattern, as well as certain genes affecting the structure of the fins, are concentrated in the sex chromosomes X and Y. Males of the guppy are heterogametic; no heterochromosomes, however, have been detected under the microscope. Schmidt (1919 a) was the first to establish that many "colour" genes are transmitted in the guppy from the male parents to the male offspring through the Y-Chromosome. Later Winge (1922, 1927) demonstrated that several of these genes are always located in the Y-chromosome, while some others may be transferred from Y to X and vice versa by crossing-over.

Until now, up to nineteen genes responsible for the colour pattern and permanently associated with the Y-chromosome have been described in the guppy; more than sixteen such genes are linked with the X- and Y-chromosomes. In addition around a dozen autosomal mutations have been described which change the general background colour of the fishes or affect other morphological traits (Table 10). Mutations affecting all types of pigment cells – melanophores, erythro- and xanthophores, as well as guanophores (iridocytes) – have been described (Schröder 1969 b).

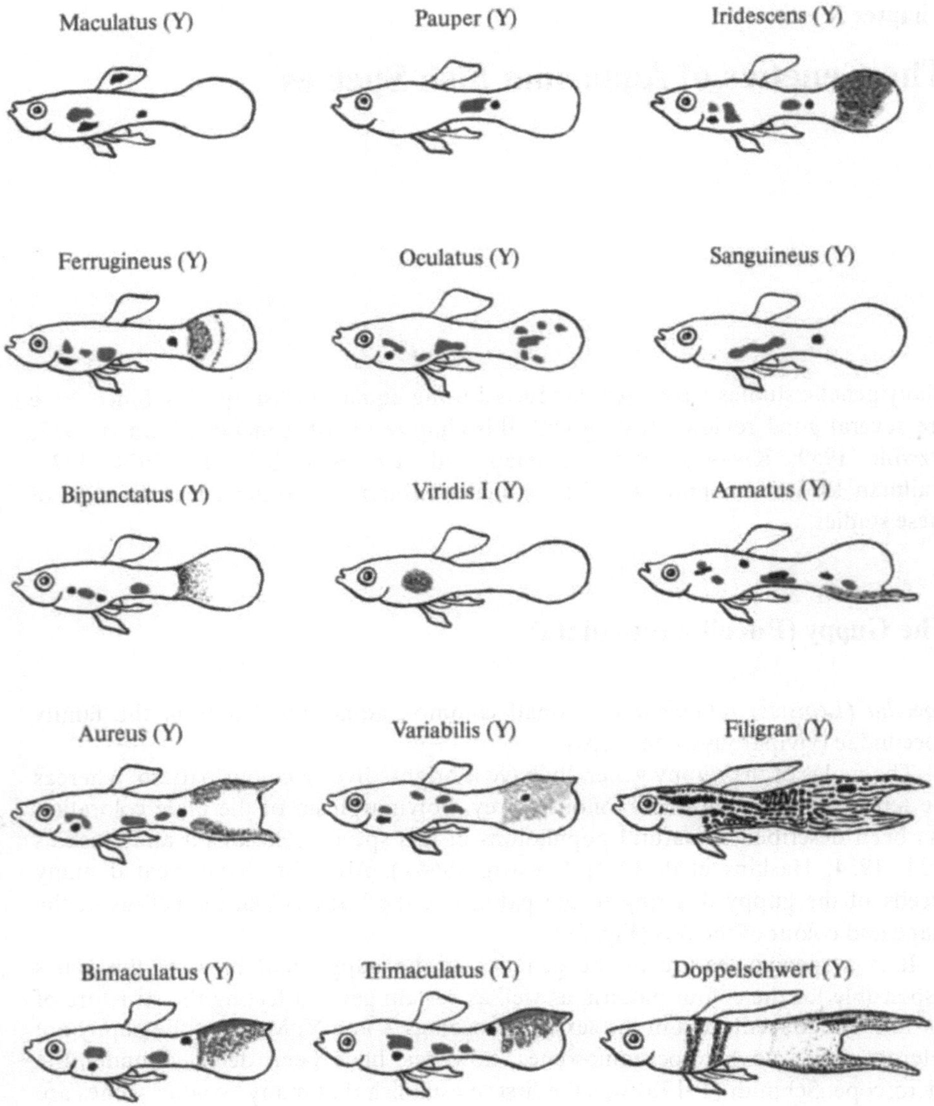

Fig. 28. Polymorphism in males of the guppy *Poecilia* (Lebistes) *reticulata* (Winge 1927; Natali and Natali 1931; Kirpichnikov 1935 b; Dzwillo 1959)

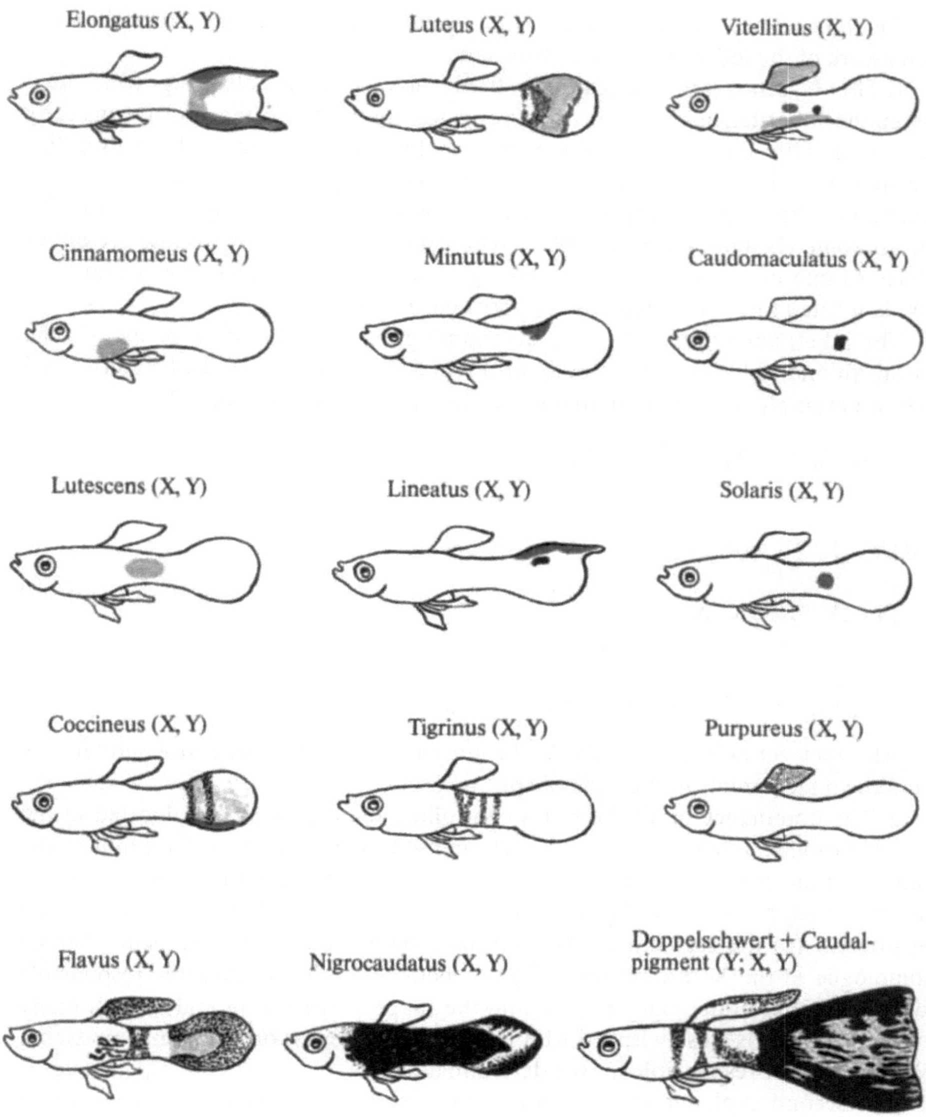
Elongatus (X, Y)
Luteus (X, Y)
Vitellinus (X, Y)
Cinnamomeus (X, Y)
Minutus (X, Y)
Caudomaculatus (X, Y)
Lutescens (X, Y)
Lineatus (X, Y)
Solaris (X, Y)
Coccineus (X, Y)
Tigrinus (X, Y)
Purpureus (X, Y)
Flavus (X, Y)
Nigrocaudatus (X, Y)
Doppelschwert + Caudal-
pigment (Y; X, Y)

The inheritance of genes in *Poecilia reticulata* can be described within the framework of the following main circumstances:

1. The Y-genes are transmitted strictly to the male offspring (one-sided masculine inheritance). No two or more genes among those that are intrinsically associated with the Y-chromosome can be located simultaneously within this chromosome. The genes Vir-I and Vir-II (green spots on the body) are an exception; these genes appear to be combined with other genes in the Y-chromosome (Natali and Natali 1931). The genes of the Y-chromosome are either alleles of one of the same locus or belong to the so-called supergene, a family of closely located loci with completely suppressed crossing-over.

The inheritance of the Ma gene determining the presence of black spot on the dorsal fin and red spots on the body of the males (Winge 1927) can be considered to be an example of the uniparental masculine transmission of "colour" genes:

1. ♂$Y_{Ma}X_o$ × ♀X_oX_o (or $X_{ch}X_{ch}$)

↓

♂♂$Y_{Ma}X_o$ + ♀♀X_oX_o

2. ♂$Y_{Ma}X_o$ × ♀$X_{Ti}X_o$

↓

♂♂$Y_{Ma}X_{Ti}$ + ♂♂$Y_{Ma}X_o$ + ♀♀X_oX_{Ti} + ♀♀X_oX_o

All other genes located in the Y-chromosome are transmitted in a similar way from the male parents to the male offspring.

2. The dominance of all the genes controlling the colour pattern located in the Y-chromosome. When both X- and Y-chromosomes are present in the genome, the manifestation of the genes located in the Y-chromosome does not depend on the genetic structure of the X-chromosome. Two explanations of such dominance can be offered: (a) the segment of the Y-chromosome containing these genes has no homologue in the X-chromosome; (b) the locus of the Y-chromosome responsible for the diversity of the pigmentation and the shape of the fins in males (with many multiple alleles) is closely linked with the gene of maleness or represents by itself a dominant factor responsible for sex determination.

The second explanation is supported by observations about differences in the sexual activity and the stability of the mechanism of sex determination in the stocks of guppy differing in alleles of the main Y-gene. The gene Ma is apparently the strongest one, the introduction of this gene into "weak" strains has resulted in particular in the disappearance of the hermaphroditism sometimes observed in this species (Spurway 1957).

Not only are the genes always located in the Y-chromosome dominant, but this applies to the genes migrating from the Y into the X and from the X- into the Y-

References: 1. Blacher (1927, 1928); 2. Kirpichnikov (1935 b); 3. Natali and Natali (1931); 4. Dzwillo (1959); 5. Goodrich et al. (1944); 6. Haskins and Druzba (1938); 7. Haskins and Haskins (1948); 8. Horn (1972); 9. Nybelin (1947); 10. Rosenthal and Rosenthal (1950); 11. Schröder (1969 c); 12. Winge (1927); 13. Winge and Ditlevsen (1948); 14. Lodi (1967, 1978 a)

Table 10. Genes identified in the guppy *Poecilia reticulata*

Gene	Trait	Reference
	Y-chromosome (males only)	
Maculatus (Ma)	Pigmentation	12
Armatus (Ar)	Pigmentation, caudal fin	12
Ferrugineus (Fe)	Pigmentation	12
Iridescens I, II (Ir)	Pigmentation	12
Aureus (Au)	Pigmentation	12
Pauper (Pa)	Pigmentation	12
Oculatus (Oc)	Pigmentation	12
Variabilis (Va)	Pigmentation	12
Sanguineus (Sa)	Pigmentation	12
Reticulatus (Re)	Pigmentation, caudal fin	1, 3
Bimaculatus (Bi)	Pigmentation, caudal fin	3
Trimaculatus (Tri)	Pigmentation, caudal fin	3
Bipunctatus (Bp)	Pigmentation	3
Viridis I, II (Vir)	Pigmentation	3
Inornatus (In)	Pigmentation	2
Filigran (Fil)	Pigmentation, caudal fin	4
Doppelschwert (Ds)	Dorsal and caudal fins	4
	X- and Y-chromosomes (males only)	
Elongatus (El)	Pigmentation, caudal fin	12
Coccineus (Co)	Pigmentation	12
Vitellinus I, II (Vi)	Pigmentation	3, 12
Luteus (Lu)	Pigmentation	12
Tigrinus (Ti)	Pigmentation	12
Purpureus (Pu)	Pigmentation	12
Cinnamomeus (Ci)	Pigmentation	12
Minutus (Min)	Pigmentation	12
Lineatus (Li)	Pigmentation, caudal fin	12
Caudomaculatus (Cm)	Pigmentation	3
Lutescens (Ls)	Pigmentation	3
Solaris (So)	Pigmentation	2
Flavus (Fl)	Pigmentation, weak action in females	13
Nigrocaudatus I, II (N)	Pigmentation, both sexes	4, 9
Caudal pigment (Cp)	Pigmentation, caudal fin, weak action in females	4
Non-coloured recessive allele (ch)		4
	Autosomes	
Zebrinus (Ze)	Pigmentation, only ♂♂	12
albino (a)	Pigmentation, both sexes	7
blond (b)	Pigmentation, both sexes	5, 6
gold (g)	Pigmentation, both sexes	5, 6
blue (bl)	Pigmentation, both sexes	4
abnormis = hunch-back (hb)	Vertebral column, both sexes	2, 5
curvatus = lordose (cu)	Vertebral column, both sexes	2, 10
Palla (Pa)	Vertebral column, lethal in homozygotes	14
coecus (cs)	Eyes, both sexes	2
Elongated (Fa)	Dorsal and caudal fins, both sexes	8
Kalymma (Kal)	All fins, both sexes	11
Supressor (Sup)	All fins, both sexes	11

chromosome as well. It is most probable that this dominance is a direct result of the selection of genes responsible for the colour pattern: in natural populations these genes represent a well-balanced polymorphic system (Haskins and Haskins 1954). The polymorphism of the colour pattern in many animal species is based on the coexistence of several dominant factors (Ford 1966). Autosomal genes of the guppy unrelated to polymorphism are generally recessive.

3. The presence of recessive lethal genes in Y-chromosomes of guppy. When two Y-chromosomes were combined in a male, homozygotes with respect to genes Ma (Maculatus), Ar (Armatus) and Pa (Pauper) turned out to be non-viable (Winge and Ditlevsen 1938; Haskins et al. 1970). Meanwhile, the males with genotypes $Y_{Ma}Y_{Ar}$, $Y_{Ma}Y_{Pa}$ and $Y_{Pa}Y_{Ar}$ are viable and in addition fertile. It appears that a special lethal gene is closely linked with each of the genes located in the Y-chromosome; the lethals associated with different genes are non-allelic. Haskins and associates were able to obtain just one male $Y_{Ma}Y_{Ma}$ which had numerous offspring. Most probably this resulted from the crossing-over between the gene Ma and a lethal; in this case one of the Y-chromosomes was devoid of the lethal gene.

The presence of lethals in the Y-chromosome contributes to the stable heterozygosity of guppy males with respect to the main colour gene. At the same time, the accumulation of lethals may be considered the first stage in Y-chromosome disruption. It may be speculated that in the guppy this process is at an early stage.

4. The possibility of crossing-over between the X- and Y-chromosomes suggested that they contain homologous regions. Crossing-over was observed by Winge (1923) and later by many other authors. Certain genes pass from X to Y and vice versa relatively frequently; for example, the crossing-over between the genes Ds and Cp is equal to about 10% (Dzwillo 1959). The data on crossing-over have been used in mapping the sex chromosomes (Winge 1934; Winge and Ditlevsen 1938):

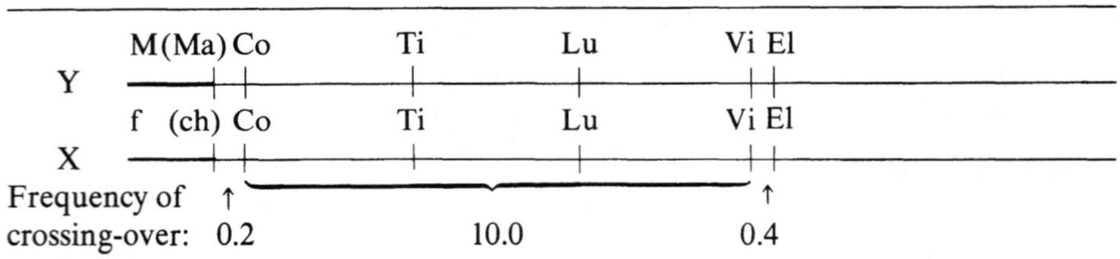

According to recent data (Nayudu 1979), loci N-II, Fl and Cp, located in X- and Y-chromosomes, are arranged in the following order: M(f) sex factor –N-II–Fl–Cp. The relation between the location of these loci and the earlier investigated genes Co, Ti, Lu and others is unknown at present. Three genes, which determine the melanistic patterns in the guppy, differ in melanophores. In Fl fish these are dendritic cells, in Cp fish – bipolar ones, and in N-II fish – corollar ones (Nayudu and Hunter 1979).

The chromosomal maps were constructed using relatively limited data and require further refinement. The location of lethals in the Y-chromosomes is particularly unclear.

5. The instability of the genetic mechanism of sex determination. The certain proportion of males in almost all strains of domesticated guppy after testing for their chromosomal structure turned out to be genetic females (XX). The selection of such spontaneously transformed males enabled Winge to construct the strain without XY males (Winge 1934):

$$♀XX \times ♂XX \text{ (transformed)} = ♀♀XX + \text{singular } ♂♂XX.$$

The proportion of transformed males in this strain has been markedly increased by selection, and the stock was maintained for a long time without the participation of normal XY males. XY males can also sometimes be spontaneously transformed into females. When such females possessing the male karyotype are crossed with normal males the proportion of males in the progeny is equal to 75%:

$$\underset{\text{(transformed)}}{♀X_oY_{Ma}} \times ♂X_oY_{Pa} = \underbrace{\underset{♂♂}{X_oY_{Ma}} + \underset{♂♂}{X_oY_{Pa}} + \underset{♂♂}{Y_{Ma}Y_{Pa}}}_{75\%} + \underset{♀♀\ 25\%}{X_oX_o}$$

The use of males with the genetic formulae $Y_{Ma}Y_{Pa}$ made it possible to obtain unisexual male offspring (Winge and Ditlevsen 1938):

$$♂Y_{Ma}Y_{Pa} \times ♀X_oX_o = \underset{(45\%)}{♂♂X_oY_{Ma}} + \underset{(55\%)}{♂♂X_oY_{Pa}}.$$

In crosses between transformed females and normal males carrying the Ma gene in the Y-chromosomes the lethal proportion 2:1 was obtained, $Y_{Ma}Y_{Ma}$ males were non-viable:

$$♂X_oY_{Ma} \times ♀X_oY_{Ma} = \underset{(67\%)}{♂♂X_oY_{Ma}} + \underset{(33\%)}{♀♀X_oX_o}.$$

In the ageing females of the guppy the activity of the female sex hormone is decreased and the expression of the "colour" genes located in X-chromosomes may be permitted. The female genotype can be determined when they are spontaneously transformed into males or when such a transformation is induced by the male sex hormone testosterone. The addition of testosterone to water or food makes it possible to obtain XX males with expressed genes of pigmentation (Dzwillo 1962, 1966; Haskins et al. 1970). Similarly, large numbers of XY females can be produced using estrone, the female sex hormone.

It is assumed that, in addition to the main genes involved in sex determination and located in the X- and Y-chromosomes, many weak additive sex genes, both male and female ones, are scattered throughout other chromosomes of the guppy (Gordon 1957; Kosswig 1964b). The overall action of these genes may become more pronounced than that of the sex factors located in the sex chromosomes, and thus lead to the transformation of a male into a female and vice versa.

6. The hormonal control of the "colour" gene expression. Almost all the dominant colour genes are inactive in females (Winge 1927; Goodrich et al. 1947),

but certain genes which escape the control of hormonal factors are an exception. For example, the gene Fl (yellow body) has a weak effect on the pigmentation of normal females; the same is true of the gene Cp (pigmented tail). The expression of genes N-I and N-II is stronger in females; this is particularly true of the second of these two genes (Dzwillo 1959). The females carrying the gene N-II ("black" according to Haskins et al. 1970) have a fairly pronounced pigmentation; in the males the pigmentation can be noted immediately after the small fishes are born.

7. Epistatic relationships have often been observed between different colour genes. Thus, in the males homozygous for the gene g (gold) the manifestation of the Y-linked gene Ma is weakened (Goodrich et al. 1947). The gene Fl in the homozygous state suppresses the action of the following genes: Cp, N-II, Ir, Ds, Ch (Dzwillo 1959; Schröder 1970, 1976).

8. Autosomal genes in the guppy are inherited strictly according to Mendel. By way of example, we shall present the segregation in the F_2 corresponding to the classical ratio 9:3:3:1 for the two genes of the background colour: blond (b) and blue (r) (Dzwillo 1959); the expected numbers are given in brackets:

grey	(B, R)	– 52	(49.0)
blue	(B, r)	– 14	(16.3)
pale	(b, R)	– 17	(16.3)
white	(b, r)	– 4	(5.4)

The strict Mendelian inheritance has also been established with respect to the dominant gene Kal, resulting in the elongation of all the fins in individuals of both sexes (Fig. 29). The manifestation of this gene requires the presence of another recessive unlinked gene Sup^+ (Schröder 1969c). The dominant allele of this last gene Sup inhibits the action of the Kal gene, and individuals with the genetic structure KalKal SupSup and KalKal $SupSup^+$ are normal.

One of the most important results of the genetic studies of the guppy was the conclusion that the Y-chromosome in this species is not destroyed and contains many genes. One of these (or a group of closely located tightly linked genes – a supergene) is responsible for sex determination. At the same time, sex determination by the sex chromosomes is imperfect, the presence of a large number of male and female genes in the autosomes and in the X-chromosome results in frequent cases of sex reversal.

The mechanisms responsible for the emergence and stable maintenance of polymorphism with respect to the colour pattern of males in natural populations of the guppy are not quite clear at present. The explanation relating this polymorphism to sexual selection, that is to the selection of the most brightly coloured males by females or to the activation of the sex game when male diversity is increased (Farr 1976) is most probable. The pressure exerted by predators may also be significant; in habitats where this pressure is high, the diversity of male pigmentation has been found to decrease, in particular almost all the genes able to migrate from the X-chromosome into the Y-chromosome and vice versa disappear (Haskins et al. 1961; Endler 1980).

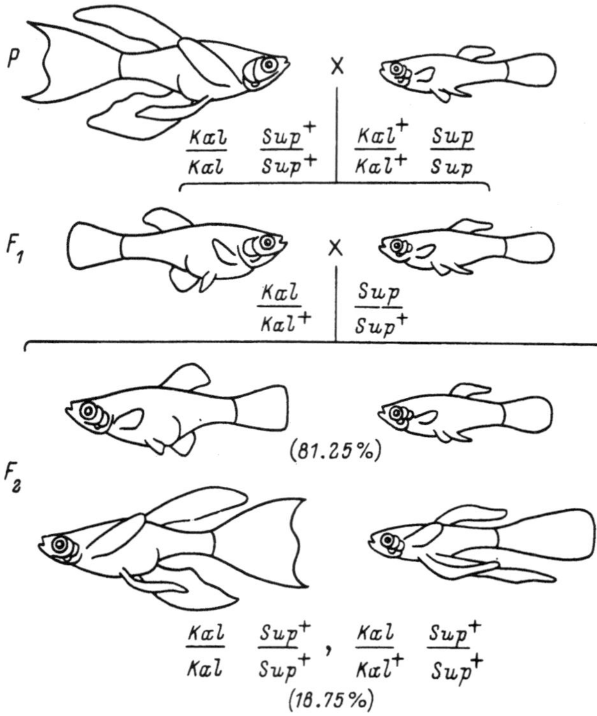

Fig. 29. Expression and inheritance of the gene Kal (Kalymma) in the guppy *Poecilia reticulata* (Schröder 1974)

The Platy (Xiphophorus maculatus)

Xiphophorus (Platypoecilis) maculatus (Poeciliidae) is a popular model of genetic studies just like the guppy. The platy does not possess such a strong sexual dimorphism with respect to pigmentation which is characteristic of the guppy. Nevertheless, the males in natural populations have much brighter pigmentation than the females (Kallman 1970b). Below we shall dwell on the most important sections of platy genetics. These include:

1. the elucidation of the mechanism of genetic sex determination in this species of viviparous fish;
2. the analysis of the inheritance of different genes and their location in chromosomes;
3. the study of the polymorphism of natural populations of the platy with respect to pigmentation;
4. the study of the genetic, biochemical and physiological mechanisms underlying the formation of malignant tumours (melanomas, erythroblastomas and other neoplasms) in the interspecific and interpopulational hybrids of the platy.

1. Sex in the platy just as in the guppy is determined by the sex chromosomes (gonosomes) which contain male and female factors (genes) (Bellamy 1923, 1928; Gordon 1937). It is a characteristic feature of the platy that natural populations

possess sex chromosomes of three types – X, Y and W. The X-chromosome contains a "weak" gene of the female sex f (recessive with respect to the male factor); in the W, it appears to be substituted for a stronger dominant female factor F; the Y-chromosome carries the gene of the male sex M. The W-, X- and Y-chromosomes are homologous (Kosswig 1954; Kallman 1973). The following combinations of sex chromosomes have been found in platyfish populations:

females	– X^fX^f, W^FX^f, W^FY^M
males	– X^fY^M, Y^MY^M

In the females carrying WY sex chromosomes factor F dominates over the factor M; in the males XY the male factor was found to dominate. The WW females have not been found in natural populations (Gordon 1947a, b, 1953, 1957; Kallman 1965a, b, 1970b, 1973).

In many populations of the platy in Honduras and Mexico all three types of sex chromosomes have been described and correspondingly females and males with different combinations of sex chromosomes are found. The W-chromosome is absent in a certain populations, and sex determination occurs according to the classical scheme: ♀♀XX, ♂♂XY (male heterogamety) (Bellamy 1936; Breider 1942; Gordon 1947a; Kallman 1975). It cannot be said at present whether the populations consisting solely of WY females and YY males (the type of female heterogamety) exist in nature.

It is remarkable that when the platy was introduced as an aquarium fish just such WY and YY individuals were brought from Honduras and gave rise to domesticated lines. Consequently, all the cultured varieties of the platyfish possess female heterogamety (Gordon 1951b).

The presence of three types of sex chromosomes allows different sex ratios to be obtained in various crosses of the platy (Anders A. et al. 1970; Kallman 1973):

♀XX × ♂XY = ♀♀XX + ♂♂XY (1:1);
♀WY × ♂YY = ♀♀WY + ♂♂YY (1:1);
♀WX × ♂YY = ♀♀WY + ♂♂XY (1:1);
♀WY × ♂XY = ♀♀WY + ♀♀WX + ♂♂XY + ♂♂YY (1:1);
♀WX × ♂XY = ♀♀WX + ♀♀WY + ♀♀XX + ♂♂XY (3:1);
♀XX × ♂YY = ♂♂XY (100%).

All the males from unisexual offspring subjected to a histological examination turned out to be genuine males and not transformed females (Chavia and Gordon 1951). The autosomal factors present in certain populations of the platy greatly affect sex determination in interpopulational crosses. Individuals from these populations yield "exceptional" offspring with altered sex. The fish with female karyotypes WX and WY are particularly prone to sex reversal. Transformed males WX and WY form a marked proportion of the offspring after the crosses of platy belonging to the two Honduras stocks N_p and C_p. Not infrequently up to 50% of the "genetic" females may be transformed into males (Kallman 1968). In the cross

♀WY × ♂WY (transformed) = ♀♀WW + ♀♀WY + ♂♂YY

females having the WW formulae which are not found under natural conditions have been observed.

The cross of the XX female from Mexico with the YY male of an aquarium strain of the platy yields only XY males in the offspring, some of these males, however, being spontaneously transformed into females. In the second generation the XX females appear again, but part of them undergoes spontaneous transformation into males (exceptional males XX). By crossing XX females and XX males with each other, purely female offspring has been obtained

♀XX × ♂XX ("exceptional") = ♀♀XX (100%).

In certain broods of this type transformed XX males are again found occasionally. The use of such males for subsequent reproduction for seven to nine generations of inbreeding resulted in an increase in the number of XX males in this stock on average to 30%, in a few crosses the number of males was equal to 50%–60% (Öktay 1959; Kosswig 1964b; Dzwillo and Zander 1967). Apparently, inbreeding resulted in the selection of additional male sex factors scattered in autosomes. These factors, acting together, could possibly inhibit the female factor present in the X-chromosome. Crosses of XX males with the females from other stocks led to a drastic decrease in the number of "exceptional" males.

It can thus be seen that two apparently incompatible mechanisms of sex determination, male and female heterogamety, appear to coexist in the platy. The genetic analysis of the coloured gene markers present in the X-, Y- and W-chromosomes allows their inheritance to be traced. In addition to the main genes responsible for sex determination and located in sex chromosomes, a large amount of weak male (and apparently female as well) genes are present in the autosomes; the combined action of these genes may alter the sex determined by sex chromosomes. Supposedly, the sex in the platy can also be altered by hormones, however, experiments of this kind have not yet been conducted.

2. In the platy many genes responsible for pigmentation are located in sex chromosomes, just as in the guppy.

The locus N (Nigra) is a dominant mutation widely used by aquarium fish culturists; this mutation results in a marked blackening of the tail part of the body and of the caudal fin. The pigmentation is due to the accumulation of large melanophores. The semidominant gene Fu (Fuliginosus) producing the black pigmentation of the whole body appears to be an allele of the gene N; the Fu homozygotes are nonviable (Öktay 1954; Gordon and Baker 1955). In the presence of the recessive allele n the macromelanophores are completely absent. The gene N is located in the Y-chromosome, but can be transferred to the W-chromosome by crossing-over (Fraser and Gordon 1928; Gordon 1937).

The locus Sp (black spots or spot-sided) in populations of the platyfish exists as many alleles such as Sp^1–Sp^{10}, Sd (spotted dorsal), Sd′ etc. (Kallman 1975). According to Gordon (1947b), this locus is responsible for the control of macromelanophore distribution over the body and fins. All the above alleles are dominant with respect to the recessive sp; the individuals with the genotype spsp have no macromelanophores. The gene Sp and its alleles in natural populations are located either in the X- or in the Y-chromosomes, but in laboratory populations they can be transmitted only through the Y-chromosome just like the gene N. The genes N and Sp are closely linked (Gordon 1937).

The Sr locus (striped body or stripe-sided), having several alleles, is also involved in the regulation of macromelanophore formation. The autonomy of this locus, however, remains doubtful, although the possibility of crossing-over between the genes Sd and Sr suggest that the latter genes are non-allelic (Anders and Anders 1963; Kallman 1970c).

The Sb gene (black bottom; Gordon 1948) is weakly investigated, but it may well be the allele of the Sp locus.

The second group of colour genes in the sex chromosomes of the platy includes the genes affecting red and yellow pigment cells – xanthophores and erythrophores. These genes belong to a very polymorphic locus (or loci) Dr (red pigmentation).

The Dr locus is involved in the control of pterine synthesis, and of the spatial arrangement of xanthophores and erythrophores (Fraser and Gordon 1928; Gordon 1937, 1956; Kallman and Schreibman 1971). Recent studies of platy populations living in nature have made it possible to detect many new variants with red and yellow pigmentation; these variants appear to be determined by the alleles of the same locus (Borowsky and Kallman 1976). Apparently, in nature the Dr locus is represented by a very large number of alleles (Dr, Ar, Rt, Mr, CP_o, V_o, Br, Fr, Nr – different types of red and orange pigmentation; T_y, CP_y, A_y – variants of yellow pigmentation). The genetic basis of natural variants of pigmentation is unfortunately unclear, but it may well be that more than one locus is involved in this variation (Kallman 1970a, 1975). For instance, the genes Ir and Iy (iris red and iris yellow) appear to belong to an independent locus (Kallman 1970b). All the alleles in this series are dominant with respect to a common recessive allele designated by the symbol dr. The Dr locus (or loci) is also located in the Y- and X-chromosomes and is closely linked with the genes N, Sp and Sr. Crossing-over and the transition of "colour" alleles into the W-chromosome are possible (McIntyre 1961) although in natural populations this chromosome generally contains only the recessive "colourless" allele dr. In contrast to the gene Sp, the expression of the Dr gene in the platy is under hormonal control and depends on gene dosage (Valenti and Kallman 1973). The genes Mr, CP_o-2, V_o, Fr, Rt, Tr and some others are manifested only in males (Kallman 1975).

The semidominant genes P^e and P^l are responsible for early and late differentiation and activation of the gonadotropic region of adenohypophysis respectively (Kallman and Schreibman 1973). The males with gene P^e display characteristic early maturation and smaller body size. The gene P^e is linked with the genes Sp, Sr, Dr, Ir, while the gene P^l is connected with the genes Br and N (Y-chromosome). It cannot be excluded, although this appears improbable, that one is dealing in this case not with different genes but rather with the pleiotropic action of the gene responsible for pigmentation (Schreibman and Kallman 1977). It was recently detected that there are more than two alleles of this gene – four or five (Kallman and Borkoski 1978). Each of these alleles affects the time of gonadal maturation in males and apparently in females of the platy. It should be added that the rate of gonad maturation in males appears to be affected not only by genes, but also by environmental conditions, in particular it may depend on the presence or absence of other individuals of the same sex (Sohn 1977).

It has been assumed that the X- and Y-chromosomes also contain regulator genes (R) and operators (O) modifying the activity of the structural loci (Anders et al. 1973). Thus, in the platy the Y-chromosome contains the genes N, Sp, Sr, Dr, P, R, O and the gene M determining development in a masculine direction. It should be added that further thorough studies of the linkage and localization of different genes in the X- and Y-chromosomes are needed.

It can be noted that the red and yellow colour genes (pterine series) are located nearer to the sex-determining part of X- and Y-chromosomes than the black colour genes (macromelanophore series). Regulating genes are arranged very closely to the respective colour genes.

In the aquarium strains of the platy with female heterogamety (♀♀WY, ♂♂YY) the genes located in the W-chromosome are inherited maternally. In other words, they are transferred from female parents to daughters. The only exception is that of individuals originating from crossing-over, but the frequency of their appearance under normal conditions does not exceed 1% (Fraser and Gordon 1928; Bellamy 1933 b; Gordon 1937; McIntyre 1961). The genes linked with the Y-chromosome are inherited as ordinary sex-linked mutations. We shall present one characteristic example of such inheritance:

$$♀W_{nDrSp}Y_{ndrSd} \times ♂Y_{NdrspY}_{Ndrsp} = ♀♀W_{nDrSp}Y_{Ndrsp} + ♂♂Y_{ndrSd}Y_{Ndrsp}$$

(DrSpSd) (N) (DrSpN) (NSd)

In this cross the genes Dr and Sp linked with the W-chromosome are transmitted to female progeny, while the genes Sd and N are inherited in a crosswise manner from female parents to sons (Sd) and from male parents to daughters (N).

Only a few genes have so far been mapped in the autosomes of the platy. Two loci control the formation and localization pattern of the small black pigment cells, the micromelanophores. The locus C (P according to Gordon) controls the pattern of micromelanophores on the tail stem and in the base of the caudal fin. Eight or more dominant alleles of this locus, M, Mc, T, Cc, O, Co, C, D (Fig. 30) and some rare variants are an example of yet another polymorphic system in the natural populations of the platy; generally no less than four to five alleles of the C locus are found in each population. All the alleles listed above are dominant with respect to the normal allele c^+, in the presence of which no accumulation of micromelanophores on the tail of such individuals could be observed (Gordon 1947 b, 1953; Kallman 1970 b, 1975; Borowsky and Kallman 1976). In some cases, the different tail spot patterns were inherited together, for example, the OT, CcO, CcD combinations (Kallman and Atz 1966; Kallman 1975). It has been suggested that the tail spots are controlled by a complex locus (supergene) as well as the body spot patterns Sp, Sd, Sr and N.

Another autosomal locus, St, controls the development of micromelanophores over the whole body surface of the platy. Substitution of this gene for its recessive allele st results in the complete disappearance of small black pigment cells and a marked decrease in the number of macromelanophores. This leads to the almost white colour of such fishes (Gordon 1927, 1957). The gene oc in the platy is not normally manifested, but cataracts are formed on the eyes of interspecific hybrids

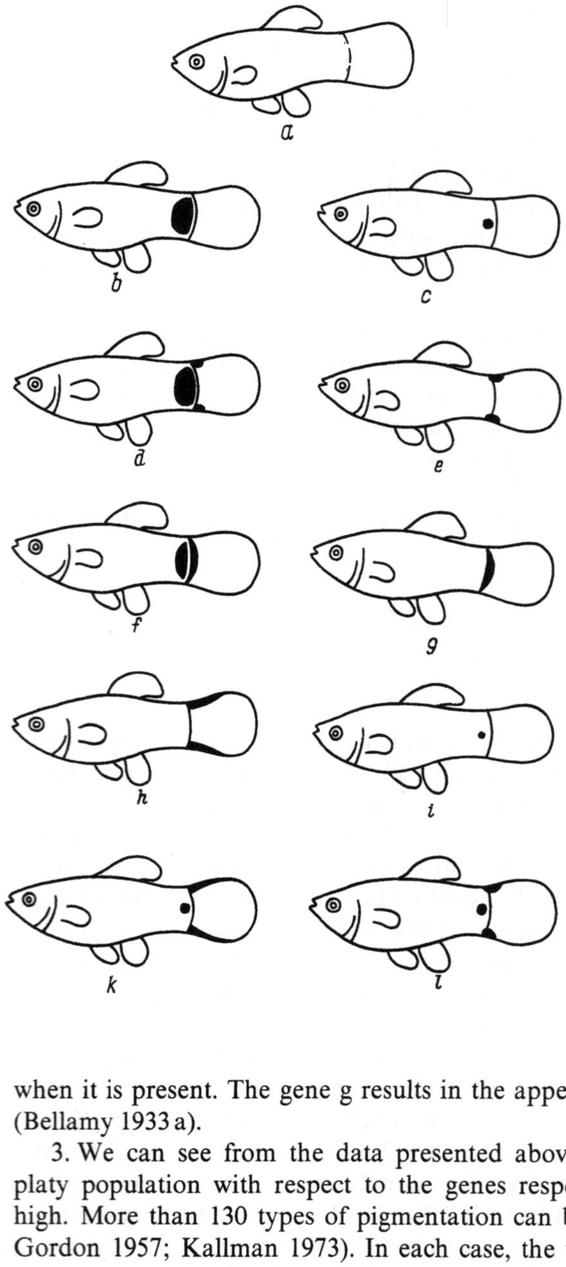

Fig. 30a–l. Alleles of an autosomal gene P affecting the distribution of micromelanophores in the platyfish *Xiphophorus maculatus* (Gordon 1947b; Gordon and Gordon 1957; Kallman 1975). **a** normal recessive allele p. Dominant alleles; **b** P^{M} (moon); **c** P^{O} (one spot); **d** P^{Mc} (moon complete); **e** P^{T} (twin spot); **f** P^{Cc} (complete crescent); **g** P^{C} (crescent); **h** P^{Co} (comet); **i** P^{D} (Dot); **k, l** heterozygotes P^{OCo} and P^{OT}

when it is present. The gene g results in the appearance of "golden" pigmentation (Bellamy 1933a).

3. We can see from the data presented above that the polymorphism of the platy population with respect to the genes responsible for pigmentation is very high. More than 130 types of pigmentation can be found in nature (Gordon and Gordon 1957; Kallman 1973). In each case, the variation is controlled mainly by two or three loci affecting the development of macro- and micromelanophores, as well as by one or two loci controlling the development of xanto- and erythrophores. Each of the loci is represented in populations by a large number of dominant or semidominant alleles. The first authors who described Mexican and Central American populations counted seven alleles of the autosomal locus C and five alleles of the sex-linked Sp locus (Gordon and Gordon 1957); more recent data have, however, demonstrated that the diversity of these loci is markedly higher

(Kallman 1971). Now I would like to mention the most essential conclusions drawn from populational investigations of the platy.

All the genes present in the polymorphic systems described above are dominant with respect to the "colourless" allele.

The frequencies of different genes vary in different populations, the frequency appears to change with time but generally speaking any changes in the concentration of the different genes over the last 70 years were only very limited. Apparently, the proportion of alleles in each population is maintained by selection (Gordon and Gordon 1957). Similar types of colour patterns in different populations are controlled by different sets of genes. Moreover, even the identical designation of an allele (Kallman 1970 a) does not automatically mean that it is identical in a different population of a given species: a common designation can mask a whole number of different alleles. All these data provide additional evidence of the selective value of polymorphism.

Most populations of the platy possess all three types of sex chromosomes (X, Y and W), but generally the W-chromosome contains only "normal" recessive alleles of colour loci determining the protective grey pigmentation. The females have much less prominent pigmentation than the males in all populations: apparently the selective pressure appears to be directed against excessively bright pigmentation of the females. The appearance of the W-chromosome with the strong female sex factor can be considered to be adaptation, contributing to the maintenance of sex dimorphism with respect to pigmentation without any hormonal control. The intensity of pigmentation of the males and females is almost identical in a few populations consisting solely of XX females and XY males (Kallman 1971).

Thus, despite certain differences in the mechanism of genetic and hormonal control, the polymorphism of pigmentation in the platy and in the guppy has many common features. In both cases, the dominant genes are known to accumulate in the populations, not infrequently they are so tightly linked with each other that one may consider them to be one "supergene". All these loci exist as a series of multiple alleles, and most of them are located in sex chromosomes. Polymorphism in the platy as well as in the guppy is stabilized by sexual selection (Ferno and Sjölander 1973), the bright colour pattern correlates with the high activity and agressiveness of the males. On the contrary, where there are predators, only more modestly pigmented and less noticeable individuals will survive. It is noteworthy that similar opposite trends between sexual selection and the selection exerted by the pressure of predators has been noted in the African fish *Notobranchius guentheri* from the family Cyprinodontidae (Haas 1976). The higher survival of heterozygotes or heterosis appears to serve as one of the mechanisms responsible for the maintenance of colour polymorphism in nature. It may be mentioned that heterozygotes for alleles of the autosomal gene C leave more offspring as compared to homozygotes (Borowsky and Kallman 1976).

4. The genetics of malignant tumours may be regarded as one of the most interesting and important chapters in the genetics of the platyfish. Hereditary melanomas were discovered almost simultaneously by three different authors (Haussler 1928; Kosswig 1929; Gordon 1931) in studies of hybrids of the platy *Xiphophorus maculatus* and the swordtail *Xiphophorus helleri*. When certain genes of the platy in particular, such genes as N, Sd, Sp, Sb, Fu and Sr, are transmitted to

F_1 hybrids, the expression of these genes is intensified (Kosswig 1937a, 1937b; Gordon 1948, 1950, 1951a). When such hybrids are crossed with the swordtail, this leads to the development of malignant tumours – melanomas or melanosarcomas affecting the fins, skin, and frequently the inner organs as well (Fig. 31). Disseminated prominent melanomas may be a direct cause of death for such individuals (Gordon 1957; Kosswig 1965; Anders et al. 1973). The development of melanomas depends on the presence of genes controlling the development of macromelanophores: these genes probably escape from the proper control in the hybrid genome (Berg and Gordon 1953; Gordon 1958; Zander 1969).

The gene Rt (red throat) also has a stronger action in the hybrids. Backcross hybrids carrying the gene Rt develop erythroblastoma, another malignant disease. The combination of Fu and Rt is accompanied by the appearance of both tumours, melanoma and erythroblastoma together, and these two compete with each other (Gordon 1957). When the gene of albinism from the swordtail is combined with the gene Sd (from the platy), a tumour still appears, but the rapidly dividing cells do not contain melanine. Gordon proposed that such a tumour should be called an amelanotic melanoma; in this case no macromelanophores are present (Vielkind et al. 1970). Finally, when the oc gene of the platy is combined with the swordtail gene a, ocular tumours appear in the homozygotes ococ in the F_2.

Genetic control of tumour formation has also been demonstrated in experiments when backcross hybrids of the two platy species (*Xiphophorus maculatus* and *X. variatus*) and the swordtail were treated with chemical carcinogens or subjected to irradiation. When the hybrids carried the genes Sr, Sd^{del}, Li, and Pu, melanomas developed; in individuals which carried the Li gene from *X. variatus*, neuroblastomas, epitheliomas and fibrosarcomas appeared (Schwab et al. 1978).

A common feature in all cases of pigment cell malignization is the disturbance of the gene balance induced by interspecies hybridization. Apparently, it is the direction and degree of cell differentiation that is changed and not the number of cells (Leuken and Kaiser 1972; Henze and Anders 1975; Vielkind 1976). A similar phenomenon was observed by Gordon in the crosses between different populations of the platy. For example, when individuals from the Yamapa population were crossed with partners from Kotzakoleas population (Honduras and Mexico respectively) the expression of the Sd gene in the hybrids became stronger. After three generations of selection for maximal expression a typical melanoma developed in the hybrids (Gordon 1950; Gordon and Gordon 1957).

The formation of melanomas is based on the inhibition of the normal pigment cell development: the sequence consisting of initial nerve crested cells – chromatoblasts – melanoblasts – melanocytes – melanophores is terminated at the stage of melanocytes (Gordon 1959; Anders and Anders 1978). Under normal conditions the melanocytes capable of further multiplication are transformed into melanophores, a significant proportion of which is then destroyed by special phagocytic cells. Under the conditions of malignization the melanocytes continue to divide without any control and this process initially results in the pronounced formation of pigment cells and is then followed by the appearance of genuine melanomas. In the platy as well as in other species prone to melanoma formation the large number of genes in the sex chromosomes and in autosomes (Anders F. et al. 1972; Anders A. et al. 1973) controls the transformation of melanocytes into

Fig. 31 a–c. Melanomas in hybrids of the platyfish and swordtail *Xiphophorus maculatus* × *X. helleri*. *1st and 2d lines* F_1; *below* F_B (for *X. helleri*) (Anders et al. 1973)

melanophores; at the appropriate time these controlling or regulatory (R) genes block the multiplication of the melanocytes and stimulate the phagocytic destruction of melanophores. A loss or substitution of part of these genes as a consequence of hybridization can be considered to be a primary mechanism of tumour formation. Malignization can be conditioned by external factors altering the course of pigment cell differentiation (non-hereditary tumours), by somatic and germ-cell mutations in regulatory genes and finally by hybridization disturbing the co-ordination between the colour genes and their regulating elements (Anders and Anders 1978).

A number of autosomal factors with strong action (Anders F. 1967) is found among the controlling genes which may be regarded as suppressors limiting melanine formation and the multiplication of melanocytes. Segregation with respect to each of these stronger factors allows the backcross hybrids of the platy and the swordtail to be classified into two distinct groups differing in the degree of malignization. For example, when the gene Sd was present in the hybrids, 1358 individuals with advanced melanomas and 1344 individuals in which the tumour was only slightly developed were found (Anders A. et al. 1973).

Each gene controlling the development of macromelanophores has its own system of modifiers/suppressor genes. These systems apparently emerged during the long natural selection for the optimal level of expression. Such systems can be destroyed not only by hybridization but also by ionizing radiation inducing mutations in the controlling loci (Anders F. 1968; Anders A. et al. 1971).

When the hybrid melanoma is transplanted to the embryos of the platy, the swordtail or of their hybrids, tumour development is most pronounced in hybrid embryos (Humm et al. 1957).

In general, the highest tendency to neoplasms has been observed when the colour genes of the platy are combined with the genome of swordtails (*Xiphophorus helleri* and *X. montezumae*). On the contrary, the transfer of swordtail colour genes into the platy chromosomal set leads to their weakened manifestation (Kallman 1975).

Recently chromosomal aberrations namely deletions of the X-chromosome and translocations of the large segment of Y-chromosome to X-chromosome has been found in the platy. Development of melanomas in hybrids has been detected only in those cases when colour genes Sd, Sr or Ar are present in the changed karyotype (Ahuja et al. 1979).

The formation of tumours in hybrids is sometimes explained by the presence of a special gene Tu (Ahuja and Anders 1976; Vielkind 1976; Anders and Anders 1978; Schwab et al. 1978).

The hypothesis postulating the presence of a specialized "Tumour" gene in the genome of all xiphophorine fish including healthy ones (Anders and Anders 1978) is insufficiently argumented from our point of view. There is no experimental proof of this hypothesis. Transplantation of skin from the individuals with the gene "Tu" to the embryos devoid of this gene, but treated with DNA prepared from tumour individuals results in the malignization of pigment cells (Vielkind et al. 1976), but in this case transformation can be due to the presence of the colour gene or its products in the graft.

In all cases of tumours associated with the abnormal development of pigment cells we observe the presence of one of the main colour genes in the fish genome. It is impossible to explain why useless and even harmful "tumour genes" were retained in the chromosomal set of the poeciliid fish.

The simplest assumption, as was supposed earlier by Gordon (1957, 1958), is that the disturbance of the stable genetic balance by hybridization, selection or by mutations should result in stimulation or weakening of colour gene activity. Melanomas and other pigment tumours can arise as a consequence of exaggeration (strong intensification) of colour gene expression.

Several authors have tried to link the formation of malignant tumours with the increased content of free amino acids (Anders F. et al. 1962; Kosswig 1965). According to Anders, the stimulation of tumour formation was observed when the fish was kept in water containing an excess of amino acids.

These are the main results of tumour studies in hybrid fishes. Tumours may appear after crosses between other poeciliid fish species (for example, between

Xiphophorus helleri and *X. variatus, X. montezumae* and *X. maculatus*), as well as in hybrids in a number of other families and orders (Ermin 1954; Kosswig 1965; Schwab et al. 1978).

Studies conducted with several fish species, but mainly with the platy, have provided an insight into the genetic mechanisms underlying the formation of melanomas, melanosarcomas and other pigmented malignant tumours. Similar genetic mechanisms may operate in the development of similar tumours in higher vertebrates, including man. An extremely important part in these processes is apparently played by the mutational alteration of genetic systems controlling the normal course of development of the pigment cells. Studies of fish have shown that different elements of the genome are adjusted to each other with a high degree of precision.

The Medaka (Oryzias latipes)

Oryzias (Aplocheilus) latipes (Cyprinodontidae) has been studied mainly by Japanese scientists. Attention has been focussed on the mechanisms of chromosomal sex determination in this species and on the development of a method of hormonal and genetic sex regulation. In the medaka the males are heterogametic (♀♀XX, ♂♂XY). The pigmentation of aquarium varieties is determined by the combination of alleles belonging to three loci. The genotypes of all the most common variants of pigmentation have been determined (Table 11). The brown pigmentation of the wild-type individuals is due to the presence of three dominant genes R, B and I. The mutation in the gene R leads to the reduction of carotinoids in xanthophores and as a result to a more or less decreased pigmentation or in combination with the gene ci to the appearance of "blue" individuals (guanine reduction). The mutations in the gene B inhibit melanophore development; the allele B′ gives variegation of pigment distribution, the allele b in combination with R results in orange-red pigmentation, and in combination with r it results in a white phenotype. The mutation i inhibits the development of all the pigment cells, ii homozygotes look like albino individuals with red eyes. The gene ci linked to the i gene (crossing-over frequency around 4%–5%) partially inhibits the action of the i gene leading to the appearance of an incomplete albino phenotype (Yamamoto and Oikawa 1963).

The X- and Y-chromosomes of the medaka were found to contain just one "pigment" locus R with three alleles which are now well known. All the alleles of this locus can pass from the X-chromosome to Y by crossing-over, but the cross-over frequency is very low, being equal to 0.2 in XY males (Yamamoto 1964) and to about 1% in XY-females, that is individuals transformed into females by estrone treatment. Normal fishes living in nature characteristically contain the R gene in the sex chromosomes X and Y. Genetic analysis has shown that Y^RY^R males obtained in crossing are practically non-viable and the lethal ratio 2:1 has been observed in crosses ♀X^rY^R × ♂X^rY^R. It has been suggested by Yamamoto (1967) that an inert region in the Y-chromosome is located in the neighbourhood of the R gene (this region is absent in the X-chromosome), and homozygotes with respect to the inert region R^-R^- are non-viable. It would perhaps be simpler to assume

Table 11. Pigmentation phenotypes and genotypes of the medaka *Oryzias latipes* (Aida 1921, 1930; Goodrich 1929; Yamamoto 1969)

Phenotypes	Genotypes of loci		
	X-, Y-chromosomes	Autosomes I	Autosomes II
Brown, wild-type	RR, RR^d, Rr	BB, BB^1, Bb	II, Ii
Brown, weakened	R^dR^d, R^dr	BB, BB^1, Bb	II, Ii
Orange (red)	RR, RR^d, Rr	bb	II, Ii
Orange spotted	RR, RR^d, Rr	B^1B^1	II, Ii
Orange spotted, weakened	R^dR^d, R^dr	B^1B^1	II, Ii
Blue	rr	BB, BB^1, Bb	II, Ii
White, spotted	rr	B^1B^1, B^1b	II, Ii
White	rr	bb	II, Ii
Albino (embryos) [a]	All genotypes	All genotypes	ii

[a] The recessive gene i is epistatic towards all alleles of the B locus; epistasis towards the R locus is observed at the embryonic stages; in adult fishes homozygous for i (ii) in the presence of R and R^d traces of pigmentation caused by these genes may be detected

that the R gene in the medaka is closely linked with a specific lethal gene. In a few exceptional Y^RY^R individuals the lethal gene is most probably located in just one of the two Y-chromosomes. The presence of just one "colour" locus in the sex chromosomes of the medaka is most probably connected with the absence of polymorphism with respect to pigmentation in the natural populations of this species.

Although the genetic mechanisms of sex determination in the medaka operate very precisely, a spontaneously transformed individual can still be observed (Aida 1936; Yamamoto 1955). Sex reversal can also be achieved by hormones (such as estrone or testosterone). Yamamoto (1955, 1958, 1959a, b, 1961, 1963, 1967) was able to develop the procedure for the large-scale transformation of genetic males into females and vice versa. Changes in the development of the gonads were achieved predominantly by the addition of hormones to the food. Transformed individuals produced unisexual offspring or offspring with an altered sex ratio (Yamamoto 1958, 1967, 1968).

1. ♀XX × ♂XX (transformed) = ♀♀ (100%),
2. ♀XY (transformed) × ♂XY = ♂♂ (75%) + ♀♀ (25%),
3. ♀XY × ♂YY (from the second cross) = ♂♂ (100%),
4. ♀YY (transformed) × ♂YY (from the second cross) = ♂♂ (100%),
5. ♀XX × ♂YY (from the second cross) = ♂♂ (100%),
6. ♀$X^rY^{R,l}$ (transformed) × ♂$X^rY^{R,l}$ = ♂♂ (67%) + ♀♀ (33%).

Since spontaneous sex-reversal is very rare, and the crossing-over frequency is low, the actual frequencies observed in the crosses followed these predictions very precisely.

Although in all the external traits the transformed fishes were virtually indistinguishable from the normal ones, their growth rate was somewhat slower: XX males grew more slowly than the XY males while XY and YY females grew more slowly than XX females. This delay can probably be explained by differences in the level of metabolism between normal females and males; these differences are probably due to the different sets of genes present in X- and Y-chromosomes (Fineman et al. 1974, 1975). In addition, it has been established that YY males are more active in copulation than normal XY males. This is probably due to the presence of two copies of the gene responsible for "masculinity" (Walter and Hamilton 1970).

The mutations affecting the structure of the vertebral column have also been found in the medaka: these are the recessive genes fu (fused) and wa (wavy) similar to corresponding genes in the guppy (Aida 1921; Yamamoto et al. 1963).

Although the sex in the medaka can be changed by hormonal treatment, the mechanism of sex determination in this species is more "advanced" than in the platy. The Y-chromosome contains a strong factor of the male sex; the effect of sex genes located in autosomes is relatively modest. The Y-chromosome is not empty in the genetic sense, but the process of its destruction involving the accumulation of chromosomal aberration is going on. One can assume that the sex chromosomes in the medaka will soon be discovered by cytological methods as well, using modern techniques of differential staining.

The Fighting Fish and the Paradise Fish (Anabantidae)

The Siamese fighting fish (*Betta splendens*) is distinguished among other aquarium fish species for its bright and variable coloration. In their tension and intensity the famous games of the fighting fish males resemble rooster fights (hence the name fighting fish). The pleasant, commonplace coloration of the males becomes extremely bright during the games, because the intensity of coloration greatly increases when the fish is excited. Although the females of the fighting fish appear more modest, they are still very impressive.

Only the inheritance of the fighting fish coloration has been studied in some detail.

We shall now list the genes described by different authors. The genes V and v (A_2 and A_1, Umrath 1939) affect the number of guanophores (iridocytes) produced in the skin of the fighting fish in marked quantities (Wallbrünn 1958). In vv individuals the pigmentation turns somewhat green, while in the VV ones it is lighter with a steel tone. The gene V is semidominant; Vv heterozygotes have a blue tone and can easily be differentiated from both homozygotes.

The gene c results in a sharp reduction of all the chromatophores leading to virtually complete albinism. Just as in other fish species, the albinotic mutation is epistatic with respect to other "pigment" genes (Umrath 1939; Eberhardt 1941; Wallbrünn 1958).

The genes B and b (M and m, Umrath 1939) affect the size of the melanophores: in bb homozygotes these are smaller and therefore xantho- and erythrophores are more prominent with the resulting light-red (almost golden) colour of the fishes. The gene B in combination with the gene C (the normal allele of the gene responsible for albinism) leads to dark-red background pigmentation (Wallbrünn 1958).

The genes L and l affect the erythro- and xanthophores; in llCB individuals the body pigmentation becomes brown-black. The gene L is semidominant, Ll heterozygotes are clearly different from homozygotes. The combination of the genes llCCbb results in the appearance of a particular dark-red pigmentation (Umrath 1939).

The gene ri, like the gene V, affects the number of iridocytes, decreasing their area density (Eberhardt 1941). The idea of the non-identity of the ri and Ri genes, on the one hand, and of the V and v genes, on the other, needs verification.

The gene nr does apparently lead to a partial reduction of the erythrophores (it affects pterine synthesis). The individuals with the gene nr acquire yellowish pigmentation instead of the red one (Lucas 1972; Royal and Lucas 1972). The gene nr is not linked to any of the other genes described above.

Among the mutants with an altered fin structure one should mention a dominant gene leading to a prominent elongation of the unpaired fins in males. The expression of this gene is under hormonal control (Eberhardt 1943); in the presence of this gene females have normal fins.

The mutant genes V (steel pigmentation), b (less prominent melanophores) and c (albinism) decrease the viability of the homozygotes. All the series of these genes are autosomal. The fighting fish belongs to protogynic hermaphrodites: the development of the sex glands in all the individuals initially proceeds in the female direction, thereafter part of them becomes transformed into males (Lucas 1968). The sex reversal may occur either spontaneously or after hormone treatment (Gordon 1957). The removal of ovaries in adult individuals results in the transformation of most of them into bona fide males. It has been assumed that a large number of genes is involved in sex determination; these genes are located in many different chromosomes; no specialized sex chromosomes have been observed in this species (Lowe and Larkin 1975).

The paradise fish (*Macropodus opercularis*), like the fighting fish, has a bright and variable pigmentation. Females also have a characteristic colour pattern, but it is somewhat more modest than that of the males. There is little information available regarding the genetic traits of the paradise fish. A recessive albinotic mutation (a) has been found. Ratios showing a good correspondence to the expected ones have been found in F_2 and in F_b (Goodrich and Smith 1937): F_2–204:62 (3:1), F_b–552:551 (1:1).

The sex ratio in this species is close to 1:1; it can be suggested that sex determination is monofactorial, although there is no direct evidence of this. In a related species (or more probably subspecies) *M. concolor* males generally predominate in the offspring, their numbers varying from 68% to 91% (Schwier 1939).

Sex determination in this case is apparently polygenic just as in *B. splendens,* in addition protogynic juvenile hermaphroditism appears to be involved.

Other Aquarium Fishes

Some information on the inheritance of certain traits in other aquarium fish species is available.

Several mutations have been found in the swordtail (*Xiphophorus helleri*), a close relative of the platy. The following genes belong to better-studied ones. The recessive genes a (albinism) and g ("golden" colour) in the homozygous state decrease the viability (Kosswig 1935b; Gordon 1941, 1942). The genes Db^1 and Db^2 (diffuse pigmentation) are dominant (Kallman and Atz 1966; Wolf and Anders 1975).

The gene Sn (seminigra or black spots) is dominant; it is homologous and apparently allelic to the genes of the macromelanophore series (Sp) in the platy (Kosswig 1961). The gene Sn when present in a hybrid genome has a stronger action, while in back crosses it leads to the formation of melanosarcomas (Breider 1956). The gene Mo (montezuma) leads to the development of orange-red colour, as well as the gene Rb (rubescens); in the latter case we have a gene from the platy genome (Kallman and Atz 1966).

The genes C and P lead to the formation of micromelanophore clusters at the base of the caudal fin and are allelic to C genes of the platy. The genotypes and phenotypes in this case are associated with each other in the following way (Kerrigan 1934): CCPP, CcPP, CCPp and CcPp – the oval black stripe on the tail (crescent); CCpp and Ccpp – two black spots on the tail (twin spot); ccpp, ccPP and ccPp – non-pigmented fishes. The gene c, as we can see, is epistatic with respect to the genes P and p. Probably the gene Gr (grave), discovered in only one swordtail population, belongs to the same locus (Kallman and Atz 1966).

The genes E and i play the part of modifiers and amplifiers with respect to the manifestation of the gene Co ("comet") and of the gene oc in the platy. By combining the genes E and Co a strain of platy with black fins has been obtained; this is the so-called wagtail platy (Gordon 1946, 1952). The combination of the genes oc and i, as we have already mentioned, results in the formation of ocular tumours (Kosswig 1965).

Red swordtails well known to aquarium fish breeders result from the introduction of the platyfish gene Rt into the swordtail genome and the combination of this gene with the albino mutation a. Repeated back-crosses were used for the construction of typical red swordtails (Kosswig 1961, 1965).

The gene Da (dorsalis alta) leads to some elongation of the rays in the dorsal and anal fins. Homozygotes with respect to the Da gene are non-viable. The heterozygous Dada males survive, but are sterile because of the alteration in the shape of the gonopodium. Their propagation is only possible by artificial insemination (Schröder 1966).

All the genes in the swordtails are inherited autosomally. No sex chromosomes have been found in this species, sex is determined polygenically, male and female factors are apparently scattered throughout all the chromosomes (Kosswig 1935a, 1954, 1964b; Dzwillo and Zander 1967). The sex of each particular individual depends on the relative abundance of male or female factors in the genome. Environmental factors also affect sex determination to a certain extent. Recently the hermaphroditic strain of this species has been found (Lodi 1979).

A few mutations have been discovered in other species related to the swordtail and the platy.

The gene Nc (Nigrocaudatum or spotted caudal) has been found in *X. montezumae;* the expression of this gene may be increased up to melanoma formation in hybrids with *X. helleri* (Marcus and Gordon 1954). In the subspecies *X. montezumae cortezy* three autosomal unlinked macromelanophore colour genes have been described – At (Atromaculatus), Cam (Carbomaculatus) and Sc (Spotted caudal), as well as one gene affecting micromelanophores (Kallman 1971). All these genes are dominant and have been detected in all the populations studied; in contrast to the polymorphism characteristic of the platy, polymorphism in *X. montezumae* with respect to the macromelanophore pattern is based on the variation of several independent loci. The Sc gene (black spots on the tail) sometimes leads to melanomas in inbred strains. Sex determination in this species is of the mixed type, i.e., determined both by sex chromosomes and by other factors present in the autosomes (Zander 1965).

The Y-chromosome of *X. milleri* carries a Gn gene (black spot on the gonopodium). The populations of *X. milleri* are polymorphic with respect to the Gn gene and its allele which defines non-pigmented gonopodium (Kallman and Atz 1966; Kallman and Borowsky 1972). The expression of the Gn gene is stimulated by androgens, and under such conditions it becomes prominent in females as well. In *X. milleri* the macromelanophore gene Sv (Spotted ventral, Y-chromosome) and three autosomally inherited patterns of tail micromelanophores – B (Bar), Pt (Point) and Ss (single spot) have been described (Kallman and Atz 1966). The females of *X. milleri* have XX sex chromosomes, while the males of this species carry XY-chromosomes.

Two genes affecting the distribution of xanthophores and resulting in the appearance of an intensive yellow pigmentation have been found in *X. pigmaeus* – these are Fl and Vfl (yellow body and yellow tail). The expression of both genes is augmented in *pygmaeus*×*maculatus* and *pygmaeus*×*variatus* hybrids, the number of xanthophores increases, and they are transformed into xanthoerythrophores: red granules appear in the pigment cells. Backcross *X. maculatus* hybrids become completely red (Öktay 1964; Zander 1968).

In *X. xiphidium* the macromelanophore genes Fl^1 (flack), Fl^2 (dusky) and Ct have been found (Kallman and Atz 1966; Schwab et al. 1978).

The following genes have been found in *X. variatus:* P or Pn (Punctatus), Li (Lineatus), R (Ruber), Rb (Rubescens), O (Orange), Pu (Purpureus); all these genes appear to be located in the X-chromosome and homologous to Rd and Sp loci of the platy (Rust 1939, 1941; Breider 1956; Anders A. et al. 1973; Borowsky and Khouri 1976). The P locus may in addition be located in the Y-chromosome as well. The Gn gene analogous to a similar gene of *X. milleri* is responsible for the presence of the black spot on the gonopodium (Kallman and Borowsky 1972) and is also located on the X-chromosome. Melanomas develop in hybrids carrying the genes Li and Pu (Anders A. et al. 1973; Anders and Anders 1978; Schwab et al. 1978). In the presence of two doses of the P^2 gene in *X. variatus* the melanomas develop even in the absence of any hybridization (Borowsky 1973).

Three types of micromelanophorous pattern, C (crescent), Ct (cut crescent) and + (absence of black spots) are determined by three autosomal alleles of one locus. This locus appears to be homologous to the corresponding locus of the platy. Sex is determined by sex chromosomes, males are heterogametic. Crossing of *X. variatus* female with males belonging to aquarium strains of the platy results in a unisexual progeny (Kosswig 1936):

$$♀XX_{var.} \times ♂YY_{mac.} = ♂♂XY\ (100\%).$$

Male heterogamety is also characteristic of *X. xiphidium* (Gordon 1957) and *X. milleri* (Kallman 1965 a).

Nineteen species of xiphophorine fish are found in freshwater basins in Mexico and Honduras. In almost all species some polymorphism could be observed for such traits as the distribution and numbers of black, red and yellow pigment cells. In many cases, the expression of the colour genes changes in response to intra- and interspecific hybridization (Zander 1974).

In the limia [*Limia (Poecilia) nigrofasciata* and other species], genes affecting the development of macro- and micromelanophores have been described; these genes are homologous to the same genes in the platyfish (Breider 1935). The hybridization of the *Limia* with related species leads to the formation of melanomas (Breider 1936, 1938).

In the molly (*Mollienesia sphenops*) there is a wide variation in the intensity of pigmentation. In nature these fishes are uniformly grey without any black pattern. The aquarium strains with a "black" pattern are, according to Schröder (1964, 1974), characterized by the presence of the additive semidominant factors N and M. The increase in the number of these genes in the genotype results in the intensification of black pigmentation. By the criterion of black colour intensity the fishes may be divided into six different genetic classes (Table 12). Segregation in F_2 leads to the appearance of all of these colour groups (Fig. 32). Chemical studies with fishes belonging to different pigmentation classes have shown that in individuals with high levels of melanine the level of free amino acids in tissues is slightly increased (Schröder and Yegin 1968).

The Ly (lyra) gene in *M. latipinna* in the heterozygous state changes the shape of the dorsal, caudal and ventral fins (Fig. 33). The LyLy homozygotes are non-

Table 12. The genetics of the black pigmentation in the molly *Mollienesia (Poecilia) sphenops* (Schröder 1974)

Group of pigmentation intensity	Phenotype		Genetic formula	Number of dominant genes
	at birth	in the adult state		
I	Grey, without spots, the iris is light-coloured	Grey, without spots, the iris is light-coloured	nnmm	0
II	Same	Dark grey, many small grey spots, the iris is light-coloured	Nnmm, nnMm	1
IIIa	Same	Almost black with a few small grey spots, the iris is generally light-coloured	NNmm, nnMM	2
IIIb	Grey, weakly spotted, the iris is light-coloured	Almost black with a few small grey spots, the iris is generally light-coloured	NnMm	2
IVa	Black with a lighter abdominal side, the iris is light-coloured	Uniformly black with the dark iris	NnMM, NNMm	3
IVb	Uniformly black with the dark iris	Same	NNMM	4

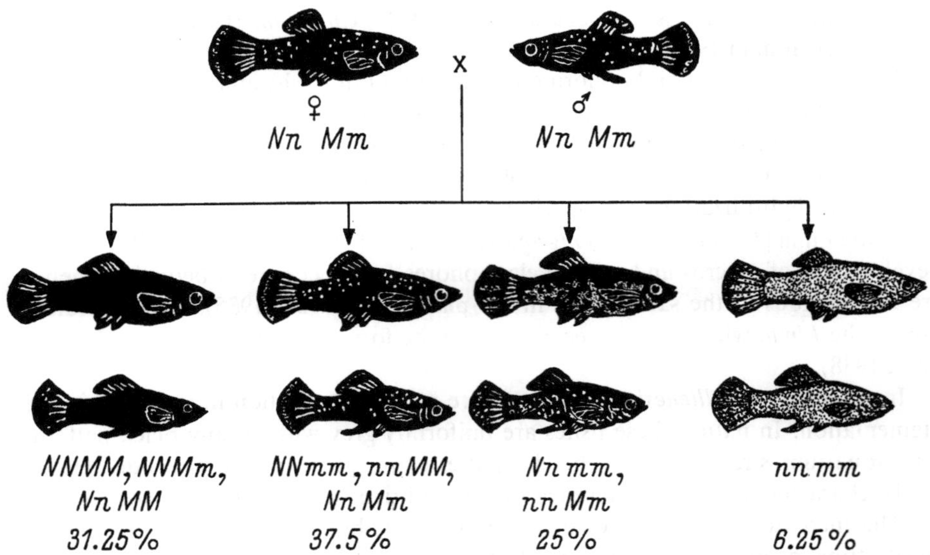

Fig. 32. Segregation of black pigmentation in crosses of *Mollienesia sphenops* (Schröder 1974)

viable perhaps because of the incompatibility of the gametes carrying the Ly gene (Schröder 1964, 1965).

Mutations have also been found in *Poeciliopsis lucida;* these include a recessive mutation stubby (curved vertebral column) (Schultz 1963) as well as a recessive autosomal mutation tr (transparency). In the tr homozygotes the pigmentation becomes variegated since certain regions do not contain iridocytes and non-contracting melanophores (Moore 1974). In *P. viriosa* the recessive autosomal mutation gr (grey) blocks the formtion of xanthophores in the scale epithelium (Vrijenhoek 1976).

In *Rivulus urophthalmus* a recessive mutation has been described which leads to the appearance of red pigmentation (Constantinescu 1928).

In *Brachydanio rerio (Cyprinidae)* the genes responsible for the distribution of the pigment cells (fr^+, fr) and the degree of pigmentation (mlr^+, mlr, modificator ur) have been described. A strain of leopard danio, named *Brachydanio frankei* (spotted danio), contains only one colour gene, the fr (Frankel 1979). The Gr gene leads to the destruction of all the xanthophores as well as of part of the guano-

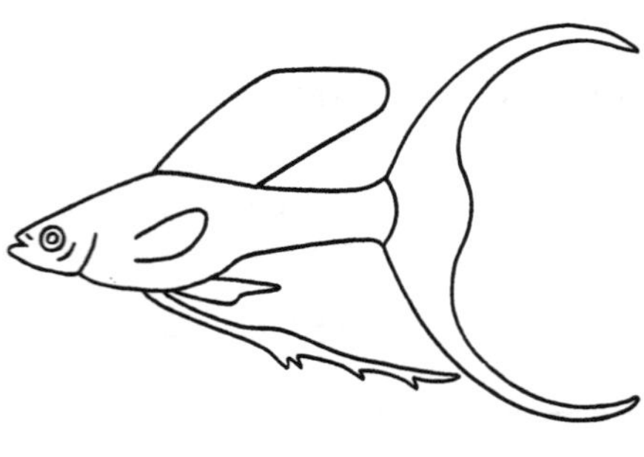

Fig. 33. Mutation "Lyra" in *Mollienesia latipinna* (Schröder 1964)

phores and melanophores (Kirschbaum 1977). An albino mutant with lowered viability in the homozygous state has been described in *Hemmigrammus caudovittatus (Characidae)* (Stallknecht 1975). In *Pterophyllum eimekei* (Cichlidae) the semidominant gene Lf leads to the elongation of all the fins (Sterba 1959). It has been suggested that this gene acts via the pituitary gland. Certain aquarium fish species, *Barbus* in particular, show surprisingly low variability; this is perhaps a consequence of strict selection for homogeneity of the external pigmentation of these fish species under natural conditions.

We shall now try to draw several conclusions.

Among tooth-carp we find fishes with very different sex determination mechanism extending from genuine hermaphrodites (*Rivulus mormoratus*) to species with strict sex determination by the sex chromosomes (*Poecilia reticulata, Xiphophorus maculatus, X. milleri, X. variatus, Oryzias latipes*). In the presence of sex chromosomes heterogamety may be present either in males or in females. In *X. maculatus* examples of male and female heterogamety are found in one and the same population; sex chromosomes of three types (X, Y and W) may be present in karyotypes. The presence of three different sex chromosomes is suspected also in the tilapia (Avtalion and Hammerman 1978; Hammerman and Avtalion 1979). But even with the most perfect mechanisms of chromosomal sex determination unpaired sex chromosomes differ from their homologues only in the presence of one gene coding for sex (M or F) or of an unpaired region; other regions of unpaired chromosomes contain many active genes. The Y (or W)-chromosome in these species is thus at the very beginning of the destruction process. At present, quite a number of fish species have been found possessing genuine heterochromosomes: in such cases, unpaired chromosomes appear to have undergone much more prominent changes. Recently well-defined X- and Y-chromosomes have been found in the males of *Xiphophorus maculatus* (Ahuja et al. 1979).

An analysis of colour polymorphism in aquarium fish varieties has shown that this polymorphism is usually associated with sex dimorphism and with the accumulation of dominant genes. The emergence and maintenance of such polymorphism in nature appears to stem from two selective factors acting in opposite directions: sexual selection leads to intensification of the pigmentation in males, thereby providing greater diversity; on the other hand, individuals with bright pigmentation are more readily eliminated by predators. The nature of this polymorphism is probably adaptive.

An analysis of the action of pigment genes in a "foreign" (hybrid) genotype has made it possible to discover a very important tendency, the fine mutual adjustment of all the genes in the genotype of a species and the loss of this adjustment induced by hybridization. One of the main mechanisms responsible for the emergence of malignant tumours in fishes (such as melanomas, melanosarcomas, erythroblastomas and some others) involves alterations in the sequence of the pigment cells' development which lead to the loss of control of their proliferation. These changes may be purely genetic ones and may emerge as a result of a disturbance of the genetic balance acquired in the course of the long history of evolution.

Almost all the varieties and strains of aquarium fishes were developed by the selection of prominent "qualitative" mutants, the selection of quantitative traits was of secondary significance in this process.

Chapter 4

The Inheritance of Quantitative Traits in Fishes. Phenodeviants

General Features of Quantitative Variation

With rare exceptions, no two completely identical individuals can be found in any fish population. The differences may appear in morphological, physiological or biochemical features; more often than not they are of a quantitative nature. The origin of the intrapopulation individual quantitative variation is two-fold: the development of an organism and of all its parts is modified (1) by changes in the genotype and (2) by environmental fluctuations.

The variation of a large number of meristic or discontinuous quantitative traits in fishes follows the law of binomial distribution, the variation in number of elements may be described by the binomial coefficients of expansion $(a+b)^n$. When n is sufficiently high, and a and b values are close to each other, such a distribution should become indistinguishable from the normal one; this is based on the combinational probabilities of large numbers of random events (Fig. 34). Indeed, the variation of continuous traits in fishes is generally described by a normal distribution. Changes in any quantitative trait which depend on many genes and environmental factors may be regarded as "random events" within the framework of biological variation. Both genes and environmental factors may affect a given trait either in the positive or the negative sense. It is easy to understand that the random combination of various genetic and environmental factors with positive and negative action will most probably result in traits with values close to the mean.

The normal distribution described graphically by the symmetrical variation curve (Gaussian curve) may be characterized by two easily calculated constants[3]: the mean:

$$\bar{x} = \frac{\sum v}{n} \tag{1}$$

and the standard deviation:

$$\sigma = \pm \sqrt{\frac{\sum d^2}{n-1}}, \tag{2}$$

3 Calculation method for statistical constants of the normal distribution and other distributions are described in numerous textbooks dealing with biological variation statistics

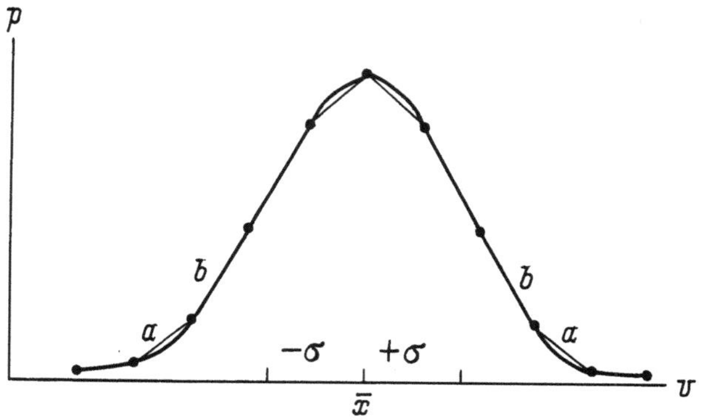

Fig. 34. The binomial distribution corresponding to binomial coefficient of the tenth power (*a*) and the normal variation curve (*b*)

where v is the individual value of a given trait in a given individual, n the number of individuals and d the individual deviation (defined as the difference between the individual measurement and the mean value of the character). Another expression of variation is the variance (V or σ^2) defined as the square of the standard deviation. In order to abolish dimensional values, the standard deviation is expressed as a percentage of the mean value (variation coefficient):

$$\mathrm{C.V.} = \frac{100\,\sigma}{\bar{x}}. \qquad (3)$$

The variation in such quantitative traits as body weight and length, exterior indices, the size of different organs, the haemoglobin content in the blood, the intensity of oxygen consumption and others is continuous. It is therefore convenient to divide the range of variation into equal intervals corresponding to different classes: individuals with close values of a trait then belong to one and the same class. The calculation of biometric constants is then made approximately, on the basis of the mean value of a given trait for all the individuals of each class.

The variation of fishes with respect to their body weight frequently deviates from the normal distribution: the variation curve becomes irregular, multimodal or asymmetric. The multimodality may stem from the intermixing of different groups of fishes or from genetic segregation with respect to a few genes affecting the rate of growth. The negative asymmetry or a shift in distribution to the right which is quite frequent, due to the strong dependence of the growth rate on the initial weight of the individual and, to an even greater extent, to the advantages gained by large fishes in the competition for food. More often than not, the positive asymmetry is associated with the presence in many populations and shoals of fishes of a few individuals with malformations, deviations from the norm and a poor growth rate.

When there is moderate asymmetry, it can be abolished if the curve is plotted in the semilogarithmic scale. The distribution after that rearrangement becomes log-normal, frequently almost identical to the normal distribution in its appearance (Fig. 35).

The variation of fishes with respect to certain other traits cannot be expressed in terms of the normal distribution; usually in such cases we are dealing with the Poisson distribution characteristic of the occurrence of rare events. We shall not, however, discuss this type of distribution, which is relatively rare among fishes.

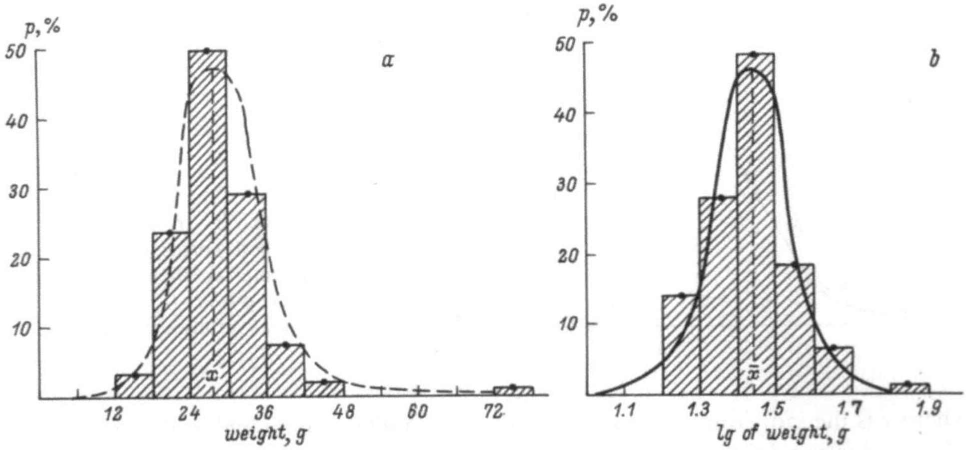

Fig. 35 a, b. Asymmetric variation of Ropsha common carp fingerlings with respect to weight. Ropsha, 1956, Pond Bystryanka. **a** linear scale; **b** semilogarithmical scale

Usually not the whole population but a more or less numerous sample of the population is studied. Mean values and the indices describing the variation of the trait determined from a random sample represent approximate values. The mean standard errors of these values can be calculated, using the following formulae:

$$m_{\bar{x}} = \pm \frac{\sigma}{\sqrt{n}}; \quad m_{\sigma} = \pm \frac{\sigma}{\sqrt{2n}}; \quad m_{C.V.} = \pm \frac{C.V.}{\sqrt{2n}}. \tag{4}$$

The magnitude of these errors is used to determine the range of the true mean and the true variation indices for the whole population. This range defined by confidence limits may be established with a certain, greater or lesser, degree of probability. The addition or subtraction of two standard errors ($\bar{x} \pm 2\,m_{\bar{x}}$ or $\sigma \pm 2\,m_{\sigma}$) corresponds to the finding of confidence limits with a probability (p) equal to 0.95. Subtraction or addition of three standard errors increases the probability to 0.997. For example, if the mean weight of fishes ($\bar{x}$) was equal to 635 g in the sample studied and the mean standard error m_x was equal to ± 5.94 g, the confidence limits would be equal to 623.12 – 646.88 g at $p = 0.95$ and to 617.18 – 652.82 g at $p = 0.997$.

The variance (σ^2) is very important in analyzing the mechanisms underlying the variation, in particular for the separation of two of its main components: the genotypic and the environmental or paratypic component. Such separation can be achieved by variance analysis: the presentation of the variance as a sum total of its components. The dispersion analysis of the quantitative variation of organisms makes use of the following equation:

$$\sigma^2_{Ph} = \sigma^2_G + \sigma^2_E + 2\,r\,\sigma_G\,\sigma_E, \tag{5}$$

where σ^2_{Ph}, σ^2_G and σ^2_E correspond to the total phenotypic, genotypic and paratypic or environmental variance respectively, while r is the correlation coefficient between the genotypic and environmental variation.

Determination of the contribution of the hereditary variation in the total variation of a given trait is associated with many difficult problems. If the environmental variance (σ_E^2) was equal to zero, in other words if all the individuals in a population or family lived their entire lives in completely identical conditions, we would be able to observe the genotypic variation in the pure form per se. In practice, however, it is impossible to make the living conditions of different fishes identical even within a single family. The isolation of a purely environmental variation represents a simpler problem. Fish populations, including individuals with identical genotypes, have been found. I refer to clones, or groups of individuals originating from one female, which reproduces by self-fertilization, parthenogenesis or gynogenesis. Hermaphroditism and self-fertilization are frequently observed in the small fish *Rivulus marmoratus.* Clones of this species consist of genotypically identical individuals (Kallman and Harrington 1964). Similar homogeneous clones have been found in *Mollienesia (Poecilia) formosa,* which reproduces by gynogenesis (Kallman, 1962b). Most probably similar clones will be found in the unisexual gynogenetic crucian carp *Carassius auratus gibelio.*

Clones are not generally found in fishes with normal sexual reproduction but genetic variation may be reduced to a certain extent by inbreeding; however, a marked residual heterogeneity usually remains even after prolonged inbreeding. It can be mentioned that the rapid production of sufficiently homogeneous stocks may be achieved by artificial gynogenesis induced by radiation or by chemical agents (see Chap. 7).

Methods of Heritability Determination in Fishes

Heritability is defined as a hereditary fraction in the total variation of a trait (Lush 1941). In the broad sense of the word heritability (h_G^2) is equal to the ratio between the genotypic and phenotypic variance:

$$h_G^2 = \frac{\sigma_G^2}{\sigma_{Ph}^2}. \qquad (6)$$

It is more important for the breeder, however, to define the fraction of the additive genetic variation, that is of the variation due to genes with the simple additive effect:

$$h^2 = \frac{\sigma_A^2}{\sigma_{Ph}^2}. \qquad (7)$$

The relative magnitude of additive genetic variation defines heritability better and is decisive for the effectiveness of selection.

Several methods are used in fish breeding to determine heritability.

Determination of Realized Heritability on the Basis of Selection Effectiveness (Response). Let us define the difference between parental pairs with respect to some trait such as S, the selection differential; let the corresponding difference between their offsprings be defined as R (selection response). The R:S ratio then

Table 13. Determination of realized heritability in the tilapia (Chan May Tchien 1971)

Sex	Mean weight at five months ($\bar{x} \pm m$) (g)						Heritability (large – small) $h^2 = \frac{R}{S}$
	Weight of spawners			Weight of offspring (separate rearing)			
	Small	Middle	Large	From small parents	From middle parents	From large parents	
Females	15.72 ±0.53	20.24 ±0.32	24.80 ±0.41	11.79 ±0.38	12.87 ±0.29	12.91 ±0.29	$\frac{12.91-11.79}{24.80-15.72}=0.123$
Males	16.84 ±0.32	23.81 ±0.22	28.37 ±0.61	19.34 ±0.56	20.30 ±0.50	20.68 ±0.39	$\frac{20.68-19.34}{28.37-16.84}=0.116$

shows which fraction of the difference between parents is retained in the offspring, that is, represents a hereditary component. This value is termed the realized heritability:

$$h^2 = \frac{R}{S}. \tag{8}$$

A good illustration of the application of this calculation technique can be found in the data of Chan May Tchien (1971) on the realized heritability of weight in *Tilapia mossambica* (Table 13).

In any studies aimed at determining realized heritability one has to weigh and measure fishes of two different generations at one and the same age, and to somehow normalize and standardize the conditions of their cultivation and growth, particularly such factors as the density and feeding conditions. Greater differences between the parents can be achieved by the selection of more contrasting pairs. If several different pairs are compared, the mean value h^2 is calculated. The calculation of realized heritability becomes much more precise if three or more sequential generations are used to generate the data.

The Determination of Heritability from the Regression Between Parents and Offspring. This approach requires calculation of the regression coefficient b; in other words, one has to establish the level of variation of a trait in the offspring when it is changed per unit for the parents. The equation of linear regression can be used in almost all cases:

$$y = a + b\,x, \tag{9}$$

where x and y are the mean values of the trait for the parents and the offspring, a is a constant, reflecting the overall differences in the expression of the trait between generations and b is the regression coefficient. If the mean values of a trait characteristic of the parents are used, then heritability becomes equal to the regression coefficient

$$h^2 = b. \tag{10}$$

If the offspring is compared with just one parent the formula is slightly modified:

$$h^2 = 2\,b. \tag{11}$$

Multiplication by two is necessary since in each case the offspring receives only half of its genes from the parent under consideration.

Data on the regression of the number of vertebrae in the viviparous fish *Zoarces viviparus* can be found in the paper by Smith (1921) (Fig. 36). The regression coefficient is equal to 0.404. The offspring were only compared with the mothers since the embryos withdrawn from the mother's body were studied. Then on the basis of Eq. (11):

$$h^2 = 2\,b = 0.808.$$

In our experiments with the common carp conducted in 1958 we obtained the following mean values for the number of soft rays in the dorsal fin of parents and their offspring:

Parents			Number of crosses	Offspring
♀	♂	$\bar{x}$		$\bar{x} \pm m$
20	18	19	12	18.91±0.037
20	19	19.5	13	19.20±0.036
20	20	20	5	19.54±0.064

The regression equation calculated from this data was as follows:

$y = 7.32 \pm 0.61\,x$; then $h^2 = b = 0.61$.

Using the regression analysis, where possible, one has to use parents with values of the trait studied close to the mean; the regression is generally linear only in the

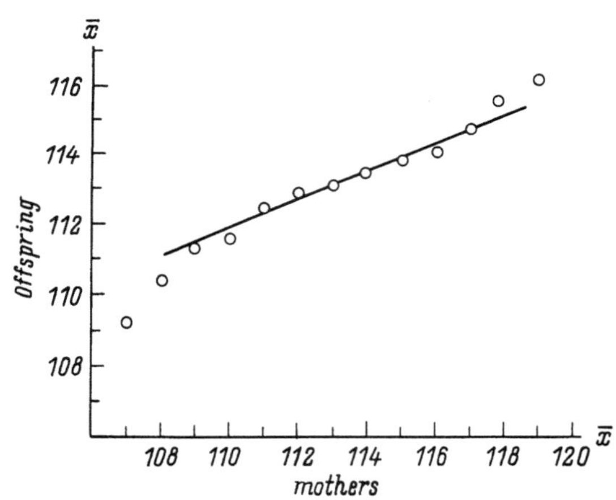

Fig. 36. Regression between mothers and offspring with respect to the number of vertebrae in the eelpout *Zoarces viviparus* (Smith 1921)

middle part of the range. Parents and offspring should be cultivated under identical environmental conditions just as in the case when realized heritability is determined.

The Determination of Heritability from the Correlation Between the Values of a Trait in Close Relatives. Three groups of relatives can be used in order to calculate the correlation: parents and offspring, brothers and sisters (full sibs) and half brothers-half sisters (half sibs).

Heritability coefficients in fish breeding are generally calculated by correlating the parents and the offspring. Both parents or just one of them can be compared with the offspring. Correspondingly, heritability is calculated from the following formulae:

$$h^2 = r \quad \text{or} \quad h^2 = 2\,r. \tag{12}$$

The method of correlation has been used by us to calculate the heritability coefficient for the ray number in the dorsal fin of the common carp. Many additional crosses have been added to the crosses employed for the regression analysis. It has been found that the heritability was rather significant (Table 14), but lower than that calculated from the regression equation ($r = 0.48$). The decrease in the heritability coefficient is explained by the effect of the extreme variants present in low numbers, but nevertheless markedly affecting the magnitude of the correlation coefficient.

The correlation between full sibs can be used when there are sufficient numbers of offspring for the comparison (no less than 25–30). It is recommended that no more than two or three pairs of fishes are taken from each brood originating from one pair of parents. The common environment for each family represents a disadvantage in calculating the heritability coefficient based on the sibs correlation. It is therefore necessary to start common cultivation of fishes from different families as early as possible. On average, half of the genotype is common to brothers and

Table 14. Correlational square for the number of the soft rays in the dorsal fin (D) in parents and their offspring. Special common carp crosses were conducted using artificial insemination of eggs (Ropsha 1958; $r = 0.48 \pm 0.11$)

Range of the variation of the mean (D) in the offspring	Mean values in parents						Number of crosses (n)
	18.0	18.5	19.0	19.5	20.0	20.5	
20.0 – 20.5					1		1
19.5 – 20.0				4	3		7
19.0 – 19.5	1	2	7	7	1	1	19
18.5 – 19.0		3	6	4	1		14
18.0 – 18.5			4				4
17.5 – 18.0			1				1
17.0 – 17.5		1					1
16.5 – 17.0							0
16.0 – 16.5		1					1

sisters, and heritability is therefore obtained from the proportion:

$$h^2 = 2\, r_{fs}, \tag{13}$$

where r_{fs} is the phenotypic correlation between the full sibs.

Half sibs or fishes with a common father but different mothers or vice versa have a genetic relatedness equal to ¼. The determination of heritability from the correlation of half sibs is less reliable and is at the same time very laborious. The correlation coefficient in this case should be multiplied by a faktor of four:

$$h^2 = 4\, r_{hs}. \tag{14}$$

The conditions of cultivation of parents and offspring should be identical in this case as well.

The Determination of Heritability from the Expansion of the Variance of Phenotypic Variation Using Variance Analysis. For this purpose one has to obtain simultaneously a sufficient number of related offspring from parents representing a certain shoal or a population of fishes. The offspring is obtained either by diallelic crosses or more frequently on the basis of the so-called hierarchic complex (Fig. 37).

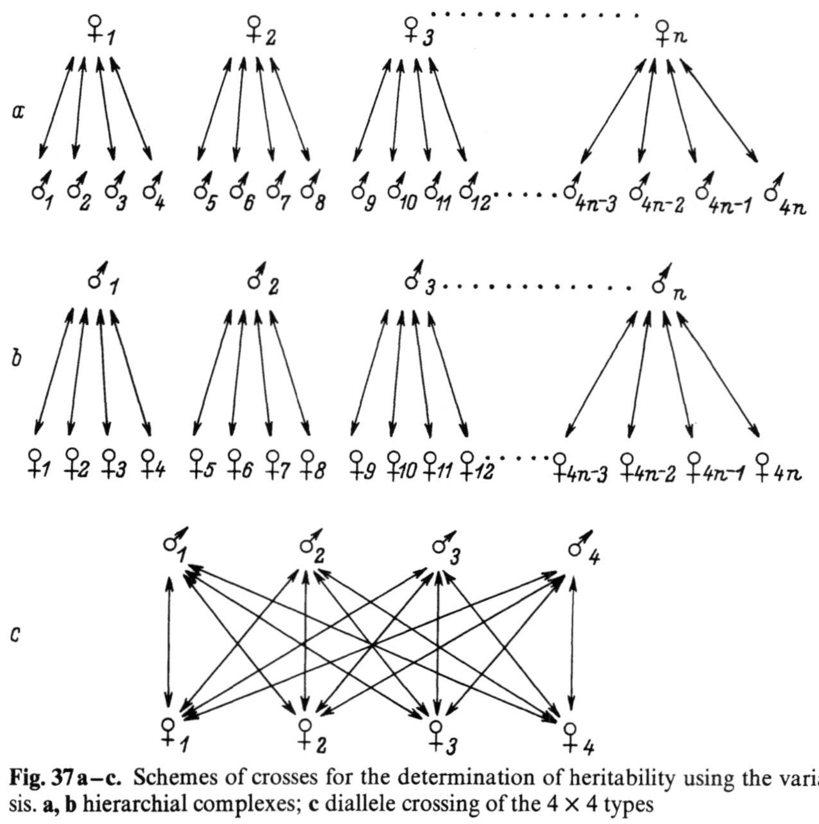

Fig. 37 a–c. Schemes of crosses for the determination of heritability using the variance analysis. **a, b** hierarchial complexes; **c** diallele crossing of the 4 × 4 types

The approach making use of the hierarchic complex has repeatedly been employed in fish breeding. The external fertilization of eggs in most fish species and the high fertility of females facilitate the simultaneous conduct of a large number of crosses. The first experiments with hierarchical complexes were performed by Nenashev (1966, 1969) with the common carp. In the hierarchical complex employed by this author (Fig. 37 a) offspring produced by one female from several different males are half sibs with respect to each other, while the individuals within each given offspring represent full sibs. An analysis of the variation of different groups of offspring makes it possible to calculate the heritability of the trait under study.

In the treatment of lengths or weight measurements for a limited number of individuals originated from each cross (usually from 5 to 20) the first task involves calculation of the sum total of square deviations from the mean (SS) separately for females, for males crossed with one and the same female, within different broods and for all the fishes measured. The calculations are made using the following formulae:

$$\begin{aligned} SS_{♀♀} &= ab\ \Sigma\ (\bar{x}_{♀}-\bar{x})^2; \\ SS_{♂♂} &= n\ \Sigma\ (\bar{x}_W-\bar{x}_{♀})^2; \\ SS_W &= \Sigma\ (x-\bar{x}_W)^2; \\ SS_{Ph} &= \Sigma\ (x-\bar{x})^2, \end{aligned} \tag{15}$$

where $\bar{x}$ is a common mean for all fishes, $\bar{x}_W$ mean values for different offspring, $\bar{x}_{♀}$ mean values for different offspring of one and the same female, x is individual measurements, n is the number of individuals in a given offspring, a is the number of females, b = the number of males crossed with one female.

The so-called observed variances (MS or V) may then be calculated. They are obtained after the division of the sum total of the squared values by the number of degrees of freedom: a−1 for $SS_{♀♀}$, a (b−1) for $SS_{♂♂}$, ab (n−1) for SS_W and abn-1 for SS_{Ph}. After this division we obtain:

1. $V_{♀♀}$ or V_D – the variance of mean values for females (dam component);
2. $V_{♂♂}$ or V_S – the variance of mean values for males crossed with one female (sire component);
   V_{random} or V_W – the variance of differences within separate offspring;
   V_{Ph} – the variance for all descendants.

Each of these observed components of the variance is non-uniform and contains a number of components of different origin (Falconer 1960):

$$V_D = \sigma_W^2 + n\ \sigma_S^2 + bn\ \sigma_D^2; \quad V_S = \sigma_W^2 + n\ \sigma_S^2; \quad V_W = \sigma_W^2. \tag{16}$$

Then it is easy to determine the values for σ_S^2 and σ_D^2:

$$\sigma_D^2 = \frac{V_D - V_S}{b\,n}; \quad \sigma_S^2 = \frac{V_S - V_W}{n}. \tag{17}$$

Each of the variances σ_S^2 and σ_D^2 contains, as calculated, one quarter of the additive genetic variation of the parents. Heritability can be obtained from the following formulae:

$$h^2_{♂} = \frac{4\,\sigma^2_S}{\sigma^2_{Ph}}; \quad h^2_{♀} = \frac{4\,\sigma^2_D}{\sigma^2_{Ph}}; \quad h^2_{♂♀} = \frac{2\,(\sigma^2_S + \sigma^2_D)}{\sigma^2_{Ph}}. \tag{18}$$

The sequence of calculations is presented in Table 15. By way of illustration, we have chosen a simplified hierarchical complex with three females, three males for each female, and three individuals examined in each cross. Heritability among females was greater than among males. This is frequently the case in studies with fishes. The higher heritability among females is usually explained by the existence of a "common environment" for all the offspring produced by a given female; this "common environment" is difficult to account for, but it increases the apparent variation "between females" and increases the value of V_D and σ^2_D.

The hierarchical complex does not allow the variance resulting from the interaction of genotypes of males and females to be singled out. This variance is an integral part of σ^2_S and σ^2_D terms. This leads to the decreased precision of heritability determination using the hierarchical scheme. The main causes of errors are, however, associated with the technical problems related to the growing of a large number of different offspring.

In the selective breeding of fishes more precise results are obtained by the complete bifactorial dispersion complex resulting from diallelic crosses of different degrees of complexity. Recently crosses have even been obtained using the scheme 20♀♀×20♂♂ with the simultaneous production of 400 offspring. The variation components are calculated, using conventional techniques and methods of bifactorial variance analysis. The sum total of the square deviations or variances "between females", and "between males", due to the interaction of females and males and random deviations, i.e., "within the offspring" are then determined (for explanations see p. 112):

$$\begin{aligned}
SS_{♀♀} &= nb\ \Sigma\ (\bar{x}_{♀♀} - \bar{x})^2 \\
SS_{♂♂} &= na\ \Sigma\ (\bar{x}_{♂♂} - \bar{x})^2 \\
SS_{♀♀♂♂} &= n\ \Sigma\ (\bar{x}_{♀♀♂♂} - \bar{x}_{♂♂} - \bar{x})^2 \\
SS_W &= \Sigma\ (x - \bar{x}_{♀♀♂♂})^2 \\
SS_{Ph} &= \Sigma\ (x - \bar{x})^2.
\end{aligned} \tag{19}$$

After the division by the number of degrees of freedom [for the deviations of the interaction it is equal to (a−1) (b−1)] we obtain the variances observed (MS or V). The transition to true or causal variances σ^2 is accomplished using the following formulae:

$$\sigma^2_D = \frac{V_D - V_{D,S}}{n\,b}; \quad \sigma^2_S = \frac{V_S - V_{D,S}}{n\,a}; \quad \sigma^2_{D,S} = \frac{V_{D,S} - V_W}{n}. \tag{20}$$

We shall present a simplified scheme for the calculation of weight heritability in a diallelic cross of the 2×2 type (Table 16). Heritability for females is somewhat too high due to the effect of a common environment and to the maternal effect. Heritability among males provides a more adequate description of the real picture. The negative interaction variance is a suggestion for a weak and perhaps even negative interaction between the female and male genotypes.

Table 15. The scheme for the calculation of weight heritability of using the hierarchical complex

		♀1			♀2		
		♂1	♂2	♂3	♂4	♂5	♂6
Offspring weight (grams)		4.0 6.0 8.0	5.0 7.5 10.0	6.0 8.0 13.0	9.0 11.0 13.0	8.0 8.0 11.0	4.0 7.0 10.0
x̄ for offsprings		6.0	7.5	9.0	11.0	9.0	7.0
x̄ for offsprings of individual females		7.5			9.0		
x̄ for all individuals		9.0					
Deviations for ♀♀	d	– 1.5			0		
	d²	2.25			0		
Deviations for ♂♂	d	– 1.5	0	1.5	2.0	0	– 2.0
	d²	2.25	0	2.25	4.0	0	4.0
Random deviations	d	– 2, 0, 2	– 2.5, 0, 2.5	– 3, – 1, 4	– 2, 0, 2	– 1, 1, 2	– 3, 0, 3
	d²	4, 0, 4	6.25, 0, 6.25	9, 1, 16	4, 0, 4	1, 1, 4	9, 0, 9
Total deviations	d	– 5, – 3, – 1	– 4, – 1.5, 1	– 3, – 1, 4	0, 2, 4	– 1, – 1, 2	– 5, – 2, 1
	d²	25, 9, 1	16, 2.25, 1	9, 1, 16	0, 4, 16	1, 1, 4	25, 4, 1

"Causal" variances:

$$\sigma_D^2 = \frac{MS_{♀♀} - MS_{♂♂}}{9} = 1.31; \quad \sigma_S^2 = \frac{MS_{♂♂} - MS_W}{3} = 0.56; \quad \sigma_W^2 = MS_W = 6.83.$$

The variance analysis as a method of heritability calculation for fishes along with its advantages possesses a number of serious drawbacks, the main ones being the following.

a) Inadequate precision in the calculation of variances and the heritability coefficient due to limitations of the technique itself. Bias associated with the introduction of correcting coefficients equal to four and two is also possible, the actual magnitude of correlations between relatives can vary to a significant extent. The formulae used in calculations with the hierarchical complex do not take into account the interaction of genotypes. It is impossible to distinguish and isolate the variance of the "common environment", and in many cases this leads to distortion of the results. The extent of genetic variation of females and males used in hierarchical complexes or diallelic crosses may be very different. In such cases, heritability calculations suffer from a much lower precision (Nikoro and Vasilyeva 1976).

b) The technical complexities associated with crossings and the cultivation of the offspring. It is particularly difficult to achieve joint cultivation of fishes from different broods. Usually one has to divide these into three or four batches and to place them in different ponds. This does, however, result in the appearance of the

♀3			SS (sum total of the square deviation)	df (degrees of freedom)	MS (mean square deviation)
♂7	♂8	♂9			
8.0	10.0	7.0			
10.5	10.0	9.0			
13.0	16.0	11.0			
10.5	12.0	9.0			
10.5					
1.5 2.25			$4.50 \cdot 9 = 40.5$	2	$MS_{♀♀} = 20.25$
0 0	1.5 2.25	−1.25 2.25	$17 \cdot 3 = 51$	6	$MS_{♂♂} = 8.50$
−2.5, 0, 2.5 6.25, 0, 6.25	−2, −2, 4 4, 4, 16	−2, 0, 2 4, 0, 4	123	18	$MS_W = 6.83$
−1, 1.5, 4 1, 2.25, 16	1, 1, 7 1, 1, 49	−2, 0, 2 4, 0, 4	214.5	26	$MS_{Ph} = 8.25$

Heritability:

$$h^2_{♂♂} = \frac{4 \cdot 0.56}{8.25} = 0.27; \quad h^2_{♀♀} = \frac{4 \cdot 1.31}{8.25} = 0.64.$$

"pond" variance which is extremely undesirable and decreases the reliability of heritability determinations. In American studies with the channel catfish *Ictalurus punctatus* (Reagan et al. 1976) the hierarchical complex has been employed using 20 males and two females for each male (two crosses were conducted immediately after each other). The larvae were kept in troughs for up to 15 months; the density of the embryos initially differed by dozens of times, and this distinctly increased the variation "between the troughs". Thereafter the density of the embryos was made equal, but new problems arose associated with feeding; the quantity of food depended on the actual weight of the fishes in a troughs and, as a result, the "best" families received an additional advantage. The factors may well explain the unusually high heritability of the weight, which was 0.61 ± 0.35 (at 5 months) and 0.75 ± 0.53 (at 15 months). As noted by the authors, the assortative cross scheme, i.e., the selection of females and males into parental pairs on the basis of their size, could also be a significant factor. The heritability of the weight and body length in fishes is generally not high and its determination depends to a particularly great extent on the conditions of the experiment.

The eggs of different females may be of a very different quality, which is in part determined by the genotype but to even greater extent by external environmental

Table 16. The scheme for the calculation of weight heritability in a diallelic cross of the 2×2 type (complete bifactorial dispersion complex)

Indices		♀1	
		♂1	♂2
Offspring weight (g)		10, 15, 20	8, 14, 20
x̄ for offspring		15.0	14.0
x̄ for families of individual females		14.5	
x̄ for families of individual males		16.5	15.0
x̄ for all individuals		15.75	
Deviations for ♀♀	d	−1.25	−1.25
	d²	1.5625 · 3	1.5625 · 3
Deviations for ♂♂	d	+0.75	−0.75
	d²	0.5625 · 3	0.5625 · 3
Deviations due to interactions	d	−0.75+1.25−0.75	−1.75+1.25+0.75
	d²	0.0625 · 3	0.0625 · 3
Random deviations	d	−5, 0, +5	−6, 0, +6
	d²	25, 0, 25	36, 0, 36
Total deviations	d	−5.75, −0,75, +4.25	−7.75, −1.75, +4.25
	d²	33.0625, 0.5625, 18.0625	60.0625, 3.0625, 18.0625

$h^2_{♂♂} = \frac{4\,\sigma^2_{♂♂}}{\sigma^2_{Ph}} = 0.16$; $h^2_{♀♀} = \frac{4\,\sigma^2_{♀♀}}{\sigma^2_{Ph}} = 0.48$; $h^2_{♀♀,♂♂} = 0.32$.

conditions and the stage of maturity. In hierarchical complexes of the type 1♀×several ♂♂ and in diallele crosses the variance "between females" is always somewhat increased. Sometimes, however, this increase is quite pronounced due to the maternal effect. In all cases it is therefore preferable for the indices to calculated on the basis of the variance for males (σ^2_S).

Determination of the Heritability of the Egg Size. Let us assume that the variation of all the oocytes in the ovary of a given female is purely paratypic. If we subtract the variance associated with this variation from the total phenotypic variance of the eggs of many different females, we shall obtain a purely genotypic component if the initial assumption is correct (Volokhonskaya and Viktorovsky 1971):

$$h^2 = \frac{\sigma^2_{Ph} - \sigma^2_E}{\sigma^2_{Ph}}, \qquad (21)$$

♀2		SS	df	MS	σ^2 (causal variances)
♂1	♂2				
13, 16, 25	11, 16, 21				
18.0	16.0				
17.0					
16.5	15.0				
+1.25 1.5625 · 3	+1.25 1.5625 · 3	18.75	1	18.75	3.00
+0.75 0.5625 · 3	−0.75 0.5625 · 3	6.75	1	6.75	1.00
+2.25 − 1.25 − 0.75 0.0625 · 3	+0.25 − 1.25 + 0.75 0.0625 · 3	0.75	1	0.75	−10.17
−5, −2, +7 25, 4, 49	−5, 0, +5 25, 0, 25	250.00	8	31.25	31.25
−2.75, +0.25, +9.25 7.5625, 0.0625, 85.5625	−4.75, +0.25, +5.25 22.5625, 0.0625, 27.5625	276.25	11	25.11	–

where σ^2_{Ph} is the variance of the egg size in a population, σ^2_E is the variance of the egg size for a single female. This formula only allows the upper limit of heritability to be determined. The mean egg size for each female not only depends on the genotype, but also on the living condition of this female during the period preceding ovulation, as well as on the extent of its maturity. The variance "between females" which is obtained by the authors of this paper by subtracting the variance "within a female" from the total variance, includes both the genotypic and the paratypic variance. It is impossible to separate these two components. The fraction of the paratypic variation "between females" (the variance of the "common environment") can be very large and therefore the genuine heritability of the oocyte diameter is less (and sometimes considerably so) than the ratio: $\sigma^2_♀/\sigma^2_{Ph}$.

Rokitsky (1974) points out another simplified approach associated with the use of hierarchical complexes. It makes use of direct comparisons of parents of one sex with their offspring. Heritability is determined from the magnitude of correlation within a class, and the values obtained are multiplied by 4:

$$h^2 = \frac{4\,\sigma_S^2}{\sigma_S^2 + \sigma_W^2}\,. \tag{22}$$

The numerator in this formula refers to the fourfold variance of the variation between males, while the denominator refers to the sum total of the variances "between males" and "between individuals" in each offspring. Heritability calculations using this technique give an approximate result. This technique has not so far been used in fish breeding.

In studies of heritability of such traits as viability and resistance to diseases using variance analysis one has to resort to no less than three replicate determinations during the cultivation period. The trait under consideration in such cases is the relative amount of survivors or dead fishes; situations of this type tend to vary greatly, depending on the numerical value of the trait, that is, percentage of survivors. In order to abolish this dependence on the date of death, a simple rearrangement may be used:

$$y = \arcsin\sqrt{x}\,. \tag{23}$$

Such a rearrangement has been used by a number of authors (Gjedrem and Aulstad 1974; Kanis et al. 1976) in experimental work with salmonids. In these cases, they recommend that the following equations be used (Bogyo and Becker 1965):

$$h_S^2 = \frac{4\,\sigma_S^2}{0.25 + \sigma_S^2 + \sigma_D^2} \quad \text{or} \quad h_D^2 = \frac{4\,\sigma_D^2}{0.25 + \sigma_S^2 + \sigma_D^2}\,. \tag{24}$$

The methods of heritability calculation outlined above predominantly take into account only a fraction of the genetic variation associated with additive genes. Elucidation of the total genetic variation, including deviations of the dominance, the effect of overdominance (the advantages of heterozygotes) and of the epistasis (the non-additive interactions of non-allelic genes) can only be achieved using special, rather complex approaches. The variation due to overdominance plays an important role in many fish species. It has been established in particular that the main part of the genetic variance in the variation of body weight, body size and total viability of cyprinids and salmonids is played by the variance associated with overdominance (Wohlfarth and Moav 1971; Holm and Naevdal 1978). A more refined analysis of the non-additive components of variance represents one of the most important problems in future studies of fish variation.

In the above paragraphs we have limited ourselves solely to the general principles of heritability determination among fishes. A convenient working formula for the calculation of heritability can be found in many textbooks. Unfortunately, the special treatise dealing with heritability published by Plokhinsky (1964) contains many serious errors. The "heritability indices" offered by Plokhinsky and based on the proportion of the sum total of deviations without any expansion of variances cannot be recommended for use. All these indices do not describe heritability in the genetic sense of the word, they do not define the fraction of the additive genetic variation in the total variation. Frequently these indices provide erroneously high estimates (Nikoro and Rokitsky 1972; Rokitsky 1974).

Common methods of heritability determination have recently been criticized by a number of experts in mathematical genetics, who have pointed out that the bias in heritability calculations is unacceptably high (Nikoro and Vasilyeva 1976). In order to obtain the relative fraction of hereditary variation these authors recommend that only the regression coefficients, the "parents-offspring" correlations and the correlations between sibs and half sibs within each class should be calculated.

Without rejecting the validity of these criticisms I would like to point out, however, that coefficients of realized heritability, as well as the indices obtained by dispersion analysis, although imprecise, do in many cases give an unequivocal picture of the level of genetic heterogeneity within a population or a shoal at least with respect to the additive variation used in mass selection of fishes.

Problems in Genetic Studies of the Quantitative Traits

Let us now list the most important problems existing for breeders and geneticists in studies of the variation of the quantitative traits in fishes.

The Analysis of the Distribution of the Character Studied and a Comparison Thereof with the Normal Distribution. The equation of the normal distribution enables one to make a precise calculation of the number of individuals within certain limits of the variation range. The mean standard deviation can be conveniently used for this purpose. The values corresponding to $\bar{x} \pm 2\,\sigma$ are generally taken as confidence limits of the variation. If the distribution is close to normal, another biometric constant, the standard error of the mean, can be used and different groups of individuals can be compared on the basis of this constant.

Determination of Trait Heritability. In planning selective breeding work and the analysis of microevolutionary processes in natural populations this problem acquires primary importance. The coefficient of heritability in the narrow sense of the word is particularly important for the breeder since the foundation of selection rests mainly on the additive variation; the higher its fraction, the more successful is the selection. If heritability is very low, the selection for relatives is preferred as compared to mass selection; this is due to the fact that the heritability of the group means is always higher than the heritability of individual differences. The main role among sources of non-additive variation in fishes is played by overdominance. The importance of dominance and epistasis (that is of the gene interaction) in the quantitative variation among fishes is apparently moderate. A high non-additive variance is evidence of the existence of well-balanced genetic systems with the advantage of heterozygotes, and, in such cases, any increase in the additive variation requires that crosses must be made.

Estimation of the Approximate Number of Genes Affecting a Given Trait and Segregating in the Population or in the Shoal. The number of interacting genes is important for the distribution of specimens in the offspring, particularly in the second and third generation after crosses of fishes with a sufficiently high degree of homozygosity. Changes in the segregation pattern in animals and plants, depending on the number of genes involved, has resulted in numerous attempts to derive formulae allowing this number to be determined. Unfortunately, none of them is suitable for the precise calculation of the number of genes. The proportion of variation in different generations depends not only on the number of genes, but also on the degree of parent heterogeneity, the level of heritability, and stability or inconstancy of environmental conditions. So far, no one has been successful in attempts to formalize the action of all these factors.

An approximate determination of the number of polymorphic genes is nevertheless possible. If this number is small, the segregation pattern frequently becomes multimodal, the variation curves become more flattened and deviate greatly from the normal ones. On the contrary, the interaction of many genes is generally manifested as a strictly normal (less frequently Poisson) distribution, so that virtually no differences appear to exist with respect to the variation magnitude between different generations.

Any conclusions regarding the number of genes involved can best be made in the cases of distant crosses, when the crossed individuals differ with respect to a relatively large number of homozygous genes. It is even better to use inbred stocks for this purpose.

It should be added in conclusion that not infrequently one or two loci in the group of interacting genes affecting the selected trait have a particularly strong impact. The finding of such loci and their genetic analysis can markedly facilitate breeding work.

Variation and Heritability of Body Weight and Length, the Age at Sexual Maturity and Fertility Among Fishes

Variation in the growth rate is characteristic of all fish species. Growth is affected by many genes since almost every change in the body structure or organ functions will affect food intake or assimilation in one way or another. The environmental factors affecting the rate of growth are also variable. The effects of environment on fish growth can be established as early as during the period of oocyte development in the female ovary; different oocytes, depending on their location in the ovary, have a different nutrient supply, grow at different rates and accumulate different amounts of essential nutrients (Meyen 1940).

It may well be that the different age of oocytes at the time of ovulation is an even more important factor (Slutzky 1971a). After ovulation and fertilization the environment imposes an additional differential effect on early fish embryos developing inside eggs. During egg incubation marked differences in the oxygen supply, temperature and illumination are inevitable. When these differences in the environmental conditions are superimposed on the differences that existed prior to ovulation, this results in the acceleration of some embryos' development and in the delayed development of others, the resulting effect providing for an increasing asynchrony between the different embryos in the course of embryogenesis. The hatching of larvae within a single batch of developing embryos generally lasts for many hours. Really, the heritability of the hatching time in the rainbow trout is not high according to the available data (McIntyre and Blanc 1973). In the wild common carp the variation coefficient of the larvae with respect to body length determined immediately after hatching does not usually exceed 5%–6% (Slutzky 1971b). In the lake whitefish (*Coregonus peled*) this variation is even less (Andriyasheva et al. 1978).

Large larvae and particularly those that hatch earlier go on to independent feeding earlier and generally grow faster than the smaller larvae from the same batch. The variation becomes even more obvious among the growing fry. Any competition for food can increase this variation even further (Nakamura and Kasahara 1955, 1957; Moav and Wohlfarth 1963; Kryazheva 1966; Wohlfarth 1977).

Later, the effect of environmental factors begins to even out when there is sufficient food, leading to a gradual secondary decrease in the indices of phenotypic variation. In the common carp such a decrease continues for a long time (Wlødek 1968):

	C.V. (%)		
	Family 1	Family 2	Family 3
0+ (fry)	49.2	36.1	55.0
0+ (one-year-old fishes)	22.7	26.8	34.3
1+ (two-year-old fishes)	13.7	20.3	19.9
2+ (three-year-old fishes)	16.0	13.8	11.2
3+ (four-year-old fishes)	11.6	12.4	10.9
4+ (five-year-old fishes)	10.6	–	12.4
5+ (six-year-old fishes)	8.8	–	11.3

Similar data have been obtained in our own experiments.

Let us now turn to the discussion of the more general principles established in studies of the variation of weight and body length in fishes.

The distribution of fishes in a population or a genetic group of uniform origin is either close to the normal one or has a greater or lesser negative asymmetry. The rearing of the carp where there is high fish density, when food is limited, sometimes results in a highly asymmetrical distribution. If the largest fishes or champions ("shoot" specimens) are removed from such a population but intensive competition for food continues, other individuals rapidly occupy their place (Nakamura and Kasahara 1957). The displacement of the variation to the right side results from the feeding advantages of the largest individuals in the population, which deprive almost all other fishes of their food. The increase of variation is facilitated by the positive correlation between the initial size and the rate of growth. Among fishes living in shoals or "schools" when the fry imitate each other's behaviour and do not appear aggressive during feeding (*Carassius auratus, Zebrias zebra* and others) the variation coefficients of the weight and body lengths do not increase to such an extent: the distribution remains close to the normal (Yamagishi 1965, 1969). It

Table 17. Heritability of weight and length

Species	Age group	Heritability: Method of calculation	Heritability: Mean values	Reference
(a) Weight				
Common carp, *Cyprinus carpio*	Larvae, 4 days	Dispersion anal.	0.20	18
	Larvae, 27 days	Dispersion anal.	0.11	18
	Fingerlings	Dispersion anal.	0.21	17
	One year old	Dispersion anal.	0.25	23
	Two years old	Dispersion anal.	0.35	23
	Spawners	Realized heritability (+selection)	0	15
		Realized heritability (– selection)	0.2–0.3	15
Rainbow trout, *Salmo gairdneri*	Fingerlings	Dispersion anal., correlation	0.01–0.29	2
	Fingerlings	Correlation (sibs)	0.04–0.18	5
	Fingerlings	Group variance	0.07–0.22	3
	Fingerlings	Realized heritability	0.06	13
	Fingerlings	Correlation (sibs)	0–0.50	16
	Fingerlings	Regression, correlation	0.26–0.29[a]	12
	Two years old	Correlation	0.20	7
	Spawners, ♀♀	Correlation	0.50	9
	Spawners, ♂♂	Correlation	0.31	9
Atlantic salmon, *Salmo salar*	Fingerlings	Dispersion anal., correlation (sibs)	0.60–0.70[b]	14
	Fingerlings	Dispersion anal.	0.08–0.15	20
	Four years old	Dispersion anal.	0.31	10
	Spawners	Families variance	0.22	21
Channel catfish, *Ictalurus punctatus*	Fingerlings	Dispersion anal.	0.61–0.75[b]	19
Guppy, *Poecilia reticulata*	Spawners	Realized heritability	0.1	22
Tilapia, *Tilapia mossambica*	Spawners	Realized heritability, ♀♀	0.12–0.32	4
		Realized heritability, ♂♂	0.12–0.29	4
(b) Length				
Common carp, *Cyprinus carpio*	Eggs, diam.	Dispersion anal.	0.24	18
	Fingerlings	Dispersion anal.	0.21	17
Rainbow trout, *Salmo gairdneri*	Eggs, diam.	Dispersion anal., correlation	0.29–0.32	8, 9
	Fingerlings	Dispersion anal., correlation (sibs)	0.03–0.37	2

Table 17 (continued)

Species	Age group	Heritability		Reference
		Method of calculation	Mean values	
Rainbow trout, *Salmo gairdneri*	Fingerlings	Correlation (sibs)	0.08–0.32	5
	Fingerlings	Correlation (sibs)	0–0.50	16
	Two years old	Dispersion anal.	0–0.3	11
Atlantic salmon, *Salmo salar*	Fingerlings	Dispersion anal.	0.12–0.17	20
	Fingerlings	Correlation (sibs)	0–0.35	11
	Four years old	Dispersion anal.	0.28	10
Peled, *Coregonus peled*	Eggs, diam.	Method of Volokhonskaya and Viktorovsky	0.41–0.65[c]	1, 24
	Larvae, 1st day	Dispersion anal.	0.27–0.58	1
Channel catfish, *Ictalurus punctatus*	Fingerlings	Dispersion anal.	0.12–0.67	19
	Fingerlings	Dispersion anal.	≧0.3	6

[a] Family heritability, h^2_f;
[b] Probably over-estimated values (include common environment variance);
[c] Over-estimated values, include non-hereditary variance.

References: 1. Andriyasheva et al. (1978); 2. Aulstad et al. (1972); 3. Ayles (1975); 4. Chan May-Tchien (1971); 5. Chevassus (1976); 6. El-Ibiary and Joyce (1978); 7. Gall (1975); 8. Gall (1978); 9. Gall and Gross (1978); 10. Gunnes and Gjedrem (1978); 11. Holm and Naevdal (1978); 12. Kincaid (1972); 13. Kincaid et al. (1977); 14. Lindroth (1972); 15. Moav and Wohlfarth (1976); 16. Møller et al. (1979); 17. Nenashev (1969); 18. Poljarush and Ovechko (1979); 19. Reagan et al. (1976); 20. Refstie and Steine (1978); 21. Ryman (1972a); 22. Ryman (1973); 23. Smišek (1978); 24. Volokhonskaya and Viktorovsky (1971)

should be pointed out that selective breeding with domesticated fish species has resulted in an accumulation of dominant genes providing higher rates of growth (Reisenbichler and McIntyre 1977; experiments with the trout).

In most studies different authors, using various techniques, reported that the indices of heritability of the weight and body length among fishes do not generally exceed 0.3, and rarely 0.4 (Table 17). The data for the salmon and the channel catfish where the coefficient of heritability was very high (up to 0.75) are apparently an exception. In the latter case, this high heritability could be due to methodological inadequacy which had led to a very high variance of the common environment and of the variance between "different batches". According to more recent data, differences between sib families in the channel catfish with respect to the body length and the dry weight do exist but are not very prominent (El-Ibiary et al. 1976). The experiments of Lindroth (1972) with the salmon also apparently suffer from certain methodological shortcomings, which had overemphasized the variance of the common environment. Particularly vulnerable in this respect is the method of calculation based on correlations between sibs. Whereas individual differences tend to be inherited to a lesser extent, the mean value (h^2_f) for the heritability of the groups turns out to be rather high (Moav and Wohlfarth 1974). This provides an argument in favour of familial selection in carp breeding.

When fishes from one and the same breed are grown separately or together in experiments designed to determine the realized heritability, the coefficients of heritability are higher when the fishes are bred together (Chan May Tchien 1971). Genetic differences increase probably as a consequence of the stronger interaction between the genotype and the environment due to the competition for food (Moav et al. 1975 b; Wohlfarth et al. 1975 a; Moav 1979).

When different families of the rainbow trout were compared for the effectiveness of utilization of different foods, genetic heterogeneity was apparent (Edwards et al. 1977; Austreng and Refstie 1979; Reinitz et al. 1979). The heritability of this trait, which is important for fish breeders, remains to be determined.

The low heritability of the weight and size among fishes should not be surprising. The rate of growth among fishes is closely related to fertility, time of maturity and viability, representing the main components of the breeding value or fitness of an individual (Falconer 1960; see also Purdom 1979). The characters affecting fitness, that is the capacity to leave sufficiently numerous and viable offspring, generally have very low heritability in all animals. Since all such traits are permanently affected by intensive selection, this results in the emergence of complex systems of interrelated allelic and non-allelic genes, increases the relative importance of non-additive genetic variation and all these factors lead to decreased heritability.

The heritability of minus-variants in the common carp and in the tilapia is greater than the heritability of plus-variants (Moav and Wohlfarth 1967, 1968, 1976; Chan May-Tchien 1971; Moav 1979). The far right part of the variation range for weight and body length is formed by the random winners in the food competition. These "champions" only exhibit minor genetic differences from other members of the community.

The group of large fishes apparently contains also heterozygous individuals, with an overdominant expression of genes affecting growth; anyway the selection for such an individual would lead to poor results.

The number of genes affecting the rate of growth in fishes remains to be established. This number is apparently rather high. Mutations inherited strictly according to Mendel have been found which drastically decrease the growth rate and result in hereditary dwarfness.

Highly accelerated growth is most frequently associated with defects in gonad development leading to deceleration or even complete arrest of sex maturation. In most cases, these defects are hereditary, the accumulation of such "sterility mutations" resulting from the selection of large fishes can present a great problem to fish breeders.

Such characters as the age of maturity and fertility are among those that are extremely important for the breeding potential of a fish. In the common carp rapid maturation depends on many almost exclusively dominant genes (Hulata et al. 1974). The variance analysis of the age at maturity of the whitefish *Coregonus peled* from the Lake Endyre population (West Siberia) cultivated in the Leningrad region has demonstrated the presence of a tremendous variation in this parameter which was reproducible from one year to another (Andriyasheva 1978 a). Throughout the years of observations the same females attained maturity early, and others much later, the curve of variation with respect to the age of maturity was generally bimodal. The coefficient of repeatability has been calculated from the formula of Rokitsky (1974):

$$r = \frac{\sigma_B^2}{\sigma_B^2 + \sigma_E^2}, \tag{25}$$

where σ_B^2 is the variance of the maturation age of different females and σ_E^2 is the variance of the maturation age of one and the same female in different years. Repeatability was very high (0.7–0.8). Although heritability cannot be determined from these data, it should apparently be high (around 0.5 or even higher).

An analysis of the whitefish fertility conducted in the same study also demonstrated high indices of repeatability with respect to the number of ovulated eggs or "working" fertility (0.83–0.84). In this case as well, unexpectedly high heritability (Andriyasheva 1978b) can be assumed. The unusually high heritability of the fertility and the age of maturity time in *Coregonus peled* may perhaps be explained by the presence in the Ropsha population of two ecologically different groups with different spawning and fertility seasons. The heritability of fertility in the rainbow trout was moderate (around 0.23) (Gall and Gross 1978) while the heritability in the age of maturity was high (Holm and Naevdal 1978; Møller et al. 1979).

The diameter of ovulated eggs in *Coregonus peled* varies within a very limited range (C.V. 3.5%–4.1%). The upper limit of heritability has been determined, using the formulae suggested by Volokhonskaya and Viktorovsky (1971); this limit was equal to 0.41–0.65 (Andriyasheva and Chernyaeva 1978). As we have already pointed out, the index of heritability calculated by this technique is erroneously high. The authors believe, however, that in the case of the Ropsha whitefish *Coregonus peled* the variance of the common environment was low (all the females were grown together) and the heritability of the egg diameter could indeed be high. Even higher values, up to 0.97, have been obtained for the chum and the pink salmon *Oncorhynchus keta* and *O. gorbuscha* (Volokhonskaya and Viktorovsky 1971); there can be no doubt, however, that these values are too high because the females taken for the study were living under very different environmental conditions and perhaps belonged to independent populations. The inheritance of differences in the size of the eggs has been established for the Atlantic salmon as well (Aulstad and Gjedrem 1973).

The age of maturation and spawning in the rainbow trout and salmon depends on the genotype (Davis 1931; Schäperclaus 1961; Gardner 1976; Naevdal et al. 1979a, b).

Although the additive genetic variance of the weight and body size in fishes is in general low, the breeding work can, nevertheless, be successful if selection is intensive, methods of group selection being used and the system of crosses being well planned. This conclusion follows in particular from the successful experience of growth acceleration of Ukrainian and Ropsha carps through five or six generation of selective breeding (Kuzema 1953; Kirpichnikov 1972a) and of the rainbow trout for three generations (Kincaid et al. 1977).

Variation and Heritability of the Common Viability and Resistance of Fishes to Various Diseases

Survival is a most important component of fitness, and it is not surprising that the first attempts to determine the heritability of this trait have shown that it is very low (Table 18). Somewhat higher values have been obtained in experiments conducted with hybrids of *Salvelinus fontinalis* × *Salvelinus namaycush* (splake); the heritability of resistance to blue sack disease and to heating varied in the range

Table 18. Inheritance of viability and resistance in fishes

Species	Trait	Age group of fishes	Heritability, h^2		Reference
			Method of calculation	Mean values	
Atlantic salmon	Viability	Eggs, larvae	Dispersion anal.	0.01–0.15	7
	Viability	Fry	Dispersion anal.	0.11–0.34	7
	Viability	Fry	Interfamilial variance	0.10–0.20	9
	Resistance to vibrio disease	Fry	Dispersion anal.	0.07–0.15	5
Brown trout	Viability	Eggs, larvae	Dispersion anal.	0.01–0.05	7
	Resistance to acid waters	Eyed eggs	Dispersion anal.	0.09–0.27	3, 4
Rainbow trout	Viability	Eggs, larvae	Dispersion anal.	0.06–0.14	7
Splake (hybrids *Salvelinus fontinalis* × *S. namaycush*)	Viability	Fry	Dispersion anal.	0.06–0.41	1
	Resistance to high temperature	Fry	Dispersion anal.	0.38	6
	Resistance to blue sack disease	Fry	Dispersion anal.	0.41–0.60	1
Chinook salmon (*Oncorhynchus tschawytscha*)	Resistance to gas bubble disease	Fry	Dispersion anal.	0.04	2
Sockeye salmon (*Oncorhynchus nerka*)	Resistance to infectious necrosis	Fry	Dispersion anal.	0.27–0.38	8

References: 1. Ayles (1974); 2. Cramer and McIntyre (1975); 3. Edwards and Gjedrem (1979); 4. Gjedrem (1976); 5. Gjedrem and Aulstad (1974); 6. Ihssen (1973); 7. Kanis et al. (1976); 8. McIntyre and Amend (1978); 9. Ryman (1972a)

0.4–0.6. This may well be associated with the hybrid origin of the experimental material: parental species might have different genes affecting the resistance which segregate in hybrid generation. Resistance to the aquatic pollution and to diseases is, in some cases, determined by a single or a few genes (Hines et al. 1974), but in most cases it has a polygenic nature. This is supported by the data showing high differentiation of the various strains and stocks of bred fishes with respect to their resistance to insecticides (Ferguson et al. 1966; Ludke et al. 1968; Macek and Sanders 1970), to low pH values (Robinson et al. 1976; Swarts et al. 1978) and to various diseases (Wolf 1954; Snieszko et al. 1959; Ehlinger 1964; Kirpichnikov and Faktorovich 1969; Kirpichnikov and Faktorovich 1972; Gjedrem and Aulstad 1974; Plumb et al. 1975; Amend 1976; Ord 1976).

Low heritability of fish resistance to diseases is also associated with the relationship between susceptibility to a disease and the general status of the organism, that is, it depends on the conditions under which a given individual was living prior to its contact with the pathogen. The heritability of resistance can be increased when fishes are grown under uniform and favourable environmental conditions.

General fish viability, particularly at the early stages of development is intimately associated with overdominance. Heterozygosity in fishes, as in other animals, contributes to the elevation of viability, while inbreeding decreases viability very quickly, more than by 10% per generation (Shaskolsky 1954; Moav and Wohlfarth 1968; Aulstad et al. 1972; Kincaid 1976). A close correlation between the general viability and heterozygosity readily explains the low heritability of the viability, depending on many genes with non-additive action.

Great differences in the resistance of fishes to elevated temperatures have been established. Heritability has, however, been determined only in the splake. The effectiveness of selection with respect to this trait has been found in studies of consecutive generations of *Salvelinus* species hybrids (Goddard and Tait 1976); it has also been determined in comparative studies of the rainbow trout from different climatic zones of Australia (Morrissy 1973). Using the sockeye salmon (*Oncorhynchus nerka*) as an example, we have been able to demonstrate that the temperature resistance of fishes can change if just one protein locus, the gene coding for lactate dehydrogenase, is varied (Kirpichnikov 1977).

Variation and Heritability of Morphological Characters Among Fishes

Exterior Indices. The highest variation of the body shape among fishes is observed among peaceful inhabitants of the water bodies with low current (the common and crucian carp and some other species). The predators such as the pike, trout, and the perch-pike, as well as species living in flowing waters and migrating over large distances, show little variation, though we have data on greater hereditary variation of body structure in the Atlantic salmon (Gunnes and Gjedrem 1978).

The variation of exterior indices is close to the normal in almost all cases. In the carp the realized heritability of the ratio of body height to the length was found to

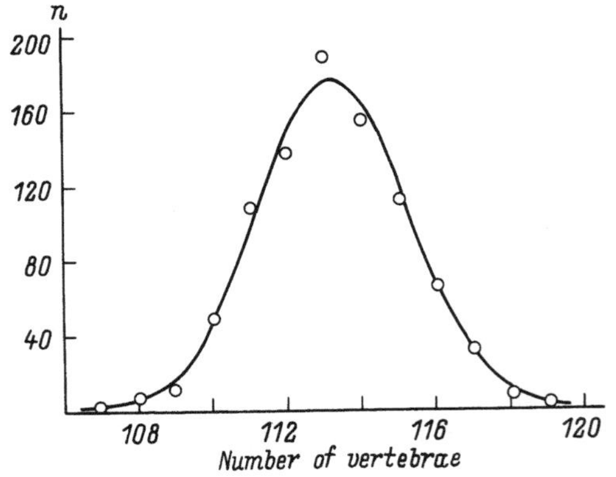

Fig. 38. Variation of the number of vertebrae in the eelpout *Zoarces viviparus.* Icefjord, Denmark, 857 individuals (Smith 1921)

be equal to 0.42 (Moav and Wohlfarth 1967). Apparently, the action of most genes affecting the shape of the trunk is additive. The selection with respect to exterior indices is effective despite the prominent dependence of these indices on environmental conditions. However, when selection with respect to exterior is conducted, some undesirable so-called "correlating changes" are frequently observed; these include deformations of the vertebral column, a slowing down of the growth rate, etc.

The Number of Vertebrae. Detailed studies of the variation with respect to the number of vertebrae have been conducted with herrings (Heincke 1898; Schnakenbeck 1927, 1931) with the eelpout *Zoarces viviparus* (Schmidt 1917, 1920, 1921), the cod and the trout (Schmidt 1930; Tåning 1952) and with several species of cyprinodonts (Kok Leng Tay and Garside 1972; Harrington and Crossman 1976).

Table 19. Heritability of the number of vertebrae in fishes

Species	Heritability, h^2		Reference
	Method of calculation	Mean values	
Carp, *Cyprinus carpio*	Realized heritability	0.86	2
	Dispersion anal., ♀♀	0.65	3
	Dispersion anal., ♂♂	0.90	3
Rainbow trout, *Salmo gairdneri*	Realized heritability	0.66	4
Brown trout, *Salmo trutta fario*	Regression "parents-offspring"	0.90	6
Medaka, *Oryzias latipes*	Interfamilial variance	0.90	1
Eelpout, *Zoarces viviparus*	Regression "mothers-offspring"	0.81	5, 8, 9
	Correlation "mothers-offspring"	0.80	7

References: 1. Ali and Lindsey (1974); 2. Kirpichnikov (1961); 3. Nenashev (1966); 4. Orska (1963); 5. Schmidt (1917); 6. Schmidt (1919b); 7. Schmidt (1920); 8. Smith (1921); 9. Smith (1922). 4.–9. our calculations of h^2

The number of vertebrae is frequently used in fish systematics as a trait allowing local subspecies and races to be distinguished. It is sufficient to mention the studies conducted with the cod (Schmidt 1930; Dementyeva et al. 1932), the smelt and the carp (Kirpichnikov 1935 a, 1943, 1967 b). Differences in the number of vertebrae between different races are apparently adaptive (Ege 1942).

The most important results of studying the variation and inheritance of the number of vertebrae can be summarized as follows.

The distribution of this trait in all the populations and shoals studied is generally close to the normal (Fig. 38). The coefficients of variation are usually low (1%–4%), but in fishes with a large number of vertebrae the extent of variation may be quite significant.

The heritability of the number of vertebrae is very high (Table 19) and most of the genetic variation is apparently additive. The variance due to the environment is low under laboratory conditions; it is, however, possible that in natural populations the environmental component of variation may be higher (Dannevig 1932, 1950). A large variation in the number of vertebrae has been observed in response to experimental changes in the temperature during embryonic development. The scope of this variation can be estimated from the variation range obtained for the rainbow trout (Orska 1963):

t °C	Number of vertebrae							
	58	59	60	61	62	63	64	$\bar{x}$
6 → 18	2	7	13	17	10	1		60.58
6 → 12				1	11	29	9	62.92

When the transfer of developing embryos to a different temperature has been accomplished at the gastrula stage, the mean number of vertebrae at the lower temperature increased by more than two vertebrae.

Similar experiments have also been performed with *Fundulus heteroclitus* (Gabriel 1944; Fahy 1972), *Engraulis anchoita* (Ciechomski and Weiss de Vigo 1971), *Gadus morhua* (Brander 1979) and several other fish species. In the majority of cases a negative correlation has been observed between the water temperature and the mean number of vertebrae; this is in agreement with the data on the geographical variation of many fish species with respect to the number of vertebrae (Fowler 1970; Gross 1977). In the trout, however, the minimal number of vertebrae was obtained upon incubation of the embryos at temperatures around 6 °C (Schmidt 1919 b; Tåning 1952). Variations in the number of vertebrae among fishes in response to changes in temperature are associated with the acceleration or retardation of embryonic development (Kwain 1975) and perhaps with the overall metabolic level. Carp with a lower number of vertebrae consume less oxygen per weight unit than carps with a higher number of vertebrae (Tzoy 1971 a). In the lake trout the metabolism appears to be most economical at 6 °C (Marckmann 1954).

The association between the number of vertebrae and the water salinity may be realized by several different ways. The direct relationship corresponding to the clinal variation within the fiords has been found in *Zoarces* (Schmidt 1921); the same is true of the herring (Hempel and Blaxter 1961). A negative correlation has been found experimentally in the case of *Fundulus* (Kok Leng Tay and Garside 1972): in freshwater the mean equal to 33.95±0.05, at 16‰ to 33.62±0.09, and at 26‰ to 33.20±0.08.

As we could see, in spite of its very high heritability, the number of vertebrae can easily be changed by such factors as temperature or salinity, which appear to be quite essential environmental factors for fishes.

The number of vertebrae in fishes is determined by a large number of genes. Experiments on the hybridization of the European domesticated and Amur wild common carp (two different subspecies) have indicated that the variation of the vertebral number in the second and third hybrid generation does not exceed that found for the parental forms:

P: C.V. = 1.45%–1.48% (n = 849 and 824);
F_2: C.V. = 1.46% (n = 911);
F_3: C.V. = 1.41%–1.52% (n = 999).

As can be estimated from these relationships, the number of vertebrae is affected by dozens of genes. Analysis of segregation in the progeny of irradiated fishes in the case of the guppy has demonstrated that the number of interacting genes is equal to 7–10 (Schröder 1969a); it is quite possible that this figure is somewhat too low. Populations of many fish species are highly heterogeneous with respect to the number of vertebrae. We attribute this heterogeneity to complex peculiar relationships between the number of vertebrae and the characteristics of fish metabolism. The variation of the number of vertebrae appears to be a consequence of the adaptive intraspecific variation of fishes with respect to the metabolic level.

The inheritance of the number of vertebrae in fishes is complicated by a distinct maternal effect which is manifested particularly after distant crosses (Table 20).

It may well be that the maternal effect is associated with the early formation of the anterior part of the vertebral column during embryogenesis at the stage of

Table 20. Maternal effect for the number of vertebrae in fish hybridization

Domesticated common carp × wild common carp (Kirpichnikov 1949)		Domesticated common carp × crucian carp (Nikoljukin 1952)	
Number of trunk vertebrae		Total number of vertebrae	
Fish group	$\bar{x} \pm m_{\bar{x}}$	Fish group	$\bar{x} \pm m_{\bar{x}}$
Domesticated carp	18.48±0.02	Common carp	36.91±0.06
Domesticated carp ♀ × wild carp ♂	18.12±0.06	Common carp ♀ × crucian carp ♂	35.10±0.06
Wild carp ♀ × domesticated carp ♂	17.80±0.04	Crucian carp ♀ × common carp ♂	34.16±0.08
Wild carp	17.43±0.02	Crucian carp	32.26±0.21

gastrulation. The father's genotype at this time has hardly any effect on the embryo's development.

The Number of Gill Rakers and Pharyngeal Teeth. The variation in the number of gill rakers in fishes is considerable. It has been established that the variation in the number of gill rakers is close to the normal (Fig. 39). This number is finally determined when carp attain a weight of 30 g or more. Therefore in the groups of one-year-old fishes studied by us the variation can be somewhat overestimated (Kirpichnikov 1943, 1967a, b). There are stable hereditary differences between different fish populations with respect to this trait (Svärdson 1952, 1957, 1970; Kirpichnikov 1967b). The heritability of this trait is rather high, although environmental conditions also affect its expression. In *Fundulus heteroclitus* grown in brackish water (16 ppm) the number of gill rakers was lower than in fishes kept in fresh water or in salt water (Kok Leng Tay and Garside 1972). This is apparently associated with the faster differentiation of the embryo in water of "mean" salinity.

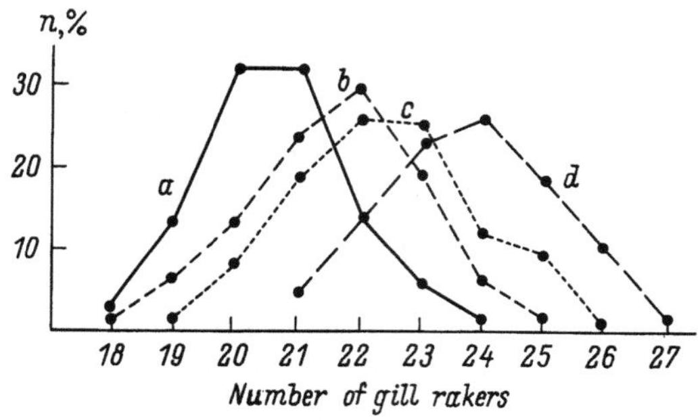

Fig. 39. Variation in the number of gill rakers in the Amur wild common carp (*a*), in two stocks of the Ropsha hybrid carp (*b, c*), and in the Galician cultured carp (*d*). Pond fish farms Jazhelbitzi and Ropsha, 1948–1960

The indices of heritability have been determined for the two fish species: in the whitefish *Coregonus lavaretus* $h^2 = 0.54–0.81$ (Svärdson 1950, 1952; the data on realized inheritance, our own calculations), in the three-spined stickleback *Gasterosteus aculeatus* $h^2 = 0.58 \pm 0.06$ (Hagen 1973; regression analysis).

The heritability of this trait in the common carp is also high. When one and the same female was crossed with two different males the number of gill rakers in the first arc was as follows (Kirpichnikow 1958a):

Cross No. 1: 22 (♀), 25 (♂); 23.21 ± 0.16 (offspring);
Cross No. 2: 22 (♀), 22 (♂); 21.49 ± 0.08 (offspring).

The heritability as we can see is close to 1.

In the carp acclimatized in Lake Balkhash the number of gill rakers increased from 23.5 to 26.9 over the 28 years from 1936 to 1964 (Burmakin 1956; Kirpichnikov 1967b). We regard this change as a direct consequence of natural selection. The main part of the variation in the number of gill rakers is apparently additive.

The number of genes affecting this trait does not appear to be very high. A definite decrease in variation in the number of gill rakers in F_1 hybrids of the domesticated and wild carp has been observed. The variation, however, has been increased in F_2. Similar data regarding the inheritance of this trait have been obtained in several generations of hybrid whitefish populations (Svärdson 1970).

The number of pharyngeal teeth in many fish species is relatively constant, but there are some species with more variable tooth formulae. For example, four species with the characteristic formulae 2.4–4.2 have been found among Cyprinidae, the variation of the total number of teeth is higher in these species (Eastman and Underhill 1973). In other cyprinids such as the wild carp as well as scaled or scattered domesticated carp any deviations from the formulae 1.1.3–3.1.1 are relatively rare, but the gene N appears to destabilize this trait and the number of teeth in linear and nude carp becomes variable, changing in a range of 6–8; the teeth in the second and third row are reduced. The heritability of the number of pharyngeal teeth has not as yet been studied.

The Number of Rays in the Fins. In many fish species including most cyprinids the number of rays (chiefly soft rays) in the dorsal fin shows some variation, but other fins such as anal, caudal, ventral and pelvic are variable as well. The structure of the fins is closely related to the peculiar features of the fish swimming pattern and their mode of life; therefore any geographical variation in the number of rays is frequently observed when a species is differentiated into subspecies and races. For example, differences in the number of rays in fins have been found for numerous local populations of *Zoarces viviparus* (Schmidt 1917), populations of *Etheostoma nigrum* (Thompson 1930; Lagler and Bailey 1947), subspecies of *Cyprinus carpio* (Svetovidov 1933; Kirpichnikov 1943, 1967b); numerous local populations of *Oryzias latipes* (Egami 1954) also differ from each other in the structure of the fins.

The distribution of the number of soft rays in fins is close to the normal with a few exceptions. In the eelpout *Zoarces viviparus* it was generally normal but in occasional populations with a decreased number of rays it was closer to the Poisson distribution (Schmidt 1917):

Sampling site	Number of rays and fish distribution											n
	0	1	2	3	4	5	6	7	8	9	10	
Limfiord				4	7	26	51	63	39	18	2	210
Gullmarfiord	24	2	1				3		1			31

The number of rays in the Limfiord population appears to have been inherited in a polygenic manner. In *Zoarces viviparus* living in the Gullmarfiord a gene (or genes) was apparently selected and then stabilized. This gene(s) completely abolished spiny rays in the dorsal fin and appeared to be epistatic with respect to all other genes of the polygenic complex. Occasional individuals with a larger number of rays (6 and 8) appear to have originated by hybridization or migrated from neighbouring populations.

More often than not, the variation coefficients of the number of rays does not exceed 6%. The variation coefficients for the dorsal fin were 1.8%–5.7% for ten species of cyprinids (Miaskowski 1957). The variation is generally lower in rapidly swimming fishes living in waterbodies with a stronger current.

Table 21. Heritability of the ray number in fins

Species	Trait	Heritability, h^2		Reference
		Method of calculation	Mean value	
Carp, *Cyprinus carpio*	Number of soft rays in the dorsal fin	Regression "parents-offspring"	0.46	3
		Regression "parents-offspring"	0.61	2
		Correlation "parents-offspring"	0.57	2
		Dispersion anal.	0.63	3
Guppy, *Poecilia reticulata*	Number of soft rays in the dorsal fin	Realized heritability	0.59	5
	Number of rays in the caudal fin	Regression "parents-offspring":		
		parents 4–6 months old	0.43–0.60	1
		parents 6–18 months old	0.80–1.00	1
Eelpout, *Zoarces viviparus*	Number of rays in the dorsal fin	Regression "mothers-offspring"	0.79	4
	Number of rays in the pectoral fin	Regression "mothers-offspring"	0.54	6
	Number of rays in the anal fin	Regression "mothers-offspring"	0.60	4

References: 1. Beardmore and Shami (1976); 2. Kirpichnikov, unpubl.; 3. Nenashev (1966); 4. Schmidt (1917); 5. Schmidt (1919a); 6. Smith (1921). 4.–6. our calculations of h^2

Data regarding the heritability of this trait are limited to three fish species, heritability coefficients are rather high (Table 21). A comparison of *Rivulus marmoratus* clones has demonstrated that interclonal differences with respect to the number of rays in all fins with the exception of the caudal one are significant, they are particularly prominent for ventral fins (Harrington and Crossman 1976). The variance between the clones is close to the genetic additive variance or perhaps somewhat higher. The maximal value of this variance has been observed for the V fin (almost unity) and in several experiments for A, P and D fins. Evidence pointing to the high heritability of the number of rays in the D and A fins has also been obtained in the medaka *Oryzias latipes* (Ali and Lindsey 1974).

The existence of a large additive component for the variation of the ray number is also supported by the data of selection experiments. Even in the guppy with its moderate variation of the number of rays in the dorsal fin the selection in the negative direction has made it possible to shift the mean value from 7 to 6.465 after five generations (Svärdson 1945b). A fixed variation of this trait in the consecutive hybrid of the common carp (Fig. 40) also provides evidence of the presence of genes with additive action. Similar conclusions may be drawn from the segregation data for this trait obtained in pairwise crosses of domesticated or wild carp and in similar crosses of the Ropsha hybrid carp (F_2-F_4) generations (Table 22).

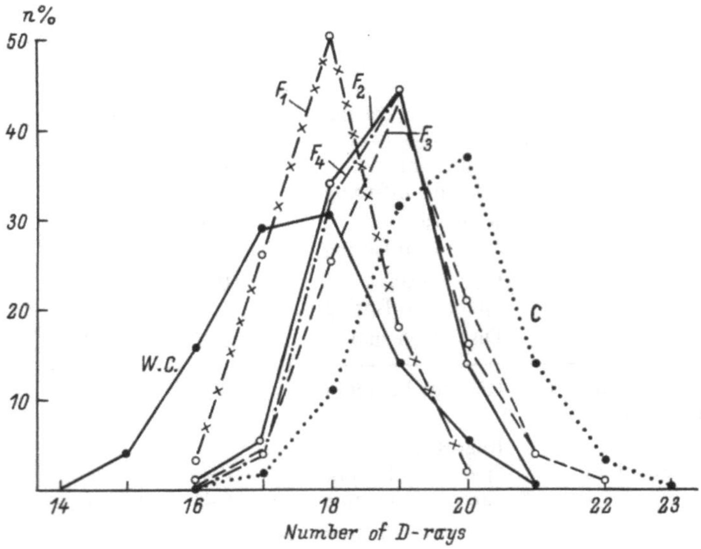

Fig. 40. Variation curves for the number of soft rays in the dorsal fin in the Amur wild common carp (*W. C.*), Galician carp (*C*) and Ropsha hybrid carps of 1st–4th selection generations (F_1, F_2, F_3 and F_4). Ropsha 1946–1964

Table 22. Inheritance of the soft ray number in the dorsal fin in pairwise crosses of the wild carp, mirror carp and Ropsha carp of the 2nd–4th generations (own data)

Mean number of rays in dorsal fins of parents	Distribution of the offspring for ray number									Mean for the offspring x̄
	14	15	16	17	18	19	20	21	22	
16.5	1	22	70	50	9	1				16.31
16.5			2	2	1					16.80
17.5		15	38	34	12	1				16.46
18.0				2	2					17.50
18.5		4	34	85	66	9	2			17.24
18.5				9	18	12	8	2		18.51
18.5				10	34	68	26	8	4	19.00
18.5				1	6	16	19	3		19.38
18.5				2	24	58	16	1		18.90
19.0			3	10	16	10	1			17.90
19.5				1	6	19	16	5		19.38
19.5					1	12	29	4	1	19.83
20.0				1	4	9	8	4		19.39
20.0					2	4	6	2		19.57
20.0			3	6	10	3	2			17.79
20.0				10	22	21	4			18.33
20.5					4	11	7	3		19.36
21.5					1	13	14	7	2	19.89

In crosses between the domesticated and Amur wild common carp the mean values of the ray number in the dorsal fins are shifted towards those characteristic of the Amur carp. The Amur carp appears to possess dominant genes controlling the formation of rays in the dorsal fin. The adaptive nature of the fin variation follows from observations with the three-spined stickleback *Gasterosteus aculeatus* (Heuts 1949). The small freshwater race of this species does not differ in the fin pattern from the larger race living in brackish water, but the number of rays in the freshwater form is modified by temperature only when the fishes are kept in the fresh water; it is changed in individuals living normally in brackish water only when they are kept in salt water. We assume that the ability to change the number of rays in the fins in response to temperature variation represents an adaptive trait.

The number of genes affecting the structure of fins in the common carp appears to be rather high. No less than ten genes are responsible for fin variation in the molly (*Mollienesia*) and in the guppy (*Poecilia reticulata*) (Schröder 1965, 1969a).

Paratypic variance plays an essential role in the phenotypic variation of the ray number. The lowest number of rays in *Fundulus heteroclitus* was observed when the embryos were kept in brackish water with a salinity equal to 16‰ (Kok Leng Tay and Garside 1972). The number of rays in the anal fin of this species is also determined by the water temperature in the course of embryogenesis (Fahy 1979). The different relationships between the number of rays in D, A and V fins and the temperature have been observed in different homozygous clones of *Rivulus marmoratus;* the negative linear correlation in certain cases is changed for the U-shaped or ∩-shaped dependence (Barlow 1961; Harrington and Crossman 1976). It is concluded from these observations that genes exist that change the character of the association between the temperature and fin development. The maximal number of rays in the brook trout is generally observed at intermediate temperatures (Tåning 1952). The drop in temperature and the lengthening of the incubation period result in an increase in the number of rays in the rainbow trout (Kwain 1975). A similar increase has been observed in the carp in years when the weather is cold (Kirpichnikov 1943).

Genes controlling the fin structure represent a component of intricate genotypic complexes affecting the whole course of fish development. The intensive selection for the changed number of rays in the guppy has been accompanied by impairments in gonad maturation (Svärdson 1945b).

The stabilizing selection for the ray number in the caudal fin of this fish results in greater fertility and viability, as well as in greater heterogeneity in several protein loci (Beardmore and Shami 1976, 1979). Selection for the number of dorsal rays is accompanied by changes in the exterior traits of the common carp (own observations) leading to other consequences that are difficult to predict.

The Number of Scales in the Lateral Line (l. l.). This trait is extremely variable in several fish species. In the whitefishes (*Coregonus* spp.) the variation in the number of scales is predominantly paratypic (Svärdson 1950, 1952, 1958). The heritability of this trait in the common carp is equal to 0.32–0.42 (Nenashev 1966; dispersion analysis, hierarchical complex). Experiments on the selection conducted with carp by Nenashev and myself did, however, yield very low heritability coefficients. The values obtained by variance analysis are apparently too high – perhaps a "common environment" could particularly affect this trait, all the more so since its determination takes place at relatively late stages of development. The variation in the number of scales is frequently asymmetric: the left half of the variation curve is extended. The mechanisms underlying this asymmetry remain unexplained.

The Reduction in the Scale Cover. In cyprinodonts from the genus *Aphanius* a geographical variation in the total number of scales on the body has been established; almost completely nude forms have been found on the borderline of the distribution range. The inheritance is distinctly polygenic; it is assumed that this polymorphism is transient in nature (Aksiray 1952; Kosswig 1965; Franz and Villwock 1972). The scale size in a related species *Kosswigichthys asquamatus* is also inherited as a polygenic trait (Villwock 1963).

The Number of Lateral Plates. In the so-called "middle morph" of the three-spined stickleback living in certain American and Canadian lakes the variation with respect to the number of lateral plates on the body turned out to be unexpectedly high. The heritability has been determined by regression analysis (Hagen and Gilbertson 1973 b). Heritability coefficients in different experimental series were 0.50–0.84. It is concluded that the development of lateral plates is controlled by additive genetic factors. The coefficient of variation for the number of lateral plates attains 18%–20% (Kynard and Curry 1976). For the European stickleback the variation in the number of lateral plates between different populations is definitely of a selective character (Gross 1977).

The Number of Intermuscular Bones. According to a number of authors, this trait is very variable in a number of fishes (Lieder 1961). The variation coefficient in the common carp, for example, attains 10%–17% (Sengbusch 1967; Kossmann 1972; Slutzky 1976). According to other authors, however, the variation in the number of intermuscular bones in the carp is markedly lower (Moav et al. 1975 a). Additional studies are required, but there can be no doubt that the variation in the number of intermuscular bones is at least in part hereditary.

The Reduction of the Eyes in Cave-Dwelling Fishes. Blindness and the pale discoloration of cave-dwelling (cavernicolous) fishes may be associated with the selection of one or a few "large" mutations. At the same time, many genes with weak manifestation are accumulated in cavernicolous forms; these genes destructively affect the body pigmentation, the structure of the eyes, and the structure of visual lobes of the midbrain (Pfeiffer 1967). The polygenic nature of eye reduction has been established in studies of hybrids between *Astyanax* and *Anoptichthys* (Characidae) (Wilkens 1970, 1971) as well as during the genetic analysis of cavernicolous representatives of *Poecilia sphenops*, Poeciliidae (Peters and Peters 1968, 1973).

Many other morphological traits in fishes show some variation. For example, it has been reported that the number of rays in the gonopodium of cyprinodonts is inherited in a polygenic fashion (Sengün 1950; Gordon and Rosen 1951). The reduction in the bones of the pelvic skeleton in the brook stickle-back *Gulaea inconstans* is also inherited in a polygenic manner, as can be judged from the data obtained by genetic analysis (Nelson 1977). The number of piloric caeca in the brown trout *Salmo trutta* and the rainbow trout *Salmo gairdneri* depends on several genes (Begot et al. 1976). Heritability of this trait determined by means of regression analysis is equal to 0.53 in the brown trout and 0.40 in the rainbow trout (Blanc et al. 1979; Chevassus et al. 1979). Differences in the number of taste tubercules between the terrestrial and cavernicolous forms of *Astyanax* depend on two or three pairs of genes (Schemmel 1974).

Variation and Heritability of Physiological and Biochemical Traits

Only very few among the physiological traits of fishes have been studied from a genetic point of view.

Polygenic inheritance has been reported for the temperature preferred (Goddard and Tait 1976). This has been established in studies of splake-hybrids of 3–5 generations between the lake and brook trout *Salvelinus fontinalis* and *S. namaycush*.

A highly significant variation between the different river strains in percent parr smoltifying in one year has been found for the Atlantic salmon. This variation was at least partially hereditary. The heritability factor for the proportion of parr transformed to smolts in a given year varied in the range 0.1–0.4 (Refstie et al. 1977a; Holm and Naevdal 1978).

The variation in the external appearance of early maturing "dwarf" males in the parr of the Atlantic salmon is partially inherited as well (Saunders and Sreedharan 1977). We suggest that a similar variation in *Oncorhynchus nerka* and some other Pacific salmon is also partially hereditary.

The ability of Pacific and Atlantic salmon and also cutthroat trout to return "their own" rivers, brooks and lakes is inherited in a polygenic manner. It has been demonstrated that this ability varies in different stocks and inbred families (Brannon 1967; Carlin 1969; Raleigh and Chapman 1971; Bowler 1975; Bams 1976). Part of this variation is additive (Ryman 1970; Ricker 1972).

The time of spawning in the rainbow trout was significantly changed after several generations of selective breeding (Lewis 1944; Millenbach 1973). Polygenic nature and the high heritability of variation at spawning time has been reported for many other species such as the whitefish *Coregonus peled* (Andriyasheva et al. 1978), the grass carp *Ctenopharyngodon idella* and the silver carp *Hypophthalmichthys molitrix* (Konradt 1973).

The ability of fish to escape fishing nets and lines depends on a large number of genes (Beukema 1969; Moav and Wohlfarth 1973a; Suzuki et al. 1978). The difference between the Chinese and European common carp with respect to this trait results from different methods of natural and artificial selection: the selection for the ability to escape fishing nets was very strict in China fish breeding while this type of selection was virtually excluded in Europe (Wohlfarth et al. 1975a, b). The selection resulted predominantly in the accumulation of dominant and semidominant genes.

The ability of hybrid trout larvae to retain gases in their air bladder also depends on many genes; the heritability coefficient was found equal to 0.20 (Butler 1968).

Polygenic inheritance has also been reported for the canibalism found in the small fishes *Poeciliopsis monacha* (Poeciliidae). This has been demonstrated by the comparison of cannibalism of diploid and triploid hybrids between *Poeciliopsis monacha* and a peaceful variety *Poeciliopsis lucida* (Thibault 1974).

The aggressive behaviour of fishes generally decreases when mutations induced by X-irradiation are accumulated. These experiments were performed with *Cichlasoma nigrofasciata* (Cichlidae). It can be assumed that aggressive behaviour is inherited as a polygenic trait (Holzberg and Schröder 1972).

Patterns of mating behaviour in viviparous cyprinodonts and in the tilapia are inherited as polygenic traits (Clark et al. 1954; Barash 1975).

In the paradise fish the active defense behaviour is a polygenic hereditary trait, with considerable non-additive component of variation (Vadasz et al. 1978). The heritability of this original trait, called "tonic immobility", is equal to 0.31–0.33 in *Macropodus opercularis* (Kabai and Csányi 1979).

We have a limited number of data on the inheritance of biochemical traits in fish. The heritability of the nitrogen and lipid content in the common carp varies from 0.14 to 0.17 (Smišek 1978, variance analysis). Heritability of the lipid content in the channel catfish is equal only to 0.08 (El-Ibiary and Joyse 1978). The genetic component in the variation of this trait is more substantial in the rainbow trout (Ayles et al. 1979). The biochemical characteristics of fishes appear to be greatly affected by the environment.

Physiological and biochemical traits of fishes depend on a large number of genes and apparently on many environmental factors as well. The genetic variation of these traits appears to include an additive component, but non-additive sources of genetic variation play an important part in the phenotypic variation of many of these traits.

The accumulation of data on the heritability of the physilogical and biochemical characteristics of fishes represents an important and urgent task in the genetics of economically essential fish species.

Phenodeviants

Phenodeviants are a group that has peculiar changes, occupying an intermediate position between qualitative and quantitative traits. This definition was suggested by Lerner (1954) in order to describe hereditary deviations from the normal state which vary greatly with respect to their manifestation and frequency of occurrence and which do not lend themselves to genetic analysis. Aberrations of this kind have been found earlier in large numbers in *Drosophila* (Dubinin and Romashov 1932; Dubinin et al. 1937).

Various aberrant forms are usually found in natural fish populations; their frequency is generally low, but in some cases may be quite pronounced. We have found massive anomalies in particular in our studies with the wild carp fry

Table 23. Types of aberration in the smolts of the Low Volga wild carp (1937–1938)

Aberrations	Delta of Volga		Akhtuba flood-plain		"Kamenny Yar"	
	Number	%	Number	%	Number	%
Displaced scales	286 (359)[a]	1.99 (2.50)[a]	7 (8)[a]	0.39	1	0.07
Malformed fins	254 (271)[a]	1.76 (1.88)[a]	11 (15)[a]	0.61	11	0.75
Absence of ventral or anal fins	5	0.03	0	0	0	0
Malformations of the caudal vertebral region	27 (28)[a]	0.19 (0.19)[a]	2	0.11	3	0.20
Bulldog-like head and distorted jaws	26	0.18	1	0.06	20	1.37
Reduction or absence of eyes	13	0.09	1	0.06	0	0
Reduction of "whiskers"	3	0.02	44 (45)[a]	2.47	0	0
Underdevelopment of the gill cover	36	0.25	9	0.50	3	0.20
Interrupted or bent lateral line	45 (46)[a]	0.31 (0.32)[a]	1	0.06	0	0
Variegated pigmentation	2	0.02	0	0	0	0
Malformations of traumatic origin	44	0.31	0	0	1	0.07
Total: Aberrant individuals	741	5.15	76	4.25	39	2.66
Aberrations[a]	833	5.79	82	4.60	39	2.66
Total number of examined fishes	14,375		1783		1465	

[a] Total numbers of each aberration including those present together with other aberrations

(*Cyprinus carpio*) in the waterbodies of the Volga estuary: up to 5% of all the fingerlings examined were aberrant, that is displayed some deviations (Table 23). Some of the aberrations found by us could results from simple recessive or dominant qualitative mutations, part of the malformations being due to traumas, but the majority of the abberrant forms was represented by phenodeviants. These primarily include various types of scale displacement, many types of fin aberrations, as well as a bulldog-like head shape, the reduction of the gill cover and frequent cases of vertebral fusion. Similar types of phenodeviants have been found in the domesticated carp; the most frequent of them will be described below (Fig. 41):

1. fin malformations – the reduction of fins down to complete disappearance, changes in the fin structure (condensed and curled dorsal fins, hanging caudal fin) (Kirpichnikov and Balkhashina 1935, 1936; Wunder 1949b, 1960; Tatarko 1961, 1963, 1966);

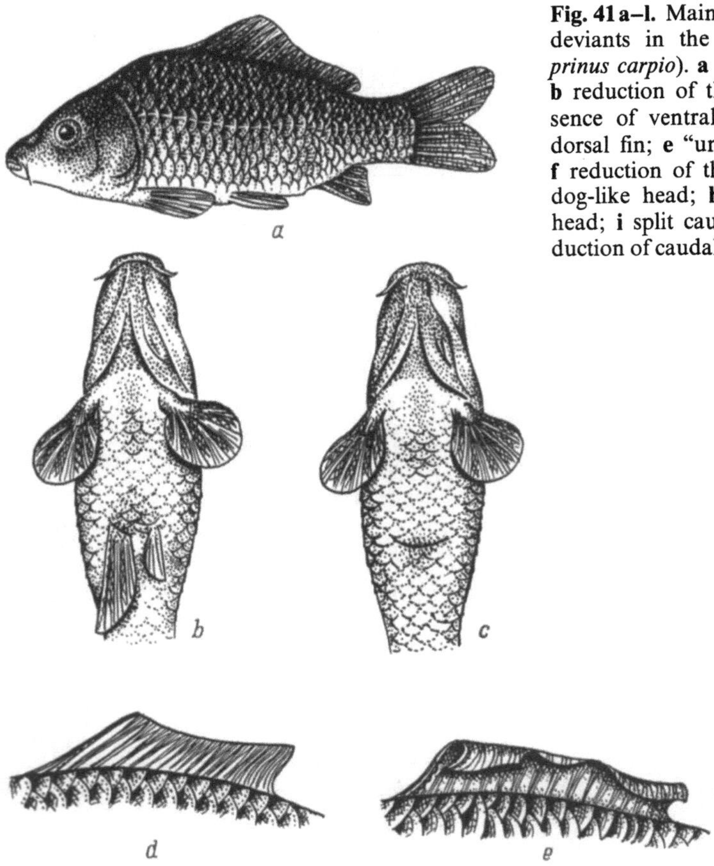

Fig. 41 a–l. Main types of phenotypic deviants in the common carp (*Cyprinus carpio*). **a** irregular scale cover; **b** reduction of the ventral fin; **c** absence of ventral fins; **d** compressed dorsal fin; **e** "undulating" dorsal fin; **f** reduction of the gill cover; **g** bulldog-like head; **h** deformation of the head; **i** split caudal fin; **k** partial reduction of caudal fin; **l** glass-like fish ►

2. scale displacement – incorrect location of the scales in different regions or on the whole trunk (Kirpichnikov and Balkhashina 1936; Probst 1953; Steffens 1966);
3. the reduction of the gill cover (Wunder 1931, 1932; Kirpichnikov and Balkhashina 1936; Schäperclaus 1954; Tatarko 1961, 1966);
4. malformations of the vertebral column (Wunder 1931, 1934; Volf 1956);
5. anomalies in the structure of the intestines (Shulyak 1961);
6. changes in the structure of integument and scales resulting in the appearance of the so-called "glassy" carps with drastically retarded growth (Kirpichnikov 1961).

This list is by no means complete. The phenodeviants may be found in most fish species. Up to 1%–2% of individuals with such changes have been found in fishes belonging to the family Bothidae (Haaker and Lane 1973). In the guppy and in the goldfish anomalies in the structure of the vertebral column are inherited in a complex fashion (Schröder 1969d; Asano and Kubo 1972). Anomalies in the structure of the lateral line are very frequent in many species (Geyer 1940). The

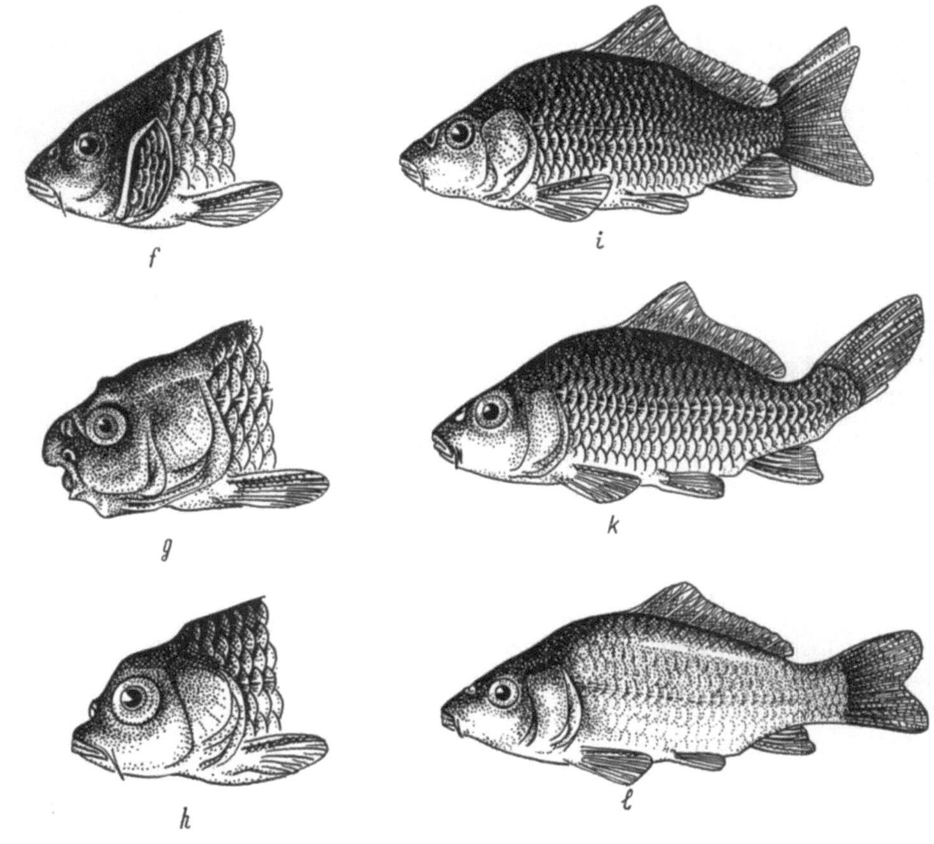

phenodeviants frequently result from inbreeding. In an inbred line of the swordtail many changes in the structure and location of certain blood vessels have been found (Baker-Cohen 1961); after two generations of strict inbreeding (crosses between brothers and sisters) in the rainbow trout, the number of malformed larvae increased by 191% (Kincaid 1976).

We shall now try to formulate certain general concepts regarding the pattern of inheritance of phenotypic deviations.

The frequency of appearance of any phenodeviant does, to a large extent, depend on the environmental conditions under which the fishes exist. In carp of common origin kept in different ponds, the number of individuals with anomalies of the gill cover and ventral fins varies from 1.4% to 18% (Tatarko 1961). Temperature and the extent of food supply for the fry represent the most important environmental factors affecting the frequency and the degree of manifestation of these defects. Phenodeviants are more likely to occur during inbreeding, sometimes achieving high levels of up to 30%–40% and in some cases up to 70%–80%, but the inheritance at any frequency of aberrations does not follow simple Mendelian laws. Generally phenodeviants are inferior as compared to normal fish by the rate of growth and viability. In the common carp the decrease of body weight associated with certain types of aberrations may be very pronounced (Tomilenko and Shpack 1979).

The incorporation of strong pleiotropic genes into the genotype, for example, N and L genes in the common carp, is accompanied by an increase in the number of phenodeviants. The frequency is much higher in nude and linear carp than in scaled or scattered ones. Big differences have been observed between different stocks, strains and varieties of fishes with respect to the occurrence frequency of phenodeviants.

The largest number of deviations in natural fish populations can be found upon the inspection of the youngest fishes. Their number in the wild carp was the highest, when very small fingerlings were released from the water bodies into the estuary. The number of aberrations among large fingerlings was twofold less. A strict, stabilizing selection appears to take place in natural water bodies: all the individuals deviating from the norm are destined to die.

The existence of phenodeviants can be considered as an index of weakening of the genetic constitution, i.e., a decrease in genetic and developmental homeostasis. Genes or combinations of genes having no visible manifestation in the case of a well-balanced genotype and optimal conditions of life may be manifested when the genetic balance is distorted and in an unfavourable environment. The manifestation of such genes depends on the remaining portion of the genotype and on many environmental factors. Therefore the inheritance of such phenodeviants is usually poorly explained by conventional Mendelian laws. Perhaps it would be more correct to speak of hereditary predisposition to malformation in such cases. Prolonged selection may convert certain aberrations into Mendelian traits, but a marked proportion of the phenodeviants is controlled by many genes and this markedly complicates the study of them. Certain aberrations and mainly the malformations of the vertebral column may be used as a very sensitive indicator of the mutagenic action of radiation upon fishes (Schröder 1979).

Phenodeviants are particularly interesting for fish breeders because they can be regarded as objective indices characterizing the general state of the population. The increase in their numbers may be taken as evidence of the undesirable consequences of breeding work, of the excessive intensity of selection or of too close inbreeding.

Chapter 5

The Biochemical Genetics of Fishes

The General Principles of Fish Immunogenetics

Various defense mechanisms including the immunological reactions protecting the body from the infection caused by pathogenic microorganisms have been described among fishes.

Such protective systems include (Corbel 1975; Marchalonis 1977):
1. The macrophages, including granulocytes and lymphocytes.
2. Non-specific substances possessing a defense function, e.g., properdin, lysozyme, complement, and interferon.
3. The presence on the surface of the erythrocytes of specific and diverse proteinaceous components – the agglutinogens (antigens) responsible for blood group variation.
4. The ability to form specific antibodies appearing in the blood and binding or destroying foreign proteins or invading organisms.

In the composition of the macrophage cell populations and in the presence of the non-specific protective substances fishes are similar to higher vertebrates, with the reservation that the structure of some of the components of the defense system may be somewhat "simpler" among them. The complement in bony fishes, for example, appears to consist of four or five components instead of the nine characteristic of man (Marchalonis 1977). The variation with respect to the blood groups in fishes is quite extensive. The capacity to respond by the synthesis of specific antibodies (immunoglobulins), in contrast, is much less characteristic of this group of animals, as compared to mammals. The immunity acquired as a consequence of contact with foreign proteins, or the immunological memory, in fishes is transient in its nature and persists for a relatively short time. It should be pointed out at the same time that the formation of antibodies in fishes even occurs at low temperatures (Ridgway 1962).

In studies of the genetic variability of fishes considerable attention has been paid to the analysis of blood groups; these studies made use of the technique of erythrocyte agglutination. This method is based on the sticking of erythrocytes when they are mixed with the heterologous sera (haemagglutination). The differences in the composition of erythrocyte antigens between different individuals may be detected using the so-called normal sera or isosera obtained from fishes of the same species (isohaemagglutination); sera from other fish species or even from other groups of animals (heterohaemagglutination) may also be used. Only in a few cases were the isosera helpful in detecting differences in blood groups, the attempts to distinguish the blood groups using non-immune heterologous sera were much more successful. In addition to the sera obtained from fishes or mammals the extracts of seeds of certain plants containing lectins turned out to be convenient differentiation agents.

Immune sera are widely used nowadays in studying blood groups in fishes. When experimental animals are injected with fish erythrocytes this leads to the formation of antibodies to the erythrocyte antigens of the donors. The sera of immunized animals are called the immune antisera. Mixing of the antiserum with the erythrocytes of the fish examined results in haemagglutination when erythrocytes contain the antigens that were present in the donor fish. Isoimmunization may be successful, but heterologous immunization, i.e., the injection of fish erythrocytes into the experimental animals and the preparation of heteroimmune antisera is generally more effective.

The antiserum can be "depleted" by mixing with erythrocytes containing one, (or several) of the antigen studied; the antibodies produced by the immunization of animals with these antigens will fall out of the antiserum together with the respective antigens. In a number of cases, such absorbed or depleted antisera allow genotypes to be differentiated more distinctly.

Precipitation, another serological reaction, is frequently used in studies of the degree of relatedness of animals. This reaction is based on the formation of precipitin – antibodies in response to the foreign proteins (antigens) present in the blood serum and in other tissues of the organism. The technique of precipitation has been successfully used in studies of fish systematics and evolution; it has been instrumental in obtaining many interesting data regarding the degree of relatedness of various subspecies, species and genera (Taliev 1941, 1946; Zaks and Sokolova 1961; Limansky 1964; Pokhiel 1969; Koehn 1971; Lukyanenko 1971; Hodgins 1972; Zhukov 1974).

The methods used in serological and immunogenetic studies have been presented in detail in a number of good reviews (Altukhov 1969a, 1974; Ligny 1969; Lukyanenko 1971; Ridgway 1971, and others). In this presentation therefore we shall limit ourselves to a brief description of the methods used in studies of the genetic variability of blood groups in fish populations and consider just a few of the most interesting examples.

The antigenic variability of fish erythrocytes is determined by genetic loci that have two or more alleles, each of which is responsible for the formation of a specific antigen. The blood groups depending on the alleles of one locus represent a single system; each species may have several such systems. Not infrequently the series of allelic genes resemble the AB0 blood system of man in the type of inheritance and the interaction of products. Such a system contains three alleles of which one is termed the null allele. The null allele does not yield any protein product or antigen; in a heterozygote where the null allele is present only one erythrocyte antigen, which is coded by the "acting" allele, is found, an organism homozygous with respect to the null allele does not possess any antigens of the given system. Therefore, individuals possessing the genotype aa and a0 have only one antigen A, fishes with bb and b0 alleles have the B antigen, the heterozygotes ab have two antigens (A and B) and the homozygotes with respect to the null allele 00 are devoid of both antigens, A and B. The specific immune antisera containing antibodies against antigen A or antigen B allow us to distinguish between these genotypes. The anti-A serum will agglutinate the erythrocytes of individuals having the genotypes aa, a0 and ab; the anti-B serum will agglutinate erythrocytes of bb, b0 and ab fishes; in the case of the 00 genotype, no haemagglutination will be observed using either antiserum. The genotypes aa and a0 can sometimes be distinguished quantitatively by the minimal titer of sera necessary for agglutination; the same is true of the genotypes bb and b0. In practice, however, in this case one is able to divide the fishes into four phenotypic classes alone characterized by the presence or absence of the antigens AB, A, B and 0. If the gene a completely dominates over b, the number of phenotypes decreases to three: A (genotypes aa, ab and a0), B (bb and b0), and 0 (00).

Simpler systems with two, three or an even larger number of active codominant alleles are frequently found among fishes. In the simplest system containing two alleles the genotypes and phenotypes correspond to each other completely:

Genotypes (alleles)	aa	ab	bb
Phenotypes (antigens)	A	AB	B.

In these cases both antigens are produced in heterozygous organisms. A similar correspondence (in the presence of the codominance and the absence of null alleles) is observed in multiple allele systems, for example with four allelic genes:

Genotypes	aa	bb	cc	dd	ab	ac	ad	bc	bd	cd
Phenotypes	A	B	C	D	AB	AC	AD	BC	BD	CD.

All ten groups may be distinguished when the appropriate antisera or other differentiating reagents are selected.

Sometimes a normal serum, an immune serum or lectin only allow one antigen to be detected. Then, in the case of a two-allelic blood group system the positive reaction (agglutination) is observed with animals of the two genotypes and the negative reaction with animals of the third genotype. The fish population may then be divided into two phenotypically different groups: +(aa and ab) and −(bb); in the presence of the null allele the plus-groups contain aa and a0 individuals and the minus-group 00 homozygotes.

The number of codominant alleles in certain systems of blood groups attains ten or twelve. Much depends on the differentiating power of the reagent used: if it only allows a rough identification, then several genotypes may be taken for one.

The correctness of one hypothesis or another regarding the inheritance of blood groups can be verified by a hybridological analysis. Such an analysis was conducted using three freshwater fish species – the common carp, the rainbow trout and the brown trout.

In experiments with the common carp (*Cyprinus carpio*) only preliminary data on the inheritance of several serum antigens, proteins and lipoproteins have been obtained (Slota et al. 1970; Rapacz et al. 1971; Slota 1973). One of these (Lpf-1) was inherited as a dominant trait. The variability of the erythrocyte antigens has also been described in the common carp (Pokhiel 1967; Yu. I. Ilyasov, pers. com.), but has not been examined genetically.

The inheritance of codominant alleles r_1 and r_2 (erythrocyte antigens R_I, R_{II} and R_I+R_{II}) has been examined in the rainbow trout (*Salmo gairdneri*). The segregation was in good agreement with the expected relationships (Sanders and Wright 1962):

$R_{I-II} \times R_{I-II} = 49R_I + 91R_{I-II} + 42R_{II}$;
$R_{II} \times R_{I-II} = 66R_{II} + 58R_{I-II}$, etc.

A more complex segregation pattern involving several alleles has also been described (Ridgway 1962).

The genetic analysis of the brown trout (*Salmo trutta fario*) has confirmed the presence of an AB0-type system with three alleles. The interpretation of segregation was, however, complicated by developmental changes in the antigenic composition of erythrocytes in certain individuals (Sanders and Wright 1962); these changes involved the conversion of B_{I-II} and B_{II} types into types B_I and B_0.

The variability of the blood groups found in natural fish populations detected by iso- or heteroimmunization or finally by normal sera may be analyzed, using the Hardy-Weinberg equation for the equilibrium state:

$$p^2\,(AA) + 2pq\,(AB) + q^2\,(BB) = 1, \tag{1}$$

where p and q are the frequencies of the alleles and AA, AB and BB the genotypes and phenotypes of individuals when two alleles of one and the same locus are codominant. When the conditions of reproduction in a population are unrestricted (panmictic), the inbreeding is low, and the impact of natural selection is little felt, the equilibrium of the frequencies described by Eq. (1) is established as early as in the second generation in any such population. Knowing the values of p and q, one can calculate the equilibrium frequencies of all three genotypes and compare them with those actually observed. We shall present one example. A large sample of the Black Sea anchovy collected near Odessa in 1963 was divided into three antigenic groups, using horse and pig antisera. According to the working hypothesis, these three groups correspond to the three genotypes of the two-allelic systems of blood groups: A_1A_1, A_1A_2 and A_2A_2. The frequencies were found to be the following (Altukhov et al. 1969):

A_1A_1	A_1A_2	A_2A_2	n
138	28	2	168.

The frequency of the A_2 allele (q_{A_2}) in the sample is determined from the equation:

$$q_{A_2} = \frac{2\sum(A_2A_2) + \sum(A_1A_2)}{2n}\,. \tag{2}$$

In all the A_2A_2 homozygotes the allele A_2 is present in the double dose; to this value one has to add the number of alleles A_2 in the A_1A_2 heterozygotes and to divide them by the total number of alleles in the sample (2n). In this case, we obtain:

$$q_{A_2} = \frac{4+28}{336} = 0.095\,.$$

It follows that the frequency of the A_1 allele (p_{A_1}) is equal to 0.905. We can substitute these values into Eq. (1):

$$p^2\,(A_1\,A_1)=0.819; \quad 2pq\,(A_1\,A_2)=0.172; \quad q^2\,(A_2\,A_2)=0.009.$$

By multiplying the frequencies of the genotypes by 168 (the number of animals in the sample), we obtain the following theoretical expected values:

$$A_1\,A_1 = 137.6; \quad A_1\,A_2 = 28.9; \quad A_2\,A_2 = 1.5.$$

As we can see, the calculated frequencies are very close to those actually determined. Using the least squares (χ^2 method), one can easily demonstrate that the difference between the empirically found and theoretically calculated frequencies does not exceed the standard error.

Similar calculations can be made in the presence of three or four codominant alleles of one locus in the population. We shall present the equations of equilibrium for these cases:

$$p^2\,(AA)+q^2\,(BB)+r^2\,(CC)+2\,pq\,(AB)+2\,pr\,(AC)+2\,qr\,(BC)=1 \tag{3}$$

$$p^2\,(AA)+q^2\,(BB)+r^2\,(CC)+t^2\,(DD)+2\,pq\,(AB)+2\,pr\,(AC)+2\,pt\,(AD) \\ +2\,qr\,(BC)+2\,qt\,(BD)+2\,rt\,(CD)=1. \tag{4}$$

Panmixia of a greater or lesser extent, the size of the population sufficient to prevent inbreeding and the relatively small selection coefficient are the conditions characteristic of the reproduction of most fish species, particularly marine ones. This leads to an equilibrium of genotypes in a population; therefore the investigator of blood groups can almost always verify his hypothesis regarding inheritance by comparing the proportions observed for the phenotypes with the theoretically expected ones. The agreement of these two groups of data provides evidence of the correctness of the selected hypothesis, although its final proof can only be obtained by hybridological analysis. The situation may be more complex, when one antigenic group contains several genotypes as is the case, for example, with the system of erythrocyte antigens of the AB0 type. Frequency of the alleles in a sample can be determined only approximately in such a case since the genotypes AA and A0 in the group, A cannot be distinguished as well as the genotypes BB and B0 in the group B. As regards the three-allelic system with four phenotypes, equations have been derived which allow the frequencies of all the alleles to be calculated fairly accurately (Bernstein 1925; Tikhonov 1967):

$$\begin{aligned} p_A &= (1-\sqrt{\bar{B}+\bar{0}})(1+D); \\ q_B &= (1-\sqrt{\bar{A}+\bar{0}})(1+D); \\ r_0 &= (\sqrt{\bar{0}+D})(1+D). \end{aligned} \tag{5}$$

In these equations $\bar{A}$, $\bar{B}$ and $\bar{0}$ are the frequencies (in fractions of unity) of the antigenic groups A, B and 0; D is the correction coefficient decreasing the calculation error of the allele frequencies and determined from the formula:

$$D = \frac{\sqrt{\bar{B} + \bar{0}} + \sqrt{\bar{A} + \bar{0}} - \sqrt{\bar{0}} - 1}{2}. \quad (6)$$

If only three phenotypes are distinguishable with the system AB0, the errors in determining allele frequencies become too large and a comparison of them with the equilibrium frequencies is of little value.

The disagreement between the frequencies observed and the theoretical ones in the analysis of blood group variability in fishes may lead to several suggestions, such as:

1. The hypothesis of inheritance is incorrect;
2. the sample used contains individuals from several populations with different frequencies of the given allele locus; the mixing of populations usually results in a decrease in the heterozygote number (the Wahlund effect) as compared to the expected value;
3. the population is very small and an important part is played by inbreeding, this also results in a decrease in the number of heterozygotes;
4. different genotypes have different selective value and as a result – great differences in the selection coefficient and the survival of fishes with different genotypes;
5. reproduction involves a pairwise assortment of individuals with certain genotypes (assortative crosses) or selective fertilization.

The testing of these possibilities requires specialized experiments.

Examples of Blood Group Variability in Commercially Important Fish Species

The Herring (Clupea harengus). The first studies dealing with the immunogenetics of the oceanic herring allowed two phenotypic groups to be detected which were characterized by the presence or absence of the specific erythrocyte antigen (Ridgway 1958; Sindermann and Mairs 1959; Sindermann 1962, 1963). Later, mainly due to studies made by Soviet scientists, this system was divided into the following three phenotypes, A_1, A_2 and A_0. It has been assumed that the inheritance of these antigens is determined by the three alleles of one and the same locus including one null allele:

Genotypes	A_1A_1, A_1A_2, A_1a	A_2A_2, A_2a	aa
Phenotypes	A_1	A_2	A_0

In order to distinguish between these three groups of the herring, normal sera of different animals, rabbit antisera to herring erythrocytes as well as lectins have been used (Altukhov et al. 1968; Zenkin 1969, 1971, 1972, 1973, 1974, 1978;

Truveller and Zenkin 1977a). Testing of the three-allele hypothesis, using the Hardy-Weinberg equation in this case, can be carried out solely with a very rough approximation. An attempt of this kind has nevertheless been undertaken (Altukhov et al. 1968), but as has already been noted (Truveller and Zenkin 1977a) the errors in calculations of theoretical frequencies were too great.

The populational analysis of different herring races has revealed important genetic differences between the populations belonging to the Atlantic and Baltic groups (Altukhov et al. 1968; Zenkin 1974; Truveller and Zenkin 1977b). North Sea herrings and particularly those of the Northwest Atlantic are generally characterized by the low frequency of the groups A_2 (0–0.06) and A_0 (0–0.08); the null phenotype is completely absent in a number of populations. In the Baltic herring the frequency of the A_2 groups is somewhat elevated (0.03–0.10); in all samples with the exception of just one the null group is present. Correspondingly, the maximum frequency of herrings with the group A_1 is characteristic of Northwest Atlantic herrings and is minimal for most of the Baltic Sea populations (Truveller and Zenkin 1977b). Within each of these large distribution areas of the oceanic herring subpopulations with characteristic frequencies of the phenotypes are found. We shall cite just one example with the herrings of the North Sea:

The Norwegian trough, spring spawning: $p_{A_1} = 1.00$ (n=249);
Fair Isle: $p_{A_1} = 0.89$; $p_{A_2} = 0.02$; $p_{A_0} = 0.09$ (n=127).

Truveller and Zenkin (1977b) have pointed out that there is an "... ordered differentiation stable in time with regard to the occurrence of erythrocyte antigens of the A-system in each of the regions studied" and they emphasize the correspondence between the herring differentiation with respect to the A-system and the pattern of the intraspecies ecological structure of this species (see p. 248). It appears that all the subspecies of the herring (Atlantic, White Sea, Baltic, Pacific) are polymorphic for the antigens of the A-system. In all the intraspecies groups only one type (A_1) predominates, but two other types are still present, however in relatively low numbers. The adaptive value of this persisting variability of blood groups which is constant in time is not clear at present, but it is difficult to attribute it to the action of some statistical random factors.

The Anchovy (Engraulis encrasicholus). Two reproductively isolated races, the Black Sea and the Azov Sea anchovy are found in the USSR aquatoria. Morphologically they are practically indistinguishable, but the Black Sea anchovy is somewhat longer than the Azov variety. Their biology, however, shows some differences: the Black Sea anchovy spawns and feeds in the Black Sea although periodic summer migrations of some shoals of this species to the Azov Sea in the summer months are possible. The Azov Sea anchovy spawns and spends all the summer in the Azov Sea, migrating to the Black Sea only for wintering, near the Caucasian and Crimean coast. In spring they return to the Azov Sea through the Straits of Kerch (Nikolsky 1950; Mayorova and Chugunova 1954). Immunogenetic studies have made it possible to detect a system of blood groups similar to the A system of herrings in this species. In the Black Sea anchovy the presence of two alleles of one locus (A_1 and A_2) corresponding to the three genotypes A_1A_1, A_1A_2

and A_2A_2 is assumed (Limansky 1964; Altukhov et al. 1969; Limansky and Payusova 1969).[4]

Two phenotypes (antigens), A_1 and A_2, have been detected using mainly normal pig and horse sera; the individuals belonging to the A_1 group can be subdivided into two further groups by the intensity of the agglutination reaction. According to the suggestion put forward by the authors, the three phenotypes correspond to the following three genotypes of the A locus:

Genotypes	A_1A_1	A_1A_2	A_2A_2
Antigens	A_1	A_1A_2	A_2
Degree of erythrocyte agglutination	Strong, by both sera	Weaker, by both sera	Only by the pig serum

The first and second group are difficult to distinguish, therefore in most studies they are combined into one heterogeneous group A_1. Control calculations have been made, using samples divided into the three phenotypes. We shall now compare the empirical and theoretical frequencies for the sum total of the samples collected during 1963–1966 in the Black Sea (Altukhov et al. 1969):

	A_1A_1	A_1A_2	A_2A_2	
Observed	427	181	25	n = 633
Expected	423.5	188.5	21.0	$q_{A_2} = 0.182$.

The small deficit of the heterozygotes appears to stem from the mixing of subpopulations differing somewhat in the frequencies of the two alleles.

The Azov Sea anchovy have a third null allele (A_0); the erythrocytes of individuals homozygous with respect to this allele are not agglutinated by either antiserum. The two-allele system is here transformed into a three-allele one of the AB0 type. Separation into the four phenotypes (A_1, A_1A_2, A_2, A_0) has been carried out by Altukhov (1974) for the samples collected in June 1965 and has shown for most cases a close correspondence between the frequencies observed and the theoretical ones. After the summation of the data for all the samples a considerable deficit of heterozygotes was observed, as was to be expected. The frequencies of the alleles for the Azov Sea race were as follows:

$p_{A_1} = 0.395$; $q_{A_2} = 0.132$; $r_{A_0} = 0.473$.

Thus, the two races of anchovy differ significantly from one another in the A system of blood groups. Within each race many smaller groups have been found,

4 The same genotypes have been detected in the anchovy near the west coast of Africa (Limansky 1969)

differing in the frequencies of two (the Black Sea race) or three (the Azov Sea race) alleles; but the significance of these groups (in a number of papers they are called the elementary populations) has not been finally elucidated. In certain publications they were taken for reproductively isolated populations, however, it appears that there are insufficient arguments in support of this. Most probably these represent just large groups (shoals) of fishes brought together temporarily by a common origin and similar size.

In some years (for example in 1966) the Black Sea anchovy migrate to the Azov Sea; this results in the formation of mixed groups containing individuals from the two races with intermediate frequencies of the A alleles (Altukhov 1974). A similar mixing of races may also take place in the Black Sea during wintering. Unfortunately, the inmmunogenetic analysis of different races of anchovy has not been accompanied by any search for morphological differences or by labelling. It is difficult to judge of the degree of isolation of the two races on the basis of the data on allele frequencies of just one locus. In this case, it is impossible to exclude the possibility of a large variation among the subpopulations within each of these races. Additional comprehensive genetic, morphological and biological studies are necessary for the solution of this problem.

The Cod (Gadus morhua). The clinal variation of the cod with regard to the frequencies of the number of genes, including the alleles of the haemoglobin locus and blood groups A and E has been well traced in the Norwegian Sea off the Norwegian coast (Møller 1966, 1967). The study of the cod biology in this region and of the structure of the otolytes found in this species has permitted two populations to be distinguished, a coastal and an Arctic one; they appear to be reproductively isolated, but cannot be distinguished morphologically. These two groups are regarded as twin species. The clinal variability is explained by the increase in the relative abundance of the Arctic codfish variety in the northeast direction (Møller 1968, 1969).

The differences between the coastal and arctic races (or between the twin species) are particularly clearly expressed in the frequency of the E^+ antigen. If each population of the cod is divided into two groups by the criterion of the otolyte structure, the frequency of the E^+ antigen will be equal to 0.09–0.34 in the arctic group and to 0.60–0.90 in the coastal group. This difference in frequencies is significant for all samples. The difference in the frequencies for the blood group A is lower – the variation in the frequencies of the A^+ phenotype ranges from 0.40 to 0.61 for the arctic cod and 0.55–0.81 for the coastal one. In spite of the lesser differences and some overlapping of the variation the difference in frequencies in this case as well is almost always statistically significant.

The incidence of the blood groups A^+ and E^+ turned out to be correlated with the allele frequencies of Hb and Tf loci (Møller 1968). Møller has suggested that the reproductive isolation of the two forms of cod which appear to spawn in the same regions and at one and the same time was determined predominantly by behavioural mechanisms, but these require special studies.

The finding of two sympatric populations of cod that did not mix with each other was of important practical significance. However, additional studies are necessary for the final proof of the existence of the twin species.

The Tuna (Thunnus thynnus, Th. alalunga, Th. albacares, Th. obesus, Katsuwonus pelamis). Big oceanic and sea fish species belonging to the suborder Thunnoidei are of great economic importance for Japan and several other Asian countries. In this connection, the population analysis of these fish species is particularly significant. The finding of dividing lines between the local populations of the tuna was greatly facilitated by studies of the blood group variability.

Four systems of the erythrocyte antigen have been found in the albacore (*Th. alalunga*) (Suzuki et al. 1958, 1959; Suzuki 1962; Keyvanfar 1962; Fujino 1970). These include: (1) C, phenotypes + and –; (2) Tg, type AB0 with four main phenotypes (three or four alleles); (3) G, type AB0, three phenotypes; (4) the Keyvanfar system also represting the AB0 type, four phenotypes. The differentiation of blood groups in the albacore has been achieved using iso- and heteroimmune antisera and lectins. The albacores of the Atlantic and Pacific differ in the frequency of alleles of several systems such as Tg (in the tunnies of the Atlantic an additional phenotype Tg-3 is present), by the Keyvanfar system and others. The Atlantic population differs from the Mediterranean one in the frequency of occurrence of serum antigens.

In the Pacific albacore the frequencies of the phenotypes of the C-system were found to differ between the northern and southern parts of its habitation area. Meanwhile the gene frequencies of different systems in tunnies were quite similar over large areas of the ocean. Apparently because the albacore is an extremely mobile fish, the Pacific subspecies is subdivided into only a very low number of not more than three or four reproductively isolated subunits.

The variability of erythrocyte antigens in the big-eye tunny *Thunnus obesus* is rather similar to that found in the albacore. Three systems have been found in this species (Suzuki and Morio 1960; Suzuki 1962; Sprague et al. 1963; Fujino 1970). These include: (1) the Tg-locus with three or more codominant alleles; (2) the C-locus probably with three alleles (phenotypes C_1, C_2, C_3 and "–"); (3) the AB0 system with four phenotypes (A, B, AB, 0), and this appears to be the three-allele system.

An analysis of the frequencies of the AB0 system phenotypes has shown a rather close correspondence between the empirical and the theoretical values with a small excess of heterozygotes (Sprague et al. 1963):

	A	B	AB	0	n
Observed	42	56	10	319	427
Expected	48.3	61.9	4.3	312.5	

The phenotype distribution according to the Tg system deviated markedly from the one expected; most probably, this was due to a discordance between the assumed hypothesis and the actual pattern of inheritance in this complex system. In a related species *Parathunnus mebachii* a polyallelic X system containing nine phenotypic groups has been found (Suzuki 1967).

A number of studies have been devoted to the examination of blood groups in the skipjack *Katsuwonus pelamis*. No less than four systems of antigens have been found in this moderate-sized widely occurring species (Cushing 1956; Sprague and Holloway 1962; Fujino 1969, 1970). These include: (1) a C-system containing phenotypes + and –; (2) an H-system, also having two phenotypes; (3) a B-system having the phenotypes K_1, K_2 and K_0 (Fujino, 1970); according to other authors up to six phenotypes can be distinguished in this system (Sprague and Holloway 1962); (4) finally, the Y-system should be mentioned with 15 phenotypes, this system appears to possess six codominant alleles.

A good review of the populational studies with the skipjack has been published by de Ligny (1969) and Fujino (1970). Differences in the frequencies of several phenotypes have been found for the blood groups B and Y (K_1 in the B-system, Y^Y in the Y-system) between the Atlantic and Pacific populations. Western populations of the Pacific are isolated and have a somewhat decreased proportion of the K_1 phenotype. The central regions of the Pacific are apparently occupied by a single poorly differentiated population. The existence of a reproductively isolated population in the Eastern part of the ocean remains doubtful. These conclusions have been confirmed in the studies on the polymorphism of a number of proteins, esterase in particular. Thus, in the case of the skipjack, no small local isolated populations appear to exist; similar blood group frequencies are found in samples collected in geographically distant regions of the ocean (Fujino 1976).

The variability of the erythrocyte antigens in other species of tunnies is generally similar to that found in *Th. alalunga, Th. obesus* and *K. pelamis.* Several systems of erythrocyte antigens (which have not, however, been sufficiently studied) have been found in the yellow-fin tunny *Th. albacares*; one rather distinct system having six or seven phenotypes including the null group has been found in the tunny *Th. thynnus* (Suzuki 1962; Lee 1965).

Many other fish species polymorphic with respect to blood groups have been found. In particular, among them mention can be made of the sardine (Sprague and Vrooman 1962), four species of Pacific salmon (Ridgway 1958, 1966; Ridgway et al. 1959; Ridgway and Klontz 1961), the rainbow and the gold trout (Ridgway 1962, 1966; Calaprice and Cushing 1967), and the roach (Balachnin and Zrazhevskaya 1969). Individual differences in the blood groups have been detected in more than 100 fish species beginning from sharks and skates up to the most evolutionary advanced families of teleosts. It may be assumed that such differences are characteristic of all fish species.

Studies of blood groups in fishes lead to a number of generalizations.

The variability in blood groups is characteristic of fishes just as of other vertebrates. Heredity of the antigenic differences is based on allelic, frequently multiple, systems of genes codominant and sometimes dominant with respect to each other. Many of these systems contain the null alleles or genes, the products of which cannot be detected.

The main technique for differentiating blood groups makes use of specific reagents inducing the agglutination of erythrocytes in the blood samples of individuals possessing a certain genotype and unable to agglutinate erythrocytes from individuals with other genotypes. The best results are generally obtained with iso- or heteroimmune sera and with lectins in a number of cases.

In many fish species populations may differ significantly from each other in the allelic frequencies of blood group loci. Within each population the frequencies of the phenotypes and genotypes generally follow the Hardy-Weinberg law. Any deviations from the equilibrium may be due to the erroneously selected genetic hypothesis, to deviations from panmixia, to the small size of the population or lastly to the relatively strong impact of natural selection. The final verification of the correctness of a specific genetic model for a given situation requires hybridological analysis.

Protein Polymorphism Among Fishes: The Background

New improved techniques of protein separation by electrophoresis in starch and polyacrilamide gels were developed in the mid-fifties (Smithies 1955). Soon after that, multiple forms of different enzymes (differing in the mobility in gels and termed isozymes) were discovered (Hunter and Markert 1957; Markert and Møller 1959). It turned out that most enzymes and many proteins devoid of enzymatic activity (haemoglobins, for example) were present in the organism in several forms. Many of these forms are genetically encoded and differ in the primary structure (amino acid sequence) and other peculiar features, retaining a common functional specificity, however. At present, the term isozymes is used to designate different genetic variants of enzymes (Salmenkova 1973; Markert 1975; Korochkin et al. 1977) in contrast to the non-genetic variants of proteins or conformeres. Another widely used word, "allozyme", is employed to designate products coded by different alleles of one and the same gene. The terms "isoforms" and "alloforms" are sometimes used for non-enzymatic proteins.

The use of electrophoresis to detect allelic and non-allelic forms of proteins has lead to a revolution in the populational and evolutionary genetics of man, animals, plants, and microorganisms. Of outstanding significance was the possibility of conducting a precise analysis of the genetic structure of populations, since in almost all cases electrophoresis made it possible to differentiate between the heterozygotes and the homozygotes and to quantitate the number of individuals with different genotypes.

In the case of monomeric proteins devoid of a quaternary structure, each homozygote shows only one protein band in the gel after specific staining. If the charge of two allozymes (alloforms) is different, then the two homozygotes differ in the location of such a protein band. In heterozygotes both products are formed (codominance) and respectively one can find two bands in the gel, each of which is stained less intensely than the bands present in the homozygotes (Fig. 42 a).

When the protein is a dimer, that is two polypeptide chains are present in the protein globule, three bands are generally found in heterozygous individuals. Two of them correspond to the pure protein species (just as in the case of homozygotes); the third species located between them represents a "hybrid" and consists of two different polypeptides coded by different alleles (Fig. 42 b). A number of dimeric proteins does not possess hybrid products; this may perhaps be explained by the different time of synthesis of these allelic forms in heterozygotes or by the different compartmentation of their synthesis (Ferris and Whitt 1978 a). In the presence of two homologous loci (resulting from gene duplication), which code for proteins or

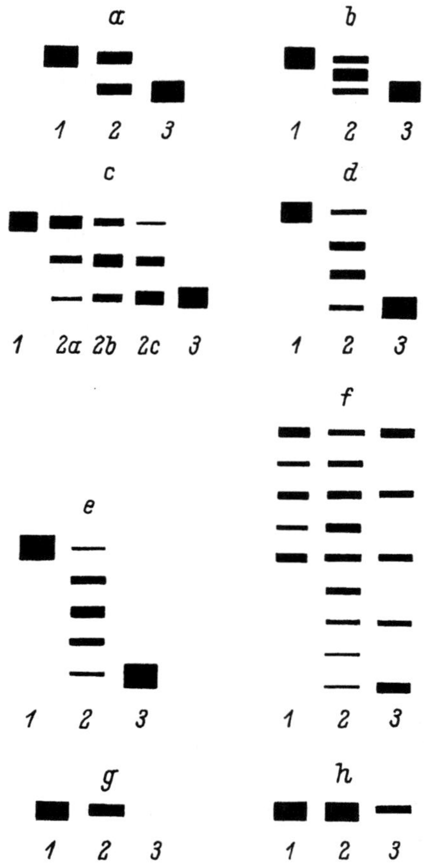

Fig. 42a–h. Most common types of electrophoregrams of polymorphic proteins in fishes (schematic). **a** monomeric protein; **b** dimer, heterozygotes with one hybrid product ($A_1^1A_1^2$); **c** dimers, two duplicated loci with similar alleles; **d** trimer, two hybrid products in heterozygotes ($A_1^1A_2^2$ and $A_2^1A_1^2$); **e** tetramer, with three hybrid products in heterozygotes ($A_1^1A_3^2$, $A_2^1A_2^2$, $A_3^1A_1^2$); **f** the same but in the presence of two non-allelic genes (one of them heterozygous); **g** variation with respect to the null allele; **h** variation for a regulatory gene. *1*, *3* homozygotes; *2* heterozygotes

isozymes possessing identical mobilities, polymorphism in similar alleles of both loci results in the appearance of five genetically distinguishable variants in the population. In the heterozygotes with respect to one of the loci the proportion between the three isozymes is equal to 9:6:1; in the heterozygotes with respect to two loci the proportion is equal to 1:2:1 (Fig. 42c).

In rare cases, when the protein is a trimer, four isozymes are present in gels with the proportion of components being equal to 1:3:3:1 (Fig. 42d).

The formation of variants of tetrameric proteins in homozygotes and heterozygotes can proceed in a different way. In the simplest case, when only one gene is present in the genome each homozygote carries just one variant (tetramers A_4^1 or A_4^2); five species of isozymes are synthesized in heterozygotes. Polypeptides coded by two alleles may be combined with each other, forming the following homomeric and heteromeric molecules:

$$A_4^1, \quad A_3^1A_1^2, \quad A_2^1A_2^2, \quad A_1^1A_3^2, \quad A_4^2.$$

The relative activity of these isozymes after electrophoresis is most frequently equal to 1:4:6:4:1, suggesting that the combination of polypeptide A^1 and A^2 occurs at random (Fig. 42 e). Such a pattern was found in particular for glyceraldehyde-3-phosphate dehydrogenase in hybrids of *Xiphophorus maculatus* and *X. helleri;* it has provided evidence of the tetrametic structure of the enzyme (Wright et al. 1972). The two parental forms had different alleles of the G3Pdh locus.

Many tetrameric proteins in a given individual are coded by two or more genes. In the case of animal haemoglobins (including many fish species) and human haemoglobin as well, the tetramer is formed by the combination of two different homodimers, the polypeptide chains of which are coded by different genes ($\alpha_2\beta_2$, $\alpha_2\gamma_2$ etc.). Each individual has several types of such molecules; as a result, the spectrum of haemoglobins after electrophoresis is very complex even in homozygotes and becomes more complex in heterozygous individuals. In the fish genome such enzymes as LDH are coded by several loci (by no less than two). The products of the two main genes, as in the case of allelic products, are combined, yielding five isozymes (less frequently only three of them are formed):

A_4, A_3B, A_2B_2, AB_3, B_4 or A_4, A_2B_2, B_4.

If one of the genes is in the heterozygous state, 15 isoenzymes are formed instead of 5 and 35 molecular species are present in double heterozygotes. The number of bands actually resolved by electrophoresis is generally lower because electric charges may be identical in some of the species (Fig. 42 f). In salmonids for example, we find in most cases only nine clearly discernible bands in heterozygotes with respect to one of the LDH loci.

A particular situation is found when one of the alleles of a protein locus does not give any product at all or when this product is inactive and cannot, therefore, be detected. In the presence of such a null allele no bands are present on the electrophoregramm in the homozygous state; in heterozygotes and in homozygotes possessing the active allele only one band is found. This band is more intense in homozygotes. Polymorphism with respect to the null allele can easily be distinguished from other types of protein polymorphism (Fig. 42 g).

A genetic analysis of allozymes (alloforms) with identical mobility in the electric field, but differing in the intensity of band staining makes it possible to detect a variation in the regulatory genes stimulating or weakening the action of one structural locus or another (Fig. 42 h). As can be judged from the *Drosophila* studies (McDonald and Ayala 1978; Ayala and McDonald 1980), the variation in regulatory genes is as common in population as the variation in the structural genes.

Recently a large number of "temperature variants" of isozymes have been found in *Drosophila.* These variants represent the allelic forms with identical mobility (and therefore a similar electric charge), but differing in the stability to heating (Bernstein et al. 1973; Singh et al. 1975; Coyne et al. 1978; Lewontin 1978). The detection of isozymes with different stability to heating is possible when electrophoresis is combined with the controlled heating of tissue homogenates or sera up to temperatures denaturing proteins or inhibiting the enzymatic activity with a subsequent hybridological analysis of the variants detected in this way. The alleles yielding products with different heat stability appear to be present in fishes as well. For example, in the sockeye salmon *Oncorhynchus nerka* the allele B^1 of Ldh-B1 locus codes for isozymes with the identical electrophoretic mobility, but different heat stability (Allendorf and Utter 1979). No genetic analysis of this situation has been conducted yet, but one can assume the involvement of two alleles with differing heat stability for the northern and southern population in the American Pacific coast. The frequency of alleles having different heat stability in fishes is apparently just as high as among insects.

The identification of genotypes on the basis of electrophoregramms in fishes just as in other organisms is generally conducted in a very simple manner. Problems may emerge when different conformational states of one and the same allozyme (alloforms) are present, as well as when bands corresponding to different alloforms are located close to one another or even coincide completely. In complex cases, one has to use immunogenetic methods, in particular the treatment of tissue homogenates by the antiserum containing antibodies to certain types of the protein investigated. Of no less significance is the hybridological analysis – the crossing of individuals with different protein phenotypes followed by the electrophoretic examination of the offspring. Such crosses are not generally time-consuming because segregation for proteins is not infrequently detectable in the embryos and the larvae.

The techniques of protein electrophoresis in starch and polyacrilamide gels have been described in great detail in a number of manuals (Shaw and Prasad 1970; Brewer and Sing 1970; Maurer 1971; Gordon A. 1975; Serov et al. 1977) and we shall not discuss these questions here. It should be pointed out, however, that chambers of different types are used for the genetic analysis of protein; the gel can be polymerized either in tubes (polyacrilamide) or in slabs (starch, polyacrylamide). In experiments with polyacrilamide gels, tubes and slabs are generally located vertically in the electrophoretic apparatus. To prevent excessive heating the buffer solution in two electrode compartments must be cooled.

The use of slabs makes it possible to analyze up to 50 or even up to 100 or more samples simultaneously, which is of tremendous significance for genetic studies. Chambers with an adjustable gel thickness have been used (Truveller and Nefedov 1974). In routine standard analyses simplified slab chambers designed by Pospelov proved to be very convenient (Fig. 43). The gel is initially polymerized between two glasses in the horizontal position. When the slab thickness does not exceed 2 mm the solution introduced by a pipette into the space between the glasses does not leak out due to capillary forces.

In a number of cases good results may be quickly obtained by electrophoresis of the proteins in the agar gel or using acetyl cellulose strips (Gauldie and Smith 1978), however, the starch and polyacrilamide gel electrophoresis provide a much higher resolving power. Even greater resolution of the proteins may be achieved by isoelectric focussing in the polyacrilamide gel (Kühnl and Spielmann 1978).

The advantages inherent in the electrophoretic separation of genetic protein variants led to the rapid implementation of this technique in population genetic studies and to the rapid development of the biochemical genetics of populations. At the same time, protein loci have become widely used as genetic markers in selective breeding.

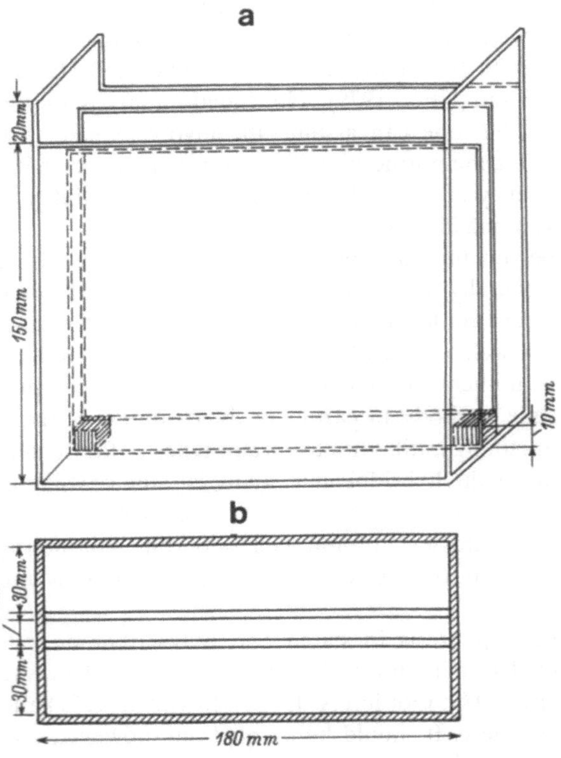

Fig. 43 a, b. Chamber for the vertical slab gel electrophoresis in polyacrylamide gells (design of V. A. Pospelov). **a** side view; **b** view from above

The General Level of Polymorphism

It has been shown in experiments with *Drosophila* that the degree of polymorphism in populations and species can be quantitatively determined simultaneously for many proteins (Hubby and Lewontin 1966; Lewontin and Hubby 1966). Soon thereafter studies of this kind were conducted with many other species. According to recent data (Lewontin 1974; Powell 1975; Selander 1976; Nevo 1978), the proportion of polymorphic loci is equal to about 45% in plants, 40–50% in invertebrate animals and 15–30% in vertebrates, including man. The proportion of heterozygous genes or the average number of genes present in the heterozygous state in a given individual related to the total number of loci was found to be equal to about 17% in plants, 12%–15% in invertebrates and 3%–8% in vertebrate animals and man.

A number of studies dealing with the estimation of the level of biochemical polymorphism has been conducted using various fish species (Table 24). Certain species such as the Oceanic herring and the smelt, the rainbow and golden trout, the eel and the cod, the killifish and the guppy, and some others show a very high variation in the loci coding for various proteins. A relatively lower variation has been observed for most species of salmonids; it should be pointed out that several authors have included in their calculation a large number of loci of non-enzyme proteins showing low variation such as haemoglobins and eye lens crystallins (Altukhov et al. 1972; Salmenkova and Volokhonskaya 1973; Omelchenko 1975 b). The calculation made for albumins, myogenes and enzymes shows higher mean values of variation, 19% for the number of polymorphic loci and 5% for the proportion of heterozygous genes.

But even with this correction the salmonids still remain a group showing a relatively low variation. The suggestion according to which this may be explained by their polyploid origin (Altukhov et al. 1972; Altukhov 1974) appears to be insufficiently argumented. In the same group of salmonids the rainbow trout displays a very high degree of variation, the increased level of variation is also characteristic of the polyploid common carp (Allendorf and Utter 1979; Paaver 1979). The relatively low level of protein polymorphism in the Pacific and Atlantic salmon, as well as in some other taxons (several species of catostomids and cobitids, cave-dwelling fishes, some specialized families of the order Perciformes, etc.), may rather be a consequence of their more narrow ecological specialization or smaller population size. According to Nevo (1978), the mean values of heterozygosity for specialized and non-specialized animal species are equal to 0.037 and 0.071 respectively.

It is interesting to mention in this connection the genus *Astyanax* belonging to the family Characidae. The terrestrial varieties belonging to this genus show high variation; their close relatives dwelling in caves show minimal levels of polymorphism – in one of the populations no heterozygous individuals were found (Avise and Selander 1972). The same results have been obtained for the epigean and cave-dwelling species of the family Amblyopsidae (Swofford et al. 1980). Such monomorphism may be the result of the inbreeding inevitable in view of the low numbers of populations living in the caves, but it is more probable that it is related to the extreme constancy of the environmental conditions in the caves. Recently it has been established that the level of polymorphism in the group of cellular membrane proteins and histones was significantly lower (Manchenko and Nikiforov 1979). It may be added that the heterogeneity of the cell proteins in the clones of cultured human cells examined by two-dimensional electrophoresis and radioactive labelling was very low (McConkey et al. 1979). But even taking into account these recent observations, we have come to the conclusion that fishes are extremely polymorphic, just like other animal species. Assuming that the average fish genome contains no less than 10,000 active structural genes and that 2.5% of these genes are in a heterozygous state in each individual, we obtain about 250 heterozygous genes in one genome. Electrophoresis is only able to detect about one third of the alleles of each locus, consequently the level of variability in natural populations should actually be higher. Populations are indeed saturated by polymorphic genes represented by two, three and sometimes even by a markedly greater number of alleles. The origin and significance of such tremendous polymorphism within the populations will be discussed after all the data related to the genetics of the protein loci in fish have been reviewed.

Table 24. Level of protein variability in fish

Family	Genera and species	Number of loci	Number of polymorphous loci, P%	Mean heterozygosity, H̄%	Reference
1	2	3	4	5	6
Clupeidae	*Clupea harengus*	33	28–34	11.3	32, 33
Osmeridae	*Hypomesus olidus*	28	31–38	11.3	32, 33
Salmonidae	*Salmo salar*	15–30	20	2.4	1, 18
	S. gairdneri	19–23	26	3.7	39
	S. gairdneri	10	50	12.1	14
	S. gairdneri	30	50	6.0	1
	S. gairdneri (hatcheries)	24	–	10.3	8
	S. clarkii, coastal	30	–	6.3	1
	S. clarkii, internal	30	50	2.3	1
	S. clarkii	23	–	5.0	23
	S. aguabonita	10	50	13.3	14
	S. apache	25	0	0	1
	Salvelinus fontinalis	13	38	–	41
	S. alpinus	34	11	3.6	33
	S. leucomaenis	38	8	2.4	33
	S. leucomaenis	–	13–15	2.5	29
	S. malma	–	13–15	–	32
	Oncorhynchus keta	20–25	18	4.5	1
	Oncorhynchus keta	46	11	2.0	2
	O. gorbuscha	30	15–17	2.9	29
	O. gorbuscha	20–25	–	3.9	1
	O. nerka	20	30	4.6	42
	O. nerka	–	14–15	4.6	29
	O. tschawytscha	30	–	3.5	1
	O. kisutch	20–25	–	1.5	1
	O. masu	–	14–17	4.4	29
	O. spp. (5)	19–23	9–13	1.2	39
	Coregonus clupeaformis	–	22–28	7.7	20
Cyprinidae	*Rhinichthys cataractae*	21	15	5.4	24
	Lavinia exilicauda	24	25	5.3	3
	Hesperoleucus symmetricus	24	25	6.8	3
	California minnows (9)	24	13	3.8	3
	Campostoma spp. (3)	19	19	5.6	10
Catostomidae	*Thoburnia rhothoeca*	34	18	5.5	9
	Th. hamiltoni	34	6	1.5	9
	Th. atripinnae	34	6	0.8	9
Cobitidae	*Cobitis delicata*	20	10	1.1	19
Characidae	*Astyanax* sp., epigean	17	29–41	11.2	4
	Astyanax sp., cave-dwelling	17	0–10	3.6	4
Amblyopsidae	*Chologaster cornuta*, epigean	19–22	–	4.0	38

Table 24 (continued)

Family	Genera and species	Number of loci	Number of poly-morphous loci, P%	Mean hetero-zygosity, H%	Refer-ence
1	2	3	4	5	6
	Ch. agassizi, facult. epigean	19–22	–	2.8	38
	Typhlichthys subterraneus, cave-dwelling	19–22	–	1.9	38
	Amblyopsis spelaea, cave-dwelling	19–22	–	0	38
	A. rosae, cave-dwelling	19–22	–	0.6	38
Anguillidae	*Anguilla anguilla*	20	65	18.1	31
	Conger conger	20	32	7.6	31
Cyprinodontidae	*Fundulus heteroclitus*	25	56	18.0	25
	Aphanius dispar	19	15	4.9	22
Poeciliidae	*Poecilia reticulata*	16	19	8.4	26
	Poecilia reticulata	16	25	10.4	34
	Poeciliopsis monacha	16	15	4.7	40
	P. occidentalis	16	11	1.8	40
	P. lucida	16	5	1.6	40
Gadidae	*Gadus morhua*	15	30	8.0	11
	Theragra chalcogramma	25	32	–	15
Macrouridae	*Coryphaenoides acrolepis*	25	16	3.3	35
Centrarchidae	*Lepomis* spp. (10)	14–15	16	5.9	5
	Pomoxis, Micropterus a. oth. spp. (9)	11–14	–	3.0	6
Pomacentridae	Differ. spp. (3)	20–29	15	8.3	37
Labridae	*Halichoeres* spp. (3)	28	–	5.7	37
Sparidae	*Chrysophrys auratus*	23	17–26	8.2	36
Clinidae	*Gibbonsia metzi*	28	18	4.3	37
Zoarcidae	*Zoarces viviparus*	32	28	8.9	12
Cichlidae	*Cichlasoma cyanoguttatum*	13	0	0	21
	Cichlasoma spp. (2)	13	15	3.5	21
	Petrotilapia tridentiger	14	14	6.9	21
	Pseudotropheus spp. (4)	15	15	6.8	21
Gobiidae	*Bathygobius ramosus*	23	–	0.5	37
	Gillichthys mirabilis	29	20	4.6	37
Notothenidae	*Trematodus* spp. (3)	21–26	9	2.1	37
Scorpaenidae	*Sebastes* spp. (3)	23	4–8	2.6	17
	Sebastolobus altivelis	20	30	4.7	35
	S. alascanus	20	20	4.9	35

Table 24 (continued)

Family	Genera and species	Number of loci	Number of polymorphous loci, P%	Mean heterozygosity, $\bar{H}$%	Reference
1	2	3	4	5	6
Mugilidae	*Mugil cephalus*	30	20	7.1	37
Atherinidae	*Leuresthes tenuis*	33	15	3.6	37
	Menidia spp. (5)	24	10	5.1	16
Pleuronectidae	*Pleuronectes platessa*	45	51	10.2	7
	Platichthys stellatus	32	56	9.7	13
	Kareius bicoloratus	32	25	8.9	13
	Mean, 88 (P) – 105 ($\bar{H}$) species	–	18.7	5.0	–
Different families	37 species	–	–	5.8	30
	51 species	–	15	5.1	28
	64 species	–	–	4.8	27

References: 1. Allendorf and Utter (1979); 2. Altukhov et al. (1972); 3. Avise and Ayala (1976); 4. Avise and Selander (1972); 5. Avise and Smith (1974); 6. Avise et al. (1977); 7. Beardmore and Ward (1977); 8. Busack et al. (1979); 9. Buth (1979b); 10. Buth and Burr (1978); 11. Cross and Payne (1978); 12. Frydenberg and Simonsen (1973); 13. Fujio (1977); 14. Gall et al. (1976); 15. Iwata and Numachi (1979); 16. Johnson M.S. (1976); 17. Johnson et al. (1973); 18. Khanna et al. (1975b); 19. Kimura (1978b); 20. Kirkpatrick and Selander (1979); 21. Kornfield and Koehn (1975); 22. Kornfield and Nevo (1976); 23. Loudenslager and Kitchin (1979); 24. Merritt et al. (1978); 25. Mitton and Koehn (1975); 26. Nayudu (1975); 27. Nei et al. (1978); 28. Nevo (1978); 29. Omelchenko (1975b); 30. Powell (1975); 31. Rodino and Comparini (1978); 32. Salmenkova and Omelchenko (1978); 33. Salmenkova and Volokhonskaya (1973); 34. Shami and Beardmore (1978a); 35. Siebenaller (1978); 36. Smith et al. (1978); 37. Somero and Soule (1974); 38. Swofford et al. (1980); 39. Utter et al. (1973a); 40. Vrijenhoek (1979a); 41. Wright and Atherton (1970); 42. Utter et al. (1980)

The Genetics of Nonenzymatic Proteins in Fishes

Transferrins. The transferrins participating in the transport of the iron necessary for the biosynthesis of the haemoglobin molecules play an important part among beta-globulins of the blood serum. The ease of detection of transferrins by electrophoresis and the surprising simplicity of inheritance of the transferrin (Tf) locus probably facilitated the appearance of many papers dealing with the polymorphism of this protein (Table 25). In the majority of fish species the Tf locus is variable, the number of alleles varies within a range of from 2 to 13 (more often than not it is equal to 3–4). Polymorphism has been found in all the cyclostomes and fish taxons examined. Transferrin represents a monomeric protein with a molecular weight of some 70,000 (Valenta et al. 1976a; this determination has been made for common carp transferrin). The electrophoresis gel patterns of the material from homozygous

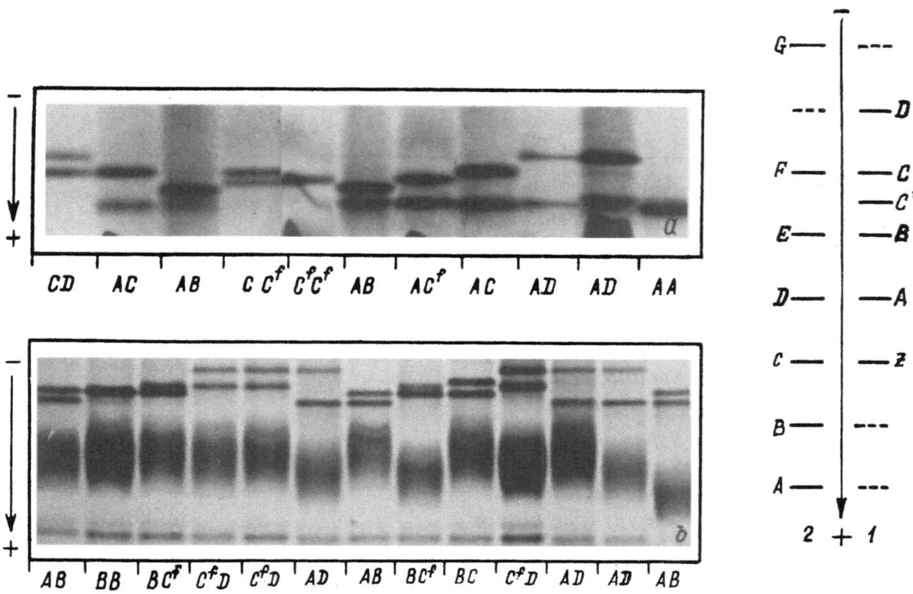

Fig. 44 a, b. Transferrins in the common carp (Cyprinus carpio). **a** the Ural carp; **b** the carp of the Moscow region (photography K. A. Truveller). Designation: *1* according to Moskovkin et al. (1973); *2* according to Valenta et al. (1976 a). (Photography of Yu. I. Tscherbenok)

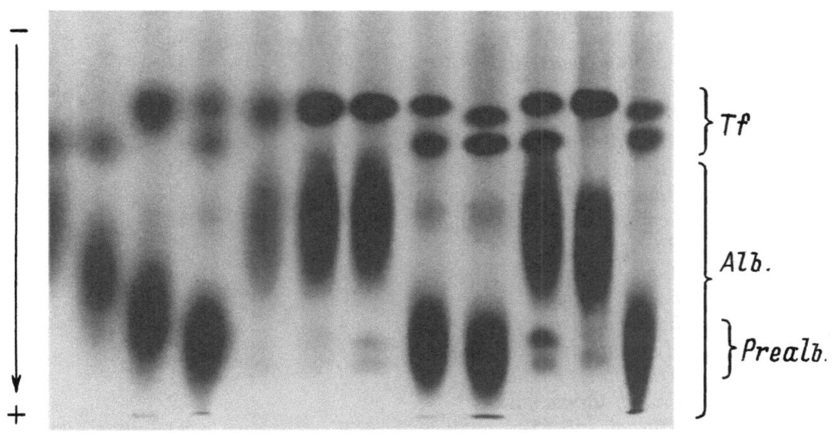

Fig. 45. Variation of transferrins (*Tf*), albumins (*Alb.*) and prealbumins (*Prealb.*) in the Ropsha common carp. (Photography of Yu. I. Tscherbenok)

animals generally contain only one band; in the case of heterozygous individuals two bands are usually present (Figs. 44 and 45). In several fish species, for example, in the Atlantic salmon (*Salmo salar*) two bands are present in the homozygotes instead of one, and four bands can be seen in heterozygotes.

No second Tf locus has been found in any of the fish species studied: only one locus is present or is active in the genome and usually it has several alleles

Table 25. Polymorphism of cyclostomes and fishes in the transferrin locus

Family	Genus, species	Number of alleles (q > 0.01)	Reference
Hagfishes (Myxinidae)	*Myxine glutinosa*	2[a]	20
Petromyzonidae	*Petromyson marinus*	2	6
Scylliorhynidae	*Scylliorhinus* spp. (2)	2	6
Acipenseridae	*Acipenser güldenstädti*	2[a]	2
	A. stellatus	3	7
Clupeidae	*Alosa aestivalis*	3	22
	Sprattus sprattus	3	50
	Clupea harengus	2–4	30, 40
Salmonidae	*Salmo salar*	4	24, 34, 51, 52
	S. gairdneri	2	44, 45
	Salvelinus fontinalis	3	53
	Oncorhynchus kisutch	3	38, 43, 45
	Coregonus lavaretus	3	33
	C. nasus	2	18
	C. albula	2	18
Argentinidae	*Argentina silus*	2[a]	27
Cyprinidae	*Cyprinus carpio*	8	1, 3, 8, 29, 46, 47, 49
	Carassius carassius	5	35, 47
	C. auratus gibelio	7	35
	Tinca tinca	3	46, 47
	Leuciscus leuciscus	9	47
	L. cephalus	7	47
	L. idus	2	47
	Abramis brama	7	47
	Rutilus rutilus	5	37
	Alburnus alburnus	9	47
	Scardinius erythrophthalmus	6	47
	Blicca bjorkna	5	47
	Aspius aspius	2	36
	Barbus barbus	10	47
	B. meridionalis petenyi	4	49
	Hypophthalmichthys molitrix	3–5	31, 47
	Aristichthys nobilis	3–5	31, 47
	Chondrostoma nasus	2	47
	Notropis spp. (5)	9	23
Catostomidae	*Catostomus commersoni*	2	5
	Ictiobus cyprinellus	2	15, 16
Anguillidae	*Anguilla anguilla*	4	9, 32
	A. rostrata	3	9
Ictaluridae	*Ictalurus melas*	4	21
Gadidae	*Gadus morhua*	13	13, 14
	G. virens	2	26
	G. polachius	2	25
	G. merlangus	2	25
	G. aeglefinus	3	25
	Merluccius productus	4	42

Table 25 (continued)

Family	Genus, species	Number of alleles (q > 0.01)	Reference
Serranidae	*Morone saxatilis*	2	28
Cichlidae	*Tilapia* spp. (2)	3	19
Thunnidae	*Thunnus albacares*	3	4, 11
	Th. alalunga	3	10
	Th. oxilunga	3	4
	Th. thynnus maccoyii	2	10
	Katsuwonus pelamis	3	11
Scombridae	*Sarda chiliensis*	2	4
Pleuronectidae	*Pleuronectes platessa*	3	17
	Hippoglossus stenolepis	4	41
Tetraodontidae	*Opsanus tau*	3–4	12

[a] The precise number of alleles is unknown

References: 1. Balachnin and Galagan (1972b); 2. Balachnin et al. (1972); 3. Balachnin et al. (1973); 4. Barrett and Tsuyuki (1967); 5. Beamish and Tsuyuki (1971); 6. Boffa et al. (1967); 7. Chikhachev and Tsvetnenko (1979); 8. Creyssel et al. (1966); 9. Drilhon and Fine (1971); 10. Fujino (1970); 11. Fujino and Kang (1968); 12. Fyhn and Sullivan (1971); 13. Jamieson (1975); 14. Jamieson and Turner (1978); 15. Koehn (1969b); 16. Koehn and Johnson (1967); 17. de Ligny (1966); 18. Lokshina (1980); 19. Malecha and Ashton (1968); 20. Manwell (1963); 21. Marneux (1972); 22. McKenzie and Martin (1975); 23. Menzel (1976); 24. Møller (1970); 25, 26. Møller and Naevdal (1966, 1974); 27. Møller et al. (1967); 28. Morgan et al. (1973); 29. Moskovkin et al. (1973); 30. Naevdal (1969); 31. Nenashev and Rybakov (1978); 32. Pantelouris et al. (1970); 33. Pavlú et al. (1971); 34. Payne (1974); 35. Polyakovsky et al. (1973); 36. Sedov and Krivasova (1973); 37. Sedov et al. (1976); 38. Suzumoto et al. (1977); 39. Tammert (1974); 40. Truveller et al. (1973a); 41. Tsuyuki et al. (1971); 42. Utter and Hodgins (1971); 43–45. Utter et al. (1970a, 1973b); 46. Valenta and Kálal (1968); 47, 48. Valenta et al. (1976a, 1977b); 49. Valenta (1978b); 50. Veldre and Veldre (1979); 51, 52. Wilkins (1971b, 1972); 53. Wright and Atherthon (1970)

corresponding to the different alloforms of transferrin. The polyploid species are no exception: in the common carp, in two species of the crucian carp, in the barbel *Barbus barbus* and in all species of salmonids and catostomids, only one locus has been found as well. This can be explained by the conservative main function of transferrin, i.e., iron transport. If the second locus emerges a result of polyploidization or regional duplication, it becomes unnecessary and is either destroyed sooner or later or ceases to be active. The presence of several allelic forms of transferrin in most fish species is not a random coincidence. Several authors relate this phenomenon to the second function of this protein, i.e., to its bactericidal or defense properties (Manwell and Baker 1970; Hegenauer and Saltman 1975); these bactericidal properties are more strongly expressed in the heterozygotes.

A genetic analysis condicted using the common carp and three species of salmonids has revealed a very simple and precise Mendelian inheritance of all the variants of transferrin. The results of several typical crosses are presented below:

the common carp:	♀CC×♂AB=98 AC+93 BC (Balachnin and Galagan 1972 a)
	♀AD×♂AA=40 AA+32 AD (Valenta et al. 1976 a) [5]
the rainbow trout:	♀AA×♂AC=32 AA+28 AC (Utter et al. 1973 b)
the coho salmon:	♀AC×♂BC=30 AB+28 AC+25 BC+29 CC (Utter et al. 1973 b)
the Atlantic salmon:	♀AC×♂AC=29 AA+52 AC+21 CC (Møller 1970)

All the alleles of the Tf locus are codominant, the proportions of phenotypes in the progeny are generally close to the expected ones, only in a few cases is there a certain deficit of individuals belonging to some classes. No null alleles have been found. Sometimes the deduction of the genotype from electrophoregramms is more difficult because of the presence of three bands instead of two; an analysis of one such case in the common carp has shown that one of the three variants results from a conformational change in the transferrin molecule. Conformational variants are apparently not rare, but their concentration is rather low, and in most cases they appear as weak "ghosts" (Valenta et al. 1977 b). The multiplicity of electrophoretically revealed molecular forms of transferrin in salmonids and some other fish species may be explained by the different number of sialic acid residues (from one to four) which can be added to the molecule (Herschberger 1970).

The proportion of different Tf alleles in fish populations is generally accurately described by the Hardy-Weinberg formula (Table 26), although deviations may sometimes be observed. Certain samples may be deficient in heterozygotes, but rarely there is excess of them. The deficit of heterozygotes can most probably be explained by the mixing of the population differing in the frequencies of the alleles (the Wahlund effect). For example, in the two samples of the Atlantic salmon from Newfoundland rivers the following equilibrium frequencies have been found (Møller 1970):

	AA	AC	CC
Indian River:			
Observed value	1	14	97
Expected value	0.6	14.8	96.6
Adis Spring:			
Observed value	21	56	43
Expected value	20.0	58.0	42.0

The mixing of these samples results in a large deficit of heterozygotes:

	AA	AC	CC
Observed	22	70	140
Expected	13.9	85.8	132.2

5 The designations of alleles have been changed in accordance with the designation of Moskovkin et al. (1973)

Table 26. Populational variation in transferrin types among fishes

Species	Site of sampling	Transferrin types and their frequencies[a]										n	Reference
		A	B	C	D	AB	AC	AD	BC	BD	CD		
Common carp	Danube	34	10	10	0	21	30	4	5	2	2	119[b]	2
Cyprinus carpio		(31.8)	(4.9)	(7.0)	(0.1)	(24.9)	(30.9)	(4.2)	(11.7)	(1.6)	(2.0)		
Atlantic salmon	Canada and USA	10	–	75	–	–	32	–	–	–	–	117[b]	4
Salmo salar	(Labrador, Maine)	(6.0)	–	(71.0)	–	–	(40.0)	–	–	–	–		
	Newfoundland	0	–	134	–	–	22	–	–	–	–	156	5
		(0.8)	–	(134.9)	–	–	(20.3)	–	–	–	–		
	England and	–	2	4288	–	–	–	–	124	–	–	4414	6
	Ireland	–	(0.9)	(4291.2)	–	–	–	–	(121.9)	–	–		
Brook trout	USA, Pennsylvania	2	54	249	–	21	36	–	240	–	–	602	8
Salvelinus fontinalis	(fish ponds)	(2)	(57)	(248)	–	(19)	(39)	–	(237)	–	–		
	Same, natural	–	1	125	–	–	–	–	14	–	–	140	8
	waterbodies	–	(0.5)	(124.9)	–	–	–	–	(14.6)	–	–		
Ocean herring	North Sea	34	79	–	–	103	–	–	–	–	–	216	7
Clupea harengus		(33.9)	(78.8)	–	–	(103.3)	–	–	–				
Cod	Norvegian Coast	27	73	1409	–	77	373	–	605	–	–	2564	3
Gadus morhua		(25)	(67)	(1405)	–	(81)	(373)	–	(613)	–	–		
Common tunny	Pacific Ocean	267	0	–	–	35	–	–	–	–	–	302	1
Thunnus thynnus		(268)	(1)	–	–	(33)	–	–	–	–	–		
Albacore	Pacific Ocean	167	25	–	–	115	–	–	–	–	–	307	1
Thunnus alalunga		(164)	(22)	–	–	(121)	–	–	–	–	–		

[a] The expected number of fishes calculated from the Hardy-Weinberg formula is shown in brackets

[b] Significant deviation from the theoretical equilibrium frequencies (the excess of homozygotes). A specimen probably heterozygous in the fifth (rare) allele was excluded from the sample of the common wild carps

References: 1. Fujino and Kang (1968); 2. Galagan (1973); 3,4. Møller (1968, 1971); 5. Payne (1974); 6. Payne et al. (1971); 7. Truveller et al. (1973a); 8. Wright and Atherton (1970)

Evaluating the pattern and the level of variation of transferrins in various fish species, the following two conclusions can be drawn:

1. The diversity of transferrins is based on the variability of one locus which exists in most fish species as several codominant alleles. Due to the simplicity of inheritance and the low coefficients of selection as can be judged from the data obtained from a population analysis, the transferrin alleles can be widely used as genetic markers.

2. The number of individuals with the heterozygous state of the Tf gene in populations attains 30%–40% and sometimes when the number of alleles is high 50%–60% and even >80%. Transferrins belong to the group of proteins with the maximal genetic polymorphism; this is perhaps related to the protective functions of this protein in fish diseases.

Haemoglobins. Of the 75 fish species investigated before 1969 only 17 had polymorphic haemoglobins (Altukhov 1969b; Altukhov and Rychkov 1972). The relative constancy of the haemoglobins has been noted by other authors as well (Bushuev et al. 1975). By 1979, the number of known cases of polymorphism had increased to more than forty. Polymorphism has been observed in the hagfish *Eptatretus stoutii* (Li et al. 1972), in several species of sharks (Fyhn and Sullivan 1974), in the sprat and Chile's anchovy (Wilkins and Iles 1966; Simpson and Schlotfeldt 1966; Naevdal 1968; Veldre and Veldre 1979), in some salmonids (chum and coho salmon, red-spotted trout, lake whitefish) (Lindsey et al. 1970; Omelchenko 1975b), in some cyprinids, namely in the tench, the barbel and the Indian big carp *Catla* and *Labeo* (Callegarini and Cucchi 1968; Dobrovolov 1972; Krishnaja and Rege 1977), in the Atlantic eel (Sick et al. 1967; Pantelouris et al. 1970), in the eelpout (Christiansen and Frydenberg 1974; Hjorth 1974, 1975), in the flatfish *Hippoglossoides platessoides* (Naevdal and Bakken 1974). Some other fish species are also polymorphic for Hb loci (Callegarini 1966; Raunich et al. 1966, 1967, 1972; Westrheim and Tsuyuki 1967; Schlotfeldt 1968; Callegarini and Cucchi 1969; Cucchi and Callegarini 1969; Sharp 1969, 1973; Manwell and Baker 1970; Hasnain et al. 1973; Kartavtsev 1975; Kimura 1976). Many gadid fishes are polymorphous for haemoglobins, ten variable species have been found including the cod, the haddock, the pollock and the ling (Sick 1961, 1965a, b; Frydenberg et al. 1965, 1969; Wilkins 1966, 1971a; Møller and Naevdal 1969; Jamieson and Jonsson 1971; Jamieson and Thompson 1972; Omelchenko 1975c).

Polymorphic forms have been found in widely different taxons and often the same taxons also contain many monomorphic species. Monomorphism is characteristic of most representatives of salmonids and cyprinids. The species-specificity of haemoglobin zymograms allows them to be used successfully for the diagnostics and analysis of evolutionary relationships in related fish species.

The polymorphism of haemoglobins is not infrequently distinguished by the presence of two "ordinary" allelic forms and 1–2 "rare" alleles observable in a few cases. The alleles are inherited codominantly. Both parental variants are formed in heterozygotes, but the hybrid products are by no means synthesized in all cases. A hybridological analysis has not yet been conducted, and the pattern of inheritance can be deduced only from population analysis.

Fish haemoglobins, like mammalian ones, frequently exist as several molecular forms transiently appearing in the course of development or coexisting simultaneously (Buhler and Shanks 1959; Koch et al. 1967; Lukyanenko and Geraskin 1971; Wilkins 1971a; the pooled data regarding multiple haemoglobins are presented in the book by Manwell and Baker (1970). Each haemoglobin molecule consists of four subunit polypeptide chains (generally present in groups of two). Different subunits are coded by different genes. Each of the tetrameric species of haemoglobin appears to be adapted to the effective transport of oxygen at certain stages of development characterized by specific conditions. The wide variety of haemoglobins provides for normal metabolic processes which must proceed in constant changes in the oxygen content in water, fluctuations in temperature and in the activity of the fishes etc. The fine, precise "adjustment" of the haemoglobin molecule to fulfil its functions lessens the likelihood of the selection of new "useful" mutations and helps in understanding why the genetic polymorphism of haemoglobins is limited in fish just as in many other animals.

Two genes coding for haemoglobin have been found so far in cyprinids (Hilse et al. 1966; Ohno 1969a); it is expected that the actual number will be greater. In polyploid species of the salmonids no less than eight structural genes have been detected for haemoglobin. Their polypeptide products combine, forming different heterotetrameres, each consisting of two, three and even four different polypeptides (Tsuyuki and Ronald 1970, 1971). The number of bands on corresponding electrophoregrams is 15–18 (Omelchenko 1973).

Haptoglobins. These proteins belonging to beta-globulins of the blood serum play a specific role in the organism – they bind free haemoglobin, but the significance of this binding is still not known (Harris 1970). The polymorphism of haptoglobins is ubiquitous in man (Efroimson 1968; Harris 1970). Hardly any studies have been made of the variation of haptoglobins in fishes. Three variants of haptoglobin (A, B and 0) have been described in two species of the Red Sea perch, *Sebastes mentella* and *S. marinus* (Nefedov 1969), but the inheritance of these variants has not been followed up. A simple two-allelic codominant system of haptoglobins has been discovered in the crucian carp *Carassius carassius* (Polyakovsky et al. 1973). Evidence has been obtained of the polymorphism of haptoglobins in the common carp (Chutaeva et al. 1975b).

Albumins and Prealbumins. Genetic variation of albumins is found in very many fish species. It has not always been possible to explain the pattern of albumin distribution after electrophoresis, since the fractions are diffused and do not always form distinct bands. This is the main explanation of the relatively few papers dealing with the genetics of albumins. The greatest number of studies has been conducted with salmonids. The polymorphic systems of albumins have been detected in the Atlantic salmon, in the rainbow trout, in the sockeye and chum salmon (*Oncorhynchus nerka, O. keta*), in three species belonging to the genus *Salvelinus* (Wright et al. 1966; Nyman 1967; Wilkins 1971b; Altukhov et al. 1972; Altukhov 1973; Salmenkova and Volokhonskaya 1973; Keese and Langholz 1974), and in four species of the whitefish (Lokshina 1980). The number of alleles varies from two to four or five. Many phenotypes, no less than seven or eight, have been observed in the common carp (see Fig. 45). Three phenotypes of albumins have been detected in the silver crucian carp from Lake Sudoblya (Byelorussia); it has been assumed that in this case two codominant alleles are present (Polyakovsky et al. 1973) with the distribution of frequencies close to equilibrium:

	A	A0	00
Observed frequencies	17	79	61
Expected frequencies	20.3	72.3	64.4

Polymorphism of albumins has also been described in *Protopterus* (Masseyeff et al. 1963), four species of sturgeon (Lukyanenko et al. 1971, 1975; Balachnin et al. 1972; Chikhachev and Tsvetnenko 1979; Tsvetnenko 1980), in the Oceanic herring and the sprat (Naevdal 1969; Veldre and Veldre 1979). It can be assumed that in very many fish species albumins are polymorphic, as is the case of the locus Tf. Related species frequently differ in alleles of the albumin locus. In the first generation of interspecific hybrids of various salmonids the fractions of parental albumins are superimposed (Nyman 1965, 1967, 1970; Haen and O'Rourke 1968).

A very simple system of prealbumins containing just two alleles has been found in the rainbow trout, and in the brook trout (Wright et al. 1966; Keese and Langholz 1974), in the common carp (Balachnin et al. 1973), and in the sculpine *Myoxocephalus quadricornis* (Nyman and Westin 1968). The variation in these proteins has been found within the genus *Protopterus* (Masseyeff et al. 1963) and among several species of the family Cyprinidae.

Unidentified Serum Proteins. The variation of unidentified serum proteins has been found in sturgeon (Lukyanenko and Popov 1969; Lukyanenko et al. 1975), in sardines (Baron 1972), and in several salmonid and cyprinid species (Dufour and Barrette 1967; Mc Kenzie and Pain 1969; Taniguchi and Ichiwatari 1972; Harris 1974; Keese and Langholz 1974). A simple apparently two-allelic system has been found in the sculpine *M. bubalis* (Nyman and Westin 1969). These authors suggest that the protein examined by them was ceruloplasmin. The variability in zone A in the sockeye and chum salmon described by Altukhov et al. (1972) was later shown to be due to the variation in the quantitative composition of the phospholipids present in lipoproteins. Alleles C and D differ in the content of lecithin and isolecithin in their products, the heterozygotes occupying intermediate position. The fractions coded by these alleles differ distinctly in their electrophoretic mobility. The alleles C and D are at equilibrium in populations of these two species (Akulin et al. 1975).

Not infrequently one can observe the variations of serum proteins in the region of the "slowest" protein fractions, i.e., of gamma globulins. Any detailed genetic and populational studies of these variable zones are difficult because of the insufficiently distinct resolution of gamma globulin fractions in gel slabs. The variation of serum protein binding Ca^{2+} has been observed in the species of American minnow *Notropis* (Buth 1979c).

Myogenes. The electrophoresis of fish muscle proteins results in the resolution of 15–20 fractions. Many of them are monomorphic within a species or a population, but in certain fish species the variation in one or two zones of muscle proteins could be observed. The number of alleles rarely exceeds two. In the common carp and in vimba (Cyprinidae) one myogen fraction is absent in part of the individuals (the null allele) and only two phenotypes are found in the sample (Fig. 46), "+" (A) and "–" (a) (Payusova and Koreshkova 1973; Truveller et al. 1973b; Payusova et al. 1976; Paaver 1979). An analysis of the crosses has shown that the plus-type is dominant and the minus-type is recessive (Truveller et al. 1973b; Cherfas and Traveller 1978):

1. aa × aa = aa (100%);
2. A × A (probably heterozygotes Aa) = 73 A (AA + Aa) + 23 a (aa).

The subspecies of the common carp differ in the presence or absence of the a phenotype. Almost all the Amur wild carp belong to the recessive type a; the type A generally predominates in Central Asian and European subspecies (Truveller et al. 1973b; Paaver 1979). Interpopulational differences have also been found in the vimba (Payusova et al. 1976). A similar pattern of variation has been observed in the hake *Merluccius capensis* (Nefyodov et al. 1973).

The presence of two allelic codominant systems of the ordinary type is characteristic of many fish species. Such a system has been traced in the populations of *Anoploma fimbria*, in this case, the frequencies of three phenotypes and genotypes are at equilibrium (Tsuyuki et al. 1965; Tsuyuki and Roberts 1969).

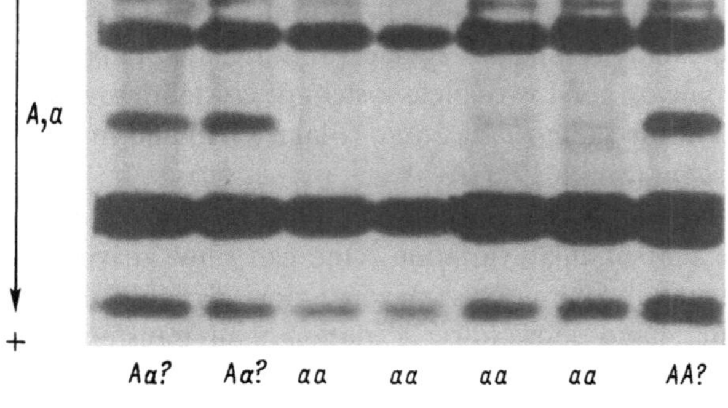

Fig. 46. Myogenes of the common carp. Variation in loci with the null allele (Truveller et al. 1973b)

In the herring two loci of myogenes show some variation (Salmenkova and Volokhonskaya 1973; Omelchenko 1975b). In salmon and whitefish one locus is present with two alleles, but in the sockeye salmon one locus with three alleles has been described (Ferguson 1975, 1980; May et al. 1975; Omelchenko 1975b; Lokshina 1980). Polymorphism has been described in the clupeid fish *Clupeonella delicatula,* in the sprat *Sprattus sprattus* (Dobrovolova 1978; Dobrovolov and Tschain 1978), in the platyfish *Poeciliopsis monacha* and *Xiphophorus maculatus* (Siciliano et al. 1973; Leslie and Vrijenhoek 1977), and in the plaice (Ward and Beardmore 1977). Two alleles of one gene have been found in the silver crucian carp (Dubinin et al. 1975; Taniguchi and Sakata 1977), in four species of catostomids (Tsuyuki et al. 1967), in the hake *Merluccius productus* (Utter and Hodgins 1971), in the perch *Stizostedion vitreum* (Uthe et al. 1966), and in the skipjack (Fujino 1970); in the two latter cases two loci appear to be polymorphic. A similar pattern of variation has been found in three of the five species of *Myoxocephalus* studied (Kartavtsev 1975) and in *Menidia menidia* (Morgan and Ulanowicz 1976). Mention should also be made of polymorphic but insufficiently studied systems of myogens in the cod (Odense et al. 1966b), in the horse mackerel *Trachurus mediterraneus* (Nefyodov et al. 1973) and in two species of the rockfish (*Sebastes*) (Tsuyuki et al. 1965). The total number of species in which myogens are polymorphic exceeds thirty at present. Muscle proteins are markedly more conservative than the proteins of blood serum. This makes it possible to use them to study the systematic relatedness of organisms.

Proteins of the Eye Lens (Crystallins). In the brook trout four out of ten bands of crystallins (probably 10 loci) turned out to be polymorphic (Eckroat and Wright 1969; Eckroat 1971, 1973). The genetic variation of three loci has been studied in great detail; the alleles of one of the loci were found to be codominant, and two others of "+" and "–" type. The analysis of crosses has shown that the segregation with respect to the two later genes is independent. It is interesting that in one of the systems of crystallins the "–" type was found to be dominant (Eckroat 1973).

Polymorphic systems of the rainbow trout (Smith 1971a), the arctic char (Saunders and McKenzie 1971), the cod and the common bonito (Odense et al. 1966b; Barrett and Williams 1967) have been studied to a lesser extent. Polymorphism in one or two loci has been found in tunnies – *Thunnus thynnus, Th. alalunga* and *Th. albacares* (Smith 1965, 1968; Gutierrez 1969; Smith and Clemens 1973). In the genus *Myoxocephalus* (Cottidae) three species were found to possess polymorphic alleles coding for crystallins (one locus, two codominant alleles – see Kartavtsev 1975). *Decapterus pinnulatus* in the family Scombridae (Smith 1969) and *Caulolatilus princeps* in the family Branchiostegidae (Smith and Goldstein 1967) were found to be polymorphic. A good two-allele system of crystallines with three phenotypes has been found in the skate *Raja clavata* (Blake 1976) and in the scorpaenid fish *Sebastolobus alascanus* (Smith 1971b).

Scarce information on the genetics of crystallins does not allow us to make any comparative evaluation of the extent of their variation. One can only arrive at a preliminary conclusion that generally they are somewhat more variable than muscle proteins, but on the other hand show lower variation than transferrins and other serum proteins.

The Genetics of Enzymes. Oxydoreductases

Lactate Dehydrogenase (LDH, 1.1.1.27). The LDH involved in the metabolism of lactate and pyruvate represents one of the better-studied enzymes in fishes. In fishes, as in mammals including man, this enzyme possesses a tetrameric quaternary structure. Four subunits of the LDH molecule may be either identical or different, therefore the tetrameric molecule may be either homo- or hetero-tetramer. In the latter case, the two types of polypeptides coded by two different non-allelic loci may be combined, forming a series consisting of five isozymes: A_4, A_3B, A_2B_2, AB_3, B_4. In certain fish species such as mackerel and others the isozymes A_3B and AB_3 are not synthesized at all (Markert and Faulhaber 1965), in many other species their levels are reduced. Sometimes heteropolymeric molecules are completely absent (e.g., in *Cynoscion regalis* and some other species; see Whitt 1970b). The number of genes coding for LDH in the genome of cyclostomes and fishes varies from one to five (Table 27). In the majority of teleosts three loci have been found, two of them, A and B, are homologous to the muscle (M) and heart (H) genes of mammals; the third gene (C) is characteristic solely of fishes (Markert et al. 1975). The gene C (otherwise called L or E) in the majority of teleost fishes is active in the eye retina. It appears to be necessary for the rapid regeneration of the visual pigments, rhodopsin or porphirin (Whitt et al. 1973c; Whitt 1975a). In cyprinids and several gadid species the products of these genes are not present in the eyes, but the liver contains isozymes with properties similar to those of the eye isozymes of salmonids; these isozymes are coded by the specific locus (or loci) which is designated at present as the C locus (Kepes and Whitt 1972; Markert et al. 1975).

Table 27. The number of LDH loci in cyclostomes and fishes (Markert et al. 1975, with additions)

Taxon	Number of loci	Loci designation
Cyclostomes		
Lampreys	1	A[a]
Hagfishes	2	A, B
Fishes		
Chondrichthyes	2	A, B
Dipnoi	2 (3)	A, B, C?
Chondrosteans	2–5	A (1–2), B (1–2), C[b]
Teleost fishes:		
Diploid species	3	A, B, C
Tetraploid species:		
Salmonids[c]	5	A_1, A_2, B_1, B_2, C
Common and crucian carp	5	A_1, B_1, B_2, C_1, C_2

[a] Any homology between the single LDH locus of lampreys and the locus of other fishes remains to be established. Lampreys possibly have the second weakly expressed Ldh locus (Dell'Agata et al. 1979)

[b] Exact number of loci in tetraploid sturgeons has not been determined

[c] The number of loci in graylings is unknown

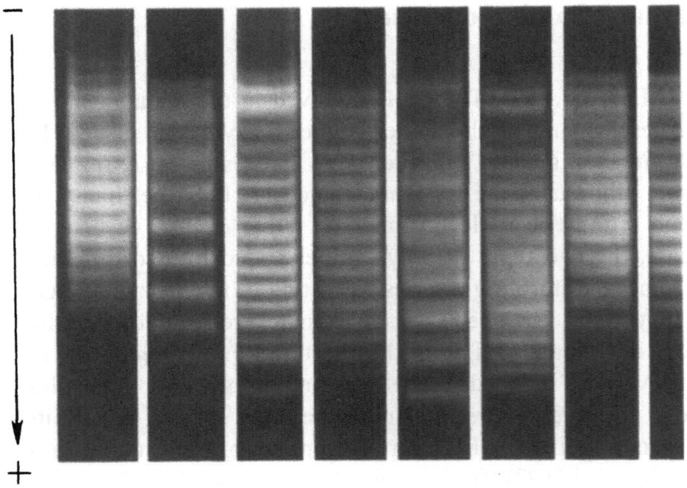

Fig. 47. Multiple isozymes of LDH in the common sturgeon *Acipenser güldenstädti* (Slynko 1976 a)

All species of salmonids studied, as well as of tetraploid cyprinids (the common carp and apparently the crucian carp), possess five genes for LDH; the genes A and B in salmonids are duplicated like the genes B and C in the common carp.

It has recently been found that the spectrum of the LDH isoenzymes is extremely rich in the sturgeon *Acipenser güldenstädti* (Fig. 47). The number of LDH loci in the sturgeon genome is five or more, i.e., it is identical to or even exceeds the number characteristic of salmonids.

In the presence of the duplicated genes A, B or C in tetraploids the heterotetramers are formed preferentially by the combination of the related polypeptides yielding the following series of isoenzymes: A_1–A_2 (A^1_4, $A^1_3A^2_1$, $A^1_2A^2_2$, $A^1_1A^2_3$, A^2_4), B_1–B_2 (B^1_4, $B^1_3B^2_1$, $B^1_2B^2_2$, $B^1_1B^2_3$, B^2_4) and C_1–C_2 (also five isoenzymes). The heterogenic isozymes (AB, AC and BC) are synthesized only in a few species and generally have a lower expression (Massaro 1972; Shaklee et al. 1973; Markert et al. 1975). It is difficult to achieve molecular hybridization of the subunits A, B and C of salmonids LDH also under the conditions of in vitro experiments. This is related to the great differences between the three types of subunits in their total amino acid composition and primary structure. For example, in the chinook salmon *Oncorhynchus tschawytscha* LDH-A and LDH-B isozymes significantly differ from each other in the content of eleven amino acids (Lim et al. 1975).

The diversity of LDH isoenzymes in the fish further increases when different loci contain different allelic variants. The series of alleles of the A, B and C loci have already been found in more than 60 species of cyclostomes and fishes extending from hagfishes (Myxinae) to flatfishes (Pleuronectidae) (Sachko 1971; Markert et al. 1975). It is not possible to provide a description of all the polymorphic Ldh loci; we shall limit ourselves to their enumeration and consideration of a few of the most interesting examples.

In the hagfish *Eptatretus stoutii*, in each of the two independently inherited loci of LDH (A and B) two alleles have been found. The frequencies of these alleles were similar to the equilibrium ones (Ohno et al. 1967b). Two alleles of the gene A have been found in the skate *Rhinobates schlegelu* (Bridges and Freier 1966).

The sterlet *Acipenser ruthenus* is polymorphic for one of the Ldh loci (Rolle 1979), other sturgeons remain to be investigated in this respect. The intrapopulational heterogeneity involving predominantly the B locus has been found in the Atlantic herring *Clupea harengus* (Odense et al. 1966a, 1971; Naevdal 1969, 1970); heterogeneity involving the A locus has been described in the anchovy *Engraulis encrasicholus* and in *E. mordax* (Klose et al. 1968; Ohno et al. 1968, 1969a; Dobrovolov 1976).

In salmonids in most cases we are dealing with a variation in the structural genes of LDH itself, but in the chum salmon, pink salmon and the arctic char cases have been found when only the degree of manifestation or the activity of the alleles was different. Therefore one can assume polymorphism in regulatory genes controlling the timing and extent of the manifestation of Ldh loci (Table 28).

Table 28. Polymorphism for Ldh loci in polyploid fish species

Species	Loci (number of alleles in brackets)	Reference
Salmo salar	C (2)	17
S. trutta	B (2), C (2)	1, 3, 26, 43
S. gairdneri	B1 (3), B2 (3), C (2)	2, 15, 33, 34, 36, 40, 43
Oncorhynchus keta	A (2), B (2)	4, 23, 29, 30
O. gorbuscha	C (2)	4, 27
O. nerka	B1 (2)	5, 13, 14, 19, 20, 21, 29, 30, 35
Salvelinus fontinalis	B1 (2), B2 (3), C (2)	7, 8, 12, 25, 26, 42
S. alpinus	B (2)	31
S. malma	C (2)	27
Coregonus clupeaformis	B (2)	6, 10, 18
C. lavaretus	B (2)	24
C. albula	B (2)	24
Cyprinus carpio	B1 (3), C1 (4), C2 (2)	9, 16, 28, 32, 37, 38, 39
Carassius carassius	B2 (2)	37

References: 1, 2. Allendorf and Utter (1973, 1975); 3. Allendorf et al. (1975); 4, 5. Altukhov et al. (1970, 1975a); 6. Clayton and Franzin (1970); 7, 8. Davisson et al. (1972, 1973); 9. Engel et al. (1973); 10. Franzin and Clayton (1977); 11. Goldberg (1966); 12. Goldberg et al. (1971); 13. Hodgins and Utter (1971); 14. Hodgins et al. (1969); 15. Huzyk and Tsuyuki (1974); 16. Ivanova et al. (1973); 17. Khanna et al. (1975b); 18. Kirkpatrick and Selander (1979); 19. Kirpichnikov (1977); 20. Kirpichnikov and Ivanova (1977); 21. Kirpichnikov and Muske (1980); 22. Krueger and Menzel (1979); 23. Kulikova and Salmenkova (1979); 24. Lokshina (1980); 25. Morrison (1970); 26. Morrison and Wright (1966); 27. Omelchenko (1975a); 28. Paaver (1980); 29. Ryabova-Sachko (1977a); 30. Sachko (1973); 31. Salmenkova and Volokhonskaya (1973); 32. Shaklee et al. (1973); 33. Utter and Allendorf (1978); 34. Utter and Hodgins (1972); 35, 36. Utter et al. (1973a, 1973b); 37. Valenta (1978a); 38, 39. Valenta et al. (1976b, 1978); 40. Williscroft and Tsuyuki (1970); 41. Wiseman et al. (1978); 42. Wright and Atherton (1970); 43. Wright et al. (1975)

Five Ldh loci have been discovered in the brook trout. The homotetrameric products of these genes may be aligned in the sequence of increasing electrophoretic mobility towards the cathode in the following way (Wright et al. 1975):

A2 → A1 → B1 → B2 → C.

Each of the series A2–A1 and B1–B2 includes five isoenzymes of which three represent heterotetramers. Heteromers may also be formed between the species B and C (Fig. 48). Allelic forms have been found for the genes B1 and B2 as well. Rare alleles of the C locus have also been found. Two species of the trout *S. fontinalis* and *S. namaycush* differ from each other in the alleles of the gene B1. In one of the back-crosses (♀*S. fontinalis* × ♂F_1) there was a deviation from independent distribution in the offspring, while in the reciprocal crossing a segregation ratio of 1:1:1:1: was observed (Morrison 1970):

$$♀\frac{\text{B1B2}}{\text{B1B2}} \times ♂\frac{\text{B1 B2}}{\text{B1}'\text{B2}'} = 495\frac{\text{B1B2}}{\text{B1B2}} + 461\frac{\text{B1 B2}}{\text{B1}'\text{B2}'} + 136\frac{\text{B1B2}}{\text{B1B2}} + 122\frac{\text{B1 B2}}{\text{B1}'\text{B2}'} \qquad (1)$$

Salvelinus fontinalis F_1

$$♀\frac{\text{B1 B2}}{\text{B1}'\text{B2}'} \times ♂\frac{\text{B1B2}}{\text{B1B2}} = 92\frac{\text{B1B2}}{\text{B1B2}'} + 113\frac{\text{B1 B2}}{\text{B1}'\text{B2}} + 122\frac{\text{B1B2}}{\text{B1B2}'} + 116\frac{\text{B1 B2}}{\text{B1}'\text{B2}} \qquad (2)$$

F_1 *Salvelinus fontinalis*

In subsequent hybrid generations all cases were found from those with a prevalence of parental gene combinations B1B2 and B1′B2′ to such ones where all genotypes were present in virtually identical amounts. This appears to be the case of so-called pseudolinkage: the genes B1 and B2 are probably located in two pairs of chromosomes related by their common origin. The presence of homologous regions in the chromosomes of these pairs provides for a non-random chromosome assortment during the meiotic divisions in males. It is furthermore possible that these chromosomes undergo temporary association, forming a complex interchromosomal associations. The presence of such multivalents has been frequently observed during meiosis in certain salmonids, in particular in the rainbow trout (Ohno et al. 1969 a), the Atlantic salmon (Barshiene 1977 a) and the sockeye salmon (Chernenko 1971). Several authors have tried to relate the appearance of chromosomal multivalents to the centric fusions of the Robertsonian type (Morrison 1970; Davisson et al. 1973).

Recently a specimen of the brook trout trisomic in the chromosomes carrying the gene Ldh-B2 was found (Davisson et al. 1972). The analysis of crosses has confirmed the hypothesis that the genome of this fish contained three alleles of the locus B2 (B2, B2′ and B2″). For example, it has been obtained in one of the following crosses:

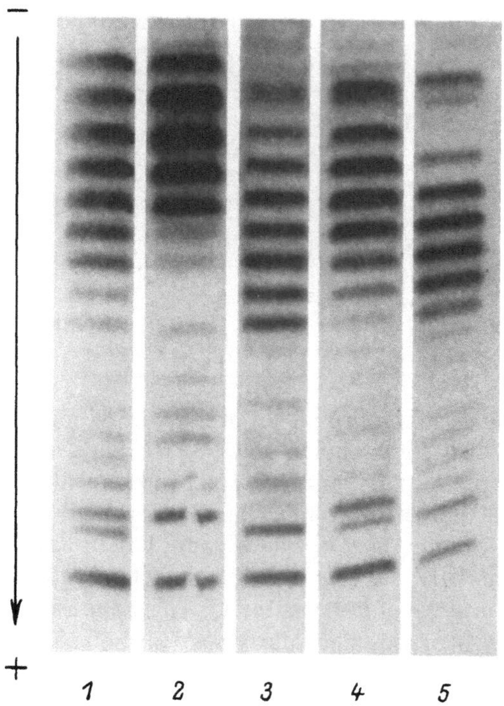

Fig. 48. The spectrum of LDH allozymes and isozymes in the brook and sea trout *Salvelinus. 1, 2 S. fontinalis; 3, 4* hybrids *S. fontinalis* × *S. namaycush* (splake); *5 S. namaycush* (Morrison and Wright 1966)

$$♀B2B2 \times ♂B2B2'B2'' = 13\ B2B2 + 17\ B2B2' + 23\ B2B2'' + 25\ B2B2B2' + 10\ B2B2B2'' + 20\ B2B2'B2''.$$

The presence of an additional metacentric chromosome (85th chromosome) has been established in the presumed trisomic individual. Probably this chromosome was obtained by the embryo after the fusion of one normal gamete and a gamete possessing an additional chromosome as a consequence of non-disjunction (Davisson et al. 1972).

Up to 27 LDH isozymes and allozymes could be observed in gels in the case of interspecific *Salvelinus* hybrid (Bouck and Ball 1968).

The inheritance of different LDH alleles in non-hybrid strains of the brook trout obeys simple Mendelian laws. The alleles of the locus B2 belong to the best-studied ones (Wright and Atherton 1970):

$$B2B2 \times B2'B2'' = 50\ B2B2' + 54\ B2B2'';$$
$$B2B2'' \times B2B2'' = 12\ B2B2 + 26\ B2B2'' + 12\ B2''B2'';$$
$$B2'B2'' \times B2B2'' = 51\ B2B2' + 44\ B2B2'' + 44\ B2'B2'' + 31\ B2''B2'' \quad \text{etc.}$$

In large fish stocks living in ponds as well as in natural populations, the proportions of different genotypes are close to the theoretically calculated ones.

Two examples with the American brook trout will be given (Wright and Atherton 1970):

	BB	BB′	BB″	B′B′	B′B″	B″B″	n
Fish farm							
Observed	280	25	25	0	0	1	331
Expected	280.7	23.2	25.0	0.5	1.0	0.5	
Brook							
Observed	196	8	6	0	0	0	210
Expected	196.4	7.7	5.7	0.1	0.1	0.04	

In almost all populations of the sockeye salmon two alleles of the Ldh-B1 locus have been found (Hodgins et al. 1969). The "slow" homzygotes B1′B1′ have five isozymes of the B series (the combinations of B1 and B2 polypeptides), after electrophoresis they are located at identical distances from each other. Zymograms of the rapid homozygotes B1B1 also contain five isozymes, but the distance between components is twice as great. In the zymograms of heterozygous specimens up to 9 bands can be counted instead of the possible 15. This appears to be related to the charge identity for several iso- and allozymes (Fig. 49). The inheritance of these codominant alleles follows the simple monohybrid scheme (Table 29).

A population analysis of the sockeye salmon has been conducted several times, and the frequencies of the genotypes were found to be close to the equilibrium ones. Two examples of the Kamchatka populations are presented below (theoretical frequencies are given in parentheses):

	B1B1	B1B1′	B1′B1′	n
Lake Azabachye (Altukhov et al. 1975 a)	379	1247	1140	2766
	(364)	(1278)	(1124)	
Lake Dalneye (Kirpichnikov and Ivanova 1977)	9	195	1063	1267
	(9)	(197)	(1061)	

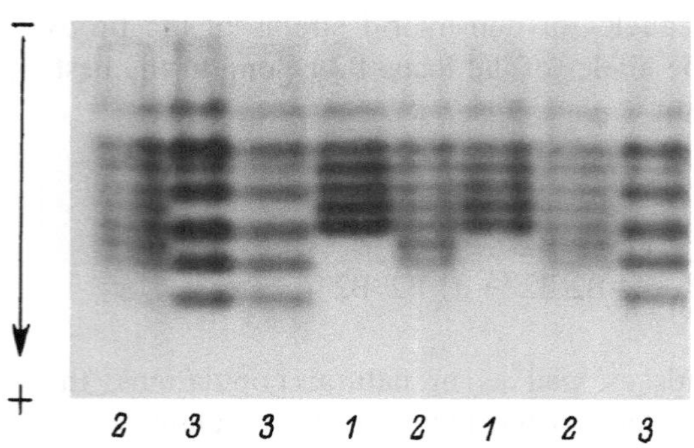

Fig. 49. LDH isoenzymes and allozymes in sockeye salmon *Oncorhynchus nerka*. *1, 3* homozygotes B1′B1′ and B1B1; *2* heterozygotes B1B1′. (Photograph G. M. Sabinin)

Table 29. Segregation for Ldh-B1 alleles in crosses of the sockeye salmon (embryonic material) (Kirpichnikov 1977)

Cross number	Parental genotypes		Number of individuals with different genotypes in the offspring			N
	♀	♂	B1′B1′	B1B1′	B1B1	
19	B1′B1′	B1 B1′	293 (308)	323 (308)	–	616
6	B1 B1′	B1′B1′	14 (15)	16 (15)	–	30
4	B1 B1′	B1 B1′	82 (72)	144 (144)	62 (72)	288
20	B1 B1′	B1 B1′	70 (69)	139 (138)	67 (69)	276

Theoretically calculated values in brackets

The small deficit of heterozygotes in Lake Azabachye may explained by the mixing of samples from many local populations differing in the frequencies of the B1 and B1′ alleles.

In the rainbow trout the alleles of the B1, B2 and C loci have been studied. In addition to active alleles, the null allele of the gene B2 has been found, it was called the b allele. In homozygotes for this allele only one isozyme, the B_4^1 was found to remain of all the B_1–B_2 series. All the five isozymes have been found in heterozygotes; the activity of these is, however, different, with B_4^1 homopolymer being most active and B_4^2 showing the lowest activity. The heteropolymers occupy an intermediate position. A pseudolinkage between the genes B1 and B2 has again been observed. It should also be pointed out that the number of recombinant forms of the enzyme in the case of pseudolinkage is genetically controlled (Wright et al. 1975).

Many cyprinid species among the more than thirty studied show polymorphism for Ldh genes (Clayton and Gee 1969; Klose et al. 1969b; Numachi 1972a; Ivanova et al. 1973; Rainboth and Whitt 1974; Avise and Ayala 1976; Valenta et al. 1976b, 1978; Valenta 1978a; Paaver 1978, 1980; Buth 1979c). Three Ldh loci out of five showed variation in the common carp; the liver loci C1 and probably C2 were found to possess the null alleles (Engel et al. 1973; Shaklee et al. 1973; Ferris and Whitt 1977; Paaver 1980). It may well be that different authors deal with different null alleles of one and the same gene. In the natural population of the common carp the genes C1 and C2 are represented by a larger number of alleles than in the strongly inbred cultivated populations; particularly heterogeneous is the population of the Amur wild carp *Cyprinus carpio haematopterus* (Paaver 1980). The gene B1 is also polymorphic both in domesticated strains and natural populations of the common carp (Valenta et al. 1978; Paaver 1980). The pattern of LDH isozymes after electrophoresis in the carp appears to be somewhat unusual: the A_4 isozyme is located between the isozymes C_4^2 and C_4^1 (Shaklee et al. 1973). The identification of the subunits present in such heteropolymers has not been completed.

In the case of heterozygosity for one of two duplicated loci of Ldh theoretically we can expect the appearance of 15 LDH isozymes. Paaver (1980) observed such a case in the population of the Amur wild common carp. The isozymes were arranged in groups of one, two, three, four and five bands on slabs:

C_4^1;
$C_3^1C_1^2$, $C_3^1C_1^{2'}$;
$C_2^1C_2^2$, $C_2^1C_1^2C_1^{2'}$, $C_2^1C_2^{2'}$;
$C_1^1C_3^2$, $C_1^1C_2^2C_1^{2'}$, $C_1^1C_1^2C_2^{2'}$, $C_1^1C_3^{2'}$;
C_4^2, $C_3^2C_1^{2'}$, $C_2^2C_2^{2'}$, $C_1^2C_3^{2'}$, $C_4^{2'}$.

As regards other cyprinid fishes polymorphism in the Ldh loci has been observed in many European and American forms including the crucian carp, the tench, the bream, the white bream, the rudd, the south barbel (*Barbus meridionalis*), and some American minnows (*Notropis* sp. sp., *Lavinia exilicauda, Hesperoleucus symmetricus*).

A variation in the structural and regulatory genes was recently found in the loach *Misgurnus fossilis* (Kusen and Stojka 1980).

Many species of tooth carp also possess polymorphic LDH (Whitt 1970b; Massaro and Booke 1971, 1972; Scholl and Schröder 1974; Mitton and Koehn 1975; Place and Powers 1978). The alleles of all the three loci, A, B and C have been found in different species. In triploid *Poeciliopsis monacha* and *P. lucida* three alleles of the C locus have been found (Vrijenhoek 1972; Leslie and Vrijenhoek 1978).

Diploid *P. monacha* is polymorphic for the A and B loci (Leslie and Vrijenhoek 1977), the killifish *Fundulus heteroclitus* for the B locus (Place and Powers 1978, 1979). In the medaka *Oryzias latipes* polymorphism of the A locus is only found in males carrying YY chromosomes. It can be assumed that the X chromosome contains genes smoothing the manifestation of different alleles of the A locus (Matsuzawa and Hamilton 1973).

Polymorphism of the genes A and B has also been detected in the characid fish *Astyanax mexicanus* (Avise and Selander 1972); specimens of this species dwelling in caves are monomorphic. The American pike *Esox americanus* shows polymorphism in the gene A, it has been found that the frequencies of all three alleles are fairly constant in the different populations of this species (Eckroat 1975).

Many codfishes are polymorphic for the genes A and B. The number of alleles is generally equal to two, more rarely to three (Markert and Faulhaber 1965; Odense et al. 1969, 1971; Utter and Hodgins 1969, 1971; Lush 1970; Jamieson 1975; Odense and Leung 1975; Johnson A. 1977; Mork et al. 1980). Variation has been observed in a few species of flatfishes (Dando 1971; Johnson A. 1977; Ward and Beardmore 1977).

Two species, the largemouth black bass *Micropterus salmoides* and the smallmouth bass *M. dolomieu,* differ in alleles of the C locus; segregation in the proportion 1:2:1 has been observed in the F_2 (Whitt et al. 1971). Among Perciformes two species of *Percina* possess polymorphic LDH (Page and Whitt 1973b), as well as *Etheostoma* sp. (Echelle et al. 1976; Wiseman et al. 1978), *Anoplarchus purpurescens* (Johnson M. 1971, 1977), *Sebastes inermis* (Numachi 1972b), two species of Cichlidae and one species of Carangidae (Kornfield and Koehn 1975; Basaglia and Callegarini 1977).

The number of genetically determined evolutionary stable LDH isozymes which are synthesized simultaneously in a given teleost fish is generally equal to three, five or more; in a number of cases it may be equal to 10–15. This variability is determined by the presence of two, three, or even five loci of Ldh in the fish genome.

In addition, in populations of many (perhaps of most) fish species the allelic variation is maintained for long periods of time. The presence of several Ldh genotypes probably increases the breeding value of population (the mean fitness) and contributes to the survival of the species as a whole.

NAD-Dependent Malate Dehydrogenase (MDH, 1.1.1.37). MDH of fish tissues exists in two forms, soluble (s) and mitochondrial (m). Each of these forms is coded by an independent gene and in several fish species two or even three genes are involved in the coding of each form. MDH is a dimeric enzyme and therefore after gel electrophoresis of the material obtained from homozygotes generally one band of activity is present; in heterozygotes three bands are seen, the intermediate band corresponding to the hybrid molecule (Fig. 50).

The polymorphic loci of MDH (predominantly of sMDH) have already been found in many fish species. Unfortunately, only in a few cases is polymorphism well documented.

In the sturgeon *Acipenser güldenstädti* the spectrum of sMDH is represented by 7–8 and in several individuals by 9–10 fractions. Polymorphism of MDH in the sturgeon is very complex (Slynko 1976a) and a thorough populational and hybridological analysis of this variation is needed. All the genes coding for MDH in sturgeons appear to be duplicated.

In the herring *Clupea harengus* no less than three phenotypes of sMDH can be found; it is assumed that a genetic system with two or three alleles is present (Odense and Allen 1971; Salmenkova and Volokhonskaya 1973). Two alleles have been found in the sprat *Sprattus sprattus* (Koval 1976).

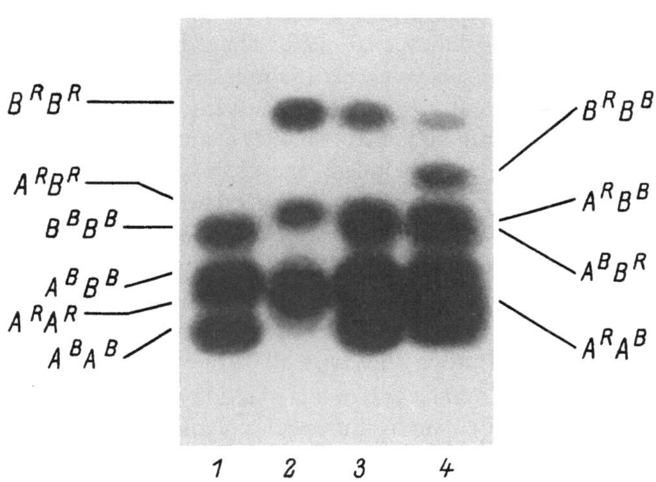

Fig. 50. Isoenzymes of MDH in sunfish species. *1 Lepomis macrochirus; 2 Lepomis microlophus; 3* mixture of homogenates of two species; *4* hybrids (the hybrid isoenzyme in distinctly seen) (Whitt et al. 1973a)

In the rainbow trout and in other salmon the gene sMdh-B is duplicated, both daughter loci (B1 and B2) being polymorphic (Numachi et al. 1972; Allendorf et al. 1975; Clayton et al. 1975; Slynko 1976b). The sMdh-A loci are also polymorphic. The identical mobility of several isozymes during gel electrophoresis complicates the interpretation of the separation patterns, although the analysis of segregation of one of the B genes in eight crosses has revealed the presence of a simple two-allelic system (Utter et al. 1973b). The finding of three or even four loci of sMDH in the rainbow trout and the brown trout (Bailey et al. 1970; Allendorf et al. 1977) is in agreement with the hypothesis about the polyploid origin of salmonids. There is some evidence that the loci of mMDH are duplicated as well, one of two of these loci being polymorphic (Clayton et al. 1975).

Two polymorphic systems of sMDH have been described in the brown trout (Allendorf et al. 1977). In one of them a null allele was discovered (May et al. 1979a). It has been suggested by a number of authors that the allelic variants of the duplicated loci of MDH in salmonids are inherited in a tetrasomic manner. When one of the loci is heterozygous, the genetic structure of the individual would be the following: $AAAA^1$ or $BBBB^1$. Consequently, when the polypeptides A and A^1 are present in a cell in equal amounts and are free to combine, the isozymes AA, AA^1 and A^1A^1 must be synthesized in a proportion of $9AA:6AA^1:1A^1A^1$. Such proportions have indeed been found after gel electrophoresis of the enzyme from heterozygous individuals of tetraploids (Baily et al. 1970; Slynko 1976a). It should be pointed out, however, that the hypothesis of tetrasomic inheritance has been criticized by many authors (Allendorf et al. 1975; May et al. 1975; Allendorf and Utter 1976). Evidence has been presented in favour of a strictly disomic inheritance for each of the loci (May et al. 1979b).

Complex patterns of polymorphism for sMDH loci have been revealed in the pink- and chum salmon (Slynko 1971a, b; Aspinwall 1974a, b; Omelchenko 1975b; Salmenkova and Omelchenko 1978; Kulikova and Salmenkova 1979). Of the four sMDH loci (A1, A2, B1, B2) at least three were polymorphic, each having two or three alleles. Polymorphic systems of MDH have also been demonstrated in the chinook, cherry and sockeye salmon (Bailey et al. 1970; Omelchenko 1975b; Grant et al. 1980); as well as in *Salvelinus leucomaenis* (Salmenkova and Volokhonskaya 1973), in two species of the genus *Prosopium* (Massaro 1972), in the whitefish *Coregonus clupeaformis* (Franzin and Clayton 1977), and in the ayu *Plecoglossus altivelis* (Sato and Ishida 1977). In all these species only one polymorphic locus has been found so far. Systems consisting of two alleles have been described in two species of smelts (Salmenkova and Volokhonskaya 1973; Omelchenko 1975b; Salmenkova and Omelchenko 1978). The variability of sMDH and mMDH has been described in many species of the family Cyprinidae both in America (Rainboth and Whitt 1974; Avise and Ayala 1976; Buth and Burr 1978; Merritt et al. 1978; Buth 1979c) and in Europe (Brody et al. 1976; Valenta 1977, 1978b). In the American minnow *Notropis venustus* two loci of sMdh and one locus of mMdh are polymorphic (Sorensen 1980). Among European polymorphic species we have the common and the crucian carp (Valenta 1978b).

Populations of the eel *Anguilla anguilla* in the Mediterranean have a large number of MDH phenotypes; it is assumed that seven alleles of one sMDH locus are responsible for this polymorphism (Comparini et al. 1975; Rodino and

Comparini 1978). The American eel is also polymorphic in MDH phenotypes (Williams et al. 1973). The variation of one locus has been described in the catostomid fish *Thoburnia rhothoeca* (Buth 1979b), in the characid *Astyanax mexicanus* (Avise and Selander 1972) and in six species of the tooth carp (Scholl and Schröder 1974; Mitton and Koehn 1975; Shami and Beardmore 1978a; Vrijenhoek 1979a; Vrijenhoek and Allendorf 1980; Turner and Grosse 1980). The killifish *Fundulus heteroclitus* was found to be polymorphic in both s- and m-types of MDH, each of the loci having two alleles (Whitt 1970a; Place and Powers 1978). The eelpout *Zoarces viviparus* was found to be polymorphic for one Mdh locus (Frydenberg and Simonsen 1973).

Among the Percidae polymorphism has been found in *Stizostedion* and *Etheostoma* spp. (Clayton et al. 1971, 1973b; Martin and Richmond 1973; Schweigert et al. 1977). Crosses between two species of perches belonging to the genus *Lepomis* have shown that the genes of sMdh-A and sMdh-B are non-linked (Wheat et al. 1972). The formation of a hybrid enzyme of MDH has been described in hybrids of *Micropterus salmoides* and *M. dolomieu* (Wheat et al. 1971).

Finally, the presence of two alleles in one of the Mdh loci has been demonstrated in *Pleuronectes platessa* (Purdom et al. 1976; Beardmore and Ward 1977), in four other species of flatfishes (Johnson and Beardsley 1975; Johnson A. 1977), in *Gasterosteus aculeatus* (Avise 1976b), *Cololabis saira* (Numachi 1970), and in the trigger fish *Novodon scaber* (Gauldie and Smith 1978). A hybridological analysis has been conducted in studies with the plaice.

Fish MDH belongs to the enzymes possessing a large number of iso- and allozymes. The systems of inheritance are generally simple in diploid species. In tetraploids such as salmonids they are more complex because of loci duplication and the superposition of the isozymes during electrophoresis.

NADH-Dependent Malate Dehydrogenase or Malic-Enzyme (ME, 1.1.1.39). Data on the allelic variation of malic-enzyme in fish are rather scarce. Intraspecies variability has been described in salmonids – *Salmo salar, S. trutta, Salvelinus fontinalis* and *Coregonus clupeaformis* (Allendorf et al. 1977; Cross et al. 1979; Kirkpatrick and Selander 1979; Stoneking et al. 1979), as well as in the cyprinid fish *Rhinichthys cataractae* (Merritt et al. 1978), in the snapper *Chrysophrys auratus,* Sparidae (Smith et al. 1978) and in pollock *Theragra chalcogramma* (Grant and Utter 1980).

Alpha-Glycerophosphate Dehydrogenase (Alpha-GPD; 1.1.1.8). The variation of the alpha-GPD loci has been studied in detail in several species of the salmon and the whitefish. In the rainbow trout the two loci are polymorphic, but for one of them only a rare allele is known ($q < 0.01$) (Engel et al. 1971b; Utter and Hodgins 1972; Utter et al. 1973a, b; Allendorf et al. 1975). Crosses have shown that the two alleles of one of this loci are inherited in a codominant fashion strictly according to Mendel (Utter et al. 1973b):

BB × AB = 87 BB + 84 AB;
AB × BB = 10 BB + 10 AB;
BB × BB = 140 BB.

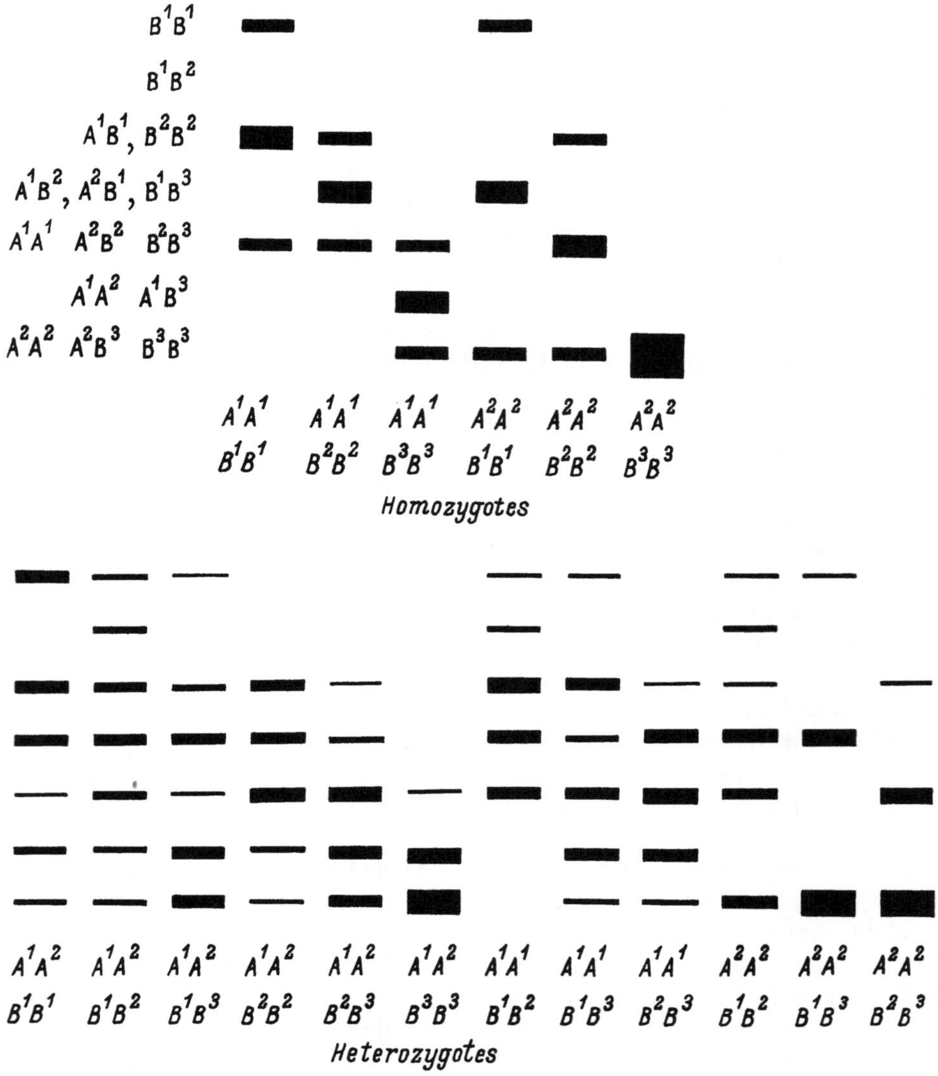

Fig. 51. The scheme of α-GPD spectra in two species of the lake whitefish *Coregonus clupeaformis* and *C. artedi*. *Left* probable isozymes; *below* genotypes (Clayton et al. 1973a)

The enzyme αGPDH (L-glycerol-3-phosphate; NAD-oxidoreductase, 1.1.1.8) is designated some times as G3PD(H), this leads to confusion because the enzyme D-Glyceroaldehyde-3-phosphate dehydrogenase (1.2.1.12) often also designated as G3PDH.

In the brown trout polymorphism in one of three loci has been found (Engel et al. 1971b; Allendorf et al. 1976, 1977; Ryman and Ståhl 1980). *Oncorhynchus keta* and *Oncorhynchus gorbuscha* are also polymorphic, and in both cases have two simple allelic systems (Altukhov et al. 1972; Aspinwall 1973; May et al. 1975; Kulikova and Salmenkova 1979). Complex diallel crossing carried out with the pink salmon (15♀♀×15♂♂, total 225 crosses) confirmed the correctness of the hypothesis postulating codominant diallelic inheritance (Aspinwall 1973). Polymorphism has also been established for three species of the genus *Coregonus, C. clupeaformis, C. artedi* and *C. peled* (Clayton et al. 1973a; Lokshina 1980). An analysis of the intrapopulational variability and of the 29 specially planned crosses of the lake whitefish *Coregonus clupeaformis* has made it possible to formulate a hypothesis about the independent inheritance of two alpha-Gpd loci, one of which (A) has two and another (B) four codominant alleles (Clayton et al. 1973a; Kirkpatrick and Selander 1979). The molecule of this enzyme consists of two polypeptide chains, the products of the two genes are free to combine with each other. The formation of 15 isozymes is possible, but experimentally one can distinguish only seven because of the identical electric charge of different dimers. Eighteen phenotypes have been detected on electrophoretic gel patterns (Fig. 51).

The populational analysis of lake whitefish living in one Canadian lake and acclimated in another has demonstrated a good coincidence between empiric and theoretical frequencies in the initial habitat and the relative excess of heterozygotes in the introduced population in the second generation (Loch 1974):

	B1B1	B2B2	B3B3	B1B2	B1B3	B2B3	n
Lake Clearwater							
Observed	24	5	20	16	37	21	123
Expected	20.7	4.4	19.7	19.2	40.3	18.7	
Lake Lyous							
Observed	19	2	6	30	43	18	118
Expected	26.1	5.7	11.3	24.4	34.4	16.1	

In the European smelt *Osmerus eperlanus* some rare alleles of two loci have been found (Engel et al. 1971b); in the Pacific smelt *Hypomesus olidus* two alleles are observed (Salmenkova and Volokhonskaya 1973). Polymorphism for αGpd is also obtained in the ayu (Sato and Ishida 1977). Some species of American minnows (Cyprinidae) have been found to be polymorphic in alpha-GPD genes (Avise and Ayala 1976; Buth and Burr 1978; Buth 1979c). Many catostomids are polymorphous for αGpd, namely *Moxostoma* spp., *Thoburnia rhothoeca, Erymyzon succetta* and other species (Buth 1977b, 1979b; Ferris and Whitt 1980). Polymorphism has also been described in *Astyanax mexicanus* (Avise and Selander 1972), in the skipjack *Katsuwonus pelamis* (McCabe and Dean 1970), in the Pacific ocean perch *Sebastes alutus* and in *S. caurinus* (Johnson et al. 1970a, 1973), in the bass *Morone americana* and *M. saxatilis* (Sidell et al. 1978), in the tooth carp *Ilyodon* (Turner and Grosse 1980), in the walleye pollock *Theragra* and in six species of flatfishes including the plaice and flounder (Johnson and Beardsley 1975; Purdom et al. 1976; Beardmore and Ward 1977; Johnson A. 1977; Ward and Galleguillos 1978). Seven species of cichlids are also polymorphic for αGpd loci (Kornfield 1978). The largest number of alleles – four – have been found in the saury *Cololabis saira* (Numachi 1971a).

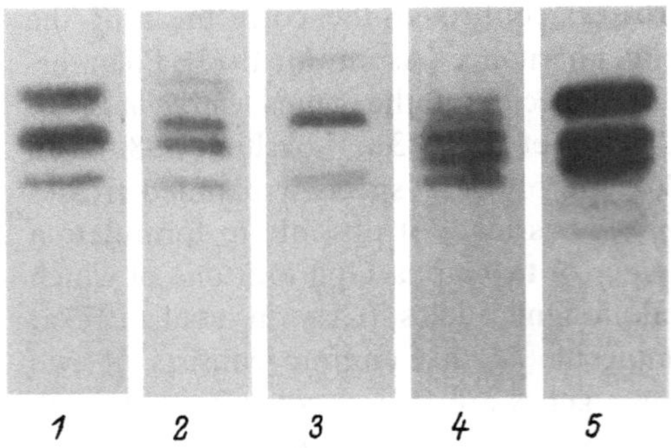

Fig. 52. Isozymes and allozymes of 6 PGD in the goldfish *Carassius auratus. 1, 3* homozygotes; *2, 4, 5* heterozygotes (Schmidtke and Engel 1974)

Thus, the alpha-GPD in fish appears to be one of most variable glycolytic enzymes. The number of alleles of the alpha-Gpd loci is frequently three or even four. In tetraploid salmonids heteropolymers consisting of products of two genes are formed. A further increase in the number of isoenzymes in this group is due to allelic variation.

A large number of fish species display polymorphism of 6-phosphogluconate dehydrogenase (6-PGDH) and tetrazolium oxidase (also termed superoxide dismutase, TO or SOD) (Figs. 52 and 53). High polymorphism of SOD may be partially explained by the peculiar aspects of the essay of this enzyme – it is easily detected in gels stained by tetrazolium salts for various enzymes. SOD inhibits the conversion of tetrazolium into diformazan and is therefore visible in the gel as distinct white bands.

The data on the variation of ADH, 6PGD, G6PD, SOD and other oxydoreductases are given below as a list of polymorphic species.

Alcohol Dehydrogenase (ADH, 1.1.1.1). *Salmo gairdneri* and *O. keta* (Allendorf et al. 1975; Kijima and Fujio 1978), eight species of catostomids (Buth 1977a; Ferris and Whitt 1980), *Ptychocheilus grandis* (Avise and Ayala 1976), *Anguilla rostrata* (Koehn and Williams 1978), *Lepomis macrochirus* (Felley and Avise 1980),

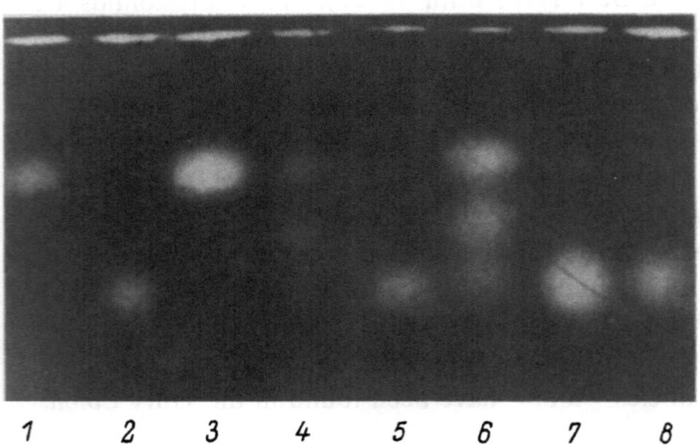

Fig. 53. Polymorphism of tetrazolium oxydase (superoxyde dismutase) in the walleye pollock *Theragra chalcogramma. 1, 3* slow homozygotes SS; *2, 5, 7, 8* fast homozygotes FF; *4, 6* heterozygotes FS (Iwata 1975)

Poecilia formosa (Turner et al. 1980a), *Rexea solandri, Girella tricuspidata* (Gauldie and Smith 1978), *Pleuronectes platessa* (Ward and Beardmore 1977).

Octanol Dehydrogenase (ODH, 1.1.1.1). Anoploma fimbria (Johnson A. 1977), *Pleuronectes platessa, Platichthys flesus* (Ward and Beardmore 1977; Ward and Galleguillos 1978).

6-Phosphogluconate Dehydrogenase (6PGD, 1.1.1.43). *Clupea harengus, Hypomesus olidus* (Salmenkova and Volokhonskaya 1973), *Oncorhynchus gorbuscha* (Omelchenko 1975b, Salmenkova et al. 1979), *Coregonus peled* (Lokshina 1980), *Barbus tetrazona, Carassius auratus* and three other European cyprinids (Bender and Ohno 1968; Klose and Wolf 1970; Schmidtke and Engel 1974), *Misgurnus anguillicaudatus* (Kimura 1976), *Ictiobus cyprinellus* and eight other catostomid species (Ferris and Whitt 1980), *Astyanax mexicanus* (Avise and Selander 1972), *Anguilla anguilla* (Rodino and Comparini 1978), *Poecilia sphenops, Poeciliopsis monacha, Ilyodon* spp. (Scholl and Schröder 1974; Leslie and Vrijenhoek 1977; Turner and Grosse 1980), *Zoarces viviparus* (Yndgaard 1972), *Cheilodactylus macropterus* (Gauldie and Smith 1978), *Katsuwonus pelamis* (McCabe and Dean 1970), *Pleuronectes platessa* (Ward and Beardmore 1977).

Glucose-6-Phosphate Dehydrogenase (G6PDH, 1.1.1.49). *Clupea harengus* (Salmenkova and Volokhonskaya 1973), *Salmo gairdneri* (Cederbaum and Yoshida 1976), *Oncorhynchus nerka* (Omelchenko 1975b), *Salvelinus fontinalis* (Stegeman and Goldberg 1971, 1972), *Cyprinus carpio, Abramis brama* (Paaver 1978, 1980), *Conger conger* (Rodino and Comparini 1978), *Fundulus heteroclitus* (Mitton and Koehn 1975), *Caranx georgianus, Cheilodactylus macropterus, Rexea solandri, Thyrsites atun* (Gauldie and Smith 1978).

Glyceroaldehyde-3-Phosphate Dehydrogenase (G3PDH, 1.2.1.12). *Xiphophorus maculatus* (Siciliano et al. 1973), *Anoploma fimbria* (Johnson A. 1977).

Superoxidismutase or Tetrazolium Oxidase (SOD or TO, 1.6.4.3). *Hypomesus olidus, Salmo trutta, S. gairdneri, Salvelinus* sp. sp., *Oncorhynchus tschawytscha, Coregonus peled, C. nasus* (Utter 1971; Locascio and Wright 1973; Salmenkova and Volokhonskaya 1973; Utter et al. 1973a; Omelchenko 1975b; Allendorf et al. 1977; Lokshina 1980), *Cyprinus carpio* (Paaver 1979) and American cyprinids, more than six species (Avise and Ayala 1976; Buth 1979c), seven species of catostomids (Ferris and Whitt 1980), Esox lucius (Healy and Mulcahy 1979), *Poecilia sphenops, P. reticulata* (Scholl and Schröder 1974; Shami and Beardmore 1978a), *Anguilla* spp. (Tegelström 1975), three species of Percidae (Martin and Richmond 1973; Page and Whitt 1973a), *Zoarces viviparus* (Frydenberg and Simonsen 1973), *Sebastes inermis* (Numachi 1972b), *Pleuronectes platessa, Limanda limanda, Hippoglossoides elassodon, Atherestes stowias, Theragra chalcogramma* (Iwata 1973, 1975; Johnson A. 1977; Ward and Beardmore 1977; Ward and Galleguillos 1978), *Merluccius merluccius* (Mangaly and Jamieson 1978), *Thunnus thynnus* (Edmunds and Sammons 1973).

Isocitrate Dehydrogenase (IDH, 1.1.1.42). *Clupea harengus* (Wolf et al. 1970), *Hypomesus olidus* (Quiroz-Gutierrez and Ohno 1970), *Salmo salar, S. gairdneri, Oncorhynchus keta* (Wolf et al. 1970; May et al. 1975; Cross and Payne 1977), *Cyprinus carpio, Carassius auratus* (Quiroz-Gutierrez and Ohno 1970), *Barbus barbus, Rutilus rutilus* (Engel et al. 1971 a), two species of American cyprinids (Avise and Ayala 1976), *Astyanax* sp. (Avise and Selander 1972), *Anguilla anguilla, Conger conger* (Rodino and Comparini 1978), four species of poeciliids (Scholl and Schröder 1974; Siciliano and Wright 1973; Leslie and Vrijenhoek 1977), *Gasterosteus aculeatus* (Avise 1976), *Zoarces viviparus* (Frydenberg and Simonsen 1973), *Chrysophris auratus* (Smith et al. 1978), *Morone americana* (Sidell et al. 1978), *Seriola grandis, Rexea solandri, Genypterus blacoides, Merluccius australis* (Gauldie and Smith 1978), *Theragra chalcogramma* (Johnson A. 1977). *Pleuronectes platessa, Platichthys flesus* (Ward and Galleguillos 1978).

Sobitol Dehydrogenase (SDH, 1.1.1.14). *Clupea harengus, Salmo salar, S. gairdneri, S. trutta, Coregonus lavaretus, C. autumnalis* (Engel et al. 1970; Allendorf et al. 1977; Ferguson et al. 1978), *Carassius auratus* (Lin et al. 1969), *Rutilus rutilus, Tinca tinca* (Engel et al. 1971 a), *Anguilla* spp. (Koehn and Williams 1978; Rodino and Comparini 1978), *Arripis trutta* (Gauldie and Smith 1978), *Pleuronectes platessa* (Ward and Beardmore 1977).

Xanthine Dehydrogenase (XDH, 1.2.3.2). *Catostomus commersoni* (Ferris and Whitt 1980), *Scomber scombrus* (Smith and Jamieson 1978), *Chrysophris auratus* (Smith et al. 1978).

Diaphorase (DIA, 1.6.9.1). *Salmo trutta* (May et al. 1979 a).

Catalase (CAT, 1.11.1.6). *Salmo gairdneri* (Utter et al. 1973 c), *Sebastes* spp. (Numachi 1971 b).

Peroxidase (PO or PX, 1.11.1.7). *Clupea harengus* (Bogdanov et al. 1979), *Salmo salar* (Nyman 1967), *Oncorhynchus nerka, O. masu* (Omelchenko 1975 b), *Myoxocephalus quadricornis* (Nyman and Westin 1968).

References: 1. Allendorf and Utter (1973); 2. Allendorf et al. (1977); 3. Altukhov et al. (1975 a); 4. Avise and Ayala (1976); 5. Avise and Selander (1972); 6, 7. Buth (1979 b, c); 8. Buth and Burr (1978); 9. Christiansen and Frydenberg (1974); 10. Echelle et al. (1976); 11. Frydenberg and Simonsen (1973); 12. Gauldie and Smith (1978); 13. Hjorth (1971); 14. Johnson A. (1977); 15. Johnson et al. (1971); 16, 17. Kimura (1976, 1978 a, c); 18. Kirkpatrick and Selander (1979); 19. Kirpichnikov (1977); 20. Kirpichnikov and Ivanova (1977); 21. Kirpichnikov and Muske (1980); 22. Kulikova and Salmenkova (1979); 23. Leslie and Vrijenhoek (1977); 24. Loudenslager and Kitchin (1979); 25. Lush (1969); 26. May et al. (1975); 27. Mitton and Koehn (1975); 28. Odense et al. (1966 a); 29. Omelchenko (1975 b); 30, 31. Paaver (1979, 1980); 32. Pantelouris (1976); 35. Roberts et al. (1969); 36. Salmenkova and Volokhonskaya (1973); 37. Sassama and Yoshiyama (1979); 38, 39. Schmidtke and Engel (1972, 1974); 40. Shami and Beardmore (1978 a); 41. Shearer and Mulley (1978); 42. Smith (1979 b); 43. Smith et al. (1978); 44. Tills et al. (1971); 45. Turner (1973 b); 46, 47. Utter and Hodgins (1970, 1972); 48. Utter et al. (1973 b); 49. Vrijenhoek (1979 a); 50. Ward and Beardmore (1977)

Table 30. Polymorphism for phosphoglucomutase and aspartate aminotransferase loci

Taxon	PGM		AAT	
	Number of polymorphic species	Reference	Number of polymorphic species	Reference
1	2	3	4	5
Clupeidae:				
Clupea harengus	1	25, 36	1	28
Sprattus antipodum	1	12	–	–
Osmeridae	1	36	1	29
Salmonidae:				
Salmo spp.	1	35, 36	3	1,2,24,26,39
Oncorhynchus spp.	3	3, 19–22, 26, 29, 46, 48	2	26, 29
Coregonus spp.	1	18	1	2
Cyprinidae:				
Cyprinus carpio	1	30, 31, 41	1 (?)	38, 39 (?)
Other European species	–	–	4	38
American species	7	4, 7, 8	4	4, 7, 8
Catostomidae *(Thoburnia)*	2	6	–	–
Characidae *(Astyanax)*	1	5	–	–
Cobitidae	3	16, 17	1	17
Anguillidae	–	–	1	32
Cyprinodontiformes				
Poeciliopsis monacha	1	23	1	49
Other species	2	27, 33, 40	1	45
Zeidae	1	12	–	–
Perciformes:				
Zoarces viviparus	1	9, 11, 13	1	11
Etheostoma spp.	–	–	1–2	10
Other species	9	12, 14, 37, 42, 43	1	14
Scorpaenidae:				
Sebastes spp.	3	14	–	–
Sebastodes spp.	2	15	–	–
Gadidae *(Gadus morhua)*	1	44	–	–
Pleuronectidae				
Pleuronectes platessa	1	14, 34, 50	1	50
Other species	7	14	9	14
Total	50	–	34	–

The number of oxidoreductases studied and especially the number of fish species investigated is still quite low. There can be no doubt that the number of polymorphic enzymes detected and of fishes studied will increase very rapidly in the future.

Transferases, Hydrolases and Other Enzymes

Phosphoglucomutase (PGM, 2.7.5.1). The number of polymorphic fish species is rather high (Table 30). In many cases, polymorphism of only one locus has been shown, but in some species two or even three varying loci have been detected. The number of alleles is mostly equal to two or three; in the common carp three allelic variants were found (Paaver 1979), in characid fish *Astyanax* even six allelic variants (Avise and Selander 1972).

It should be pointed out that usually 2–3 bands of PGM activity are revealed by electrophoresis, they probably correspond to two or three independent genes. Hybrid products are not formed in heterozygotes. This provides a basis for the opinion that the enzyme is a monomer devoid of a quaternary structure. Crosses conducted with flatfishes and the sockeye salmon have confirmed that the alleles of Pgm are inherited according to Mendel (Purdom et al. 1976; Kirpichnikov 1977).

Two of three Pgm loci, the products of which are detected by electrophoresis, are polymorphic in the eelpout (Hjorth 1971; Christiansen and Frydenberg 1974;

Table 31. The proportion of PGM genotypes in pairwise crosses and natural populations of the sockeye salmon *Oncorhynchus nerka* (Kamchatka 1975–1976)

State of cross or sampling	Fish age, years	Number and type of cross	Distribution between genotypes				Frequency of the A allele (q_A)
			AA	AB	BB	total	
Crosses							
Lake Kurilskoye	Embryos and larvae (0+)	16 AB×AB	55 (54)	115 (108)	48 (54)	216	–
		19 AB×AB	106 (123.5)	264 (247)	123 (123.5)	493	–
Populations							
Lake Kurilskoye (Eastern spawning site)	Spawners (2+ –5+)	–	8 (7.6)	36 (36.8)	45 (44.6)	89	0.292
Same (West Bank)	Fingerlings (0+)	–	14 (9.6)	57 (65.8)	117 (112.6)	188	0.226
Lake Dalneye	Smolts (1+ –3+)	–	55 (54)	333 (335)	520 (519)	908	0.244
Lake Azabachye[a]	Spawners (2+ –6+)	–	108 (130)	975 (932)	1650 (1671)	2733	0.218

Theoretical frequencies in brackets
[a] According to Altukhov et al. (1975a)

Christiansen et al. 1976). The allelic frequencies of each of the two loci show relatively moderate variation between different populations. In the sockeye salmon all the populations studied are polymorphic with respect to the Pgm loci, however, differences in frequencies are not very pronounced (Utter and Hodgins 1970; Altukhov et al. 1975 a, 1975 b; Kirpichnikov and Ivanova 1977; Kirpichnikov and Muske 1980). In the sockeye salmon the number of heterozygous individuals among spawners in some populations is greater than expected. A small excess of heterozygotes after crossing heterozygous individuals with each other is already noticeable at the larval stage (Table 31). In the smolts of Lake Dalneye (Kamchatka) the proportions of the genotypes are close to those of the equilibrium ones, while among the fry captured near the shore in Lake Kurilskoye (the same region) a certain deficit of heterozygotes has been noted. The deviation in the latter case could be explained by the mixing of populations with different allele frequencies.

Aspartate Aminotransferase or Glutamate Oxalate Transaminase (AAT or GOT, 2.6.1.1). This enzyme is found in the cell in the soluble (s) and mitochondrial (m) forms. The variation of the soluble AAT (sAAT) has been found in the herring (Odense et al. 1966 a; Naevdal 1970) and in many representatives of salmonids and cyprinids as well as in ten species of flatfish and in a few other forms (Table 30). Aspartate aminotransferase of all salmonids is coded by two unlinked loci (Table 32). The genetic interpretation of AAT segregation in the chum salmon is based on the assumption that homologous dimeric isozymes possess identical mobility:

$$A_2^1 = A_1^1 A_1^2 = A_2^2;$$
$$A_1^1 A_1^{1'} = A_1^1 A_1^{2'} = A_1^{1'} A_1^2 = A_1^2 A_1^{2'};$$
$$A_2^{1'} = A_1^{1'} A_1^{2'} = A_2^{2'}.$$

Assuming that the polypeptides coded by these two genes are free to combine, the total amounts of these three isozymes are to be present in the ratio 9:6:1 in individuals with one heterozygous gene and in the ratio 1:2:1 in double heterozygotes and in homozygotes for different alleles. Such proportions have indeed been demonstrated by enzyme electrophoresis (Fig. 54).

Table 32. Inheritance of AAT variants in the chum salmon (May et al. 1975)

Assumed parental genotypes				Phenotypic segregation in the offspring			
♀		♂		Group I AAAA	Group II AAAA′	Group III AAA′A′	Group IV AA′A′A′
A_1A_1	A_2A_2	A_1A_1'	A_2A_2'	44	85	43	–
A_1A_1	A_2A_2	A_1A_1'	A_2A_2 [a]	32	31	–	–
A_1A_1	A_2A_2	A_1A_1	A_2A_2	20	–	–	–
A_1A_1'	A_2A_2 [a]	A_1A_1'	A_2A_2'	16	40	44	19
A_1A_1	A_2A_2	A_1A_1'	A_2A_2 [a]	138	120	–	–

[a] Or A_1A_1 A_2A_2'

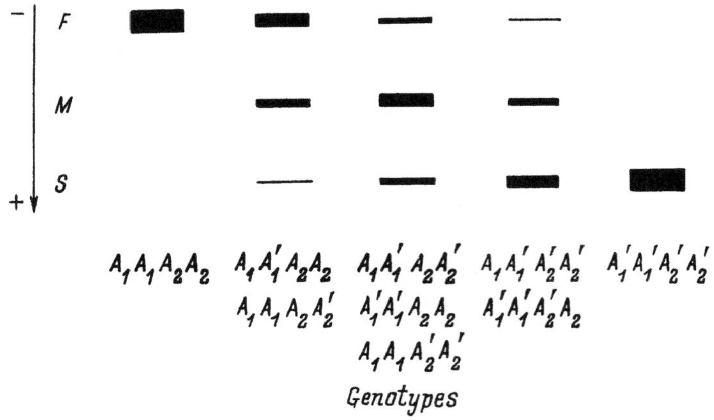

Fig. 54. Spectra of AAT isoenzymes in the pink and chum salmon *Oncorhynchus keta* and *O. gorbuscha* having different genotypes (schematic). Variation in two loci. *Left* fractions (isozymes); *below* genotypes (May et al. 1975)

In three cyprinid fish species, the roach, the tench and the barbel *Barbus tetrazona* one locus with two or three alleles appears to be present while in the common and crucian carp the presence of two duplicated loci has been suggested (Schmidtke and Engel 1972). Three bands of enzyme activity are present in the gels of the common carp homozygous with respect to sAAT alleles, and six bands can be discerned in individuals heterozygous for one of these genes.

Glutamate Pyruvate Transaminase (GPT, 2.6.1.2). The sockeye salmon (*Oncorhynchus nerka*) is polymorphous for this enzyme, two codominant alleles have been found in Alaskan populations (Grant et al. 1980).

Esterases (EST, 3.1.1.1, 3.1.1.2, 3.1.1.6, 3.1.1.7, 3.1.1.8, 3.1.1.10). Esterases are enzymes having different functions, able to cleave ester bands between carboxylic acids and naphthol (Korochkin et al. 1977). Generally these enzymes are divided into four main groups: carboxy esterases, aryl esterases, acetyl esterases and acetylcholine esterases, including pseudocholine esterases (Holmes and Masters 1967; Manwell and Baker 1970; Korochkin et al. 1977). Both serum esterases (predominantly carboxy esterases) and tissue esterases in fish generally are products of several loci. Individual variations have been found in many of them, this variation has been detected in species belonging to very different families. The number of species in which polymorphic systems of esterases have been described is almost one hundred. We are unable to present all the literature dealing with fish esterases here. Since a part of this literature has been reviewed (Ligny 1969; Nyman 1971; Kirpichnikov 1973a; Altukhov 1974), only a few of the most illustrative studies will be discussed.

In the oceanic herring *Clupea harengus* at least five loci of esterases are represented by several genetic variants, the number of alleles for each locus is equal to four or five (Naevdal 1970; Ridgway et al. 1970; Simonarson and Watts 1971; Salmenkova and Volokhonskaya 1973; Omelchenko 1975b; Zenkin 1976, 1979).

Most frequently the one "dominating" allele is present in greater abundance, its relative concentration attains 0.7–0.9 or more. In the natural populations of the Atlantic herring *Clupea harengus harengus* the proportions of alleles are stable in time and across the habitat. We shall cite as an example the data on the populations examined in the North Atlantic (Table 33; rare alleles are excluded from these calculations).

In the pilchard *Sardinops ocellata* the system of the Est-II locus which codes for muscle esterases consists of nine alleles; 24 phenotypes have been found in the population (Thompson and Mastert 1974). The other species, *Sardina pilchardus,* is also polymorphic for at least three loci (Krajnović-Ozretić and Zikić 1975) as well as *Sardinella* (Baron 1973) and the sprat *Sprattus sprattus* (Howlett and Jamieson 1971). In the capelin *Mallotus villosus* the system of serum esterases includes 10 alleles (Nyman 1971).

Eight bands of esterase activity were found after the electrophoresis of preparations from the Atlantic salmon. Six bands corresponding to the liver enzyme show polymorphism (Khanna et al. 1975 a). It has been pointed out by the authors that carboxy-, aryl- and cholin-esterases appear to belong to enzymes showing polymorphism. Polymorphism of the esterase locus has been found in the rainbow trout (Kingsburg and Masters 1972) and the arctic char (Henricson and Nyman 1976).

Many species of cyprinids show polymorphism in the Est loci. Of the five species belonging to the genus *Notropis* which have been studied, at least three possess polymorphic systems (Koehn et al. 1971). The number of alleles varies from two to eight (Menzel 1976) and null alleles are sometimes found.

The system of "fast" serum esterases with six alleles (including the null allele) has been studied in the common carp (Moskovkin et al. 1973; Tscherbenok 1973; Truveller et al. 1974; Paaver 1979; Tscheglova and Iljasov 1979). In the fish homozygous with respect to each allele two bands are present on electrophoregrams, the main and the "ghost" one; in heterozygotes three bands are visible. The "ghost" bands probably correspond to conformational variants. The alleles are inherited in Mendelian fashion (Truveller et al. 1974; Tscherbenok 1976; Cherfas and Truveller 1978; Paaver 1980). The "slow" serum esterase is coded by the locus with three alleles; one of them being the null variant. The Est system in erythrocytes (also having the null allele) is possibly identical to the slow esterases of the serum (Brody et al. 1976). A system of the Est locus with three alleles has been

Table 33. Polymorphism for a locus of liver and heart esterase in populations of the Atlantic herring (Ridgway et al. 1970)

Sampling site	Genotype frequencies						Concentration of alleles		
	mm	ms	ss	mf	sf	ff	q_m	q_s	q_f
Georges Bank	218	117	19	7	1	0	0.77	0.22	0.011
Cape May	54	37	4	1	0	0	0.76	0.23	0.005
Western Maine	61	48	9	2	2	0	0.71	0.28	0.016
Nova Scotia	82	64	12	2	0	0	0.72	0.27	0.006

found in a bisexual population of the silver crucian carp (Polyakovsky et al. 1973), two allelic systems have been found in several other cyprinids such as the bream, the roach, the bleak, and the ruff *Acerina cernua* (Nyman 1965, 1969). Serum and liver esterases of the eel exhibit a complex pattern of variability (Pantelouris and Payne 1968; Pantelouris et al. 1976).

Multiple bands possessing esterase activity have been found in the killifish and in two species belonging to the genus *Cyprinodon* (Cyprinodontidae). One of the polymorphic systems of the killifish has five alleles (Kempf and Underhill 1974). In the *Cyprinodon* at least three of the ten bands of liver esterases correspond to different alleles in different species (Turner 1973 a). Polymorphism of the Est locus have also been described in other representatives of the tooth carp (Hodges and Whitmore 1977; Leslie and Vrijenhoek 1977; Shami and Beardmore 1978 a).

Polymorphic esterases coded by two loci in the eelpout have been studied in detail (Simonsen and Frydenberg 1972; Christiansen et al. 1973, 1974; Hjorth and Simonsen 1975). The presence of two condominant alleles of the Est-III locus has been confirmed by hybridological analysis. The frequencies of different genotypes are at equilibrium; similarly an equilibrium has been observed in the populations of the ruff (Nyman 1975).

It has been found in the case of the horse mackerel *Trachurus trachurus* that the variation for the "slow" and "fast" alleles of the two esterase loci, EstS and EstF, is closely correlated (Zenkin 1979). The relationship between the allelic variability of the different esterase loci was observed in some other fish species.

The variation of serum and liver esterases has been reported for several species of the tunny, three or four alleles being described for each locus (Sprague 1967, 1970; Fujino and Kang 1968; Fujino 1970; McCabe and Dean 1970; Serene 1971). In the tunny serum esterases are coded by no less than three loci.

Polymorphism for esterases has also been recorded in the catfish *Ictalurus nebulosus* (Skow 1976), in two species of *Anoplarchus* (Johnson M. S. 1977), in many flatfishes (de Ligny 1968; Koval and Bogdanov 1979), in *Macrurus rupestris* (Alekseev et al. 1979; Dutschenko 1979), in *Pagrus major* (Taniguchi and Tashima 1980). Among the 125 species inhabiting coral reefs (68 families) many were found to be polymorphic with respect to the Est loci (Leibel and Markert 1978). In actual fact, polymorphic esterase systems appear to exist in the majority of fish species belonging to all the familis independent of their taxonomic state.

Phosphoglucoisomerase (Glucose Phospate Isomerase, PGI, 5.3.1.9). Polymorphism of this enzyme has been found in the Atlantic herring (1 locus, three alleles) and the brown trout (two or three loci, one of them polymorphic with two alleles) (Schmidtke and Engel 1974; Schmidtke et al. 1975 b). Polymorphic systems predominantly having two alleles have been detected in more than ten species of cyprinids. Two loci of PGI have been found in diploid species, while in tetraploids such as the barbel *Barbus barbus* and the silver crucian carp *Carassius auratus gibelio* three or four loci have been detected; the number of alleles does not generally exceed two (Avise and Kitto 1973; Schmidtke et al. 1975 b; Avise and Ayala 1976; Merritt et al. 1978; Buth 1979 c). In two characinid species both loci of Pgi were polymorphic, each having four or more alleles (Avise and Selander 1972; Kuhl et al. 1976). Diallelic systems has been described in the killifish (Place and

Powers 1978) and in the platyfish *Xiphophorus maculatus* (Siciliano et al. 1973); in the latter case there are two variable loci of Pgi. Polymorphism of this enzyme has also been detected in cobitid fishes *Botia macracantha* and *Cobitis delicata* (Ferris and Whitt 1977 b; Kimura 1978 b). Two polymorphic loci, each having three or four alleles, have been described in the eels *Anguilla anguilla* and *Conger conger* (Williams et al. 1973; Comparini et al. 1975, 1977; Rodino and Comparini 1978).

A comparison of mothers and their progeny (embryos) has been made in the eelpout in which no less than two loci of Pgi are polymorphic (Yndgaard 1972). A similar pattern of variation has been found in the plaice where two loci have been shown to have allelic variants (Purdom et al. 1976); in the latter case a hybridological analysis has been performed. The cod has also been found to be polymorphic in Pgi (Cross and Payne 1978), the number of alleles being equal to six.

The Pgi, catalyzing the interconversion of glucose-6-phosphate and fructose-6-phosphate is a dimeric enzyme, therefore hybrid molecules are formed in heterozygotes. In gels one can see one enzyme band in homozygotes or three bands in heterozygotes for each of the loci. The spectra of the isozymes in tetraploid species having secondary duplicate Pgi loci have not yet been completely interpreted. Polymorphous systems of the Pgi loci have also been described in the following fish species: *Oncorhynchus nerka* (Grant et al. 1980), *Campostoma,* two species (Buth and Burr 1978), *Gasterosteus aculeatus,* for two loci (Avise 1976 b), more than ten species of the order of Perciformes (Gauldie and Smith 1978; Kornfield 1978; Felley and Avise 1980), *Merluccius australis* and *Novodon scaber,* fam. Balistidae (Gauldie and Smith 1978). Pgi may be regarded as one of the most variable enzymes in the family of glycolytic enzymes.

So far, there is very little information available on the genetics of other fish enzymes. Cases where polymorphism has been described will be listed below.

Purine Nucleoside Phosphorylase (PNP, 2.4.2.1) – lamprey, *Lampetra planeri* (Ward et al. 1979). The PNP has an uncommon trimeric structure in this fish.

Creatine Phosphokinase (CK, 2.7.3.2) – the shark *Carcharinus springeri,* the rainbow and brown trout, the common carp, American cyprinid fish *Notropis,* the suckers *Maxostoma macrolepidotum* and *Catostomus discobolus, Argentina silus,* the cobitid fish *Botia modesta,* the tooth carp *Xiphophorus helleri* and *Ilyodon* sp., the catfish *Ictalurus melas,* the rockfish *Prionotus tribulus,* the bass *Morone saxatilis* (Scopes and Gosselin-Rey 1968; Eppenberger et al. 1971; Perriard et al. 1972; Allendorf et al. 1976, 1977; Ferris and Whitt 1978 a; Fisher and Whitt 1978 b; Buth 1979 a; May et al. 1979 a, b; Guse et al. 1980; Turner and Grosse 1980).

Hexokinase (HEX, 2.7.1.1.) – the eelpout (Frydenberg and Simonsen 1973).

Adenylate Kinase (AK, 2.7.4.3.) – the sockeye salmon, *Notropis* spp., the eelpout (Frydenberg and Simonsen 1973; Omelchenko 1975 b; Buth 1979 c).

Amylase (AMY, 3.2.1.1.) – the swordtail *Xiphophorus montezumae* (Herrera 1979).

Peptidase or Leucinaminopeptidase (PEP or LAP, 3.4.11.1.) – the lake whitefish *Coregonus clupeaformis,* American cyprinid fish *Campostoma* (3 sp.), and *Rhynichthys cataractae, Poeciliopsis monacha,* characid fish *Astyanax,* rockfish *Sebastes* (3 sp.), plaice and other flatfish, *Glyptocephalus zacharius* (Avise and Selander

1972; Johnson et al. 1972; Johnson A. 1977; Leslie and Vrijenhoek 1977; Ward and Beardmore 1977; Buth and Burr 1978; Merritt et al. 1978).

Adenosine Deaminase (ADA, 3.5.4.4.) – eel *Anguilla anguilla,* plaice (Ward and Beardmore 1977; Rodino and Comparini 1978).

Inorganic Pyrophosphatase (IPP, 3.6.1.1.) – striped bass *Morone saxatilis* (Guse et al. 1980).

Inosine Triphosphatase (ITP) – striped bass (Guse et al. 1980).

Aldolase (ALD, 4.1.2.13) – *Coregonus clupeaformis* (Kirkpatrick and Selander 1979).

Carboanhydrase (CA, 4.2.1.1.) – rainbow trout (Diebig et al. 1979).

Fumarase (FUM, 4.2.1.2.) – *Xiphophorus maculatus* (Siciliano et al. 1973).

Triosephosphate Isomerase (TPI, 5.3.1.1.) – brown trout, American cyprinid fish *Hesperoleucus symmetricus* (Avise and Ayala 1976; Allendorf et al. 1977).

Phosphomannose Isomerase (PMI, 5.3.1.8.) – sockeye salmon (Grant et al. 1980).

General Conclusions

Polymorphic fish proteins can be divided into several groups by the nature of the genetic variation. This division is, of course, rather arbitrary. Transferrin occupies a particular place among all the fish proteins studied. Only one locus, Tf, is present in almost all fish species and it almost always has three or more alleles. Most fish species appear to be polymorphic with respect to the Tf locus. A considerable proportion of all the individuals in fish populations, not infrequently up to 40%–50% or more, is heterozygous with respect to Tf alleles. Most probably albumins are similar to transferrins in this respect, but so far they have been insufficiently studied. From our standpoint, the peculiar aspects of transferrins and albumin genetics are related to their monomeric structure and probably to the high specialization in their function. The question, why only one locus of transferrin is active even in polyploids such as salmonids or cyprinids, remains unresolved.

The second group includes esterases. The variation in esterases is rather high, and this is achieved by an increase in the number of homologous loci present in the genome and of the number of alleles in each locus. The esterases are generally monomeric proteins, less frequently their molecules are dimeric. Most esterases are active towards many substrates. This does perhaps explain the increased genetic variation of the esterase loci and the increase in the number of the loci themselves.

The fish lactate dehydrogenases comprise the third group. The main peculiarity of the LDH is its ability to form "hybrid" molecules – the heterotetramers, subunits of which are coded by different genes. In diploid fish species these are the genes A and B, similar to the genes M and H in mammals; in tetraploids these are duplicated loci A1 and A2, B1 and B2, C1 and C2. The large number of these loci results in the formation of a large number of isozymes, up to 15–18 in homozygotes and up to 25–30 in heterozygotes. In spite of the wide isozyme spectrum present in homozygotes the intrapopulational genetic variation of LDH in fishes is quite extensive, although it appears somewhat decreased in certain polyploid species. The diversity of LDH species probably reflects the apparently high differentiation of the function of this enzyme, both in the different organs and during embryonic development. The different isozymes of LDH have significant differences in the number of their properties. The pattern of malate dehydrogenase variation does in many respects resemble the variation of LDH.

A rather high variation is characteristic of such enzymes as 6-phosphogluconate dehydrogenase (6-PGD), superoxide dismutase (SOD), phosphoglucomutase (PGM) and perhaps phosphoglucose isomerase (PGI). Isocitrate dehydrogenase poorly studied in the fish also appears to belong to this group. Phosphoglucomutase occupies a somewhat different place, although the number of species with polymorphic PGM is large, (more than 50), polymorphism is limited to just one locus, which usually has two codominant alleles.

Further we have a large group of moderately variable enzymes (G6PDH, SDH, alpha-GPDH, AAT and probably some others) not yet examined to a sufficient extent. Eye lens proteins, crystallins, also appear to belong to this group.

The last group is comprised of a number of enzymes with low variability and includes haemoglobins and myoglobins. It is clear that the inclusion of an enzyme into one of these two last classes is rather arbitrary and depends to a large extent on the method used to assay the particular enzyme. The situation appears to be more unequivocal for haemoglobins and myogens; most fish species are monomorphic with respect to these enzymes, in a few cases of polymorphism the number of alleles is limited to two (without taking rare alleles into account). Nevertheless, the number of loci and of corresponding protein variants may be rather large in fish. Haemoglobins and myogens, probably like many nonenzymatic cell proteins, show only limited variation, just as in other animal species. It appears that mutations in the loci Hb and My somehow interfere with the normal work of the proteins smoothly tuned to perform their function.

As regards other enzymes such as ADH, AMY, AK, CAT, as well as gamma-globulins and haptoglobins, the information on their genetic variation in fish is extremely limited. Any mechanisms underlying the different variation of various groups of proteins in fishes are far from being clear. Most authors attempt to relate this to the functional characteristics of the proteins themselves. In the case of enzymes the type and number of the substrate for a given enzyme may be an important factor. The number of subunits is another essential determinant, the tetramers are less variable than dimers and particularly monomers (Harris et al. 1976; Ward 1977, 1978).

We have already pointed out that probably all the fish species are heterogeneous with respect to blood groups. The significance and particularly the possibilities offered by these data for solving many problems regarding evolution and for improving methods of fish selection emphasize the need for further intensive research in this area.

In the overwhelming majority of cases the alleles of protein loci in fishes are codominant. Heterozygotes have isozymes (or isoforms) characteristic of both homozygotes. Inheritance is generally strictly Mendelian. Only null alleles and certain alleles of the regulatory genes are recessive. However, individuals carrying the dominant gene in the homozygous state may be frequently differentiated from the heterozygotes by the quantity of the synthesized product. The problem is more complex with blood group antigens, but even here the selection of more sensitive reagents often makes it possible to distingish between the homozygous and heterozygous genotypes by the intensity of the agglutination reaction.

Gene linkage has only been established in a few cases. There is some evidence that the locus of one of the muscle proteins (My) and the locus of fast esterases (Est-F) in the common carp are linked (Moskovkin et al. 1973). Linkage has also been found for the genes of G6pd and 6Pgd in the sunfish (the genus *Lepomis*) where the crossing-over frequency is 15%–22% (Wheat et al. 1973).

Two linkage groups were described in *Xiphophorus* spp. (Poeciliidae). In the first of them the crossing-over frequency between the genes Ada, G6pd and 6Pgd is equal to 6% and 24% respectively, in the second group the crossing-over frequency between the genes Est-2, Est-3, Ldh-1 and Mpi is equal to 43%, 26% and 19% (Morizot and Aravinda 1977; Morizot et al. 1977; Morizot and Siciliano 1979).

The linkage between the genes Ldh-B and Ldh-C has been shown in the killifish *Fundulus heteroclitus* (Whitt 1969), the linkage between the genes Ldh-1, Est-5, Idh-2 and Est-4 was demonstrated in *Poeciliopsis monacha* (Leslie 1979). It is also suggested that the two genes coding for haemoglobin in the cod are linked (Sick et al. 1973), as well as the genes sMdhA1 and A2 in the pink salmon (Aspinwall 1974a) and the genes Aat1 and Aat2 in *Salmo clarkii* (Allendorf and Utter 1976). In the latter cases we are probably dealing with the homologous genes, which originated as a result of tandem duplication.

In most cases, the genes coding for different enzymes and non-enzyme proteins in fishes are not linked. Taking into account the large number of chromosomes in fish genomes, generally 24–25 pairs or more, the probability of detection of linked loci is quite low.

The autosomal character of inheritance has been demonstrated for many protein systems. In the polyploid fish species such as the common carp, the crucian carp, the barbel, all the salmonids, catostomids and others the number of loci per enzyme is generally elevated although the same number of gene codes for many enzymes in diploid and tetraploid species (Engel et al. 1975; Ferris and Whitt 1977a, b). When duplicated genes are present in polyploids they are almost always inherited disomically and independently of each other.

The variation of regulatory genes is poorly understood at present in fishes, it can only be suggested that it may be as extensive as the variation of the structural loci.

Phylogenetically related fish species frequently possess different alleles of one and the same locus. When multimeric proteins are analyzed in interspecific hybrids, characteristic hybrid bands can usually be observed in gels.

In conclusion, we shall list the polymorphic genetic systems detected until now in the three most extensively studied fish species. We did not include the loci controlling blood groups in this list (Table 34). The number of genetic systems examined is still quite small, but data is rapidly being accumulated.

The genetics of fish proteins is generally based on studies of natural polymorphism. Useful alleles have been selected by nature for hundreds of thousands and

Table 34. Genetic variation of proteins in the carp *Cyprinus carpio*, the rainbow trout *Salmo gairdneri* and the Atlantic herring *Clupea harengus*

Cyprinus carpio		*Salmo gairdneri*		*Clupea harengus*	
Loci	Number of alleles	Loci	Number of alleles	Loci	Number of alleles
Tf	7–8	Tf	2	Tf	2–4
Hp	2	Postalb	2	Alb	2
Prealb	2	Alb	2	My-1	2
Alb	4	Lp (lens protein)	2	My-2	2
My	2	Ldh-B_1	3	Ldh-A	2
Ldh-C_1	2–3	Ldh-B_2	2	Ldh-B	2–3
Ldh-C_2	2–4	Ldh-C	2	sMdh	2–3
Ldh-B_1	2–3	sMdh-A	2	G6Pd-1	2
sMdh	2	sMdh-B_1	3	G6Pd-2	2
G6Pd	2	sMdh-B_2	4	6Pgd	2
sIdh-1	2	mMdh	4	sIdh	2
sIdh-2	3	G6Pd	2	Sdh	3
Pgm	2	αGpd	2	Pgm	3
Est-F	2	sIdh-1	2	Pgi	3
Est-S	2	sIdh-2	2	Est-I	4
Aat	2	Sdh	2	Est-II	4
Ck	2	Adh	2	Est-III–V	2–3
To	2	Pgm	2	Aat	2
		Est-I, Est-II	2	Px	2
		Aat	2		
		Ck	2		
		To	3		
		Ca	2		
		Cat	2		

millions of years, these were the alleles which did not cause an appreciable lessening of viability. By this criterion the alleles of biochemical loci differ fundamentally from the majority of mutant genes investigated which affect the qualitative morphological traits and generally lead to a deviation from the normal state.

The use of biochemical markers in genetic studies and in selective breeding work is therefore particularly important and opens up impressive prospects.

Chapter 6

The Use of Biochemical Variation in Embryological, Populational and Evolutionary Studies of Fishes

Gene Expression During Embryogenesis

The discovery of a large number of polymorphic biochemical systems in fishes was instrumental in elucidating certain general aspects of gene action during embryogenesis in animals. Below we shall consider the most interesting results of these studies conducted with fishes.

The General Characteristics of Gene Action During Development. It is a general characteristic of oogenesis in most fish species that the mass of the oocyte increases one hundred or even a thousandfold as a consequence of the accumulation of reserve substances such as yolk. At the stage of rapid growth (during vitellogenesis) the chromosomes take on the appearance of the so-called lampbrushes due to the formation of numerous lateral DNA loops. Such loops are the sites of active mRNA synthesis; protein synthesis increases as well at this stage, and proteins accumulate in the oocytes in large quantities. Part of the mRNA is conserved in the mature egg in the form of stable mRNP (ribonucleoprotein) complexes which are being actively studied at present. Such mRNA, which is inactive during protein synthesis in the oocyte was called masked mRNA; the corresponding mRNP was called the informosome. Protein synthesis is virtually absent in mature eggs, which can be retained in fish ovaries for long periods of time, e.g., for several months or even years. Protein synthesis begins again after fertilization.

The genes of the embryo appear to be inactive up to the stage of the blastula and sometimes even up to gastrulation. Protein synthesis at the early stages of development makes use exclusively of the template (or mRNA) molecules which had been accumulated in the form of informosomes during oogenesis; it follows that only maternal proteins are synthesized at this period. This conclusion follows from numerous observations of various fish species (Whitt 1970b; Neyfakh and Timofeeva 1977; Shaklee and Whitt 1977; Neyfakh and Abramova 1979). The amount of enzymes, in particular the enzymes involved in glycolysis and in carbohydrate metabolism, remains almost constant in the developing fish embryo or slowly decreases (Milman and Yurovitzky 1973). These enzymes include among others LDH (the B locus, homologous to the heart locus H of mammals), MDH and PK (Shaklee and Whitt 1977). A particularly important part in this case is played by the enzymes synthesized on maternal RNA templates; these enzymes are only gradually substituted by embryonic (zygotic) proteins.

The products of the alleles of certain loci coming from the male parent can be detected for the first time in developing fish embryos at the stage of the late blastula or early gastrula; however, an appreciable fraction of the genes becomes active even later, sometimes after embryogenesis is already completed (Champion et al. 1975; Ivanenkov 1979; Neyfakh and Abramova 1979). The differentiation of genes by the time of their activation and by the tissue specificity of expression is typical of fishes (Markert and Ursprung 1971; Korochkin 1977; Shaklee and Whitt 1977).

Classification of Enzymes by the Time of Their Appearance During Development, Tissue Specificity and Functional Characteristics. All the known fish enzymes can be classified into several groups. The first group includes the enzymes of universal or almost universal occurrence present in almost all tissues and organs of the developing and adult individuals. These enzymes are associated in one way or another with the most important intracellular pathways such as glycolysis, carbohydrate metabolism or the interconversion of phosphorus compounds. Usually this group of enzymes is called the "housekeeping" enzymes (Neyfakh and Abramova 1979). These enzymes include LDH, MDH, cytochrome oxydase, G6PDH and 6PGDH, as well as aldolase, certain esterases, creatine phosphokinase, phosphoglucomutase and others (Shaklee et al. 1974, 1977; Frankel and Hart 1977; Neyfakh 1979; Pontier and Hart 1979). It is a characteristic feature of all these enzymes that they are present in fairly constant amounts in developing embryos. The embryos appear to conserve the enzyme molecules synthesized during oogenesis for long periods of time. The synthesis of new or "embryonic" enzymes belonging to this group begins at relatively late stages (Shaklee and Whitt 1977).

The second relatively small group consists of enzymes characteristic only of the embryonic period; they disappear by the end of embryogenesis or at the larval stage.

The third and most numerous group includes enzymes intimately associated with tissue differentiation and appearing as a rule at definite, specific stages of development, sometimes even after embryogenesis is completed (Shaklee et al. 1974; Champion et al. 1975; Ivanenkov 1976; Timofeev 1979).

Many fish enzymes are coded by several loci, their number varying from two to four in diploid species to five to six loci in tetraploids such as salmonids, certain cyprinids and catostomids. The increase in the number of loci is achieved either by the tandem duplication of genes or is due to the duplication of the whole genome (Ohno 1970a, b; Markert et al. 1975). The appearance of two or more homologous genes during evolution creates a basis for their evolution involving structural changes and the functional divergence of the protein products. "The division of labour" between isoenzymes (or variants) controlled by duplicated loci contributes to the optimization of the function of each enzyme (or a non-enzymatic protein such as haemoglobin) in different tissues and organs at different stages of development and in different environmental conditions (Hochachka and Somero 1973).

The differentiation of LDH loci has been studied in some detail. The loci A, B and C in diploid fishes have undergone a very pronounced functional divergence (the A and B loci are homologous to the M and H loci of higher vertebrates). The isoenzyme A_4 is generally found in the muscles and appears at relatively late stages of development. The locus B is generally active throughout embryonic development; later the products of this locus can be detected in many organs and tissues, primarily in the heart, spleen, liver and blood. The B_4 LDH isoenzyme can thus be included in the group of "housekeeping" enzymes. Its role, however, can sometimes be taken over by the enzyme A_4 (Philipp and Whitt 1977). The C locus in most teleostean fishes is specifically expressed in the eye retina; the beginning of the C_4 isozyme synthesis in larvae corresponds to the period when the differentiation of receptor cells in the eye retina is already completed. In other words, the appearance of this enzyme corresponds to the beginning of the active visual function (Nakano and Whiteley 1965; Whitt 1969, 1970b, 1975a; Whitt et al. 1973c; Markert et al. 1975; Miller and Whitt 1975). In Cyprinidae, Gadidae and certain other fishes no specialized eye-specific Ldh locus is present, the gene Ldh-C being active in the liver, where it plays the part of Ldh-B locus (Whitt and Maeda 1970; Markert et al. 1975).

Differences in the tissue distribution and the time of activation of duplicated Ldh genes have been found in *Brachydanio* (Frankel and Hart 1977), *Lepomis cyanellus* (Miller and Whitt 1975), *Erymyzon* (Champion et al. 1975), *Oryzias latipes* (Philipp and Whitt 1977), and other species.

Distinct and sometimes very significant differences between the Ldh-B1 and B2 loci products of secondary duplication due to polyploidization, have been observed in tetraploid fishes, where the functional differences between Ldh A, B and C loci are retained at the same time. In spite of the relatively short period of time that has elapsed since the moment of duplication of the number of chromosomes in the evolutionary ancestors of salmonids (probably more than 20 million years) the B1 and B2 loci appear to play different roles (Bailey and Lim 1975; Lim et al. 1975). The isozymes $B1_4$ and $B2_4$ in Pacific salmon differ from each other in their heat stability and activity (Kirpichnikov 1977; Kirpichnikov and Muske 1980). In the case of Ldh genes, the divergence has proceeded so far that in many fish species, including diploids, the formation of heterotetramers consisting of A and B subunits (AB_3, A_2B_2 and A_3B) or of A and C subunits (AC_3 etc.) became difficult or even impossible.

Below we shall mention just a few other loci belonging to the class of duplicated ones which underwent this type of functional divergence. Each of the enzymes such as glucose phosphate isomerase (GPI), malate dehydrogenase (MDH) and creatine phosphokinase (CK) in the sunfish *Lepomis cyanellus* is coded by two loci. One of these loci codes for the isozyme present in all the tissues of the adult fishes; the other isozyme is activated at the stage of hatching and is only active in the white or skeletal muscles. Similar, although somewhat less distinct differentiation is characteristic of such loci as Ldh, Pgm, G6pdh, Ak and Est; in each of these cases, at least one locus has a broad action coding for the synthesis of a housekeeping enzyme (Champion and Whitt 1976a). Quite frequently the time of the appearance and localization of isozymes produced by different homologous loci are very similar even in systematically distant fish species. Similarity in the ontogenic pattern of isozymes has been established for 15 enzymes (30 loci) in *Lepomis cyanellus* (Centrarchidae) and *Erymyzon succetta* (Catostomidae) (Champion et al. 1975).

The loci of the soluble cytoplasmic malate dehydrogenase sMdh-1 and sMdh-2 (Valenta 1977) and three loci of creatinekinase Ck-A, Ck-B and Ck-C (Pontier and Hart 1979) have undergone prominent divergence in the group of cyprinids. The tissue specificity of creatine kinase isozymes has been studied in various species of teleostean fishes; recently the phylogenetically youngest fourth locus of this enzyme Ck-D has been found in the testes of most evolutionary advanced teleostean fishes (Fisher and Whitt 1978a, b; Ferris and Whitt 1979).

A number of enzymes is present in fishes in the two forms, mitochondrial and soluble ones; such enzymes include aspartate aminotransferase (AAT), isocytrate dehydrogenase (IDH), superoxide dismutase (SOD) and a few others in addition to MDH. Probably the duplication of these loci occurred hundreds of million years ago, and they are generally unable to form hybrid molecules.

Fish haemoglobins also exist as multiple variants. The change in haemoglobin types due to changes in the activity of different loci during development has been observed in the coho salmon *Oncorhynchus kisutch*; this phenomenon is similar to the one described for mammals (Giles and Vanstone 1976).

Calculation of the number of isozymes of a given polymeric enzyme can be performed using the following formula:

$$C = \frac{(n+k-1)!}{n!\,(k-1)!} \quad \text{(Shaw 1965)},$$

where n is the number of loci and k the number of polypeptides in the enzyme molecule.

For lactate dehydrogenases of salmonids and certain cyprinids such as *Cyprinus carpio, Carassius* sp. sp., *Barbus barbus,* C is equal to 70 where there are five loci in the genome and four subunits in the protein molecule. In reality, however, many heteropolymeric isozymes are not formed because of the major differences in the structure of A, B or C polypeptides or because of the discordance in the time and location of their synthesis. In salmon, for example, on about 20 electrophoretically separable LDH isozymes can be observed in the homozygotes; in the heterozygotes this number increases to 30 or even more.

Expression of the Maternal and Paternal Alleles in Hybrids. The alleles of "biochemical" loci received from the female and male parent are usually activated at one and the same time during embryonic development in fishes (Korochkin 1976a). The synchronous expression of the alleles of the female and male parent is not infrequently observed in cases of interspecific hybridization, particularly when closely related species are mated. For example, the time of appearance of the male parent's aldolase in various reciprocal hybrids between the loach *Misgurnus fossilis* and small warm-water cyprinids is similar (Glushankova et al. 1973; Neyfakh et al. 1976). These conclusions are deduced from measurements of the heat stability of the hybrid aldolase, differing for the two parental forms (Fig. 55). The synchrony of activation of the alleles of several other enzymes such as Est, G6PDH, cytochrome oxidase, GlutDH and LDH has been demonstrated in these crosses as well. It has been revealed in crosses of two species of *Brachydanio,* that the male parent's and female parent's alleles of LDH (Frankel and Hart 1977), Mdh, Xdh, Aat and Ck-B (Pontier and Hart 1978, 1979) are activated simultaneously. Complete synchrony in the activation of Gpi alleles and apparently of a few other genes as well has been demonstrated in hybrids between *Lepomis cyanellus* and *L. gulosus* (Champion and Whitt 1976 b).

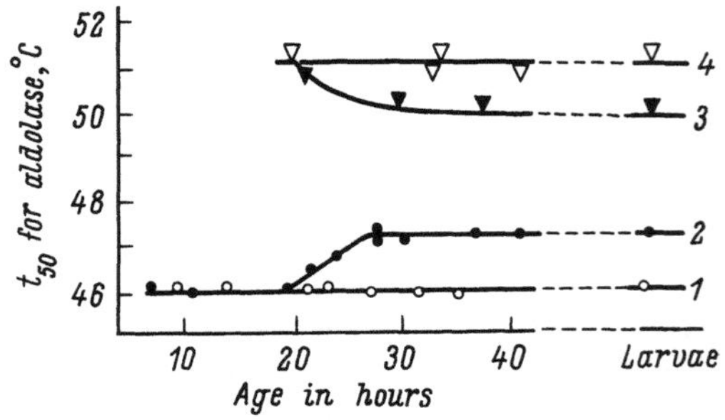

Fig. 55. Variation of aldolase heat stability in the loach *Misgurnus fossilis*, danio *Brachydanio rerio* and their reciprocal hybrids in the course of development. *1* loach; *2* hybrid loach × danio; *3* hybrid danio × loach; *4* danio (Glushankova et al. 1973)

In more distant crosses the expression of male and female parents' alleles is frequently asynchronous. Hybrids have been described, in which the paternal alleles were expressed at later stages and to a lesser extent than the maternal ones. Suppression of the male parent's alleles of 6Pgdh, α-Gpdh, Ldh-B and Adh loci has been recorded in experiments on hybridization of the rainbow trout (*Salmo gairdneri*) and the brown trout (*Salmo trutta*) (Hitzeroth et al. 1968; Klose et al. 1969 a; Ohno 1969 b); a similar observation has been made for the Pgi locus in hybrids of the brown and brook trout (*Salmo trutta* × *Salvelinus fontinalis*) (Engel et al. 1977). Repression of the paternal allele of Ldh-B and α-Gpdh loci has been observed in hybrids of the brook and lake trout (Goldberg et al. 1969; Yamauchi and Goldberg 1974; Wright et al. 1975; Schmidtke et al. 1977). The male parent's alleles of the Est and Adh genes show delayed expression in hybrids of *Brachydanio* (Hart and Cook 1977; Frankel 1978). In hybrids of two sunfish species *Lepomis cyanellus* and *Eupomotis* (*Lepomis*) *gibbosus* the activation of the paternal alleles of Gpi-B, Ck-A and Ck-B, Mdh-A and Mdh-B as well as of Ldh is delayed (Whitt et al. 1977). It should be pointed out, however, that in reciprocal crosses yielding non-viable hybrids, the beginning of expression of the male and female parent's alleles in the embryo appears to be identical. Difference in the expression of paternal and maternal alleles of several enzyme loci in sunfish hybrids shows a positive correlation with the genetic distance between crossed species (Philipp et al. 1980).

Sometimes the maternal alleles are subject to repression. This has been established for G6Pdh and Est (liver) loci in the hybrids *Micropterus salmoides* × *Lepomis cyanellus* (Whitt et al. 1972, 1973 a) and for the 6Gpdh locus in experiments on the hybridization of the brown and brook trout (Schmidtke et al. 1976 a). Finally, in the crosses of poeciliids *Xiphophorus maculatus*, *X. variatus* and *X. montezumae* differing in alleles of the gene Amy, the allele of *X. maculatus* is always more intensively expressed in F_2 heterozygotes (Kallmann 1975; Herrera 1979).

Here, just as in other cases, the asynchrony of activation and the differing extent of expression of the paternal and the maternal genes appears to result from altering the balance of nuclear-cytoplasmic relationships; in more general form this can be interpreted as a consequence of the maladjustment of the complex regulatory processes controlling the activity of genes in parental forms (Champion and Whitt 1976a; Frankel 1978; Herrera 1979). It may also well be that the regulation of gene activity is impaired at the posttranscriptional level during translation (Korochkin 1977).

Under certain circumstances the male or female parent's allele may not be expressed in hybrids at all (Ohno 1969b; Vrijenhoek 1972; Whitt 1975b). The greater the evolutionary distance between parents in a given cross, the higher is the probability of such deviations from the normal pattern.

I should specifically like to mention an extraordinary case where no expression of alleles from one of the parental species was found in certain hybrids. In the hybrids of *Salmo salar* and *S. trutta* the spectra of esterases, peroxidases, haemoglobins and a number of serum proteins are intermediate between those of both parents; in the F_2, F_3 and F_b, however, they become identical to the spectra of *S. trutta* (Nyman 1967; Cross and O'Rourke 1978). The authors explain this phenomenon by postulating "selection" favouring the genome of *S. trutta.* This selection must be exceptionally intense since no variation has been observed with respect to a large number of protein systems either in the F_2 or subsequent generations.

Functional Differences Between Isozymes and Between Allelic Forms of Proteins

Any increase in the number of loci coding for one and the same enzyme or non-enzymatic protein provides an oportunity of differentiating these loci and their protein products. Isozymes and protein variants coexisting in fishes frequently differ in their charges, K_M values, temperature stability, optimal substrate concentration, sensitivity to inhibitors, and sometimes even in the substrate specificity.

Several examples of differences of this type can be listed. The most abundant evidence is available for LDH isozymes.

Experiments conducted with the mackerel and some salmonids have demonstrated that the homopolymeric isozymes of LDH can be presented as the following series

$$A_4 — B_4 — C_4$$

on the basis of their kinetic characteristics and stability. In this series the A_4 homopolymer shows the least stability to heating. This isozyme is adapted to perform well under anaerobic conditions, the concentration of pyruvate optimal for its activity is high and the insensitivity to inhibitors is maximal. Quite opposite characteristics are found for the C_4 isozyme, the B_4 isozyme occupying an intermediate position (Hochachka 1967; Somero and Hochachka 1969; Whitt 1970b, 1975a; Wuntch and Goldberg 1970; Shaklee et al. 1973; Bailey and Lim 1975; Lim et al. 1975). All three isozymes differ in their tissue specificity: the A_4 isozyme being found predominantly in the muscle tissue, the B_4, as has already been mentioned, is characteristic of most fish organs and tissues (the housekeeping enzyme) and the C_4 is synthesized in the eye retina and in liver in some cases.

It has already been mentioned above that the genes A, B and C (L) are present in duplicate copies in fish species of tetraploid origin. The daughter isozymes $B1_4$ and $B2_4$ are functionally different in many species. The $B1_4$ isozyme shows a lower heat stability as compared to $B2_4$ in *Oncorhynchus tschawytscha* and in *Oncorhynchus nerka* (Lim et al. 1975; Kirpichnikov 1977). The $B1_4$ isozyme is inhibited at higher pyruvate concentrations and the K_M value is also changed (Lim et al. 1975). No differences between $A1_4$ and $A2_4$ have been observed (Lim and Bailey 1977).

According to Ryabova-Sachko (1977 a), the B group isozymes of salmonids are more sensitive to changes in the oxygen concentration in water, while the group A isozymes are more sensitive to temperature variation. When the killifish *Fundulus heteroclitus* is transferred to warm water, a marked stimulation of the activity or content of A_4 and A_3B isozymes is observed; these isozymes appear to be more effective in catalysis where there is intensified muscle glycolysis and an increased pyruvate concentration (Bolaffi and Booke 1974).

Similar data are available for other proteins as well. PGI isozymes differ in the temperature sensitivity in *Lepomis* hybrids (Champion and Whitt 1976 b). A similar difference has been noted between αGPDH isozymes in the lake whitefish species (Clayton et al. 1973 a).

Haemoglobin loci have also undergone functional divergence in fishes. Two variants of haemoglobins were found in the eel *Anguilla japonica,* one of these variants having a lower affinity to oxygen. Haemoglobin with a decreased capacity to transfer oxygen functions during the sea period of the eel's life, the more active haemoglobin works during the freshwater phase, when the fishes frequently happen to be in water where there is a shortage of oxygen (Poluhowich 1972). The cathodic and anodic fractions of haemoglobins in salmonids differ in their temperature stability, sedimentation constant, and other characteristics (Bushuev 1973).

The isoforms of haemoglobin in two sympatric catostomid fishes correspond in their functional characteristics to the conditions of life. In haemoglobins of *Catostomus clarkii,* which prefer a swift water habitat, there are the fractions devoid of the Bohr effect, these fractions are completely absent in the slow water fish *C. insignis* (Powers 1972).

It appears that in poikilotherms (more specifically in ectotherms) which include all fishes, the presence of only one variant of an enzyme or a non-enzymatic protein is frequently insufficient to ensure normal functioning throughout the temperature range (Hochachka and Somero 1973). The presence of several proteins with a similar function becomes understandable, if one takes into account the fact that many other external and internal environmental factors vary in addition to temperature. These factors include the degree of water saturation by oxygen, salinity, the speed of the current, the composition of the intracellular medium in different tissues and organs of fishes, etc. The set of isozymes and variants adapted for every occasion in life is advantageous for the individual and for the species as a whole (Hochachka and Somero 1973; Johnson G. 1976; Kirpichnikov 1979).

Considerable information has been accumulated regarding the functional differences between allelic variants, such as "allozymes" or "alloforms", of one and the same protein locus. The most interesting observations made in fishes will be mentioned below. A marked proportion of these observations refers to LDH allozymes.

LDH preparations isolated from fishes with three different genotypes in the minnow *Pimephales promelas* (Cyprinidae) differ in the K_m value (Merritt 1972). The $A2'_4$ allozyme in the common carp (probably the liver allozyme $C2'_4$) more temperature-resistant than the $A2_4$ allozyme (Rolle 1979). The purified B^b_4 allozyme in the killifish is more active at lower temperatures (around 10 °C) while B^a_4 allozyme is more active at higher temperatures (Place and Powers 1979). Intracellular levels of ATP are different in specimens of the killifish with different LdhB alleles. There appears to be a physiological correlation between the function of lactate dehydrogenase and haemoglobin (Powers et al. 1979). Several papers have dealt with LDH allozymes among salmonids. The allozymes $B2_4$, $B2'_4$ and $B2''_4$ in the rainbow trout possess different K_m values after purification (Kao and Farley 1978a, b). There are also differences in the reaction of genotypically different fishes to low oxygen concentrations (Klar et al. 1979). It should be pointed out that the $B2''_4$ allozyme is able to catalyse the pyruvate transformation into lactate more rapidly than the $B2_4$ allozyme; this means that in basins with a rapid current B2″ fishes have a certain advantage (Tsuyuki and Williscroft 1973; Huzyk and Tsuyuki 1974). Allozymes coded by the LdhC locus may also show differences in activity (Northcote et al. 1970). Trout carrying different alleles of the Ldh-H-alpha locus (apparently corresponding to the B1) differ in their tolerance of rapid current (Tsuyuki and Williscroft 1977). It cannot be excluded in this case, however, that the difference depends on the action of other, possibly regulatory genes linked to the Ldh-B1 locus because no genetic differences with respect to this trait were observed in the steelhead trout.

In the brook trout *Salvelinus fontinalis* differences in the limiting concentration of the inhibitor (substrate) have been observed between allozymes coded by different alleles of Ldh-B1; the temperatures of enzyme inactivation also showed some variation. These latter differences in temperature stability were quite significant (Wuntch and Goldberg 1970): the upper limit of the $B1_4$ allozyme was 70°–75° while for the $B1'_4$ allozyme these temperatures were 65°–70°.

In the sockeye salmon *Oncorhynchus nerka,* just as in other salmonids, the B1 and B2 subunits of LDH form a series of five isoenzymes in individuals homozygous with respect to two loci. Heating of homogenates prepared from the fry, homozygous in the slow or in the fast alleles of the gene B1 at 65 °C prior to electrophoresis, resulted in almost complete destruction or inactivation of the $B1_4$ homopolymer; the homopolymer $B1'_4$ ("southern") did not respond to heating. Heating at 70 °C was not only followed by the destruction of the $B1_4$ homopolymer, but also by the inactivation of two heteropolymers $B1_3B2_1$ and $B1_2B2_2$. Similar heating only slightly decreased the activity of the $B1'_4$ homopolymer, while the heteropolymers $B1'_3\,B2_1$ and $B1'_2\,B2_2$ hardly responded by any activity changes. The staining of gels at temperatures close to 0 °C has demonstrated that at such temperatures the activity of the $B1_4$ tetramer is even somewhat higher than that of the $B1'_4$ tetramer. Similar differences in stability between allozymes of LDH-B have also been detected in experiments with urea. These differences appear to be of a non-specific nature (Fig. 56).

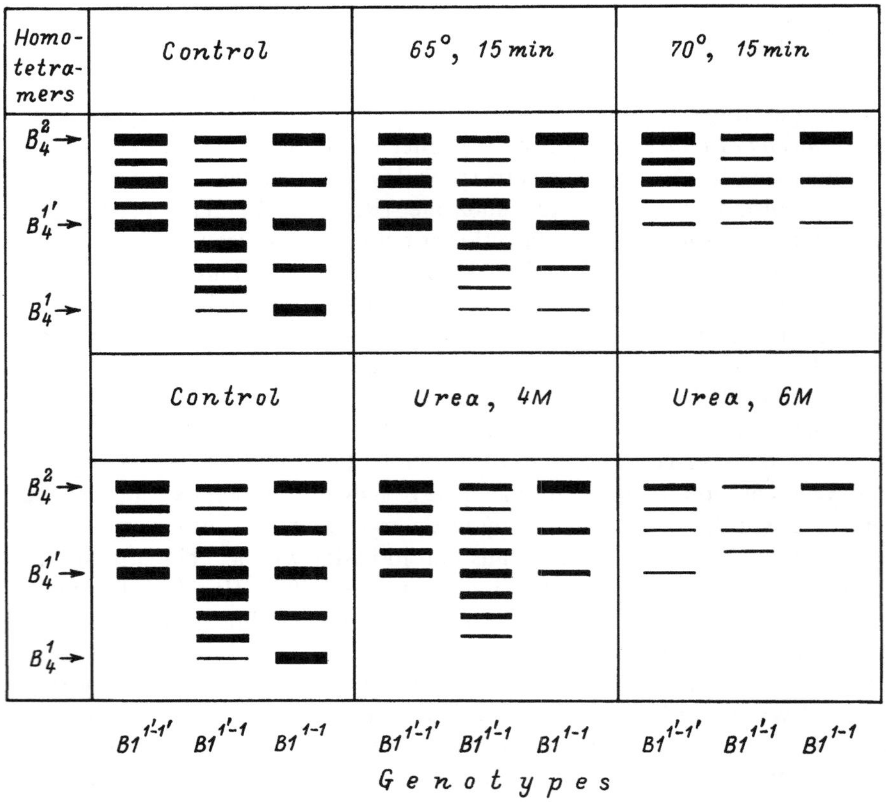

Fig. 56. Differences in heat stability and activity of LDH isoenzymes and allozymes in the sockeye salmon *Oncorhynchus nerka* (own data). *Left* types of homotetramers; *below* genotypes. *B1*^{1} and *B1*$^{1'}$ "northern" and "southern" alleles of the B1 locus.

Differences in the response of the two allelic LDH forms to heating exceed 10 °C. Final conclusions regarding the magnitude of this difference require the purification of LDH allozymes of *Oncorhynchus nerka,* since the general stability of enzymes may be affected by the complex processes of the interaction of enzymes with other substances and the formation of complexes with changed properties.

It has recently been demonstrated that closely related species of *Sebastolobus* (family Scorpaenidae) living at different depths, differ in the sensitivity of muscle LDH to pressure; in *S. altivelis* living at a lower depth this sensitivity appears to be decreased (Siebenaller 1978; Siebenaller and Somero 1978). We suggest that these two species of *Sebastolobus* possess different alleles of the Ldh-A locus, these alleles appear to code for allozymes with very different functional characteristics.

It is interesting to discuss the relevant evidence available for esterases. The curves showing the esterase activity as a function of temperature in *Catostomus clarkii* are very different for the enzymes coded by two alleles of the Est-1 locus (Fig. 57a). The esterase present in fishes homozygous with respect to the allele b, the frequency of which is highest in the northern part of the distribution range, rapidly lost their activity in response to a rise in temperature; the behaviour of the esterase present in individuals homozygous with respect to the southern allele a was quite the opposite; its activity was stimulated under these conditions; the activity of esterase from heterozygotes was maximal at intermediate temperatures (Koehn 1968, 1969a, 1970). Differences in the temperature stability of esterase allozymes were also found in the small American minnow *Notropis stramineus* (Koehn et al. 1971). In this case, heterozygotes differed greatly from homozygous individuals (Fig. 57b). The allozymes of esterases in *Fundulus heteroclitus* differed both in the substrate specificity and in the sensitivity to inhibitors (Holmes and Whitt 1970).

In *Notropis lutrensis* three alleles of the Mdh-B locus have very pronounced differences with respect to the temperature optimum of activity (Richmond and Zimmerman 1978). The allozyme present in B^{ff} homozygotes shows maximal activity at 20 °C, for B^{mm} homozygotes the maximal activity is observed at 25 °C, while for B^{ss} homozygotes the optimal activity is at 30 °C.

Differences in the temperature stability have been found with PGM allozymes of sockeye salmon even after purification of the corresponding enzyme preparations isolated from homozygotes and heterozygotes (Kirpichnikov and Muske 1980).

Functional differences between alloforms of certain non-enzymatic proteins have also been established. In the common carp, the bream *Abramis brama,* the rainbow trout *Salmo gairdneri* and the coho salmon *Oncorhynchus kisutch* individuals carrying different alleles of the Tf locus coding for transferrin differ in their viability, their resistance to disease and certain morphological and physiological traits (Tscherbenok 1973, 1978; Smišek and Vavruska 1975; Haberman and Tammert 1976; Reinitz 1977b; Saprykin 1977, 1979; Suzumoto et al. 1977). These relationships are not always stable and, in a number of cases, they were shown to depend on the genetic background (Iljasov and Shart 1979; Shart and Iljasov 1979).

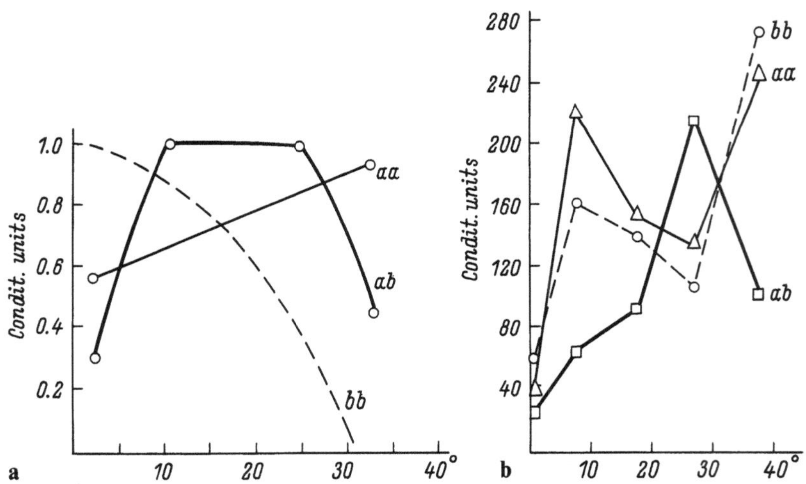

Fig. 57a, b. Functional differences between esterase allozymes. **a** the sucker *Catostomus clarkii* (Catostomidae); **b** the minnow *Notropis stramineus* (Cyprinidae), *aa, bb* homozygotes; *ab* heterozygotes; *abscissa* temperature; *ordinate* **a** activity, **b** optical density (in arbitrary units) (Koehn 1968; Koehn et al. 1971)

In the case of Tf^A and Tf^D alleles differing distinctly in the resistance of the corresponding fishes to an oxygen shortage (Saprykin 1979), one can, nevertheless, assume that the functional differences are directly associated with the A and D structural genes themselves.

The examples listed above appear to be quite convincing. The functional differentiation of alleles is a widespread phenomenon among fishes; it may well be that most isozymes and allozymes differ from each other in their specific functional characteristics. These peculiarities for such proteins as esterases, lactate dehydrogenases and haemoglobins are intimately correlated with the fishes' living conditions. Differences in the temperature stability and activity are regarded as one of more conventional indices of the divergence of allozymes and alloforms; differences of this type were observed in almost all cases. These differences are sufficiently pronounced to ensure the effect of natural selection on alleles of a polymorphic locus.

The Clinal Variation of Protein Loci

In many cases of biochemical polymorphism among fishes the allele frequencies of certain genes display progressive variation, depending on the actual geographical location within the distribution range of the species. This leads to the formation of the geographical cline or to a gradual change in gene frequencies as a function of latitude or simply as a function of distance. A number of more interesting examples will be given below.

Populations of the small sucker *Catostomus clarkii* (Catostomidae) have been studied along the Atlantic coast of America for about 800 km. The concentration of the Est^b allele in the North was 0.82, while in the South it decreased to zero (Koehn and Rasmussen 1967; Koehn 1968, 1969 a, 1970). Related species of the genus *Catostomus*, living in the North or migrating far upstream in rivers (high above sea level) have only one allele with the electrophoretic mobility of its product similar to that controlled by the northern (b) allele of the small sucker. In a *Pantosteus* sp., sympatric with *C. clarkii,* two bands of esterase activity are found after electrophoresis, these bands corresponding to A and B allozymes of *Catostomus clarkii.* It is assumed that this species possesses a kind of "fixed heterozygosity" (Koehn and Rasmussen 1967). In more southern species *C. santaanae* the band of esterase activity corresponds to the allozyme A of *C. clarkii.* Hardly any variation in frequencies has been observed in time (Koehn 1970).

Clinal variation of the esterase locus alleles has been found in the Atlantic herring *Clupea harengus* (Zenkin 1978). Parallel clines have been observed in all three main areas of the vast distribution range of this species in the northwest Atlantic, and in the North and Baltic sea. Clines associated with other protein polymorphic systems have also been described in the herring, these loci include Ldh-B, Aat (Odense et al. 1966 a) and the blood group system A (Altukhov et al. 1968; Truveller and Zenkin 1977 b; Zenkin 1978).

Extensive materials have been collected for the eelpout *Zoarces viviparus.* The concentration of one allele of the Est-III locus changes from 0.02–0.08 in the south to 0.30 in the north of the Baltic Sea in the 46 populations studied (16,000 fishes examined). The frequency of alleles of the Hb-1 locus changes in parallel, increasing from 0.138 to 0.806 (Frydenberg et al. 1973; Christiansen and Frydenberg 1974; Hjort and Simonsen 1975). The frequencies of these two loci correlate with each other; particular clines have also been observed in the fiords (Christiansen 1977; Christiansen and Simonsen 1978). The clines within fiords as well as extending throughout the range of the eelpout were previously demonstrated for a number of morphological traits, too (Schmidt 1917, 1919b, 1921; see Chap. 4).

The variation of Ldh-1 and Est-M loci in the crested blenny *Anoplarchus purpurescens* (Blennidae) is noteworthy. This variation extends in a latitudinal direction, allele frequencies in Puget-Sound bay change in a north-south direction:

Ldh-A′ from 0.02 to 0.26
Est-M^f from 0.08 to 0.27.

The correlation of frequencies of the two loci attains 0.89. Allele frequencies of the Ldh-A locus depend on the temperatures in the respective habitats and on the state of the oxygen supply. A related species *A. insignis* is monomorphic, containing only the northern allele; correspondingly, it lives in colder areas (Johnson M. 1971, 1977). According to new data the pattern of allelic geographical variation in the crested blenny is more complicated. The best correlation has been observed between allelic frequencies and the magnitude of temperature fluctuations (Sassama and Yoshiyama 1979). The clines for Ldh, Mdh, Est-4 and Gpi loci have been found in the killifish *Fundulus heteroclitus.* Although no distinct clines have been detected for other loci, nevertheless the correlation of gene frequencies and differences in environmental conditions can be detected under such conditions; the variation also appears to be associated with the sex and age of the fishes (Mitton and Koehn 1975). The correlation of frequencies of Ldh, Mdh and Gpi genes with the mean water temperature is particularly clear and distinct. In the regions where there is a large annual temperature fluctuation the overall heterozygosity appears to be increased (Fig. 58). The ranges of frequency are considerable: they vary from

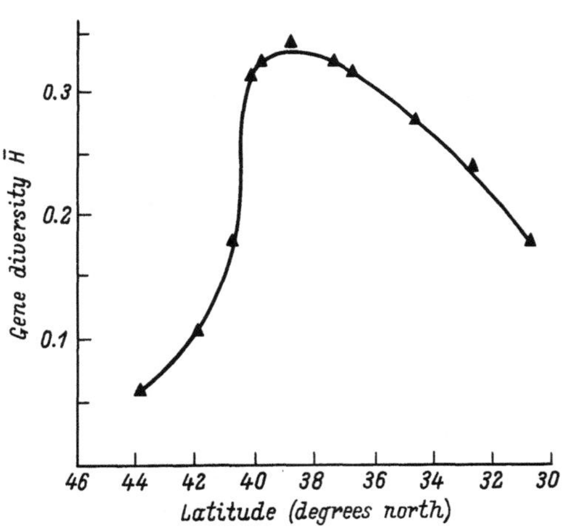

Fig. 58. Geographical changes of mean heterozygosity (four loci) in the killifish *Fundulus heteroclitus* (Place and Powers 1979)

0.05 to 1.00 for Ldh, from 0.03 to 1.00 for Mdh and from 0.5 to 0.75 for Gpi (Powers and Place 1978). In discussing the mechanism responsible for the appearance of clines in the killifish and attempting to make a choice between the drift hypothesis, the migration hypothesis and selection, same authors tend to recognize selection as the main factor responsible for the formation of geographical latitudinal clines. An important factor underlying the clinal variation does, according to the same authors, involve adaptation to the thermal conditions of life. The temperature variation can also be invoked to explain the formation of a cline for the Est locus in populations of the Arctic char *Salvelinus alpinus.* The concentration of one allele of this locus varies from 0.17 to 1.00 in different populations in a north-south direction (Nyman and Show 1971).

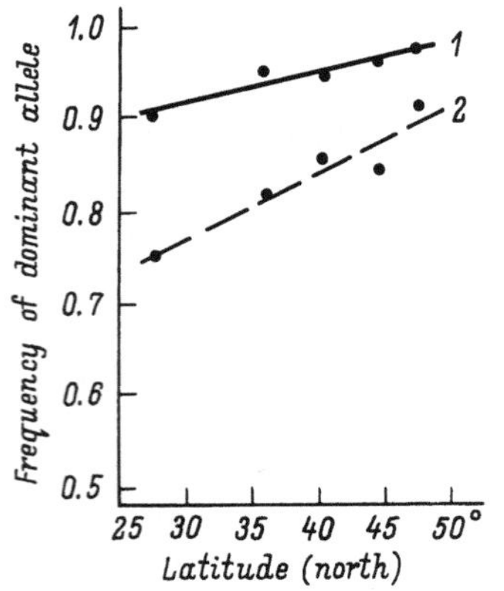

Fig. 59. The geographical variation (clines) of allele frequencies for Sdh (*1*) and Phi (*2*) loci in the American eel *Anguilla rostrata* (Williams et al. 1973)

The clinal variation for Sdh and Phi loci was demonstrated for the American eel *Anguilla rostrata* (the cline for Adh claimed earlier has not been confirmed) (Fig. 59). It has been initially suggested that during the migration of eels from spawning grounds the shoal undergoes differentiation due to some differences in behaviour (Williams et al. 1973). Later it was found that the cline for the Pgi locus is detectable solely in the permanent residents of given waterbodies; it cannot be detected for newly migrated eels. It may be suggested that the initial genetic differentiation emerging in the course of migration is intensified by the selection of certain genotypes during the freshwater period of life (Koehn and Williams 1978). The clines for both loci show no variation in time.

A cline for alleles of the Tf loci has been detected in Atlantic salmon *Salmo salar* migrating into the rivers of southern Canada and the northern United States (Table 35). The picture is complicated because of large fluctuations in the gene frequencies in different rivers within each of the subranges. The release of salmon smolts by commercial fish-hatcheries can explain these fluctuations at least in part (Møller 1970, 1971; Payne 1974).

The frequencies of two alleles of the haemoglobin locus change in the cod *Gadus morhua* living in the Norwegian and Baltic seas in a south-west to north-east direction (Frydenberg et al. 1965; Sick 1965a, b). A mixture of the two forms is found in shoals of cod near the Norwegian cost, the Arctic and the coastal one

Table 35. Frequencies of the Tf^A gene in populations of Atlantic salmon (North America coast) (Møller 1971)

Sampling site	Number of samples	Frequencies	
		Mean values	Variation of the mean
Labrador	1	0.085	–
Newfoundland	3	0.151	0.07–0.24
The rivers of St. Lawrence			
Northern region	10	0.440	0.24–0.65
Same, river Miramichi	24	0.351	0.22–0.48
Rivers of New Brunswick	7	0.318	0.23–0.42
Nova Scotia	6	0.391	0.29–0.54
Maine, USA	5	0.548	0.46–0.60

differing from each other in the structure of the otholiths (Møller 1968, 1971). Møller explains the clinal variation of the gene Hb frequencies in the cod by an increase in the relative abundance of the Arctic form in northwestern regions of the Norwegian sea. The existence of such a cline cannot, however, result from simply mixing two forms with different frequencies of the Hb allele (Fig. 60). The frequency of the Hb^1 gene varies according to Møller in two "twin" species in the following way (Møller 1968):

Arctic species: 0.03–0.21
Coastal species: 0.13–0.43.

As we can see, cod populations from the Skagerrack and the Baltic, as well as populations from the North Sea, cannot be classified as either race. It should also be added that the Hb cline is quite distinct for the Baltic cod where no Arctic form is found (Sick 1965a; Jamieson and Otterlind 1971). The frequency of the Hb^1 allele decreases, when one moves from the south-west to the north-east. It may well be, of course, that two different races, mixing within a narrow zone, populate the western and the eastern part of the Baltic Sea. It is more probable, however, that in this case as well the cline is supported by selection.

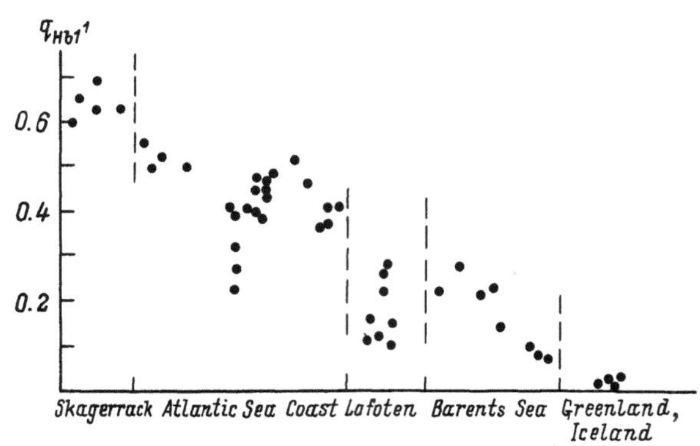

Fig. 60. Geographical variation of the Hb-1^1 allele frequency in the cod *Gadus morhua.* (de Ligny 1969, on the basis of material published by Frydenberg et al. 1965; Møller 1968)

Table 36. The frequencies of Ldh-B1 and Pgm-1 genes in the sockeye salmon from Kamchatka and the American Pacific Coast

Region	Fish age, race and year of sampling	Frequencies of alleles				Reference
		Ldh-B1		Pgm-1		
		q_{B1}	n	q_A	n	
Kamchatka						
Lake Kurilskoye	Spawners, 1976–77	0.171	281	0.333	204	4
Lake Nachikinskoye	Spawners, spring race, 1978	0.241	116	0.356	80	4
Lake Blizhneye	Spawners, 1976	0.074	74	–	–	3
Lake Dalneye	Smolts, 1973–1978	0.087	2108	0.256	1758	3
Lake Kronozkoye	Spawners, bentophage, 1976	0.235	80	–	–	2
	Same plus larvae, 1978	0.213	445	0.370	158	4
	Spawners, planctophage, 1976	0.078	243	–	–	2
	Same plus larvae, 1978	0.082	164	0.269	121	4
Lake Ushkovskoye	Larvae, 1978	0.081	480	0.200	140	4
Lake Azabachye	Spawners, 1971–1973	0.362	2766	0.218	2733	1
American Coast						
British Columbia and the State of Washington	Spawners	0.002	591	0.080	87	5
Same		–	–	0.163	95	5
South Western Alaska	Spawners	0.022	90	0.205	90	5
Same, Cooper River and Bristol Bay	Spawners	0.119	865	0.292	406	5

References: 1. Altukhov et al. (1975a); 2. Chernenko (1977); 3. Kirpichnikov and Ivanova (1977); 4. Kirpichnikov and Muske (1980); 5. Utter et al. (1973b)

Distinct clinal variation in Ldh-BI and Pgm-1 loci has been reported for the sockeye salmon of the American Pacific coast (Hodgins et al. 1969; Hodgins and Utter 1971; Utter et al. 1973b); the frequencies of rare alleles increase in a South-North direction (Table 36). A more complex distribution of frequencies is observed around the Kamchatka peninsula. The frequency of the northern allele Ldh B1 is high in the northern Lake Azabachye and attains minimal values in the southern Lakes Dalneye and Blizhneye, which are well heated by the sun. In Lake Kurilskoye to the far south where the water is, however, deep and cold, these alleles are found with a rather high frequency, their frequency in the population of Lake Nachikinskoye (Southwest) is even higher. The population of Lake Ushkovskoye differs drastically from that of the nearby Lake Azabachye, while two races with different gene frequencies coexist in Lake Kronotzkoye (Table 36).

The variation of allele frequencies of the Pgm-1 locus is also rather complex. In deep water forms (bentophage of Lake Kronotzkoe), and in some other populations the frequencies of the Northern allele Pgm-1^A are increased.

It should be pointed out that the northern isozyme of lactate dehydrogenase $B1_4$ has decreased heat resistance and exhibits high activity at temperatures around 0 °C (Kirpichnikov 1977). The northern isozyme of phosphoglucomutase A also exhibits a decreased temperature optimum of activity. The frequency variation of both genes appears to correlate with the temperature conditions for the embryonic development and subsequent life of the sockeye salmon larvae and fry in fresh water prior to their migration to the sea. At the same time, allele frequencies are affected by other factors as well, primarily by the oxygen supply and the accumulation of metabolic waste products at the spawning grounds.

Clines for different genes have also been found in the minnow *Rhinichthys cataractae* for Est locus (Merrit et al. 1978), in the suckers *Hypentelium nigricans* and *Moxostoma macrolepidotum* for Ck and Adh loci respectively (Buth 1977a, 1979a); in the stickleback *Gasterosteus aculeatus* for Est locus (Raunich et al. 1972), in several species of the atherinid fish *Menidia* for Est, Phi and other loci (Johnson M. 1974, 1976), in the largemouth bass *Micropterus salmoides* for Mdh-B locus (Childers and Whitt 1976) and in the whiting *Gadus merlangus* for the Hb locus (Wilkins 1971a); clines have also been detected in several other fish species. The clinal variation in fishes appears to be a common phenomenon. The most important characteristics of the clinal variation in fishes are mentioned below. Frequently, the variation is a function of the geographical latitude, the gene frequencies change from north to south. In several cases, an unequivocal relationship between the allele frequency and the temperature conditions of life in different parts of the range has been established. In two cases the differences in the temperature stability of the allozymes coincided with the direction of the clines: this was the case with the small sucker (Est locus) and the sockeye salmon (Ldh-B1 locus) (Koehn 1969a; Kirpichnikov 1977). Meanwhile certain loci of a given species produce distinct geographical clines, while for other loci no differences in allele frequencies across the range are observed, or no polymorphism is present at all in any population. The pattern of frequencies characteristic of the species as a whole is generally very stable in time: sometimes similar values are retained for many years. Finally, it should be mentioned that parallel clines for different loci appear to exist in different subspecies and the races of a given species, and correlated changes in the frequency of several different genes do, apparently, occur as well.

The different mechanisms responsible for the appearance of the geographical clinal variation in biochemical loci can be visualized. The clinal variation can result from a secondary contact between two populations which were previously isolated; in this case, the different alleles gradually migrate into the zone of mixing. It may well be that alleles of some loci are linked to inversions or to other genes having selective value. In the case of the emergence of a positive mutation and its gradual spread throughout the range, it can be suggested that there is a transient clinal variation. There can be no doubt, however, that among the possible reasons for clinal variation (Aronstam et al. 1977) selection is the most important. Generally, the cline can be visualized as a direct response on the part of the species to varying environmental conditions in different regions of the distribution range.

Monogenic Heterosis in Protein Loci

The survival advantages of heterozygotes within natural fish populations are frequently calculated by comparing the number of heterozygotes observed in the sample with the expected number, calculated using the Hardy-Weinberg equation. If the observed number of heterozygotes is significantly higher than the expected one, the presence of monogenic heterosis for the locus in question can be expected.

An excess of heterozygotes in different loci has been described in populations of several fish species. Among salmonids these cases include the char (Ldh locus), the sockeye salmon (Pgm locus), and the Ireland relic whitefish pollan *Coregonus pollan* (several loci) (Wright and Atherton 1970; Ferguson 1974, 1975; Altukhov et al. 1975a; Ferguson et al. 1978). The number of heterozygotes observed for the Pgm-1 locus in the sockeye salmon in almost all cases was higher than the one expected both in studies of spawning subpopulations and in individual crossings (Table 37). An excess of heterozygotes for the Hp and Alb loci (haptoglobin and albumin) has been found in the crucian carp *Carassius auratus gibelio* from Lake Sudoble, in Byelorussia (Polyakovsky et al. 1973) as well as in the goldfish (Sdh locus) from Lake Erie (Lin et al. 1969). A similar excess of heterozygotes has been found for alleles of the Hb locus in the catfish *Ictalurus nebulosus* (Raunich et al. 1966). An identical situation was observed in the cod for the Ldh locus (Jamieson 1975) and in the skipjack for the Tf locus (Fujino and Kang 1968).

The data obtained in studies of the population of the Pacific perch *Sebastes alutus* should also be mentioned. Abyssal populations had a characteristic excess of heterozygotes for the loci Pgm and α-Gpdh. There was, moreover, a significant correlation between allele frequencies in these two loci. In shallow-water populations neither excess of heterozygotes nor any correlation between loci has been observed (Johnson A. et al. 1971).

Data pointing to an excess of heterozygotes in natural populations cannot in themselves serve as proof of the presence of monogenic heterosis. The increased number of heterozygotes may be a consequence of certain mating patterns (preferential combination of different parents with different genotypes); it may also result from the so-called linkage disequilibrium, that is the tight linkage of each of

Table 37. Frequencies of PGM genotypes in individual crosses and spawning populations of the sockeye salmon

Sampling site	Year	Material	Numbers[a]	
			Heterozygotes	Homozygotes
Lake Kurilskoye[b]	1976	Larvae, cross No 16	115 (108)	101 (108)
		Larvae, cross No 19	264 (247)	229 (247)
Lake Dalneye[b]	1976	Spawners (dwarfs)	41 (34)	47 (53)
	1977	Spawners (dwarfs)	136 (108)	92 (120)
Lake Azabachye[c]	1972	Spawners (anadromous)	258 (245)	490 (503)
	1973	Spawners (anadromous)	366 (357)	638 (647)
	1974	Spawners (anadromous)	351 (330)	630 (651)

[a] Theoretical frequencies in brackets
[b] Kirpichnikov and Muske (1979)
[c] Altukhov et al. (1975 a)

the alleles with different recessive genes lowering the viability. It may also be due to an erroneous hypothesis of inheritance, in particular it can be a consequence of the presence of non-electrophoretic alleles.

The data obtained in experimental crosses of fishes are generally more convincing. Crosses followed by subsequent analysis of the sufficiently numerous offspring have been conducted with the sockeye salmon (Kirpichnikov 1977, 1979). A small excess of heterozygotes with respect to the Pgm-1 locus has even been noted during the examination of larvae (Table 37). In crosses of the common carp heterozygotes for the locus Tf (AB and to a lesser extent AC) showed an accelerated growth rate and increased resistance to dropsy (Tscherbenok 1973).

Advantages with respect to the growth rate have been observed for Tf^{BC} heterozygotes in the rainbow trout (Reinitz 1977b). In both cases, however, heterosis depended, to a large extent, on the general genetic background and its manifestation was not observed in all cases.

When crosses with different species of the family Centrarchidae have been performed, on excess of heterozygotes has been observed in F_2 and F_b in a number of cases. This was the case, for example, in crosses of two *Micropterus* species for the loci Ldh-C, Mdh and Idh (Whitt et al. 1971; Wheat et al. 1974); a similar situation has been discovered in hybrids of two species of *Lepomis* for Mdh, Est and To genes (Whitt et al. 1973b), in hybrids of *Pomoxis* species for gene Est (Metcalf et al. 1972) and in crosses of different genera, *Lepomis* × *Chaenobryttus* for the gene Hb (Manwell et al. 1963). It has been detected that the two species of *Lepomis* differed from each other in the alleles of two Pgi loci, A and B. In F_2 hybrids the number of double heterozygotes for these loci has been drastically increased (Whitt et al. 1976):

Homozygotes	42 (expected 64)
Simple heterozygotes	122 (expected 128)
Double heterozygotes	92 (expected 64)

Experiments involving crosses cannot completely exclude the linkage of two allelic genes with different semilethals: a lengthy and complex hybridological analysis is required to eliminate this possibility. The most direct proof of the presence of monogenic heterosis would include the assessment of the functional advantages (better function) of a hybrid protein in a heterozygous individual as compared to the corresponding protein molecules present in a homozygote. Unfortunately, data of this kind are at present very limited.

A specific hybrid haemoglobin is formed in the hybrids of the sunfishes *Lepomis cyanellus* × *Chaenobryttus gulosus* in addition to parental types. Examination of this haemoglobin has demonstrated that it has a higher affinity to oxygen and correspondingly a greater ability to transfer oxygen than the parental haemoglobins (Manwell et al. 1963). Transferrin isolated from the brook trout *Salvelinus fontinalis* heterozygous in A, B and C alleles binds iron better than that isolated from homozygous individuals (Herschberger 1970) (Fig. 61). The activity of esterase-1 in the heterozygotes of the small sucker is greater than in the homozygotes. This is observed over a broad temperature range extending from 10° to 30 °C (Koehn 1969a). Lactate dehydrogenase in heterozygotes of the sockeye salmon with respect to Ldh-B1 gene possesses a characteristically increased capacity to NADH binding; the K_m value is minimal at 15 °C (experiments of Guilbert, cited by Allendorf and Utter 1979).

In all these cases, heterogeneous mixtures of protein molecules have been subjected to examination. The relative contribution of individual isozymes or variants to the phenomenon of monogenic heterosis remains to be established. In the case of tetrameric haemoglobin the advantage of the heterozygotes could be attributed to the formation of additional hybrid molecules. For monomers such as transferrin and esterase a more plausible explanation of heterosis may lie in the

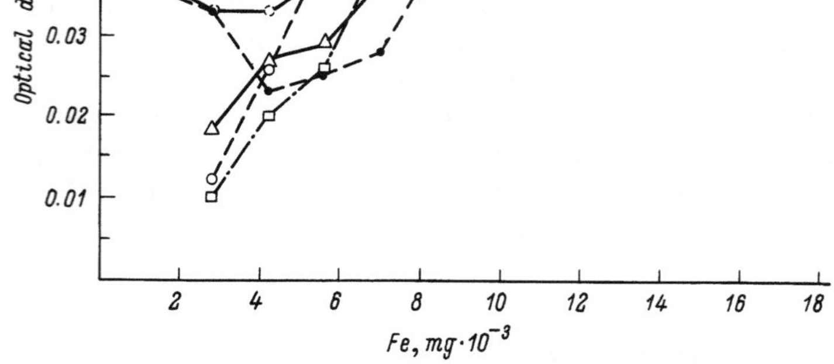

Fig. 61. Variation in the optical density of partially purified samples of transferrin (in response to the addition of various amount of iron) in the brook trout *Salvelinus fontinalis*. *1, 2* homozygous genotypes BB and CC; *3, 4, 5* heterozygotes BC, AC and AB (Herschberger 1970)

presence of two different protein molecular species in heterozygotes; then the normal function of this protein may be feasible in a wider range of environmental conditions and of fluctuations of the inner medium.

Monogenic heterosis appears to be a fairly rare phenomenon in natural fish populations. More and more data are becoming available in support of the assumption that the main part in the better performance of an individual is played not by monogenic heterosis but by heterozygosity in many loci conferring more significant advantage (Johnson G. 1976; Beardmore and Ward 1977; Beardmore and Shami 1979). Unfortunately, little is known about the mechanisms of the gene interaction responsible for such advantage.

Natural Selection with Respect to Individual Alleles

Allele frequencies in fish populations can be directly modified by the action of natural selection. In this section we shall not consider any data regarding the increase in the number of heterozygotes in hybrids, some of these data have already been discussed above. Several other contributions will be analyzed below.

Indirect evidence in favour of selection follows from determination of gene frequencies in different age groups of one and the same fish population. Age-dependent changes in the frequency of Tf alleles were observed in the plaice *Pleuronectes platessa* (de Ligny 1966), and in the skipjack *Katsuwonus pelamis* (Fujino and Kang 1968). Similar changes involving Est alleles were found in the bleak *Alburnus alburnus* (Handford 1971). The total number of heterozygotes in Mdh, Ldh and Est-4 loci increased as a function of age in populations of the killifish; it should be pointed out, however, that the frequency changes for each of these loci measured separately were insignificant (Mitton and Koehn 1975). In contrast, in the eelpout the selection operating at the early stages of development is directed against heterozygotes for the Est-III gene; since the eelpout is a viviparous fish, this allows us to calculate precisely the fitness of these three genotypes (Christiansen et al. 1974, 1977; Christiansen 1977, 1980):

	Genotypes		
	1/1	1/2	2/2
w	1.07	1.00	1.04

In the plaice caught at the site of habitation at an age of 100 to 300 days the overall frequency of heterozygotes for five loci (Pgm-1, Mdh-2, Ada, alpha-Gpdh, Phi) initially decreases and then rises again, attaining values characteristic of adult individuals (Beardmore and Ward 1977). The authors suggest that heterozygotes develop faster and are the first to occupy feeding areas.

Evidence of the selection favouring heterozygotes also follows from the data obtained in studies of populations of the guppy *Poecilia reticulata.* The overall

Table 38. Mean heterozygosity ($\bar{H}$) for Est-2, Mdh-1, Pgm and To (Sod) loci in a population of *Poecilia reticulata* (Shami and Beardmore 1978 b)

Fish age (weeks)	Number of scales (1.1.)					
	23	24	25	26	27	28
3.7	25.0	45.1	48.3	50.7	49.6	37.5
60.0	–	–	48.0	57.3	45.8	–

heterozygosity in four loci was found to be associated with the middle, most numerous phenotypes by a morphological criterion (number of scales along the lateral line); since the extreme groups do not survive during growth, we have stabilizing selection, which acts in a population simultaneously with the selection for high heterozygosity (Table 38).

Selection begins to operate even before young fishes are born, the size of the brood shows a positive correlation with the degree of heterozygosity (Shami and Beardmore 1978 b).

A comparison of the wild and hatchery populations of the chum salmon *Oncorhynchus keta* has demonstrated that the frequency of heterozygotes in the Ldh-A2 locus is greater in nature than in fish hatcheries (Ryabova-Sachko 1977 a). It has already been mentioned that in abyssal populations of the Ocean perch *Sebastes alutus* the fraction of heterozygotes in Pgm and alpha-Gpdh loci is higher than in shallow-water populations (Johnson A. et al. 1971). All this provides additional evidence of the existence of selection favouring heterozygotes.

In populations of the snapper *Chrysophrys auratus* (the New Zealand coast) allele frequencies of the Est-4 locus vary from one generation to another. It has been established that the frequency of the common allele Est-4^2 correlates with the temperature conditions during fry development varying from 0.425 in the warmest year to 0.504 in the coldest year. These differences are not large but statistically significant (Smith 1979 a).

In the area affected by the hot water effluents of a Swedish nuclear power station the frequency of the F allele of the Est locus was found to be significantly increased in the ruff *Acerina cernua* (Nyman 1975). Shifts in the frequencies of several loci in zones affected by the effluents of electric power-stations have also been observed among sunfishes (Yardley et al. 1974).

Even more direct evidence of selection is provided by experimental results. In experiments with killifish populations one group of fishes was kept in a warm basin: this resulted in changes in allele frequencies for three loci towards values characteristic of southern populations of this species (Mitton and Koehn 1975). When the coho salmon was infected with a bacterial kidney disease, the mortality of fishes with three different Tf genotypes, AA, AC and CC, was equal to 34.4%, 18.8% and 10.0%, respectively (Suzumoto et al. 1977). Shifts in the allele frequencies of Tf and Est-F loci were observed after the infection of carp with dropsy (Tscherbenok 1973; Kirpichnikov et al. 1976, 1979; Shart and Ilyasov 1979). In experiments with an acute oxygen shortage the frequencies of the Tf alleles were undergoing modification in the samples of the common carp examined as well as after crosses (Tscherbenok 1978; Saprykin 1979). The survival of fingerlings in the common carp, the rainbow trout *Salmo gairdneri* and the sterlet *Acipenser ruthenus* with different Ldh genotypes was very different in experiments where the temperature was rapidly increased (Rolle 1979).

The fry of the crested blenny obtained after crosses was grown at different temperatures. Frequencies of Ldh alleles at the end of the growing period differed significantly in warm and cold populations. Differences in the frequencies followed the pattern found for populations living in nature at different mean temperatures (Johnson M. 1971).

The larvae of the sockeye salmon obtained from parents heterozygous with respect to Pgm-1 and Ldh-B1 loci were subjected to raised temperatures and an acute oxygen shortage. The allele frequencies in both loci were different in resistant and susceptible fishes, in several experiments these differences were statistically significant (Table 39).

We come to the conclusion that there is considerable information on selection acting directly upon alleles. Such selection can be convincingly demonstrated when allozymes have specific characteristics relevant to certain environmental conditions (Allendorf and Utter 1979). This condition has been fulfilled only in a few cases so far, but all the observations provide convincing bases in support of the assertion that natural selection is an important factor contributing to the emergence and

Table 39. The effect of high temperature and oxygen deficit on the allele frequency of Ldh-B1 and Pgm-1 loci in the sockeye salmon (Kirpichnikov 1977; Kirpichnikov and Muske 1980)

Experimental conditions	Year and experiment number	Allele frequencies		Frequency differences
		Control	Survivors	
Rapid temperature shift	Ldh-B1, $q_{B1'}$			
(6° → 14°) during hatching	1976, 1	0.49	0.60[b]	+0.11
Slow temperature shift				
(6° → 28°), larvae	1976, 2	0.47	0.38	−0.09
(6° → 28°), larvae	1976, 3	0.30	0.27	−0.03
(6° → 28°), larvae	1977, 4	0.51	0.63	+0.12
(6° → 28°), larvae	1977, 5	0.30	0.23	−0.07
(6° → 28°), larvae	1978, 6	0.09	0.06	−0.03
Oxygen deficit, larvae	1976, 7	0.29	0.19[c]	−0.10
Oxygen deficit, larvae	1976, 8	0.28	0.24	−0.04
Oxygen deficit, larvae	1978, 9	0.096	0.068	−0.028
	Pgm-1, q_A			
Oxygen deficit, larvae	1976, 7	0.49	0.40[a]	−0.09
Oxygen deficit, larvae	1976, 8	0.57	0.55	−0.02
Oxygen deficit, larvae	1978, 9	0.15	0.10	−0.05
Oxygen deficit, larvae	1978, 10	0.23	0.15	−0.08

[a] Significance of the difference $p < 0.1$
[b] Significance of the difference $p < 0.05$
[c] Significance of the difference $p < 0.01$

maintenance of biochemical polymorphism in nature. It should be added that in natural populations selection should always involve a large number of polymorphic systems.

The Evolution of Fish Proteins

Nowadays biologists focus their attention on evolutionary mechanisms and the evolution rates of proteins in living organisms. Different approaches to the solution of this most important problem have been exercised. The following experimental techniques were used with fishes:

1. Direct determination of amino acid sequences and of the relative content of amino acids in the polypeptide chains of a given protein in different fish species.
2. Hybridization in vitro of polypeptide chains of homologous, but different polymeric proteins after their separation into subunits. The most common techniques involve the dissociation of subunits, using freezing-thawing and subsequent mixing of proteins of different origin.
3. The serological (immunogenetic) determination of relatedness between homologous proteins of different fish species and between isozymes of a given protein in one species.
4. Comparison of homologous enzymes (isozymes and allozymes) by catalytic and kinetic characteristics.
5. Analysis of the duplicated loci using the degree of divergence of their structure as well as the degree of functional divergence of the proteins coded by them as a criterion. Particular attention has been paid to studies of duplicated genes in polyploid fish species and specifically to the rate of loss of expression or "silencing" of one of the duplicated genes.
6. The construction of evolutionary trees for the better-studied fish proteins.

The data on the primary structure and the amino acid composition of fish proteins are rather limited. The information available on haemoglobins and immunoglobulins is extremely interesting and relevant in this respect.

The monomeric haemoglobin of the lamprey differs from human haemoglobin (alpha-chain) in 75% of all the amino acid residues (Rudloff et al. 1966). Forty percent of the amino acids are different in the haemoglobins of the lamprey and the common carp (Braunitzer 1966). Haemoglobins of the carp and the common sucker differ in 18 amino acids (their divergence occurred around 50 million years ago) while two genera of the lamprey (*Lampetra* and *Petromyzon*) differ in five amino acids (Goodman et al. 1975).

In the cod haemoglobins coded by two alleles of a single gene differ in one peptide (Rattazzi and Pik 1965); only one amino acid appears to be different in the haemoglobins coded by these alleles. By comparing the age of a taxon with the number of amino acid substitutions in a haemoglobin molecule, one can determine the approximate rate of evolution of the protein. One substitution was fixed in the molecule on average every 5–8 million years; after the genome polyploidization which repeatedly occurred in fishes, albeit at a different time in different taxons, the rate of substitutions appeared to increase. The evolution of haemoglobins at the early stages of the divergent evolution of agnathans and fishes, that is around 400–500 million years ago, took place much faster than among higher vertebrates (Goodman 1976). Goodman attributes the increased rate of accumulation of mutations to the need for the functional improvement of the haemoglobin molecule; later in fishes, as well as in other vertebrates, the selection predominantly protected this protein from functional deterioration, i.e., played a stabilizing role.

Heavy chains of immunoglobulins were compared in six species of agnathans and true fishes belonging to five taxons; Cyclostomata, Chondrichthyes, Chondrostei, Holostei and Teleostei (Marchalonis 1972). The degree of divergence was determined on the basis of differences in the content of different amino acids in the protein molecule. The maximal value was obtained in a comparison between the lamprey (Cyclostomata) and the testaceous pike (Holostei) ($\Sigma d = 70$); great differences have been discovered between immunoglobulins of the lamprey and of two chondrostean fishes ($\Sigma d = 40$ and 46). Differences between teleostean fishes and other taxons were relatively moderate (Σd from 12 to 26). The rate of evolution of immunoglobulins appeared to decrease in relatively more advanced teleost fishes. The complement of immunoglobulins present in fishes is much more limited when compared with that of mammals and birds: the heavy chains are represented solely by IgM (IgG, IgA, IgD and IgE are absent). A second component, IgN with a lower molecular weight appears only in Dipnoi (Marchalonis 1977).

The hybridization of proteins in vitro has been employed in a number of studies to elucidate the relatedness between multiple forms of fish LDH (Massaro and Markert 1968; Whitt 1970b; Massaro 1972; Sensabaugh and Kaplane 1972). A high degree of homology has been established for B and C subunits, while the homology for A and B subunits was lower up to the complete loss of their ability to form hybrid dimers. A similarity has been found between the C subunits from the eye retina of salmonids and those from the liver of gadids (Horowitz and Whitt 1972; Shaklee et al. 1973; Whitt et al. 1973c, 1975; Markert et al. 1975).

Using molecular hybridization based upon reversible denaturation, hybrid isozymes of dimeric creatine kinase were obtained that had not been detected in the heterozygotes in vivo; these studies were conducted with the actinopterigious fishes. In this case, the absence of heterodimers in vivo was not due to the incompatibility of the polypeptides, but was a consequence of the spatial and (or) temporal isolation of the synthesis of the two allelic protein variants (Ferris and Whitt 1978a). It may well be that two identical CK polypeptides have become associated into a dimer molecule precisely in the course of translation or immediately thereafter, similar to the case of haemoglobin assembly ($\alpha_2\beta_2$ and other haemoglobins). CK subunits from different, even relatively distant, species also yield hybrid isozymes; this observation has proved that the evolution of creatine kinase does not appear to proceed faster than in the case of other enzymes.

Serological and immunoelectrophoretic techniques have been widely used in studying the interrelationships between different fish species, as well as in establishing the degree of homology between multiple isozymes or the allozymes of various proteins. The problems of serological systematics will be treated below. An analysis of the degree of similarity or dissimilarity between isozymes and allozymes requires the use of specific antibodies against certain purified isozymes during electrophoretic analysis. To produce such antibodies experimental animals must be immunized with sufficiently homogeneous isozyme preparations isolated from fish tissues. After checking for the presence of antibodies, the antiserum containing these is usually added to tissue homogenates prior to electrophoresis. The isozymes against which the antibodies were raised are eliminated by this treatment and this allows precise identification of the protein fractions to be made. Progress in the classification of the isozyme composition of LDH, MDH and several other enzymes particularly in polyploid fish species may be directly attributed to the use of immunoelectrophoresis, frequently in combination with the molecular hybridization of polypeptides.

The data related to the functional and kinetic characteristics of isozymes and allozymes have already been presented above. Here I would like to emphasize that even at the early stages of the evolutionary divergence of enzymes, when the polymorphism of the allelic variants of a given gene had just emerged, in most cases one can observe marked differences both in the kinetic characteristics and in the enzyme activity as a whole. Variants of a number of non-enzymatic proteins, such as transferrin and haemoglobin, are also easily discernible.

Of late considerable attention has been paid to the analysis of the divergence of the duplicated genes. In their studies of the system of lactate dehydrogenase loci Markert and coworkers (Markert et al. 1975) developed a general scheme of the divergent evolution of the gene (Fig. 62). They postulate the presence of at least four steps in such a process.

Step 1: The formation of two identical genes as a consequence of either consecutive (tandem) duplication of a part of the chromosome or of aneuploidy and polyploidization.

Step 2: Primary divergence in the structure of the duplicated genes themselves or in regulation of their action accompanied by the appearance of differences in the kinetic characteristics, activity or stability of corresponding proteins.

Step 3: Continuation (intensification) of the divergent evolution of the structural and corresponding regulatory genes, the appearance of developmental, tissue and metabolic specialization of the proteins.

Step 4: The independent evolution of daughter genes ("divergent specialization") characterized by the progressive divergence in the functions of proteins and in the mechanisms responsible for regulating the activity of the corresponding genes.

Duplication followed by the divergence of the daughter genes is believed by the authors to represent the main and perhaps even the only mechanism leading to the emergence of new variants of the initial protein or even of a new protein. This viewpoint is shared by many other investigators (Ohno et al. 1968, 1969a; Engel et al. 1970; Ohno 1970a; Schmidtke and Engel 1972, 1974).

More recent data appear to suggest that the differences at the early stages of divergent evolution are the consequences mainly of the evolution of regulatory mechanisms; as a result differences in the degree of expression (activation) of homologous isozymes appear in various organs and tissues.

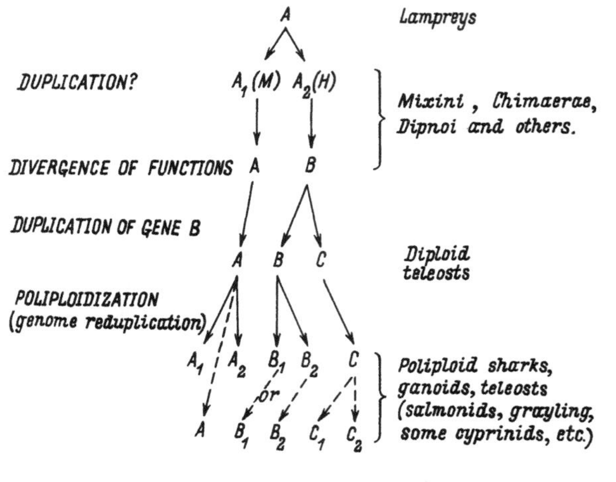

Fig. 62. The evolution of LDH loci in fishes (Markert et al. 1975)

In the family Catostomidae, which underwent polyploidization more than 50 million years ago, the bulk of the duplicated genes is at the early stage of divergent evolution: both copies of the gene are either expressed identically in all tissues (14% of all loci) or the differences in activity are unidirectional (67% loci) (Ferris and Whitt 1979) everywhere. It can be assumed with a high degree of certainty that the main part here is played by the divergent evolution of regulatory mechanisms. A relatively minor fraction of the duplicated loci is tissue-specific: one gene is more active in one type of tissue or organ, while another gene or locus is more active in different tissue types (19% loci). In such cases, the divergent evolution appears to proceed much further and the structural loci themselves have already been markedly changed.

A large divergence in regulatory mechanisms has been demonstrated recently in experiments with hybridization of sunfish species. Evolution of regulatory genes appears to play an important part in the arising of reproductive isolation between species (Philipp et al. 1980).

The evolution of loci coding for lactate dehydrogenase provides a good illustration of the divergent evolution of the gene (Markert et al. 1975; Whitt et al. 1975). In the early vertebrates that were the evolutionary ancestors of fishes and terrestrial animals only one Ldh locus was present (Fig. 62). Only one Ldh locus is found at present in the lamprey; it is usually designated as the A locus, although immunochemically the A_4 isozyme of the lamprey has an identical homology with both the A_4 and B_4 isozymes of teleost fishes (Horowitz and Whitt 1972). The first duplication of this ancestral gene appeared to take place when true fishes branched off from Agnatha; in hagfishes two Ldh loci A and B are already present. The finding of a minor LDH isozyme in certain tissues of the lampreys enables one to suggest that the duplication of the LDH locus had occurred during the earlier period of the evolution of vertebrates (Dell' Agata et al. 1979). Chimaeroid fishes, dipnoan and all cartilaginous fishes also possess two loci. In actinopterigian fishes three loci are already found, namely A, B and C loci. The second duplication event involving the duplication of the gene B took place at the earliest stages of actinopterigian evolution: in cartilaginous ganoids the daughter gene C is already present. Both in ganoids and in primitive teleostean, in the order Anguilliformes in particular, the isozyme C_4 is present in many tissues just like the isozyme B_4 (Whitt et al. 1975). Furthermore, in all relatively advanced fish groups the Ldh-C locus is active almost exclusively in the eye retina (Whitt et al. 1973c). But in several families such as Gadidae, Cyprinidae, Characidae, Catostomidae, Cobitidae the activity of the gene C is restricted to the liver.

The last stage in the evolution of Ldh loci involves their duplication resulting from the genome polyploidization that has occurred relatively recently in a number of fish taxons. In salmonids the duplication occurred more than 20 million years ago: LDH in this group is coded by five loci, A_1, A_2, B_1, B_2 and C. In cyprinids again five genes A, B_1, B_2, C_1 and C_2 are present, polyploidization took place around 50–80 million years ago. In both cases, one of the six daughter loci appeared to become silenced in the course of subsequent evolution or has lost its activity, or was even destroyed completely (Fig. 62).

It should be pointed out that all the daughter LDH genes in fishes retain their original function in controlling pyruvate and lactate metabolism (Holmes 1973; Markert et al. 1975). The divergence of Ldh loci has not yet attained the advanced stage reached by the loci coding for myoglobin and haemoglobin; the divergence of these loci did apparently occur at the very beginning of vertebrate evolution (Ingram 1963; Ohno 1970a).

Duplications with subsequent divergent evolution of the duplicated loci were characteristic of many other proteins as well. For example, the loci of creatine kinase have passed through two or three stages of duplication in fishes. The first duplication took place at the dawn of vertebrate evolution. It has resulted in the appearance of the genes A and C. In the ancestors of teleostean fishes the gene C was apparently duplicated again, and this has led to the appearance of the gene B, wich is at present strictly specialized. The isozyme A_2 in teleostean fishes predominates in the skeletal muscles, the B_2 is located in the eyes and brain, the C_2 in the intestinal muscles and the D_2 in the testes (Fisher and Whitt 1978b). The tissue specificity of the Ck loci in fishes is very remarkable as we can see, however the functional characteristic of the CK isozymes in different fishes has unfortunately not been examined.

A marked increase in the number of loci coding for esterases has been observed in many families of fishes living in the reefs of the Caribbean Sea (Leibel and Markert 1978). The esterase versatility can possibly emerge quite quickly and independently in different taxons. The mechanisms responsible for the appearance of such versatility are unclear.

The evolutionary dynamics of the loss of activity or of the silencing of one of the two homologous genes which had appeared as a result of polyploidization of the chromosomal complement has been studied in the family Catostomidae (Ferris and Whitt 1979). The rate of silencing a homologous daughter gene depends on the length of time that has elapsed since the polyploidization of the genome and also depends on the rate of speciation in a given taxon. A markedly greater number of active duplicated loci (50%–65%) was retained in primitive suckers as compared to more advanced groups (35%–45%) (Ferris and Whitt 1977b; Ferris et al. 1979). As many as 52% of the duplicated loci in the common carp are active (Ferris and Whitt 1977a). In some highly specialized polyploid cobitid species from the genus

Botia this proportion is decreased to 15%–30% (Ferris and Whitt 1977c). A comparison of the variability for 20 enzyme loci in 19 species of catostomids has been conducted recently. The loci present in these species in the double dose are significantly less variable than the single-dose loci (Ferris and Whitt 1980). The higher level of polymorphism characteristic of the "single-dose" loci compensates for the silencing or inactivation of one of the duplicate genes and is in our opinion undoubtedly selective in its nature.

Thus, the evolution of a given gene in fishes may proceed in the following two ways. The daughter genes created by duplication may be conserved and may undergo gradual divergent evolution; their protein products will then acquire different functions. Frequently, however, one of the two genes loses its ability to express itself or is even destroyed; this is called gene silencing. Both these processes can be traced in studies of different proteins in different fish taxons. The destruction of unnecessary genes may be the consequence of the random accumulation of null mutations (Ferris et al. 1979), however, this phenomenon may also be due to selection. A comparison of the rate of silencing of duplicated genes with the rate of their activity loss (which was calculated with the aid of a computer on the basis of purely stochastic processes) has indicated that the actual silencing occurs much more slowly than the calculated one (Bailey et al. 1978). This provides evidence in favour of selection, which probably works to prevent the appearance of defective individuals. It is noteworthy that the ancient polyploids possess a double dosage of ribosomal genes (Engel et al. 1975), although in such criteria as the protein content in the tissues, enzyme activity, cell size and others, these polyploid species closely resemble diploid ones.

The difference between the diploids and polyploids (in their evolutionary origin) in the order Clupeiformes with respect to a number of duplicated loci is

Table 40. The loci number of certain enzymes in diploid and tetraploid species of fishes. Clupeiformes and Cyprinidae according to Engel et al. (1975); Cobitidae according to Ferris and Whitt (1977b)

Enzymes	Clupeiformes		Cyprinidae		Cobitidae	
	2 n	4 n	2 n	4 n	2 n	4 n
LDH A, B	2	4	2	3, 4	2	3, 4
sMDH	1	2	1, 2	2, 3	2	2
mMDH	1	2	1, 2	2	1	1
6PGD	1	1	1	2	1	1, 2
αGPD	3	3	1	2	–	–
sIDH, mIDH	2	4	3	4	–	–
sAAT	1, 2	2	1	1, 2	1	1
SDH	1	2	1	1	–	–
PGI	1, 2	2, 3	2	3, 4	2, 3	2, 3
TO (SOD)	–	–	–	–	1	2
CK	–	–	–	–	2	2, 3
ADH	–	–	–	–	1	1, 2
LDH-C, XDH, G3PDH, PGM, AK, ALD, AP	–	–	–	–	7	7

greater than the analogous difference in the family Cyprinidae (Table 40). However, even in salmonids many genes are not duplicated. In the rainbow trout, for example, only around 50% of the loci are duplicated (Allendorf et al. 1975). Meanwhile, in Clupeiformes in contrast to Cyprinidae the differences between diploid clupeids and tetraploid salmonids persist when measured by such criteria as the DNA and RNA content in cell nuclei, the activity of several enzymes such as LDH, PGI and 6-PGD or the heat stability of reassociated non-repetitive DNA (Schmidtke et al. 1975 a, 1976 a, b, 1979).

Many other genes in addition to Ldh and Ck are also duplicated in diploid fish species. This is true of such genes as Pgi, Mdh, Aat and others. Sometimes the divergence of duplicated loci is very pronounced; for example, this is true of the genes sMdh and mMdh, as well as of sAat and mAat: these are the soluble and the mitochondrial forms of the corresponding enzymes. They do not form hybrid heteropolymers and they carry out very specific functions.

The evolution of genes and the corresponding proteins in fishes proceeded generally very slowly. For hundreds of millions years only a small number of amino acid residues underwent changes in fish proteins (around 30–40 residues in each protein).

This calculation refers to the differences that can actually be observed, however the genuine rate of protein evolution must be somewhat higher (Goodman 1976). Nevertheless, only in a few cases, in the case of LDH, for example, has a marked functional divergence of the duplicated genes been noted. Most enzymes and other proteins have been diverging relatively slowly, the main result of such divergent evolution including the appearance of the tissue specificity of action among multiple loci and the appearance of a certain temporal pattern of their expression.

Biochemical Genetics and Systematics

A large number of papers dealing with the biochemical genetics of cyclostomes and fishes is aimed at obtaining a better definition of the taxonomic rank of species and other taxons, as well as at analyzing the phylogenetic relationships of related species and genera. Such studies have been conducted with fishes belonging to different families; only some of the results of more general interest will be presented below.

Lamprey (Petromyzonidae). Two lamprey species of the genus *Mordax* (*M. mordax, M. praecox*) had identical spectra of the haemoglobins; in this criterion, however, they differed from the species of another genus, *Geotria australis* (Potter and Nicol 1968).

Sturgeon (Acipenseridae). A strict specificity of haemoglobin fractions is characteristic of each of the four species of the sturgeon studied. These include the common sturgeon (*Acipenser güldenstädti*), the sevruga (*A. stellatus*), the sterlet (*A. ruthenus*) and the great sturgeon or beluga (*Huso huso*) (Geraskin and Lukyanenko 1972).

Herring (Clupeidae). The herrings *Alosa pseudoharengus* (alewife) and *A. aestivalis* (blueback herring) are very close to each other serologically in the precipitation reaction and in the criterion of the blood group typing. The American shad *A. sapidissima* and particularly the Atlantic menhaden *Brevoortia tyrannus* differ significantly from the two species mentioned. The Atlantic herring *Clupea harengus* is unique because of its serological features (Sindermann 1962). The spectra of myogens are highly specific even in the related species *Alosa pseudoharengus* and *A. aestivalis* (McKenzie 1973).

Salmon (Salmonidae). Two or even three groups of species can be set apart in the genus *Oncorhynchus* on the basis of studies of the myogens, haemoglobins and serum proteins. The pink salmon *Oncorhynchus gorbuscha* is close to the sockeye salmon *Oncorhynchus nerka,* the chinook salmon *Oncorhynchus tschawytscha* to the coho salmon *Oncorhynchus kisutch,* the position of the chum salmon *Oncorhynchus keta* has not yet been finally clarified (Tsuyuki and Roberts 1966; Omelchenko et al. 1971; Utter et al. 1973a, 1975).[6] The salmon trout (cherry salmon) *O. masu* = *O. rhodurus,* in the characteristics of its haemoglobins, myogens and certain enzymes, is close to the salmon of the genus *Salmo,* in particular to the rainbow trout *Salmo gairdneri* (Utter et al. 1975; Slynko 1976b). All the species of the genus *Salmo* have specific spectra of the myogenes, the Atlantic salmon *S. salar* differing significantly from the two species of the trout *S. trutta* and *S. gairdneri* (Grag and McKenzie 1970). The rainbow trout resembles brook and lake trout belonging to the genus *Salvelinus* in the criterion of the erythrocytes' antigenic composition (Wright et al. 1963). It should, however, be mentioned that at the same time high heterogeneity has been observed within the genus *Salvelinus* itself: in the spectra of the haemoglobins and myogens *S. fontinalis* and *S. namaycush* can easily be distinguished from the Dolly Varden char *S. malma* and the East Siberian char *S. leucomaenis;* the chars living in Kamchatka are similar to the Dolly Varden char (Yamanaka et al. 1967; Omelchenko 1975a).

Sympatric sibling species differing in their allele frequencies of the Est locus have been found in the complex species such as *S. alpinus* (Nyman 1972; Henricson and Nyman 1976).

Cyprinids (Cyprinidae). Phylogenetic relationships between several species of the genus *Notropis* (subgenus *Luxilus*) have been elucidated on the basis of a comparison of enzymes and non-enzymatic proteins (up to 17 loci); up to four phyletic groups have been detected (Rainboth and Whitt 1974; Menzel 1976; Buth 1979c). The common carp differs from the crucian carp in the spectra of the peptidase loci (Pep 1, Pep 2), the glucose-phosphate isomerase (Gpi) and the malic enzyme (Me) (Shearer and Mulley 1978); the carp differs distinctly from the group of phytofagous fishes (Ctenopharyngodon, Hypophthalmichthys) and from Aristichthys when such criteria as the Hb and My spectra are compared (Jurča 1974; Jurča and Matei 1975). Three subspecies of Japanese crucian carp belonging to the species *Carassius auratus* possess different spectra of the muscle proteins and

6 The study of the chromosomal complements of Pacific salmon has demonstrated that these groups of species also have a different structure of the karyotype (Gorschkova 1979, 1980)

differ from the European crucian carp *Carassius carassius* in this criterion (Taniguchi and Sakata 1977). Closely related species of Indian carp *Labeo rohita* and *L. calbasu* possess different alleles of the Hb locus (Krishnaja and Rege 1977). Fifteen species of European cyprinids can be separated into three groups on the basis of the LDH spectra: one of these includes the ide *Leuciscus idus,* the chub *L. cephalus,* the dace *L. leuciscus,* and other fishes from the subfamily Leuciscini, as well as the gudgeon (*Gobio*); the second group includes the roach *Rutilus rutilus,* the owsianka *Leucaspius* from the subfamily Leuciscini, as well as the bitterling *Rhodeus* (subfamily *Rhodeini*) and the undermouth (*Chondrostoma,* subfamily Chondrostomini). The third group includes the tench (*Tinca tinca,* Leuciscini) and the silver carp *Hypophthalmichthys,* Hypophthalmichthyini) (Valenta et al. 1976 b).

The daces living in Lake Issyk-Kul and usually regarded as different species (*Leuciscus schmidti* and *L. bergi*) do not differ in the spectra of the haemoglobin and serum esterases or in their allele frequencies. One can assume that these two forms are closely related to each other or even belong to one and the same species (Payusova 1979).

The data of biochemical genetics, as we can see, are not always in agreement with classification of Cyprinidae which is common at the present time.

Sucker (Catostomidae). In the studies of twelve species belonging to this family a complete correspondence has been found between the spectra of myogens and the taxonomy based on morphological traits (Huntsman 1970).

Characids (Characidae). Cavernicolous and terrestrial representatives of the *Astyanax-Anoptichthys* complex only differ in the frequencies of several of their biochemical loci and appear to belong to one and the same genus and species (Avise and Selander 1972). A cavernicolous population has been found which occupies an intermediate position between the terrestrial forms possessing normal visual function and the genuine blind cavernicolous forms; fishes of this population appear to represent intermediate forms in the course of adaptation to cave life (Peters et al. 1975).

Eels (Anguillidae). A rare allele of the Hb locus absent in the European population has been detected in the American population of eels (Sick et al. 1967).

Different allelic frequencies for serum Est and Mdh loci have been found for three different forms or species of eels, the American, European and Japanese ones (*Anguilla rostrata, A. anguilla, A. japonica*) (de Ligny and Pantelouris 1973), homologous fractions of LDH and MDH also had different mobilities for these three forms (Taniguchi and Morita 1979). These data suggest that the three forms of eels belong to independent species.

Pike (Esocidae). Biochemical studies were instrumental in elucidating the taxonomic relationships between different species of the pike (Eckroat 1974).

Tooth Carp (Poeciliidae, Cyprinodontidae, Goodeidae). The relationships between several species of platyfishes have been established (Greenberg and Kopac 1968; Scholl and Anders 1973 b; Scholl and Schröder 1974). Five species belonging to the genus *Cyprinodon* are similar in the patterns of myogens and haemoglobins, but pronounced differences were observed with regard to the pattern of esterases (Turner 1973 a, 1974).

Two morphologically distinct forms of *Ilyodon* (Goodeidae) have identical protein spectra and therefore belong to one species (Turner and Grosse 1980).

Codfish (Gadidae). The immunogenetic investigation of parvalbumins in nine species of Gadidae made it possible to divide the family into two phyletic groups (Piront and Gosselin-Rey 1975). Five species belonging to the genus *Merluccius* which presented certain problems, when their classification was attempted on the basis of greatly transgressing morphological traits such as the number of vertebrae or the number of gill rakers, can easily be differentiated on the basis of the spectra of water-soluble muscle proteins (Mackie and Jones 1978).

Perch (Percidae). The comparison of 68 species belonging to the subfamily Etheostomatini in the spectra of their LDH, MDH and TO (SOD) allowed us to confirm the monophyletic origin of the genera *Percina* and *Etheostoma*; phylogenetic relationships between the species were also established (Page and Whitt 1973 a, b). Six species of *Etheostoma* (darters) can be divided into two groups according to the criteria of gene frequencies and the spectra of LDH, PGM and My (Wolfe et al. 1979).

Cichlid Fishes (Cichlidae). Seventeen species of this family can be divided into six groups on the basis of LDH patterns; these groups do not always correspond to the classification which is now generally accepted (Scholl and Holzberg 1972). At the same time, the variation, examined using 27 protein loci for comparison, was identical in several morphologically distinct local races of cichlids in one of the American lakes. This provided a basis for the conclusion that only one species with a single panmictic population is present in this lake (Sage and Selander 1975). Biochemical markers made it possible distinguish different species of *Tilapia* in mixed populations (Avtalion et al. 1976).

Horse Mackerel (Carangidae). It has been demonstrated using the precipitation reaction that the large and small horse mackerel of the Black Sea appear to belong to autonomous independent species; the serological differences between them are no less than the differences between the "good" Mediterranean species *Trachurus trachurus* and *T. trecae* (Altukhov and Apekin 1963). On the basis of studies of esterases and myogens, seven species of mackerel were divided into several phyletic groups (Alferova and Nefyodov 1973; Nefyodov et al. 1973).

Goby (Gobiidae). Two closely related species of the goby (*Bathygobius*), living on different sides of the Panama isthmus differ with respect to a large number of the 26 loci studied. The age of these species (2.5 million years) determined on the basis of biochemical differences, that is of the mean number of amino acid substitutions in the proteins, corresponds to the geological age of the isthmus (Gorman et al. 1976).

Platycephalids. Two sibling-species in this family are easily distinguishable on the basis of their myogen pattern (Taniguchi et al. 1972).

Sculpin (Cottidae). Three species belonging to the genus *Myoxocephalus* are easily distinguished by their haemoglobins but have identical patterns of myogens and crystallins; the genera *Enophrys* and *Hemitripterus* differ from *Myoxocephalus* in these latter proteins as well (Kartavtsev 1975).

In many other investigations the data on biochemical genetics allowed us to arrive at more unequivocal conclusions in the field of systematics (Taniguchi and Nakamura 1970; Gonzalez et al. 1974; Mester and Tesio 1975; Wanstein and Yerger 1976; Chernyshev 1980, and others). Meanwhile, many works of this type suffer because an inadequate number of genetic systems has been analyzed. Biochemical studies can indeed be helpful to taxonomists, but only if sufficiently large numbers of proteins have been studied and when the results of genetic investigations are thoroughly compared with the morpho-physiological and ecological characteristics of the forms studied.

A comparison of protein patterns is frequently helpful in solving the difficult question of identification of natural hybrids between closely related fish species. For example, hybrids of cyprinids may be easily determined by their EST, LDH and GPI spectra (Child and Solomon 1977). Distinct differences in the electrophoretic pattern of albumins, transferrins and blood serum esterases have been described for the Atlantic salmon *Salmo salar* and the brown trout *S. trutta.* It is easy to identify the hybrids possessing both sets of these proteins (Solomon and Child 1978). The pink and chum salmon may cross with each other and give offspring, but, as can be concluded from the study of various enzymatic systems, these hybrids are either non-viable or sterile because no hybrids have been found in natural populations of either species (May et al. 1975). The hybrids of five species of salmonids – *Salmo salar, S. trutta, S. gairdneri, Salvelinus fontinalis* and *Oncorhynchus kisutch* – are easily identifiable by the patterns of ADH, MDH, EST, PGM and PGI (preferably using the latter enzyme) as well as on the basis of myogen patterns (Guyomard 1978). Hybrids of *Xiphophorus maculatus* and *X. helleri* can also be easily distinguished, this can be determined most readily from the esterase spectra (Ahuja et al. 1977).

The quantitative determination of the genetic similarity and genetic distance between related fish taxons has been attracting ever increasing attention in genetic studies of fishes over the last few years. Various methods of determining the indices of "identity" and the "distance" are being used. The identity index of populations, subspecies, species or higher taxons can be calculated by the widely known method of Nei (1972) with the aid of the following formula, making use of the frequency determinations of a single genetic locus

$$I_j = \frac{\sum X_i Y_i}{\sqrt{\sum X_i^2 \sum Y_i^2}}, \qquad (1)$$

where X_i and Y_i are the frequency of the i-th allele of a given locus in the two populations of fishes compared. The mean identity index for all the loci is then determined as:

$$\bar{I} = \frac{\sum(\sum X_i Y_i)}{\sqrt{\sum(\sum X_i^2)\sum(\sum Y_i^2)}}. \tag{2}$$

The index of distance (or of the difference) D is calculated, using the formula:

$$D = -\ln \bar{I}. \tag{3}$$

The identity index varies over a range of 0 to 1; when the genotypes are completely identical, $I=1$ and $D=0$. The index of distance D shows a non-monotonous variation with respect to I; when the identity is low ($I=0.25$ or less) it is greater than one, at very low values of I it approaches infinity.

The indices suggested by Nei can be substituted by the distance and similarity indices (D and S) of Rogers (1972) and Prevosti et al. (1975):

$$D_R = \frac{1}{m}\sum\sqrt{\frac{\sum(X_i - Y_i)^2}{2}}, \tag{4}$$

$$S_R = 1 - D_R \quad \text{or} \tag{5}$$

$$D_P = \frac{1}{m}\sum\sum\frac{|X_i - Y_i|}{2}, \tag{6}$$

$$S_P = 1 - D_P. \tag{7}$$

In these cases both S and D vary in a range of from 0 to 1. The similarity index suggested by Rogers represents the geometric mean of the similarity indices for individual loci and is generally close in its magnitude to I. A comparative evaluation of these three methods of index calculation as well as of Hedrick's formulae (1971), which are used less frequently, has been presented by Pudovkin (1979). He has demonstrated that when the forms compared are closely related, the similarity indices of Rogers and Prevosti give somewhat underestimated values, whereas the values of the distance indices appear to be too high. When the differences between the taxons compared are significant (that is when more distant species, genera and taxons are compared) the method of Rogers is to be preferred, since the coefficient D calculated according to Nei increases out of all proportion to the degree of divergence.

Simplified techniques for similarity determinations have been used repeatedly. One of these methods made use of the calculation of the number of protein fractions with similar electrophoretic mobilities, then this number was divided by the mean number of all the visible fractions (Shaw 1970):

$$Q = \frac{n}{0.5\,(N_1 + N_2)}, \tag{8}$$

where n is the number of identical fractions, N_1 and N_2 are the number of all fractions on the zymograms of the two species. In a number of studies the presence

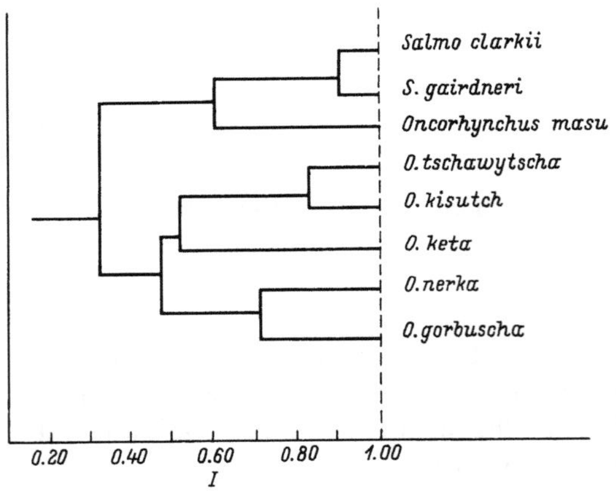

Fig. 63. The dendrogram for some salmonid fishes. *Abscissa* identity index (Utter et al. 1973 b)

of identical alleles of a locus was coded as 1 and the presence of different alleles as 0; when frequencies of one of the alleles showed variation, intermediate values from 0 to 1 were employed, these were inversely proportional to the magnitude of difference. The sum total of the similarity indices was divided by the number of loci compared (Utter et al. 1973 a).

Identity and similarity indices were calculated for many fish groups. For eight species of the Pacific salmonids they were on the average equal to 0.46, however, a very high similarity was observed for several pairs of species (S = 0.90 for *Salmo gairdneri* – *S. clarkii;* 0.82 for *Oncorhynchus kisutch* – *O. tschawytscha;* 0.72 for *O. gorbuscha* – *O. nerka*). A dendrogram based on similarity indices has been constructed (Fig. 63). Comparison of the similarity indices and peculiar characteristics of the protein spectra generally lead to identical results.

It has been established, using the index of identity calculated for 27 loci, that the lake whitefishes *Coregonus pollan* and *C. autumnalis* belong to one and the same species, in this case I = 1 (Ferguson et al. 1978).

Many studies have been conducted with cyprinid fishes. Identity indices for the nine species of American minnows studied belonging to different genera varied quite significantly (Table 41). Two species, *Hesperoleucus symmetricus* and *Lavinia*

Table 41. Similarity indices for protein loci in nine species of Cyprinidae. Calculated according to Nei (Avise and Ayala 1976)

	1	2	3	4	5	6	7	8
1. *Hesperoleucus symmetricus*	×							
2. *Lavinia exilicauda*	0.95	×						
3. *Mylopharodon conocephalus*	0.91	0.86	×					
4. *Ptychocheilus grandis*	0.82	0.81	0.88	×				
5. *Orthodon microlepidotus*	0.60	0.54	0.58	0.58	×			
6. *Pogonichthys macrolepidotus*	0.49	0.47	0.55	0.55	0.34	×		
7. *Richardsonius egregius*	0.65	0.60	0.64	0.59	0.46	0.60	×	
8. *Gila bicolor*	0.78	0.70	0.84	0.72	0.60	0.51	0.64	×
9. *Notemigonus crysoleucus*	0.41	0.40	0.45	0.37	0.34	0.33	0.38	0.41

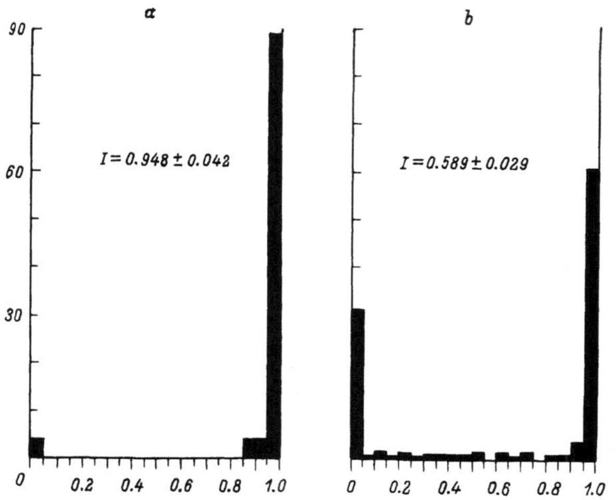

Fig. 64 a, b. The distribution of enzyme loci in Californian cyprinids by the degree of their identity index (*I*) in different species upon pairwise comparison. **a** closely related species *Hesperoleucus symmetricus* and *Lavinia exilicauda;* **b** pooled data for 9 species (and genera); *abscissa* similarity index; *ordinate* percentage of loci (Avise et al. 1975; Avise and Ayala 1976)

exilicauda turned out to be quite close to each other in spite of significant differences in their morphology and biology (Avise et al. 1975). It has been assumed that this is related to the relatively recent divergence of these two forms.

Identity indices may be used in constructing graphs reflecting the distribution of the loci according to the degree of their identity in different species (Fig. 64). Generally the bulk of the loci (up to 90% or more) is found in the two extreme classes I = 0 and I = 1; the other loci are distributed between them (Ayala 1975).

In catostomids the values of I and D were detected for three species of *Thoburnia.* One of these species, *Th. atripinne* differs significantly from the other two (Buth 1979 b).

A study of the local races of the loach *Misgurnus anguillicaudatus* (Cobitidae) has demonstrated that the difference between them (D) calculated according to Nei was equal to 0.23–0.40 (Kimura 1978 a). These races appear to be subspecies. The index of distance between populations was only minimal (0.0003–0.013).

In the family Cyprinodontidae numerous species of the subfamily Aphanini are greatly differentiated, I < 0.20 (Scholl et al. 1978). Comparison of five species from the family Poeciliidae made it possible to distinguish two pairs of related forms, *Poecilia reticulata* and *P. vittata* (I = 0.72), *Xiphophorus maculatus* and *X. helleri* (I = 0.78). *Poecilia (Mollienesia) sphenops* differs greatly from these four species (Scholl and Schröder 1974) and may easily be divided into several distinct intraspecies groups (Brett et al. 1980). Species of the genus Cyprinodon are generally very closely related (I = 0.89) (Turner 1974).

The interesting studies conducted with cichlid fishes are worthy of mention. A very large number (up to 300) of different species of this family appeared in a relatively short period of time in the African lake Malawi, which appears to have existed for only about 5000 years (Fryer 1977). A comparative study of seven species (16 loci) has demonstrated that the interspecific differences in the proteins in this case are not very great, the identity index being equal on average to 0.934 (the range from 0.86 to 0.99) (Kornfield 1978). The divergence was more significant in the Galilean Lakes (I varied from 0.25 to 0.95). The high value of the identity

Table 42. Similarity indices between different species of the horse mackerel (Carangidae) calculated on the basis of spectra of esterases and myogenes (Alferov and Nefyodov 1973; Nefyodov et al. 1973; calculated according to Shaw (1970)

Species	Esterases						
	1	2	3	4	5	6	7
1. *Trachurus trachurus*	×	0.875	0.824	–	0.631	–	–
1. *T. picturatus*	0.522	×	0.667	–	0.471	–	–
3. *T. trecae*	0.750	0.640	×	–	0.555	–	–
4. *T. sp.*	0.880	0.462	0.741	×	–	–	–
5. *Caranx rhonchus*	0.518	0.429	0.482	0.467	×	–	–
6. *Decapterus punctatus*	0.296	0.357	0.276	0.267	0.250	×	–
7. *Vomer sitipinnis*	0.207	0.267	0.258	0.187	0.412	0.353	×
	Myogenes						

index was observed for just one pair of species belonging to the genus *Tristramella* (Kornfield et al. 1979).

Comparatively high values of the similarity indices have been obtained in a pairwise comparison of four species of mackerel belonging to the genus *Trachurus* (Carangidae). Three other genera of this family differed greatly from *Trachurus* (Table 42).

In flatfishes (Pleuronectidae) the interspecific differences determined for 35 loci were moderate for the plaice and the flounder (*Pleuronectus plalessa – Pl. flesus*) (D=0.45 according to Nei); great differences were, however, found when these two species were compared with another flatfish species, the dab *Limanda limanda* (D=1.01 and 1.29). The genetic distances were proportional to morpho-anatomical differences (Ward and Galleguillos 1978).

Calculations of the identity indices and of the distance were also conducted for other fish species as well, in particular for ten species of the genus *Lepomis* (Centrarchidae) (Avise and Smith 1974) and for several species of *Myoxocephalus* (Cottidae) (Kartavtsev 1975).

By summarizing all the calculations of genetic identity indices for different fish taxons, we are able to evaluate the average degree of identity for the taxons of different ranks (Table 43). The similarity between populations is generally very high, the index being equal to at least 0.97–0.99 in all cases. The data for the subspecies are not very numerous, the indices for the species show great variations, perhaps reflecting the extent of divergence of different forms in various groups. Two extreme cases are noteworthy: a very low index for the genus *Aphanius* and very high values for the Cichlidae from Lake Malawi. In the first case, the speciation appeared to proceed at a high rate and was accompanied by intensive specialization; the very short time of divergent evolution of the species in the second case appears to be the decisive factor. Identity indices between genera do not usually exceed 0.40, they show higher values only for cyprinids.

The genetic and morphoecological divergences, as can be seen, do not always proceed in parallel. A hypothesis has been suggested according to which quantitative genetic changes accumulate as a function of time and do not appear to be

Table 43. Genetic similarity for taxons of different rank

Family and genus	Similarity indices				Reference
	Populations	Subspecies	Species	Genera	
Salmonidae	–	–	0.46	–	16
Cyprinidae, 9 genera	0.99	–	–	0.57	2
Cyprinidae, Lavinia-Hesperoleucus	–	–	–	0.95	2
Catostomidae, Moxostoma	0.99	–	0.81	–	6
Catostomidae, Campostoma	0.98	0.84	0.84	–	8
Catostomidae, Thoburnia	0.99	–	0.76	–	7
Cobitidae	0.99	0.68	–	–	9
Cyprinodontidae, Cyprinodon	–	–	0.89	–	15
Cyprinodontidae, Aphanius	–	–	0.20	–	13
Poeciliidae	–	–	0.57	0.39	12
Gadidae, Merluccius	–	–	0.65	–	11
Centrarchidae	0.97	0.85	0.53	0.29	3, 4
Carangidae	–	–	0.70	0.34	1, 11
Sciaenidae	–	–	–	0.17	14
Cichlidae, L. Malavi	–	–	0.94	–	10
Pleuronectidae, Pleuronectes, Limanda	–	–	0.64	0.32	17
Pleuronectidae, Hippoglossoides	–	–	0.73	–	5

References: 1. Alferova and Nefyodov (1973); 2. Avise and Ayala (1976); 3. Avise and Smith (1974); 4. Avise et al. (1977); 5. Bogdanov et al. (1979); 6. Buth (1977b); 7. Buth (1979b); 8. Buth and Burr (1978); 9. Kimura (1978a); 10. Kornfield (1978); 11. Nefyodov et al. (1973); 12. Scholl and Schröder (1974); 13. Scholl et al. (1978); 14. Shaw (1970); 15. Turner (1974); 16. Utter et al. (1973a); 17. Ward and Galleguillos (1978)

associated with the rate and extent of speciation. A comparison of the identity indices in *Leuciscini* (Cyprinidae) and *Lepomis* (Centrarchidae) probably supports this hypothesis (Avise 1976a) although the author points out that no final conclusion can be made on the basis of the data available. A detailed analysis of Table 43 enables one to conclude that the rate of mutation accumulation not only depends on the time factor, but is also associated with the specific features of speciation in different evolutionary branches. Intensive specialization appears to be accompanied by the accelerated selection of mutational changes in proteins (genus *Oncorhynchus* among salmonids, *Aphanini* among tooth carps and a few others).

The particularization of phylogenetic relationships between the species compared was the most important result of studies with closely related fish species, when the identity indices were calculated for a sufficiently large number of loci (more than 30). Matrices containing I (S) and D values can be used in constructing dendrograms (Fig. 63) and evolutionary trees representing the most probable pathways of phylogenesis in graphic form. The most common method of constructing dendrograms and phylogenetic trees makes use of the gradual addition of "nodes", beginning from the most closely related pairs of species for the populations with maximal I (S) values and correspondingly minimal D values (Farris 1972). In constructing a dendrogram, the values of I or D are plotted along the abscissa and, when constructing a phylogenetic tree, these values are usually plotted along the ordinate. Two or more closely related pairs of forms with minimal D values are then connected with each other by a line. The lengths of the nodes are determined by the calculation of the mean arithmetic values of all the possible distances:

$$D[(A, B), (C, D)] = \frac{D(AC) + D(AD) + D(BC) + D(BD)}{4}. \tag{9}$$

The phylogenetic tree constructed should be parsimonious, the sum total of all the nodes being minimal (Farris 1972).

The dendrogram for polyploid fish taxons can be constructed, using the number of loci differing in their state, that is duplicated or non-duplicated,instead of genetic distances. Such a dendrogram constructed for the family Catostomidae correctly represented the phylogenetic relationships between the species of the family (Ferris and Whitt 1978 b).

It should be pointed out that the dendrograms and phylogenetic trees constructed only taking into account the index of the genetic distance or the number of duplicated loci silenced appear to be somewhat arbitrary; they can only provide a general schematic view of the evolutionary relationships within a taxon. There can be no doubt, however, that the approach is still useful, if such schemes are considered together with phylogenetic trees constructed on the basis of morpho-anatomical traits.

An original method of determining the homology of DNA of different species was recently used in fish systematics. This technique included the construction of hybrid DNA molecules and the comparison of their heat stability. The stability of hybrid duplexes of DNA was found to be proportional to the degree of homology of the initial sequences (Mednikov et al. 1973 b). The fraction of homologous DNA decreases as the evolutionary distance of the fish species compared increases. When different classes of fishes are compared, the degree of homology is less than 20%; a comparison of different orders gives a value varying from 15% to 45%. A comparison of families yields a value around 50%–70%, while for the genera and species the DNA homology increases up to 75%–100% (Mednikov et al. 1973 a). The authors believe that the use of the technique based on the determination of the thermal stability of hybrid DNA duplexes may allow certain traditional concepts in systematics to be evaluated. In particular, cartilaginous fishes (Chondrichthyes), cartilaginous ganoids (Chondrostei), and Teleostei should now be regarded as different classes (Popov et al. 1973).

The use of DNA molecular hybridization in studies of salmonids permitted three autonomous groups to be found in the subfamily Coregoninae; these are (1) *Stenodus* and *Leucichthys,* (2) *Coregonus,* (3) *Prosopium* (Mednikov et al. 1977). The rainbow trout *Salmo gairdneri* and the mykish *Salmo mykiss* should probably be viewed as one and the same species (Mednikov and Akhundov 1975). The chars of Chukotka *Salvelinus alpinus* and *Salvelinus malma,* which populate the same area and are considered by taxonomists to be two different species, showed almost complete genetic identity as can be judged from similar melting curves of DNA hybrids; these forms appear to belong to one and the same species (Mednikov and Maximov 1979).

Biochemical Genetics and the Population Structure in Fish Species

Practically all species of animals or plants exist as a greater or lesser number of completely or partially isolated populations – geographical subspecies (races) and more minor groups. Free crosses between all individuals or panmixia can theoretically be observed solely within these groups. In reality, however, panmixia may not be complete even in small populations, it is limited by many factors, primarily by the isolation caused by geographical distances; there is a greater probability that individuals within a close locality will mate.

The differentiation of species among fishes is fairly prominent, but the degree thereof varies in different forms. Until recently the populational structure of the species was studied, predominantly using quantitative morphological traits, such as the number of vertebrae, the fin rays, the gill rackers or various exterior indices, etc. Unequivocal representations of the populational structure of the species could be obtained in a number of cases. In studies of the eelpout *Zoarces viviparus* the solution of the problem was facilitated by experiments confirming the hereditary nature of the differences in the quantitative traits. In studies with the wild common carp the work was simplified by such a favourable factor as the ease of carp breeding and the possibility of controlled genetic experiments. In the work with other species, in particular with the cod and the herring, the experimental systematics could not be used because of technical difficulties; therefore the analysis of the populational structure of many of these species became a practically insoluble task. Numerous studies of the various races of the herring, the geographical groups of the cod, Caspian roach and many other economically important fish species, led to no perceptible results. The conclusions of the authors were disputable and unconvincing.

The use of techniques of biochemical genetics has led to a new and fruitful stage in fish population studies. Studies of this kind are aimed at solving several tasks, the following being the most important ones:

1. the establishment of the pattern of population structure of the species (the type of population structure);
2. the analysis of the genetic differences between populations and of the adaptive nature of such differences;
3. the drawing of boundaries between adjacent populations and determination of the extent of their isolation;
4. the study of evolutionary processes in populations;
5. the analysis of the consequences of human activity such as commercial fishing, protective measures, artificial reproduction and the transfer of fish from other waterbodies, on the populational structure and the numbers of the species;
6. the development of rational ways of reproduction and protecting the resources of the most commercially valuable fish species.

The solution of these problems is facilitated by the existence of numerous populations in the case of most fish species, by the ease of collecting materials for analysis, by the ease of crosses and cultivation of the fish fry, the poikilothermal nature of the group and finally the habitation of individuals belonging to a single fish species in many different, frequently very contrasting environments (Allendorf and Utter 1979).

Before we go on to discuss the results obtained in studies of fish populations, we would like to point out several more general principles inherent in such studies.

The main technique for analyzing the genetic structure of fish populations makes use of the calculation of numbers of different phenotypes (blood groups and other proteins) in the sample studied, the formulation of a working hypothesis for the inheritance and the verification of this hypothesis, using the Hardy-Weinberg formulae. These formulae reflect the equilibrium state of the population; it may be recalled that in the presence of two alleles the formulae for the equilibrium of genotype frequencies are the following (see also p. 146):

$$p^2\,(AA) + 2\,pq\,(AB) + q^2\,(BB) = 1,$$

where p and q are the frequencies of the alleles A and B and AA, AB and BB are the genotypes of three classes of individuals.

Where the observed and theoretically calculated frequencies of the phenotypes correspond well, one can conclude that the population is at equilibrium, the hypothesis of inheritance is therefore recognized as a satisfactory one. The final verification of such a hypothesis is, however, only possible after the necessary crosses have been achieved, that is, such verification requires hybridological analysis. The crosses allow the loci to be identified precisely and the non-inheritable protein modifications, including the conformational changes, which may be quite frequent, to be rejected.

The next step involves a comparison of the different populations of a given species with respect to the frequency of blood groups and protein loci, determination of the boundaries of populations and the finding of clines for the frequencies of different genes; frequently population students seek relationships between allele frequencies and various environmental factors.

Certain problems and complications associated with the use of the Hardy-Weinberg formula should be mentioned in this connection.

1. The mixing of nonuniform groups of fishes (supopulations and different age groups) differing in their allele frequencies results in the Wahlund effect (Wahlund 1928), that is, in the increase in the number of homozygotes and the decrease in the heterozygotes. According to Wahlund the frequencies of the genotypes in the mixed population (in the system with two alleles) may be expressed as follows:

$$r_{AA} = p^2 + \sigma^2_{q_1}$$
$$r_{AB} = 2\,pq - 2\,\sigma^2_{q_1}$$
$$r_{BB} = q^2 + \sigma^2_{q_1},$$

where p and q are the frequencies of the alleles A and B in the mixed population, and the q_i is the frequency of the allele q in each subpopulation (Li 1976).

A deficit of heterozygotes is indeed frequently observed in studies of fish populations since pure panmictic groups are quite rare. Even when a population is uniform, but different age groups have different gene frequencies, the number of heterozygotes may be less than expected.

2. Inbreeding is the second important factor responsible for deviations of the frequencies from the equilibrium state and acting towards the decrease in the number of heterozygotes in the sample. Generally, if the inbreeding coefficient equals F[7], the frequencies of AA, AB and BB genotypes in the equilibrium population may be written as follows:

$$r_{AA} = p^2 + Fpq$$
$$r_{AB} = 2\,pq - 2\,Fpq$$
$$r_{BB} = q^2 + Fpq.$$

If the inbreeding is not tight (for example $F = 0.01$), the increase in the number of homozygotes is insignificant, at $p = q = 0.5$ it will only correspond to 1% and will remain virtually unnoticed during the analysis of the small samples taken from natural populations. The inbreeding coefficient is known to depend on the size of the population

$$\Delta F = \frac{1}{2\,N_e},$$

where ΔF is the increase in the inbreeding per generation and N_e is the effective size (reproductive fraction) of the population [8] (Falconer 1960). It follows that even when $Ne \geqq 50$ the increment of the inbreeding coefficient is equal to 0.01 per generation and will hardly affect the proportion of genotypes. When the number of reproductively active individuals is low, for example, if N_e equals 10, the deviations will be more pronounced. Fish populations are generally numerous, N_e being equal to thousands or even tens of thousands of individuals. Supopulations of certain migrating salmon may serve as an example of relatively small fish populations. These populations are generally "anchored" to their own spawning grounds due to the homing instinct. If, however, the exchange of individuals or migration between these subpopulations becomes possible, the effect of inbreeding will be drastically reduced and should hardly be taken into account.

3. The assortative crossing that is the preferential crossing of individuals with similar genotypes or vice versa with different genotypes; this results in the first case in a decrease in the number of heterozygotes or in the second case in an increase in the heterozygous individuals. The assortative crosses involving biochemical alleles appear to be rare among fishes if one does not take into account the genes tightly linked to the unpaired sex chromosome (Li 1976). It may, however, well be that during the selection of pairs in the spawning season involving size selection and

---

7 The inbreeding coefficient F may acquire values from 0 to 1 and shows which fractions of the genes in a given group of individuals became homozygous as a result of breeding. Methods of of calculating the inbreeding coefficient have been listed in a number of treatises (Falconer 1960; Li 1976)

8 The effective size of a population is determined by the proportion of females and males participating in reproduction (spawning) using the formula:

$$N_e = \frac{4\,N♀♀ \cdot N♂♂}{N♀♀ + N♂♂}$$

observed, in particular, in the sockeye salmon (Konovalov 1979), the selection of similar genotypes in the sense of protein loci appears to take place. When differences between age or size groups with respect to allele frequencies are great, this may result in a decrease in heterozygosity.

4. Natural selection represents a permanently acting and very important factor responsible for the shift of the genetic equilibrium in populations. If we define the number of heterozygotes in the sample as 2 B and the allele frequencies as p and q in a two-allele system, the proportion $\frac{B}{pq}$ may be taken as an index of the deviation from the panmixia (Nikoro 1976):

$$Y = \frac{B}{pq}$$

When selection acts against one or both homozygotes or against one homozygote and the heterozygotes (when the selection coefficient is low) $Y > 1$, i.e., an excess of heterozygotes is observed. For instance if the survival of the three genotypes is equal to 100% (AA), 90% (AB) and 60% (BB), then in the presence of the initial allele frequencies equal to $p = 0.2$ and $q = 0.8$ the following genotype frequencies will be obtained:

	AA	AB	BB	p	q
Before selection (equilibrium)	0.040	0.320	0.640	0.2	0.8
After selection, observed data	0.056	0.405	0.539	0.23	0.77
After selection, expected data	0.053	0.354	0.593	0.23	0.77
Difference	+0.003	+0.051	–0.054	–	–

If selection operates against the heterozygotes ($Y < 1$), it means there is a shortage of them (Nikoro 1976). The equilibrium will only be restored in the next generation. Two important conclusions can be drawn from this analysis: (1) the excess of heterozygotes does not necessarily provide evidence of their selective advantage (monogenic heterosis); it may be a consequence of the negative assortative crossing or, what is more frequent, it may result from selection against one of the alleles; (2) the deficit of heterozygotes may be the outcome of the mixing of populations, inbreeding, positive assortative crossing or finally, a consequence of the selection operating against the heterozygotes.

Since the selection coefficients in nature are generally low, deviations from the equilibrium due to selection frequently remain unnoticed. No conclusions about the presence or absence of selection favouring one allele or another as well as about monogenic heterosis can be drawn on the basis of an analysis of the ratio of the genotypes. This can be achieved either by comparing the different age groups of one and the same population (preferably in the space of one generation) or by direct selection experiments.

Numerous data dealing with the populational genetics of various marine, anadromic and freshwater fishes have been obtained over the last few years.

Several reviews dealing with these questions have been published (de Ligny 1969; Altukhov 1974; Utter et al. 1975; Allendorf and Utter 1979; Ferguson 1980). Only a few of the most interesting studies will be discussed below.

Most fish species are differentiated into subspecies (races), populations and semi-isolated subpopulations. Large geographical groups within a species known as subspecies may be divided into local races (according to Altukhov 1973, 1974: local shoals; according to Konovalov 1979: isolates). According to Altukhov and Rychkov (1970), these races or shoals form populational systems that are relatively stable in the genetic sense. The total gene frequencies in such shoals are almost unchanged and remain stable for long periods of time. Subpopulations or subisolates can be distinguished within each race, these subpopulations being more or less isolated from one another, and differing in their allele frequencies in one or several loci; they may also show some variation in their genetic structure. Population variability as a function of time is associated with the relatively small size of such "elementary" or Mendelian populations, and is due to the joint action of three main evolutionary factors: random genetic drift (changes in the allelic frequencies due to purely random causes such as fluctuations in the number of spawners and low N_e value), migration (the mixing of fishes from various subpopulations) and selection (the preferential survival of individuals adapted to local conditions). The hierarchy of intraspecific groups may be more complex, having three or even four levels of separation, the appearance of seasonal races, dwarf forms etc.

The principle governing the hierarchic differentiation of species and the relative stability of larger entities (races and subspecies) (Altukhov and Rychkov 1970) was extended to all fish species and to other animals (Altukhov 1973, 1974, etc.). Several authors have pointed out (Aronstam et al. 1977) that this principle generally appears to be valid. It should, however, be mentioned that the stability of a system of a higher order is conserved only when environmental conditions in the area occupied by this system (such as race etc.) are constant. Any directed change in the environment results in a vectorial and sometimes very essential shift in the pattern of gene frequencies. Such shifts have already been considered, when we presented examples of changing gene frequencies for fish populations living in waterbodies affected by the warm effluents of electric power stations. There can be no doubt that in nature frequency fluctuations occur also in the absence of any human participation, but they proceed slowly and are therefore difficult to detect.

Numerous studies with a wide variety of fish species deal with the problem of discerning the genetic differences between populations. Such differences are particularly apparent in the presence of any discontinuity between populations, such a discontinuity may be due to natural barriers or to the ecological separation of forms. For example, the Atlantic salmon *Salmo salar* and the flatfishes *Hippoglossoides platessoides* of the European and American Atlantic coast differ in the frequencies of a number of genes (Nyman and Pippy 1972; Naevdal and Bakken 1974). A New Zealand ling *Genypterus blacoides* (Genypteridae) has two populations separated by the underwater range off the coast of New Zealand; these populations significantly differ in the genes Gpi and Pgm (Smith 1979b). Large races or shoals of migrating fish, salmonids in particular which spawn in different lakes and rivers, generally have different allele frequencies for many genes (Utter et al. 1973b; Altukhov 1974;

Kirpichnikov and Ivanova 1977). Significant differences between seasonal races of the sockeye salmon have been detected in the frequencies of the Ldh-B1 alleles (Altukhov et al. 1975a; our calculations); any mixing of these races appears to be very limited.

In many cases, when natural boundaries between populations are absent genetic differences acquire a clinal character (see p. 210). It has already been demonstrated that the clinal variation appears to be associated with the gradient of environmental conditions and is probably adaptive.

Genetic differences between fish populations are generally moderate. The identity index (I) calculated according to Nei or Rogers is generally more than 0.97–0.99 (Table 43). When these indices are calculated for just one or two polymeric genes without taking into account the monomorphic loci, they may be even somewhat underestimated. Attempts to construct dendrograms showing the genesis of different populations have been undertaken, using such selected indices (Utter et al. 1975). It should be taken into account, however, that such schemes are markedly less reliable than dendrograms or evolutionary trees constructed for species and higher taxons using many randomly selected loci.

Determination of the boundaries between populations on the basis of differences in gene frequencies and the establishment of the degree of their isolation involves serious problems. To achieve this, one has to collect and analyze a large number of fish samples in different parts of the distribution range. In addition, repeated studies are necessary.

Only a few studies made in this direction will be mentioned. Studies of the blood groups and of several esterases in the Pacific skipjack have shown that the boundary between the western and eastern populations or systems of populations is dynamic and migrates periodically from West to East and vice versa (Fujino 1976, 1978). It can be assumed that similar mobile boundaries exist for populations of other fishes migrating over long distances, in particular for other tunas of the Pacific and Atlantic ocean. The brown trout *Salmo trutta* in one of the Scandinavian lakes exists as two sympatric populations differing in the alleles of the Ldh locus (Allendorf et al. 1976; Ryman et al. 1979). The absence of heterozygotes with respect to these alleles in this lake provides evidence of the isolation of these populations. Coastal and inland populations of the steelhead *Salmo gairdneri* also differ distinctly in the locus Ldh-4 and appear to be reproductively isolated (Utter and Allendorf 1978).

The evolutionary processes occurring in fish populations at the present time have hardly been studied. Several papers, however, deal with the problem of the origin of interpopulational differences. An interesting paper has been published dealing with the Canadian lake whitefish *Coregonus clupeaformis* (Franzin and Clayton 1977). Two main populations of this fish can be distinguished on the basis of αGpdh and Ldh allele frequencies. The western population lives in lakes near the Pacific coast, while the central population inhabits the lakes of the Central Canadian plateau up to the border between the provinces Manitoba and Ontario. An analysis of the allele frequency distribution has led to the hypothesis that Canada became populated by whitefishes during the post-glaciation period. These fishes originated from two glacial retreats – the Mississipi retreat, which was the main one, and the Behring retreat; these differed from each other in the genetic structure of the whitefish populations living therein.

Unfortunately at present we have hardly any information on the effects of human activity on the genetic structure of fish populations. A marked genetic depletion of the common carp strains in the USSR fish ponds can be mentioned (Paaver 1980). In the chum salmon the fishes cultivated at the hatcheries are less heterozygous for Ldh-A2 loci than the fishes produced by natural spawning (Ryabova-Sachko 1977 b). Noticeable changes in αGpdh and Cpk allelic frequencies have been obtained in hatchery strains of the brown trout (Ryman and Ståhl 1980).

Good examples of directed shifts of frequencies of several genes are provided by the above-mentioned rapid frequency changes in populations affected by the warm effluents from electric power stations (Nyman 1975).

Studies of the populational structure of economically important fish species conducted so far allow a number of recommendations to be made regarding the reproduction and pretection of the resources of these species. When capturing individuals for subsequent reproduction, one must attempt to use the reproductively active fishes of different ages and various origin (from all the main subpopulations), including different seasonal and ecological races. The number of individuals should be sufficient to avoid inbreeding (no less than 50 and preferably more than 100) and they should represent all the genetic groups present in the populations. When planning commercial fishing, it is important to take into account the population composition of the species and plan the permissible catch size separately for each population.

Now we shall discuss in brief the main types of intraspecies structure characteristic of fish species (see also Allendorf and Utter 1979).

The Type of Zoarces viviparus. The low mobility of these fishes living over a vast range results in differentiation into many local reproductively isolated groups adapted to life in relatively small aquatoria and mixing little with one another. These groups differ in the number of morphological traits and simultaneously in the number of protein loci and blood groups. The differences in the loci of esterase, phosphoglucomutase and haemoglobin have been described in the eelpout (Hjorth 1971; Christiansen and Simonsen 1978). The clinal variation in Est-III and the Hb loci has also been established (see review of Christiansen 1977).

It is a characteristic feature of the intraspecies structure of this type that it is extremely highly differentiated in the absence of any marked geographical barriers, isolation being provided by the distance factor. Reliable isolation underlies the emergence of a large number of local groups with different gene frequencies and specific morpho-physiological adaptations. The clinal variation easily appears as well and involves both morphological quantitativ traits and protein loci frequently even in a small waterbody, for example, in a sea bay showing a gradient of salinity and temperature.

The Type of Cyprinus carpio. The differentiation of species in the case of semi-anadromous common carp or purely lake and river fishes (bream and others) generally follows a similar pattern. Freshwater fishes from different lakes and river systems are isolated from each other by impassable geographical barriers, mainly by the presence of land masses and the salinity of the seawater. Detailed genetic studies have been conducted with the sucker *Catostomus clarkii* (Koehn 1968, 1969a), the crested blenny *Anoplarchus purpurescens* (Johnson M. 1971), the lake whitefish *Coregonus clupeaformis* (Franzin and Clayton, 1977) and the bluegill *Lepomis macrochirus* (Avise and Felley 1979; Felley and Avise 1980). There are insufficient data available on other fishes to make any generalizations. Even in the studies mentioned above, however, only the variation in one or two and in the case of *Lepomis* in four loci has been traced. Therefore only preliminary conclusions can be made. Each geographical subspecies (or race) and each isolated population of freshwater fish differs from others in the allele frequencies of certain genes. For example, in the Amur wild carp *Cyprinus carpio haematopterus* Tf^D and My^- (null allele) appear to predominate, while these alleles are rare in European populations (Balachnin and Romanov 1971; Paaver 1979). These differences, however, apply to a few loci, the identity indices of populations for protein genes generally exceed 0.97–0.99 or more (Table 43) (see also Avise 1976a). The isolation between local populations of freshwater fish species generally does not lead to any increased geographical variation of morphological traits, although there are exceptions to this rule. The only plausible explanation of this apparently paradoxical fact is that the ecological conditions are similar for fishes living in different isolated rivers and lakes but belonging to one and the same species. These populations occupy very close ecological niches in different water bodies. One would rather expect a more essential genetic differentiation in different parts of one river basin, and this has indeed been observed in a number of cases, for example, in the rainbow trout in the upper and lower reaches of one and the same river (Northcote et al. 1970; Utter et al. 1975).

The differentiation of small southern freshwater fish species belonging to the families Characidae, Cyprinodontidae and Poeciliidae should probably be considered a special case. Such fish species exist as numerous specialized populations frequently of small size living in Africa, Central and Southern America, and Asia. It can be assumed that the genetic differences between the local groups in such fishes may be prominent. Unfortunately, no studies of this kind have been made so far with the exception of investigations on intraspecific karyotype variability (Chap. 1).

The Type of Anadromic Salmon (Salmo salar, the Steelhead Salmo gairdneri, the Genus Oncorhynchus). A distinct homing instinct is a characteristic of all migrating salmon. This homing instinct makes them return to the same river or lake and even to the small region of the freshwater basin where their parents spawned, even after a prolonged stay in the sea and frequently after mixing at feeding and wintering grounds. This surprising instinct is responsible for the good reproductive isolation of local groups within each species. This frequently leads to the formation of a rather complex hierarchical structure. The subspecies living over a vast territory are differentiated into shoals originating from one river or one lake (races or isolates). Frequently, seasonal races are present within these shoals (in the Pacific salmon these are the spring and summer races and in some other fishes the summer and autumn ones; in the Atlantic salmon and among sturgeon these are the winter and spring ones). Individuals belonging to these races enter the freshwater basins at

Table 44. Gene frequencies in even and odd populations of *Oncorhynchus gorbuscha* (Aspinwall 1974b; Salmenkova et al. 1979)

Locus	Number of subpopulations	Frequencies of the more abundant allele	
		Even population	Odd population
Mdh-A	3–4	0.992–1.000	0.927–0.974
Mdh-B	3–4	0.910–0.963	0.717–0.865
alpha-Gpd	3–4	0.807–0.928	0.889–0.966
Pgm	2–3	0.998–0.999	0.943–0.944
6Pgd	2–3	0.806–0.826	0.892–0.902

different times and correspondingly have different spawning seasons. Finally, each of the races from one and the same water body consists of several, sometimes of many subpopulations reproductively isolated to a greater or lesser extent.

Genetic studies have demonstrated the presence of a marked differentiation of populations of Salmonidae for blood groups genes (Ridgway et al. 1958; Ridgway and Utter 1963, 1964; Ridgway and Klontz 1971) as well as for various protein loci (Altukhov 1974; Allendorf and Utter 1979; Grant et al. 1980). Different river and lake shoals, as well as subpopulations within the shoals, may differ from each other. The isolation due to the homing instinct is a factor which is no less important in differentiation than the "barrier" isolation in freshwater species and the isolation by distance in the eelpout. In the pink salmon significant differences have been found between the "even" and "odd" populations, that is between populations spawning during even and odd years; these have different allele frequencies for the genes Mdh-A, Mdh-B, α-Gdph, 6Pgdh and Pgm (see Table 44). These generations are completely isolated in the reproductive sense because of the presence of a strictily biennial life and maturation cycle; all individuals die after the first spawning.

The sockeye salmon from Lake Dalneye (Kamchatka peninsula) has a four- and sometimes a five-year cycle of generation changes; in other words, a numerous

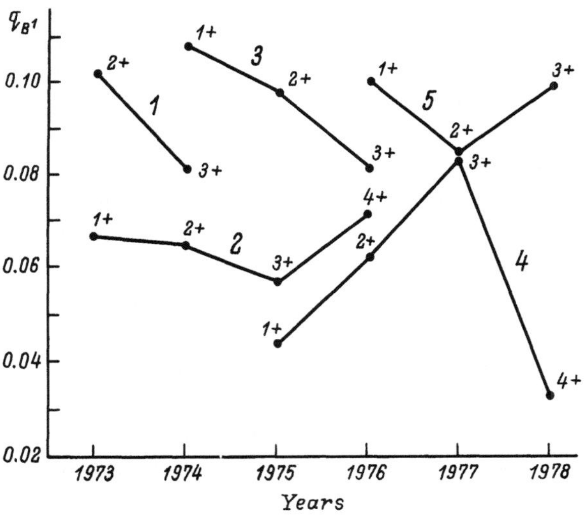

Fig. 65. Frequencies of Ldh-B1[1] allele in smolts of successive generations in the sockeye salmon *Oncorhynchus nerka* (Kamchatka, Lake Dalneye). Generations: *1* 1970; *2* 1971; *3* 1972; *4* 1973; *5* 1974. *1*+, *2*+, *3*+, and *4*+ age-groups of fishes (Kirpichnikov and Ivanova 1977; with additions)

dominating group of fishes comes to spawn once in four to five years. Each generation has a characteristically specific frequency of the Ldh-B1 gene (Fig. 65). The actual differences in frequencies correspond to the cycles. If one takes into account the fact that the alleles of this gene have different adaptive values (see p. 221) and selection with regard to these alleles appears to operate at the larval stages (Kirpichnikov 1977), one may conclude that genetic differences between generations are most probably adaptive in nature.

The exchange between different generations of the sockeye salmon is rather extensive, being more than 10% (Krogius 1975, 1978). The finding of stable genetic differences between generations is even more surprising in view of this. The exchange of individuals between local neighbouring subpopulations and probably also between seasonal races is markedly lower and does not appear to exceed 2%–3% according to some authors (Hartman and Raleigh 1964). Such an exchange probably cannot be regarded as a serious obstacle to the genetic differentiation of populations.

Unfortunately, there is hardly any information on genetic differences between the local groups and the seasonal races of the Atlantic salmon and the sturgeon. The freshwater perch *Roccus chrysops* (Serranidae), which possesses the homing instinct, has several subpopulations differing in the spectra of the serum proteins within one and the same lake (Wright and Hasler 1967).

The Herring Type (Clupea harengus). Long-term studies of the morphological traits in the various races of the herring have not provided any clear-cut picture of the intraspecific structure of this rather numerous species of marine fishes. It has only been possible to distinguish several subspecies (Atlantic, Baltic, White Sea, Pacific) and ecological groups within the subspecies such as oceanic and coastal herrings, or herring populations differing in their spawning season. Genetic studies aimed at the examination of blood groups, serum proteins, or enzymes were not very helpful either. In the case of the herring, one can only say with certainty that there are large geographical and ecological groups (Bogdanov et al. 1979; Zenkin 1979). Numerous subpopulations usually exist within such groups, but the mixing of the subpopulations appears to be very pronounced. Using better-studied systems of esterases (three loci) and the A system of blood groups (one locus), Truveller and Zenkin have been able to distinguish from three to four reproductively isolated populations within each of the three large geographical groups of the Atlantic herring (Northwest Atlantic, North Sea, Baltic Sea); parallel clines have been found in all three regions (Truveller 1978; Zenkin 1979).

A similar structure of the species appears to be correct fo the cod *Gadus morhua* (Møller 1966, 1969; de Ligny 1969; Cross and Payne 1978; Jamieson and Turner 1978), the sprat *Sprattus sprattus* (Aps and Tanner 1979; Veldre and Veldre 1979), the saury *Cololabis saira* (Utter et al. 1975), the hake *Merluccius merluccius* (Mangaly and Jamieson 1978) and many other species of marine fishes. Their common features include the absence of any physical barrier between populations and the very large population size. It should be pointed out that in many of these fish species, primarily in the herring, populations are extremely heterogeneous (Altukhov et al. 1972; Salmenkova and Volokhonskaya 1973; Salmenkova and Omelchenko 1978).

The Tuna Type (Thunnidae). The tunas and related species of oceanic fishes which are distinguished for their extreme mobility and sometimes able to span thousands of kilometres, may be considered to be a peculiar type of species organization. Only the populations living in different oceans appear to be completely isolated, only two, at maximum three groups or populations can be distinguished within each ocean (Fujino 1970, 1976). The differences in genes between the groups are moderate, and it cannot be excluded that the groups undergo extensive mixing. It may be assumed that sharks, on which there are, however, hardly any data, also belong to this type.

The nomenclature of types of the population structure presented by us is, of course, very preliminary and does not include all the existing variants. The collection of new data on the genetic variation of fish species will permit us to improve the classification presented above. At present, we are probably correct in stating that the differentiation of fishes into subspecies, races, shoals and Mendelian populations can be very diverse. With the exception of such mobile fishes as herrings, tunas, sharks, and several others, the populations of most species show distinct differentiation, which is generally associated with their reproductive isolation.

A better understanding of speciation patterns in fishes and the solution of certain practical problems related to fish breeding may significantly depend on the relative importance of different evolutionary factors in the intraspecific differentiation. These evolutionary factors include mutations, genetic drift, migration, and selection. The role of selection has already been discussed (p. 219). Attempts to make a comparative assessment of three of these factors (without taking mutations into account) have been undertaken with the populations of the sockeye salmon from the Lake Azabachye (Kamchatka peninsula) (Altukhov et al. 1975b). Using the extensive data collected from more than 2700 individuals and employing the methods of analyzing population structure proposed by Wright (1951), the authors were able to show that differences in the allele frequencies of Ldh-B1 and Pgm genes depend on the interaction of all three factors mentioned above. The role of selection in the case of Ldh-B1 was less prominent, but nevertheless quite noticeable. These conclusions follow from the graphs of allele frequencies constructed on the basis of the assumption that there are three interacting factors or only two factors without selection. Another important argument in favour of the adaptive nature of the intrapopulational variability in the sockeye salmon for the Ldh-B1 locus is provided by our own observations of the differences in the reproductive value (fitness) of two allelic variants and the frequency shifts caused by environmental stress factors (Kirpichnikov 1977; Kirpichnikov and Muske 1980).

We assume that natural selection is the main factor responsible for the genetic differentiation of fish populations; genetic drift, migration and other factors are only of secondary importance. The genetic drift based on the fluctuation of the population's size may be essential only when the reproductively active part of the population is very small. Certain fish populations have definitely passed through the "bottleneck" – the decrease of N_e to 50 individuals or less – but the incomplete isolation of a population later contributed to their mixing and the elimination of the consequences of this drift. Taken as a whole, all these observations made in studies of the genetic structure of species and of its constituent populations provide convincing evidence that the genetic differences between fish populations, as well as the genetic polymorphism within populations, are adaptive in nature.

The Adaptive Nature of Biochemical Polymorphism

According to the neutralist theory of evolution which appeared in the sixties, the evolution of proteins proceeds by random replacement (without any selection at all or with selection playing only a subordinate role) of one allele by others as a result of the interaction between the mutational process and the random changes of the allele concentration due to genetic drift. This hypothesis supported by Kimura and others (Kimura 1968a, b, 1977; Kimura and Ohta 1969, 1974) has been repeatedly criticised (Richmond 1970; Kirpichnikov 1972b; Lewontin 1974; Ayala 1975, 1976 and others) and will not be discussed here. We shall consider just one specific problem, which forms the cornerstone of the neutralist hypothesis that biochemical polymorphism is neutral or almost neutral in nature. Only if this is the case, can the fixation of the alleles be random; if alleles are not neutral and have different selective value, selection will operate markedly faster than the random stochastic processes.

The data on fish polymorphism provide evidence of the varying fitness of different alleles and do not support the view postulating the neutral nature of allelic variation. Prominent and distinct functional differences have been observed between many isozymes and allozymes as well as between the different variants of non-enzymatic proteins. These differences are expressed in the varying activity and the stability of the protein variants, in the different affinity to the substrates and in other peculiar characteristics. They appear to be correlated with the conditions under which a given variant of a polymorphic protein exerts its function.

A clinal, most frequently latitudinal, variation of allele frequencies has been observed in many species of fish. It is difficult to explain these trends from the standpoint of the secondary mixing of two forms with different gene frequencies, introgression or the spreading of a new mutant form. In a number of cases (e.g., Est genes in suckers, Ldh in sockeye salmon) a direct relationship has been established between the trend of the cline, environmental conditions and the thermal stability of the corresponding allozymes: the concentration of the alleles coding for the "northern" allozyme with a decreased optimal temperature increases in a northward direction.

Several cases of monogenic heterosis with respect to protein loci have been established. The advantage of heterozygotes provides direct evidence of the selective nature of polymorphism, no matter which specific mechanism underlies such heterosis.

A comparison of the different age groups of fishes belonging to one and the same population with regard to allele frequencies has made it possible to establish differences that can only be explained by the action of natural selection. A seasonal variation in the concentration of alleles has been extablished, these appear to be due to selection changing direction during the year. A relationship between heterozygosity and the level of variation of environmental factors, as well as the variation of morphological traits, has been established (Johnson and Mickevich 1977). Under extreme conditions such as high temperature or oxygen deficiency different genotypes possessing varying viability.

Most fish species possess polymorphic loci with alleles the frequencies of which vary from population to population (the clinal variation, differences between subspecies and races), but there are loci with allele frequencies similar throughout the range. This phenomenon is very difficult to understand if one attempts to explain it from the neutralist point of view (Efroimson 1971). Relationships between biochemical variability and environment appear to be very complex in nature. This is supported by evidence available for better-studied polymorphic systems such as LDH and Hb. The variation in these proteins may depend on fluctuations in many environmental factors – water temperature, oxygen content, pH, salinity and others (Powers 1980).

Finally, it should be pointed out that the level of enzyme variation among fishes, just as among other organisms, depends on the structure of enzymes as well as on the part they play in the organism. For example, the monomeric enzyme appears to possess an increased variation as compared to dimeric and particularly the tetrameric ones. Differences of this kind have been described in various fish species (e.g., in the plaice) and in other vertebrate and invertebrate animals as well as in man (Table 45). The decreased variation of polymeric proteins can be explained in two ways. The mutational changes in the polypeptide chains of the latter proteins will frequently lead to alterations in their quaternary structure and therefore such changes in general will be more harmful than the mutations in monomeric proteins. At the same time, in the presence of two or more polypeptides in the molecule, when these polypeptides are the products of two different genes or

Table 45. The heterozygosity of loci, coding enzymes with different quaternary structure and functional characteristics (Harris et al. 1976; Ward and Beardmore 1977; Ward 1977)

Taxonomic group	Enzyme class	Heterozygosity ($\bar{H}$)		
		Monomers	Dimers	Tetramers
Invertebrates	Glycolytic	0.164	0.099	0.067
	Others	0.201	0.152	–
	Total	0.186	0.124	0.067
Vertebrates	Glycolytic	0.119	0.048	0.013
	Others	0.111	0.033	0.024
	Total	0.113	0.040	0.015
Including:				
The flatfish:	Glycolytic	0.483	0.218	0.003
	Others	0.079	0.015	0.431[a]
	Total	0.164	0.132	0.057
Man:	Total	0.096	0.071	0.050

[a] One locus

of two alleles of one gene, heteropolymeric isozymes will be synthesized and such variants may to a certain extent replace the allelic protein variants.

The enzymes participating in glycolysis and the tricarboxylic acid cycle in invertebrate organisms are less variable than others (Gillespie and Kojima 1968), although there are exceptions to this rule (Nevo 1978). The lower variation of glycolytic enzymes characteristic of invertebrates may sometimes be explained by the existence of just one substrate for such enzymes (Gillespie and Langley 1974), but this hypothesis has not been confirmed (Selander 1976).

Although certain objections can be made with regard to some of the observations presented above which testify in favour of the adaptive nature of biochemical polymorphism, still taken as a whole, they provide evidence contradicting the neutralist hypothesis, even if one admits that neutral genes are linked with genes that have a selective value. Biochemical polymorphism is quite essential for the existence of animal and plant populations; in general, polymorphism provides for the better survival of a population in constantly changing environmental conditions, which are undergoing various trends in time and space. The constantly fluctuating conditions of intracellular metabolism may also play an important part (Johnson G. 1976).

While the adaptive nature of biochemical polymorphism appears to be evident to most naturalists, any mechanisms responsible for its maintenance in populations are far from being clear. Nevertheless, one can point out the two main mechanisms responsible for the continuing variation in protein loci. One of these mechanisms makes use of the advantages of heterozygotes. A good example is provided by the Pgm locus of the sockeye salmon. The excess of heterozygotes with respect to this locus has been observed in all populations and individual crosses. All populations are polymorphic but allele frequencies do not differ too much. An excess of heterozygotes in the Pgm locus is either due to monogenic heterosis or the linkage disequilibrium, that is to the presence in the same chromosomes of two genes lowering the viability in the homozygous state:

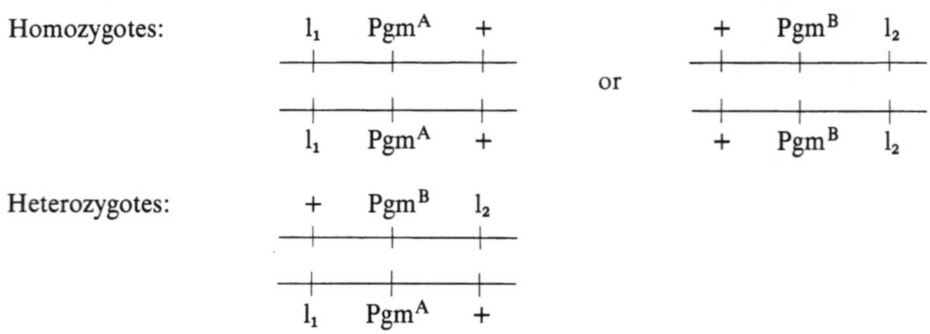

Homozygotes with this chromosomal structure will have a decreased viability, in heterozygotes, however, the semi-lethal effect of the recessive genes l_1 and l_2 is abolished. The suggestion about the presence of linkage disequilibrium in the case of Pgm is favoured by a comparison of the kinetic characteristics of the purified preparations of phosphoglucomutase obtained from three genotypic variants of *Oncorhynchus nerka,* Pgm^{AA}, Pgm^{AB} and Pgm^{BB}. The kinetic characteristics of the purified phosphoglucomutase from heterozygous individuals were intermediate, when compared to isozymes isolated from homozygotes (Kirpichnikov and Muske 1980). The preferential survival of heterozygotes appears to be associated with the formation of complex protein aggregates containing the products of several genes and functioning in a better fashion in heterozygous individuals.

No matter what the precise mechanism of heterosis is in this case, the advantage of heterozygotes with respect to viability ensures the conservation of polymorphism in the Pgm locus which appears to be maintained for a long time throughout the distribution area of the species. It is interesting to point out that a similar mechanism regarding the conservation of a polymorphic system appears to exist for this gene in several other fish species, in particular, in the eelpout, the allele frequencies of the Pgm loci in this species showing a surprisingly low variation between the different populations (Christiansen et al. 1976). Polymorphic systems of this kind appear to be common in fish species, but the increased viability of the heterozygotes is not always as prominent as in the case of the Pgm locus in the sockeye salmon; the advantage of heterozygotes can generally be detected only in more or less extreme environmental conditions or in specially designed experiments.

The second, perhaps even more important mechanism responsible for the maintenance of polymorphism is based on the variable selective value (fitness) of two or more alleles coexisting in a given population. Such "shift adaptation" of two alleles (Timoffeef-Ressovsky and Svirezhev 1966) applies to the change in the direction of selection, depending on environmental conditions, the conditions of the inner medium or a development stage; under such circumstances selection may favour one allele or another intermittently.

The differences between the alleles very frequently involve differences in the heat stability or activity of the allozymes coded by them. The heat stability of the tissues and tissue proteins is in most cases very different even for taxonomically close fish species or subspecies. This has been established in particular for various races of the lake whitefish *Coregonus autumnalis* and the grayling *Thymallus thymallus* (Ushakov et al. 1962), various species of the gadid fish (Kusakina 1967; Andrejeva 1971), several related species of the sea perch and the mackerel (Altukhov 1967; Altukhov et al. 1967). The temperature conditions of life for most species of animals, plants and microorganisms correlate with the heat stability of proteins (Alexandrov 1975); speciation appears to be accompanied by hereditary changes in the heat stability of the protein molecules. According to the hypothesis advanced by Alexandrov, a critical role is played not by direct changes in the protein heat stability since such changes are frequently outside the limits of the temperature range observable in nature, but by the change in the conformational flexibility of the protein molecules, which is coupled to protein stability with respect to heating and the action of many denaturants. The optimal functioning of a protein molecule requires a certain balance between its flexibility and stability, or between stability and activity; this is achieved by the selection of the alleles affecting both these characteristics (Alexandrov 1975). Similar suggestions have recently been put forward by Johnson G. (1976). The simultaneous evolution of heat stability and the activity of the protein molecule may probably proceed in other, perhaps more complex, ways (Hochachka and Somero 1973). These may involve changes in the ability of the protein to form complexes with other proteins and non-proteinaceous cytoplasmic components, changes in the affinity to substrates etc. There can be no doubt, however, that direct changes in the polypeptide primary structure accompanied by changes in the conformational flexibility of the protein molecules appear to have played a very important part in evolution. An increase in the protein "rigidity" may lead to a decrease in its activity over a given temperature range and vice versa.

Many observations involving the differences of this kind existing between allelic variants of proteins have been collected with fish species. An example can be provided by the Ldh-B1 locus in the sockeye salmon. In the absence of any manifested advantage of heterozygotes there is a marked difference between alleles $B1^1$ and $B1^{1'}$ in both heat stability and the activity of the allozymes coded by these alleles (p. 208; see Kirpichnikov 1977). As far as we can see, the northern allozyme (homotetramer $B1^1_4$) is unstable to heating, but retains all its activity at lower temperatures. Two alleles in the sockeye salmon populations, whose temperature optima regarding the activity of the corresponding allozymes differ, complement each other and increase the total fitness of the population.

In the presence of variable allele fitness during different seasons, in different localities and at different periods of development selection may contribute to the survival of individuals possessing different genotypes. The stability of such a system is increased if the heterozygotes finally receive some selective advantage over homozygotes. We propose that such an advantage can easily emerge as a secondary phenomenon due to the linkage or to the selection of modifier genes.

These are the two main mechanisms responsible for the prolonged persistence of many polymorphic genetic systems in fish populations; the diversity of the existing mechanisms is not, however, limited to these two. In my opinion, biochemical polymorphism is undoubtedly adaptive in its nature and cannot be a consequence of the automatic accumulation of neutral mutations.

Chapter 7

Gynogenesis in Fishes [9]

Natural Gynogenesis and Hybridogenesis

Several species of teleostean fishes are represented in nature almost exclusively by females. Such unisexual forms are reproduced by gynogenesis and hybridogenesis.

Gynogenesis is a rare type of sexual reproduction requiring insemination, when the nucleus of the sperm which has penetrated into the egg undergoes inactivation in the egg plasm, and the development of the embryo is controlled solely by maternal heredity. The chromosomes of the sperm are eliminated soon after insemination. Gynogenesis therefore requires the combination of two types of hereditary changes: (1) mutations leading to the genetic inactivation of the sperm and (2) mutations preventing the reduction of the female chromosomes in the course of oocyte maturation.

The precise role of the sperm in gynogenetic reproduction is unclear. Mature eggs of gynogenetic species do not develop in the absence of the sperm. It is not unlikely that the penetration of the sperm induces a kind of "physiological heterosis" (Golovinskaya 1954). Nor can it be excluded that the sperm's main role involves the donation of the centrosome, an organelle necessary for cell division.

Hybridogenic forms exist in nature as permanent hybrids of two closely related bisexual species. At the early stages of germinal cell maturation, all "male" (by origin) chromosomes are eliminated in the female oocytes, and only the maternal genome remains in the gametes. When such females are crossed with males of closely related species the hybrid constitution is restored in each new generation. Among vertebrate animals such a mode of multiplication is found among fishes. Hybridogenesis exists also in amphibia such as *Rana esculenta* and in the lizard *Lacerta* (Uzzell et al. 1975; Borkin and Darevsky 1980).

In fishes, gynogenesis has been described in the crucian carp *Carassius auratus gibelio* (Cyprinidae), as well as in several species of small viviparous fishes from the family Poeciliidae (*Poecilia* and *Poeciliopsis*). Hybridogenesis has only been found in *Poeciliopsis.*

Gynogenetic and hybridogenic forms are represented by unisexual populations of females. The reproduction of gynogenetic and hybridogenic forms is accomplished with the participation of males of closely related bisexual species.

---

9 This chapter was written by N. B. Cherfas

Natural gynogenesis was described for the first time in *Poecilia formosa* living in rivers of the Mexican Gulf (Hubbs and Hubbs 1932). In southern Texas and northeastern Mexico, i.e., the natural range of *Poecilia formosa,* it is generally found with other closely related fishes *P. latipinna* or *P. mexicana;* both bisexual species are found in the coastal lagoons of east Mexico together with *P. formosa.* The males of *P. latipinna* and *P. mexicana* support the gynogenetic reproduction of unisexual female populations of *P. formosa.* When the females of *P. formosa* were crossed under laboratory conditions with males of 50 different species, subspecies and races of *Poecilia,* as well as with males of another genus of the same family, the progeny always consisted solely of females identical to the maternal forms (Hubbs 1946 a).

In the natural populations, *P. formosa* is represented by the diploid form with 46 chromosomes. The same chromosomal number has been observed in the closely related bisexual species. The chromosomal complement of diploid females of *P. formosa* contains one genome of *P. latipinna* and one genome of *P. mexicana.* The first unequivocal evidence in support of this has been obtained in comparative studies of the electrophoretic spectra of serum albumins in the females of *P. formosa* and several related bisexual species (Abramoff et al. 1968; Balsano 1969). The electrophoretic data were confirmed by karyological studies (Prehn and Rasch 1969), making use of the difference in the size of the largest chromosomes in *P. latipinna* and *P. mexicana.* Occasionally populations of *P. formosa* also contain triploid females with a chromosomal number of 69 (Prehn and Rasch 1969; Rasch et al. 1970). The analysis of chromosomal complements and electrophoretic data (Balsano et al. 1972) has indicated that the triploid form originated from crosses of diploid *P. formosa* females with males of *P. mexicana.* According to morphological criteria, the triploid females occupied an intermediate position between the diploid forms of *P. formosa* and *P. mexicana,* but they were much more similar to *P. formosa* (Menzel and Darnell 1973). The triploid females represent a persisting component in the populations of *P. formosa.* The proportion of diploids to triploids varies in different basins, the diploid form being very rare in certain regions (Menzel and Darnell, 1973).

The investigation of the triploid females has demonstrated that after crosses with males of bisexual species they yield gynogenetic triploid progeny (Rasch and Balsano 1973; Strommen et al. 1975). In one case, the triploid offspring was obtained from a pregnant diploid female captured in nature; this offspring was similar to that of triploid females and had normally developed ovaries. The triploid females in *P. formosa* populations appear to emerge by two different mechanisms: (1) by reproducing themselves and (2) as a result of crosses of certain diploid females of *P. formosa* with the males of closely related species.

Aspects of gonad maturation in *P. formosa* have not been examined in detail. It has been assumed that premeiotic endomitosis involving chromosome duplication without subsequent cell division could take place in oocytes of both forms (Schultz and Kallman 1968; Rasch et al. 1970). As regards the diploid forms, there is unequivocal evidence that meiosis is arrested in them and maturation proceeds according to the ameiotic pattern. All the offspring of a given female is clone or a community of individuals genetically identical to the mother. Experiments to study tissue compatibility have demonstrated to clonal nature of different broods and of natural populations of *P. formosa* (Kallman 1962 a, 1962 b; Darnell et al. 1967).

Throughout the history of studies only a few dozen males have been found in the populations of *P. formosa* (Darnell and Abramoff 1968). Among the few fishes examined specially, only occasional males gave progeny with females of the same species, most of the males being reproductively incompetent. The males of *P. formosa* resemble the females of the same species, which were transformed into males by hormonal treatment; they also resemble males from the F_1 hybrid progeny of crosses between *P. latipinna* and *P. mexicana*. The appearance of rare males in the natural populations of *P. formosa* (and in certain laboratory stocks) most probably results from phenotypic sex transdetermination in gynogenetic females or from the overlooked hybridization of bisexual related species (Hubbs et al. 1959). Nor can it be excluded that the male parent's chromatin remains partially conserved, and this leads to the manifestation of traits characteristic of the male sex (Haskins et al. 1960).

Gynogenesis in *P. formosa* apparently emerged comparatively recently in evolution. The young phylogenetic age of diploid females of *P. formosa* is inferred from their phenotypic resemblance to hybrids of *P. mexicana* and *P. latipina,* established on the basis of morphological and biochemical criteria (Hubbs and Hubbs 1946 b; Balsano 1969). The discovery of triploid hybrids in laboratory crosses between the diploid females of *P. formosa* and the males of other species (the mean frequency of such hybrids is equal to about 1%; see Schultz and Kallman 1968), as well as the existence of a triploid progeny from a diploid female (Strommen et al. 1975), means that no complete reproductive isolation exists at present between the diploid gynogenetic females of *P. formosa* and the bisexual species. The absence of *P. formosa* in certain regions of the distribution range of the parental species can also be explained by the recent evolutionary origin of the species (Darnell and Abramoff 1968).

The gynogenetic forms of the genus *Poeciliopsis* belong to the unique unisexual-bisexual complex of fishes living off the Pacific coast of Mexico. As well as the usual bisexual species and gynogenetic forms, this complex also includes forms reproducing by hybridogenesis (Miller and Schultz 1959; Schultz 1961, 1969, 1977). Four hybridogenic forms of different origin have been found in the basins of northwestern Mexico, each of these forms being associated with its "own" bisexual species. The hybridogenic form *P. monacha-lucida* (P_{m-l}) has been studied in the greatest detail; several clonal variants have been described (Schultz 1967). The diploid females of P_{m-l} are hybrids of *P. monacha* and *P. lucida.* They are found in a common population with *P. lucida* and are reproduced, making use of males of this species (Schultz 1961, 1966, 1969).

The genome of a diploid P_{m-l} contains haploid chromosomal complements of both parental species; at the same time, the "monacha" complement of hybridogenic females P_{m-l} and the haploid complement of individuals from bisexual *P. monacha* populations are identical (Leslie and Vrijenhoek 1978). When the females P_{m-l} are crossed with *P. lucida* males or with males of more distant species, they yield unisexual hybrid offspring. Furthermore, no intensification of the traits of the father species can be observed in the progeny from back-crosses. The hybrid origin of P_{m-l} has also been confirmed by the data of biochemical studies (Vrijenhoek 1972).

The reproduction of unisexual populations of P_{m-l} appears to involve the elimination of the male (by origin) genome in the females. In other words, *lucida* chromosomes are selectively eliminated; this is confirmed by crosses making use of such genetic markers as Ldh alleles (Vrijenhoek 1972). The elimination of "male" chromosomes takes place during early meiosis prior to the onset of vitelogenesis; thereafter the haploid female complement (24 chromosomes of the *"monacha"*) undergoes one equational division, forming a single polar body. Twenty-four "female" chromosomes remain in the maturing egg. During insemination, the sperm cell brings in twenty-four *"lucida"* chromosomes, and the hybrid karyotype is restored.

The emergence of three other hybridogenic forms also required the participation of *P. monacha* females, but the actual mechanisms appeared to differ (Schultz 1977). The P_{m-0} form (*P. monacha-occidentalis*) apparently originated from crosses of *P. monacha* females with *P. occidentalis* males in the areas where these two bisexual species live in the same habitats. The P_{m-lat} females (*P. monacha-latidens*) were apparently formed when the genome of *P. lucida* was substituted for the genome of *P. latidens* in crosses of hybridogenic females of P_{m-l} with the males of *P. latidens.* Finally, the trihybrid hybridogenic form of $P_{m(v)-l}$ has been found in certain populations of *Poeciliopsis.* This form bears the traits of *P. monacha, P. viriosa* and *P. lucida* (Vrijenhoek and Schultz 1974). The genome of this form includes certain chromosomes or genes of *P. viriosa.* The introgression of *P. viriosa* genes might occur as a result of crosses of *P. monacha* females with *P. viriosa* males. Meiosis normally occurs in hybrids resulting from a cross of this type.

Hybridogenic forms in nature appear in those regions where the bisexual species occupy the same ecological niches.

Even at the present time the populations of these species include females with a hereditary tendency to hybridogenesis. This is confirmed in particular by the successful syntheses of hybridogenic P_{m-l} females under laboratory conditions (Schultz 1977).

The hybridogenic forms are regarded by certain authors as the ancestors of the gynogenetic forms found in the genus *Poeciliopsis.* The mechanism of unisexual reproduction could perhaps emerge in hybrids as a result of elimination of the "male" chromosomes at the beginning of meiosis or could result from a non-random assortment of the male and female (by origin) complements in females during reductional division (Vrijenhoek 1979b). The transition from hybridogenesis to gynogenesis is associated with the impairment of the hybridogenesis mechanisms; this includes (1) loss of the reduction of the male complement during egg maturation and (2) elimination of the sperm chromosomes soon after insemination. Three gynogenetic forms of *Poeciliopsis* have been described: P_{2m-l}, P_{m-2l} and $P_{2m(v)-l}$, which is the least studied. All of them are triploids and, like the hybridogenic forms, they have no independent taxonomic nomenclature. The arbitrary designation (2m–l etc.) refers to the number of genomes of each of the parental species in the chromosomal complement. It is believed that the hybridogenic P_{m-l} females are the ancestors of the first two forms and the females $P_{m(v)-l}$ are the ancestors of the third form. The gynogenetic forms coexist sympatrically with the bisexual host species: the P_{2m-l} lives together with *P. monacha* and P_{m-2l} is found together with *P. lucida.* They differ from each other in the structure of their teeth, the number of vertebrae and in several other morphological traits. As regards the biological characteristics, each gynogenetic form is characterized by a preferential resemblance to its "own" bisexual species. Gynogenesis has been established in experiments involving special crosses with males of several bisexual sympatric and allopatric species (Schultz 1967).

The hybrid nature of the gynogenetic individuals of *Poeciliopsis* has been proved beyond doubt by the data obtained from a morphological analysis (Schultz 1967, 1969), as well by biochemical marker analysis (Vrijenhoek 1972). The gynogenetic P_{2m-1} females have 72 chromosomes in the complement. The triploidy has been established by counting the number of chromosomes (Schultz 1971) directly and by indirect techniques: on the basis of the size of the erythrocyte nuclei, cytophotometrically, by measuring the DNA content in the cell nuclei and on the basis of LDH isozyme spectra (Vrijenhoek 1972, Cimino 1973, 1974). In the course of the maturation of gynogenetic females (Cimino 1972a, 1972b) the primary oogonia undergo endomitosis. The hexaploid nucleus passes through meiosis in which the bivalents are formed by sister chromosomes. This mechanism leads to a genetic similarity between the mother and her offspring and results in clonal reproduction.

Diploid hybridogenic and triploid gynogenetic individuals are generally found together in most unisexual populations of *Poeciliopsis,* the relative proportion of diploids varying but usually being present in greater abundance. The admixture of triploids is extended owing to their higher fertility at the end of dry season. In mixed unisexual populations triploids can be differentiated from outwardly similar diploids by the criterion of the erythrocyte diameter (Thibault 1978).

The only known case of obtaining a male in the progeny from P_{2m-1} female is explained by the loss of one genome of *monacha* (Cimino and Schultz 1970). In its external appearance this male is a typical hybrid P_{m-1}; this observation provides additional evidence of the absence of random segregation in the course of maturation of gynogenetic females P_{2m-1} (Schultz 1971).

Apparently, the genome of *P. monacha* played a central part in the origin of all the hybridogenic and gynogenetic forms of *Poeciliopsis;* this genome is generally conserved during the reduction of chromosomes preceding hybridogenesis. The presence of unisexual forms possessing clonal reproduction in populations of bisexual *P. monacha* contributed to the survival of this special despite the dramatic variations in its numbers which are characteristic of it. At the same time, when purely clonal reproduction was impaired (probably quite often), the exchange of genes between unisexual individuals and bisexual forms of *P. monacha* became possible. This provided for a sufficiently high level of variability in bisexual populations, decreasing due to a strong gene drift (Vrijenhoek 1979a, b).

The last known case of natural gynogenesis among fishes was observed in the crucian carp *Carassius auratus gibelio* (Cyprinidae) (Golovinskaya and Romashov 1947). The gynogenetic unisexual form in *Carassius auratus gibelio* is known to exist alongside with the usual bisexual form of this species, and these two forms are morphologically indistinguishable (Golovinskaya et al. 1965). *Carassius auratus gibelio* is distributed over a vast territory stretching from the Far East to Western Europe. Generally, bisexual populations are found in the Eastern part of this territory including certain basins in Siberia. In such populations females are more numerous; this is, apparently, due to the mixing of unisexual and bisexual forms; the balance between these two forms determines the overall proportion of males and females in the population. The percentage of males in the populations of *Carassius auratus gibelio* decreases in a westward direction and in the European part of the range this species is universally represented almost solely by the unisexual gynogenetic form. The appearance of bisexual populations in certain basins of the European part of the Soviet Union is due to the occasional release of "bisexual" *Carassius auratus gibelio* from the river Amur. Both forms are found in Europe (Romania, Bulgaria, the GDR and the FRG), but the unisexual form predominates. In mixed populations the gynogenetic females are reproduced with the participation of males of the bisexual form of the same species; in the unisexual female populations males of closely related species participate in reproduction, this species included the wild common carp, the roach, *Carassius carassius* etc.). In pond fisheries, males of the carp are most frequently used in breeding *Carassius auratus gibelio.*

All the crosses conducted so far have confirmed the presence of gynogenesis in this species (Table 46). The results of the experiments on the insemination of eggs of the unisexual and bisexual forms of *Carassius auratus gibelio* by the irradiated sperm of the carp were particularly impressive (Golovinskaya et al. 1965). The radiation damage to carp chromosomes, lethal for the bisexual form, did not inhibit the development of the offspring of the unisexual form; the paternal chromosomes do not participate in the development of the unisexual form of *Carassius auratus gibelio.* The early development of gynogenetic forms was examined cytologically as far as the stage of two blastomeres (Golovinskaya 1954; Kobayashi 1971). The head of the spermatozoa which has penetrated into the egg does not undergo transformation into the male pronucleus and does not participate in the first cleavage division.

Table 46. Results obtained in crosses between the females of the unisexual form of the crucian carp and the males of other fish species

Males	Offspring composition	Reference
Cyprinidae		
Cyprinus carpio	The crucian carp (*Carassius auratus gibelio*), females	4
Same, but sperm irradiated	The crucian carp (*Carassius auratus gibelio*), females	5
Carassius auratus	The crucian carp (*Carassius auratus gibelio*), females	4
C. auratus gibelio:		
unisexual form	The crucian carp (*Carassius auratus gibelio*), females	2
bisexual form	The crucian carp (*Carassius auratus gibelio*), females	3
C. carassius	The crucian carp (*Carassius auratus gibelio*), females	4
Tinca tinca	The crucian carp (*Carassius auratus gibelio*), females	4
Rutilus rutilus	The crucian carp (*Carassius auratus gibelio*), females	4
Hemibarbus labeo	The crucian carp (*Carassius auratus gibelio*), females	6
Cobitidae		
Misgurnus fossilis	The crucian carp (*Carassius auratus gibelio*), females	5
Salmonidae		
Salmo gairdneri	The crucian carp (*Carassius auratus gibelio*), females	1

References: 1. Tcherfas (1971); 2. Golovinskaya (1954); 3. Golovinskaya (1960); 4. Golovinskaya and Romashov (1947); 5. Golovinskaya et al. (1965); 6. Kryzhanovsky (1947)

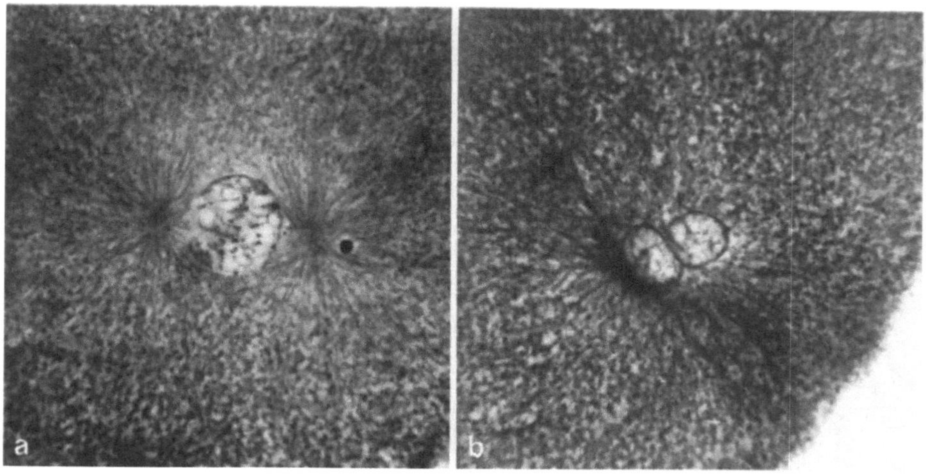

Fig. 66 a, b. Preparation to the first cleavage division in silver crucian carp *Carassius auratus gibelio*. **a** the unisexual form: the triploid female pronucleus and the sperm head; **b** the bisexual form: female und male pronuclei

During this period it looks like a dense chromatin body and is apparently eliminated at later stages (Fig. 66).

The number of chromosomes in the diploid forms of *Carassius auratus gibelio* is equal to 94–100 (Cherfas 1966 a; Ojima et al. 1966). A great similarity has been noted between the karyotypes of *Carassius auratus gibelio* and those of the carp (Ojima and Hitotsumachi 1967). The gynogenetic females of *Carassius auratus gibelio* from both European and Japanese populations are triploid. The tetraploid gynogenetic form has been found in Japanese populations. The chromosomal number is 135–146 in triploid females from the European population (Cherfas 1966 a), in triploids of the Japanese crucian carp *Carassius auratus gibelio* it is 156 and in tetraploids of the same species it is 206 (Kobayashi et al. 1970, 1977; Kobayashi and Ochi 1972). Recent studies of Japanese *Carassius* species (Muramoto 1975) have resulted in the discovery that individuals may vary in the number of chromosomes: from 153 to 165 chromosomes have been found in different females (microchromosomes have been detected in several individuals). Triploidy of *Carassius auratus gibelio* was accompanied by a proportional increase in the cell size just as was observed in gynogenetic forms of Poeciliidae. Cytometry data (the size of erythrocytes, for example) can be used in analyzing fishes in mixed populations of *Carassius auratus gibelio* (Cherfas 1966 a; Cherfas and Shart 1970).

Meiosis of *Carassius* triploids has been studied in great detail (Cherfas 1966 b). The conjugation of homologous chromosomes, crossing-over and reductional division did not occur in the course of maturation. The egg undergoes two maturational divisions. The three-pole spindle is formed during the first abortive division (Fig. 67), and later it is transformed into a bi-polar one. The univalent chromosomes are distributed between the two poles, either equally or in the proportion 1:2. Subsequently, all the univalent chromosomes become associated in a united triploid metaphase of the second (and actually of the first) division, representing conventional mitosis. When the oocyte is at this stage, ovulation takes place (Fig. 68). Mei-

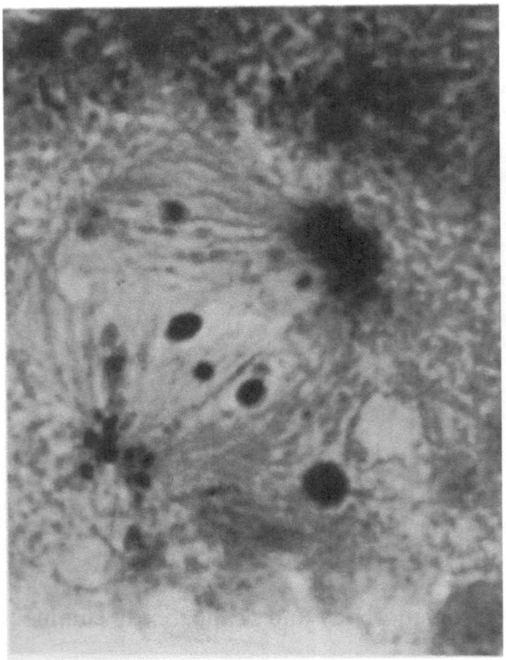

Fig. 67. The first cleavage division in a gynogentic female of *Carassius auratus gibelio* (mitosis with three poles may be seen)

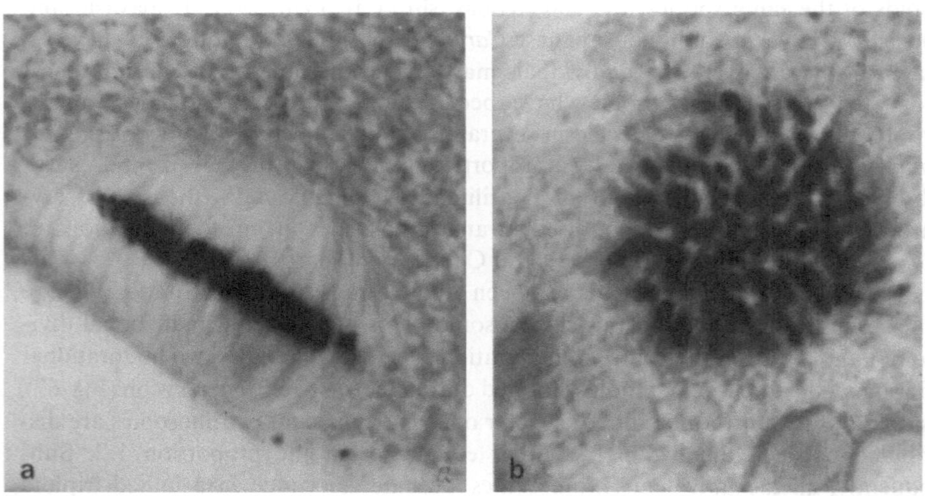

Fig. 68a, b. The oocyte of the unisexual silver crucian carp *Carassius auratus gibelio* at the moment of ovulation. **a** the spindle of the metaphase II; **b** the triploid plate

osis is completed soon after the penetration of the sperm, when the only polar body is separated; the triploid group of female chromosomes remaining in the egg is then transformed into the female pronucleus and subsequently forms the metaphase plate of the first cleavage division. These peculiarities of meiosis in triploid females of *Carassius auratus gibelio* result in the clonal reproduction of this species; in this respect, *Carassius* is similar to other gynogenetic forms of fishes.

Triploid females of *Carassius auratus gibelio* do not differ from the females of bisexual forms in their morphological features, and this would suggest that gynogenesis is not of a hybrid nature. Prominent differences in erythrocyte antigens in the unisexual and bisexual forms (Pokhill 1969) may be the consequence of the mutations accumulated in the unisexual individuals. Nevertheless, we cannot completely exclude the hybrid origin of gynogenetic forms; some evidence in favour of this hypothesis is provided by the peculiar distribution of the chromosomes during the first meiotic division; these may reflect the cytological and genetic non-uniformity of the genomes constituting the triploid chromosomal set. Other evidence is also available in favour of the hybrid origin of gynogenesis. The finding of the triploid hybrids in the offspring from the backcrosses of F_1 females (common carp × crucian carp) with males of the common carp has aroused considerable interest. The number of chromosomes in these triploids was equal to 145–160 just as in triploids of the crucian carp from natural populations; the major proportion of hybrids obtained in the back-crosses was represented by females (Ojima et al. 1975). In experiments with induced gynogenesis we discovered the ability of F_1 females resulting from crosses of females from the bisexual race of the crucian carp with carp males to produce predominantly diploid eggs (Cherfas and Ilyasova 1980a, b). This is in good agreement with the observations of Japanese scientists and explains the emergence of triploids in their experiments. If the genome of the unisexual crucian carp *Carassius auratus gibelio* incorporates chromosomes of the common carp, one has to assume that for some reason all the genes on these chromosomes are repressed and inactive. The possible repression of the genes of one of the initial species follows from the results of the studies of gynogenetic females P_{m-21} in *Poeciliopsis* (Schultz 1977). When the loss of certain traits of *P. monacha* has been observed in certain clones, the similarity of the resulting gynogenetic individuals to *P. lucida* became more pronounced.

The males are even more rare in the European gynogenetic populations of *Carassius auratus gibelio* than in *P. formosa.* During the examination of many tens of thousands of fishes for more than 22 years of study only two males were found; one of these gave offspring with its "own" gynogenetic females, but turned out to be reproductively deficient in the cross with the carp female; another male turned out to be completely sterile (Golovinskaya 1960). Triploid males have been found in Japanese populations (Muramoto 1975); their reproductive potential has not been examined.

We may conclude that in fishes, just as in parthenogenetic and gynogenetic forms of other animals, natural gynogenesis is intimately associated with the phenomena of hybridization, apomixis and polyploidy. The emergence of triploid parthenogenesis is usually preceded by the formation of interspecific (or more related) hybrids and by diploid ameiotic parthenogenesis, including the loss of chromosomal reduction in the course of oocyte maturation. Triploid ameiotic parthenogenesis and gynogenesis may occur as a result of back-crosses of parthenogenetic diploid females with males of related bisexual species (Astaurov 1971).

This scheme smoothly explains the origin of the gynogenetic forms belonging to Poeciliidae. The first step in the sequence of events resulting in triploid gynogenesis in this family involved interspecific hybridization. The hybridization was facilitated by easy crosses between the species and by the shortage of males in natural populations caused by their lowered viability.

The hybrids must have lowered fertility as a consequence of the absence of balance between the genomes of the original species.

The second most important step included the loss of the reductional division in hybrids and apparently the simultaneous appearance of the ability to develop parthenogenetically (diploid ameiotic parthenogenesis); this resulted in the hybrids recovering their fertility. In the presence of mechanisms inactivating the male nucleus after the penetration of the sperm into the egg, parthenogenesis could easily be changed for diploid gynogenesis. Finally, from time to time the back-crosses of diploid gynogenetic females with the males of bisexual species resulted in the appearance of the triploids, genuine hybrids also reproducing by gynogenesis.

The consecutive stages of the evolutionary transformations leading to triploid ameiotic gynogenesis can be followed in the most complete form on *Poecilia formosa.* It is quite probable that in some species of *Poeciliopsis* the gynogenetic form emerged via hybridization and hybridogenesis, although there is no direct evidence of this.

The data of the Japanese authors suggest that triploid gynogenesis in the crucian carp could possibly occur in a shorter sequence of events. Almost all the hybrids obtained in the back-crosses described by them appear to be sterile, but single females can be at least fertile, developing rare diploid gametes as a result of loss of reduction. The triploid zygotes can give rise to gynogenetic clones. In this case, ameiotic gynogenesis developed directly on the basis of the triploid karyotype, in the possible absence of the stage of diploid gynogenesis.

It is quite possible that the ability to spontaneously undergo the gynogenesis typical of *Carassius auratus gibelio* was advantageous since it prevented the hybridization of females of the species with males of other species when females were occasionally transferred to foreign habitats (Golovinskaya and Romashov 1947). Gynogenesis appears to contribute to better isolation in other species as well. This suggestion is supported by findings of gynogenetic individuals of the maternal type in many cases of distant hybridization of fishes. It cannot be excluded that hybridization in the case of *Carassius auratus gibelio* took place at later stages in the development of triploid gynogenesis.

There is, however, still some difficulty in directly relating gynogenesis in *Carassius auratus gibelio* to hybridization, because of the absence of morphological differences between the unisexual triploid and bisexual diploid forms. Detailed karyological and biochemical genetic studies of both forms of the crucian carp, the common carp, and their hybrids are needed if this difficulty is to be overcome.

The evolution of gynogenetic forms apparently included chromosomal rearrangements. In *P. formosa* females, a chromosome possessing a subterminal centromere has been found. The chromosomal complement of related bisexual species of this family is represented exclusively by acrocentric chromosomes (Prehn and

Rasch 1969). The DNA content per nucleus in all triploid gynogenetic females of the family Poeciliidae is lower than that expected, apparently because of the loss of part of the DNA (Cimino 1974). The ameiotic type of maturation makes such changes possible.

It is not easy to evaluate the significance of gynogenesis and hybridogenesis in fishes. The transition to these types of reproduction is associated with the loss of a component of genetic variability due to a recombination of the chromosomes and genes during conventional sexual reproduction. The genetic diversity of the gynogenetic forms is limited by the diversity of the clones. New clones can emerge in populations as a result of the repeated appearance of gynogenetic females and also as a consequence of mutations. A study of the populations of P_{2m-1} females of *Poeciliopsis* allowed us to follow the segregation of a single clone into two as a result of mutations. Generally, however, the clonal variation in the gynogenetic populations studied was low. The genetic variation in hybridogenic females is markedly greater. The paternal genome "borrowed" from the bisexual species is an additional source of genetic variation. New alleles can also be incorporated into the maternal genome as a result of periodic crosses of the hybridogenic females with the males of *P. monacha* (Vrijenhoek et al. 1977; Angus and Schultz 1979).

The advantages of parthenogenesis are usually attributed to the two-fold acceleration of the reproduction rate which contributes to the wider propagation of this species. All the unisexual female forms of fishes differ from the parthenogenetic animals in the fact that they are obligate "parasites" and depend completely on the presence of closely related bisexual species. This is primarily true of viviparous forms of the family Poeciliidae. A highly complex interaction between the unisexual members of the complex and bisexual species is established in mixed unisexual-bisexual populations of *Poecilia* and *Poeciliopsis* (McKay 1971; Moore and McKay 1971; Schultz 1971). This interaction involves the competition for males as well as for ecological niches and food resources. The males of the bisexual species prefer to fertilize their own females; a similar phenomenon has been observed in the mixed population of salamanders (Borkin and Darevsky 1980). The selectivity of fertilization particularly characteristic of representatives of the genus *Poeciliopsis* apparently results from the selection of the genes controlling the mating behaviour. Studies of P_{m-21} females have demonstrated that the adaptive changes increasing the probability of fertilization of the gynogenetic females by the males of the bisexual species (*P. lucida*) are also possible.

Usually a marked proportion of the gynogenetic females remains unfertilized; the relative amount of unfertilized females increases in proportion to the abundance of the unisexual form in the population (Moore and Bradley 1979). Because of external fertilization this tendency is less pronounced in *Carassius auratus gibelio*, and this can explain the wide spreading of the unisexual form of the crucian carp and its greater abundance in the main part of its distribution range.

The genetic advantages of the gynogenetic forms can be attributed to their hybrid nature, ameiotic type of maturation and triploidy. These characteristics are conductive to the formation of adaptive, stabile heterozygous genetic systems (Vrijenhoek et al. 1977; Turner et al. 1980). Under laboratory conditions the gynogenetic females of *Poeciliopsis* show higher viability than the females of closely related bisexual species. Heterosis, characteristic of the unisexual forms, provides

X-rays (about 100 kR) are used for this purpose. Since the irradiation selectively destroys chromosomes, even with high doses fish spermatozoa retain the ability to penetrate into the egg; the apparatus of cell division contributed by them functions normally. Since inactivated sperm chromosomes do not participate in development, only female chromosomes remain active in the cells of the embryo. There is a small but important difference from natural gynogenesis, namely, the sperm head is transformed to the pronucleus. Later, however, paternal chromosomes as well as under natural gynogenesis, undergo picnosis and do not participate in the first cleavage division (Romashov and Belyaeva 1964, 1965 a).

Mature ovulated fish eggs are at metaphase II of meiosis and contain a reduced number of chromosomes. The embryos originating from the insemination of eggs by the genetically inactivated irradiated sperm represent haploids with a certain complex of developmental deviations which can be called a haploid syndrome. The most typical defects include shortening and the characteristic deformation of the body, excessive hydration of the pericardium, and impairment of the pigmentation. The haploids pass through embryonic development relatively well, but die during the hatching and in the course of the next few days.

If the problem of experimental diploid gynogenesis is to be solved, methods of diploidizing the female chromosomal complement must first be developed. The frequency of spontaneous diploidization varies in different species of fishes and is generally very low. It can be increased by treating the eggs at extreme sublethal temperatures, which affect the normal course of the second meiotic division. Such temperature shocks can be employed prior to the insemination of the eggs at the metaphase II stage, but the treatment administered soon after insemination at the anaphase II stage is more effective. The conditions in which the temperature shock is used may vary, depending on the particular species used.

The behaviour of the female chromosomes after shock treatment has been followed in the case of the loach (Romashov and Belyaeva 1965 b). It has been demonstrated that diploidization results from a union of two haploid complements produced in the second meiotic division. A similar mechanism of diploidization operates in the case of the common carp, as can be inferred from the genetic analysis (Golovinskaya and Romashov 1966). If such a "diploidized" egg is fertilized by a normal unirradiated sperm, a triploid embryo appears. Diploidization in carp is followed by the appearance of many (up to 25%) aneuploid embryos (Gervai et al. 1980).

The frequency of diploidization of the female chromosomes in different fishes and after different shock treatments varied, but was very high in certain experiments (Table 47). However, massive gynogenetic stocks containing thousands of larvae have so far been obtained only in the common and grass carp (Cherfas 1975; Stanley 1976 a; Nagy et al. 1978).

Quite recently a great ability to diploid gynogenesis was observed in hybrids of the crucian carp *Carassius auratus gibelio* (bisexual form) and the common carp. The yield of viable diploid larvae in gynogenetic broods from hybrid females was equal to almost 100%, when calculated relative to the number of embryos that had started to develop. The easy transition to gynogenesis in interspecific hybrids can be attributed to the reduction of meiosis in the course of maturation of the hybrid eggs. An increased tendency to induced gynogenesis, up to 1.7% of the diploid embryos,

Table 47. The yield of triploids and gynogenetic diploids after the application of the heat shock during the second meiotic division

Organism	Temperature conditions		Frequency of diploids and triploids, %		Reference
	Shock temperature (°C)	Duration (min)	Experiments	Control	
Acipenseridae	3–4	180	4	1.5	7
	34[a]	3	52	0	10
Cyprinus carpio	8–9	210	8	0.1	1
	7.5	330	22	1.0	2
	4.0	60	36	0.2	3
Misgurnus fossilis	0.5–3.0	180	60	0.3	6
	34	4	17	0.3	6
Tilapia aurea	11[a]	60	75	–	9
Gasterosteus aculeatus	0–1[a]	90–180	56	0	8
	33.5–34.0[a]	5	50	0	8
Pleuronectidae	0–0.5	120–240	100	–	4
	0.5	240	94	3.5	5

[a] In these experiments triploids were obtained.

References: 1. Cherfas (1975); 2. Cherfas and Ilyasova (1980a); 3. Nagy et al. (1978); 4. Purdom (1972); 5. Purdom and Lincoln (1974); 6. Romashov and Beljaeva (1965b); 7. Romashov et al. (1963); 8. Swarup (1959); 9. Valenti (1975); 10. Vasetzky (1967)

has also been recorded in hybrids resulting from crosses of the European carp with the Japanese decorative carp (Cherfas and Ilyasova 1980a).

The gynogenetic nature of the diploid offspring can be proved either by a morphological examination of such fishes or, and this gives more precise data, using a special genetic analysis. Morphological criteria can be used in the case of distant crosses, i.e., when the sperm of a species distinctly different from the maternal form is used for insemination. In all cases of heterogeneous insemination the offspring had characters typical of the mother species (Table 48). A marker genes analysis may be used, when there is detailed genetic information on the system. The finding of scattered carps (ss) in broods resulting from a cross of the scattered female of the carp with males having the formula SS (Golovinskaya et al. 1963) can serve as an example. In other experiments with common carp, genes affecting the character of the N and n, the pattern D and the light pigmentation, L have been used.

Genes showing the expression at the early stages of development, in particular, the recessive duplicate genes of orange pigmentation are especially convenient for this work (Katasonov, 1978). Biochemical markers with codominant expression may also be used.

The Cytogenetic Characteristics of Induced Gynogenesis. The cytological mechanisms underlying the diploidization of the egg chromosomal complement determine the most important features of artificial gynogenesis; segregation in the offspring

Table 48. The gynogenetic offspring obtained in distant crosses of fishes

Species crossed		Reference
Females	Males (sperm irradiated)	
Acipenser ruthenus	*Huso huso*	5
H. huso	*A. ruthenus*	5
Cyprinus carpio	*Carassius carassius*	1
Ctenopharyngodon idella	*C. auratus*	7
Ctenopharyngodon idella	*Cyprinus carpio*	6
Misgurnus fossilis	*C. carpio*	4
Misgurnus fossilis	*Carassius auratus gibelio*	4
Misgurnus fossilis	*C. carassius*	4
Pleuronectes platessa	*Hippoglossus hippoglossus*	3
Pleuronectes platessa	*Platichthys flesus*	2

References: 1. Golovinskaya et al. (1963); 2. Purdom (1969); 3. Purdom and Lincoln (1974); 4., 5. Romashov et al. (1961, 1963); 6. Stanley and Jones (1976); 7. Stanley and Sneed (1974)

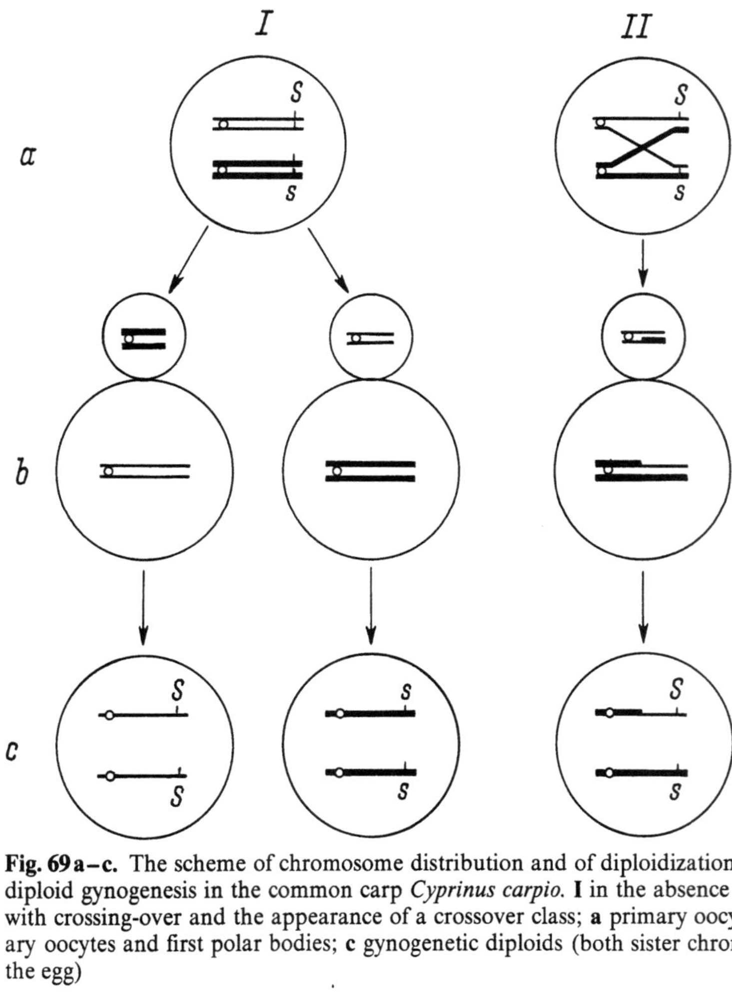

Fig. 69 a–c. The scheme of chromosome distribution and of diploidization of embryos during diploid gynogenesis in the common carp *Cyprinus carpio*. **I** in the absence of crossing-over; **II** with crossing-over and the appearance of a crossover class; **a** primary oocytes (2n); **b** secondary oocytes and first polar bodies; **c** gynogenetic diploids (both sister chromosomes remain in the egg)

Table 49. The number of heterozygous individuals in gynogenetic offspring of the carp (from heterozygous females)

Locus and trait	Number of heterozygotes	
	%	Confidence limits at p = 0.05
S (pattern of scales)	4.8	4.5– 5.1
Tf (transferrin)	5.0	0.5– 9.5
Est-S ("slow" blood serum esterase)	9.1	2.5–15.7
Est-F ("fast" blood serum esterase)	28.4	17.9–38.9
D ("pattern")	70.0	63.9–76.1
L (light pigmentation)	73.9	66.4–81.4
N (reduction of scales)	97.9	96.6–99.2

and the high homozygosity of gynogenetic individuals. Both these characteristics differ fundamentally for induced diploid meiotic gynogenesis, on the one hand, and natural gynogenesis on the other.

The analysis of segregation in gynogenetic offspring has been conducted in the carp (Golovinskaya and Romashov 1966; Cherfas 1977; Cherfas and Truveller 1978). It has been established that the gynogenetic offspring obtained from heterozygous females does, as a rule, contain a certain percentage of heterozygotes (Table 49). The appearance of heterozygotes can best be explained by the hypothesis of interchromatid meiotic crossing-over in the region between the gene and the centromere (Fig. 69). This hypothesis suggested by Golovinskaya and Romashov (1966) is now support by many authors. The number of heterozygotes for lethal and co-dominant genes may be established directly from the experimental results. In the case of complete dominance, it is determined as the difference between the number of offsprings in the dominant (AA + Aa) and recessive (aa) classes. This calculation is based on the postulate of the random segregation of alleles in the first meiotic division. If the survival of both types of homozygous organisms in the offspring is identical, the proportion between them in gynogenetic stocks must be equal to 1:1. The analysis of segregation with respect to certain biochemical traits confirms this suggestion (Table 50). The number of heterozygotes varies greatly in different genes.

Table 50. Segregation for biochemical loci in gynogenetic carp offspring (Cherfas and Truveller 1978; Nagy et al. 1978)

No.	Locus	Female pheno-type	Number of individuals		Analysis of the pro-portion of the homo-zygous offspring	
			Homozygous	Heterozygous	χ^2 (1:1)	p
1	Est-F	BC	28 (B) 25 (C)	21 (BC)	0.16	> 0.50
2	Est-S	Bb	38 (B) 32 (b)	7 (Bb)	0.52	> 0.25
3	Tf	AC	79 (A) 76 (C)	10 (AC)	0.06	> 0.75
4	Tf	AD	56 (A) 72 (D)	5 (AD)	2.00	< 0.10
5	Tf	AB	69 (A) 65 (B)	9 (AB)	0.12	> 0.50

Crossing-over represents a factor limiting the homozygosity of the gynogenetic offspring in the carp. This limitation may be quite essential with respect to a number of genes (Tables 49 and 50). Nevertheless, meiotic gynogenesis in fishes results in a faster increase in the inbreeding coefficient as compared to any other breeding system, including self-fertilization (Nace et al. 1970). Even in the first gynogenetic generation the inbreeding coefficient in the carp is equal to about 60% (Cherfas, 1977).

The General Properties of the Gynogenetic Offspring. In all the species of fishes studied the survival of gynogenetic offspring during the first year of life decreases greatly, particularly at the larval stages. This has been established in experiments with the sturgeon, the trout, the flatfish, the loach, and the peled whitefish (Romashov et al. 1963; Romashov and Belyaeva 1965 b; Purdom 1969; Tzoy 1972). High mortality has frequently been observed during hatching and transition to active feeding. The survival rate of gynogenetic carp during the first two weeks of postembryonic development is about 50% (Golovinskaya et al. 1963; Cherfas 1975). Organism with slight external deviations from the norm generally die during this period. Similar impairments are also characteristic of the gynogenetic offspring in the flatfish and the trout.

The survival rate of carp at later stages of development varies greatly; during the first year of life it varies from 5% to 66%, and during the wintering of fingerlings from 0% to 87%. Broods obtained from different females differ greatly in this respect. In older age groups the survival rate improves (80%–90%) (Cherfas 1975). When the gynogenetic and control offspring of the carp were grown together, the survival rate of the gynogenetic offspring (expressed as a percentage of the control fish survival) during the first two months of life was 9%, 13%, 21% and 95% in four different experiments (Nagy et al. 1978). There is only limited information on the survival of the gynogenetic grass carp: 24 diploids out of 34 were alive one week after hatching and six were alive after three years (Stanley 1976 b). The survival of gynogenetic fingerlings of the peled whitefish was about 90% (Tzoy 1972).

The mean weight of the gynogenetic fingerlings of the common carp determined in the autumn also varied greatly. The weight gain during the summer season in two-year-old, three-year-old and four-year-old fishes was quite satisfactory and attained one kilogram (Cherfas 1975). Nor has any growth depression been observed in the gynogenetic grass carp (Stanley and Sneed 1974). Unfortunately, however, no systematic studies regarding the effect of gynogenesis on the growth of other fishes have been conducted at present.

The sexual composition of gynogenetic offspring is determined by the chromosomal constitution of the females. In the case of female homogamety (XX), the offspring should contain only females, in the case of female heterogamety (WZ) one can expect the appearance of ZZ males; the WW females are apparently non-viable. Normal females (WZ) can appear solely as a result of a crossing-over between the sex chromosomes. In the common and grass carp no males have been observed in the progeny (Golovinskaya et al. 1974 a; Stanley 1976 b; Gomelsky et al. 1979). In the flatfish *Pleuronectes platessa* both males and females have been found; it has been suggested that the female sex is heterogametic in this species (Purdom 1972), but the data of Purdom require confirmation.

In older gynogenetic common carp great familial and individual differences in the state of the gonads have been observed (Golovinskaya et al. 1974a; Gomelsky et al. 1979). As many as half of the fishes may have severe defects of the ovaries, most frequently a great diminution thereof. A large number of females has traits of intersexuality. Apparently, the genome of gynogenetic carp carries recessive genes negatively affecting fertility after the transition to a homozygous state, just as in the case of inbreeding.

It should be pointed out that females with normal fertility are also found among gynogenetic common and grass carp; and the second (in the common carp the third) generation of induced gynogenesis has been successfully obtained from them (Cherfas 1975; Stanley 1976b; Nagy et al. 1978; Cherfas and Ilyasova 1980a). In the common carp the spontaneous yield of gynogenetic diploids in the second and third generation of gynogenesis was on average more than one order of magnitude higher than in the gynogenetic offspring of ordinary females: in two experiments it was equal to 2.0% and 3.5% of the number of embryos that had started to develop (Cherfas and Ilyasova 1979). The increased yield of diploids is the overall result of the higher frequency of diploidization of the female chromosomal complement and of the increased survival of the offspring at the early stages. Both these factors are apparently the consequence of the selection processes occurring in the primary gynogenetic offspring. The intensive selection in the presence of high homozygosity in the series of gynogenetic generations may result in a considerable lessening of the inbreeding depression.

The Practical Application of Gynogenesis

Gynogenesis can be widely used in genetic and breeding work with fishes (Golovinskaya 1968; Stanley and Sneed 1974; Cherfas 1978). Being a highy sophisticated technique used in genetic analysis, gynogenesis provides an opportunity to detect crossing-over during the first meiotic division if this crossing-over involves sister chromatids. Gene mapping with respect to the centromere does, therefore, become possible. In the first approximation such work has been done for seven loci of the carp (Table 51). The distances between the loci and the centromere varied greatly. It is noteworthy that of the two non-allelic genes determining the type of scaling of the common carp (genes S and N), one gene (S) is located at a minimal distance from the centromere while the other (N) is located at maximal distance. Gynogenesis is also helpful in solving such important problems of genetic theory as determination of the share of paratypic variation of traits, precise estimation of the extent of inbred depression in fishes, detection and analysis of the recessive genes affecting the viability, determination of the chromosomal constitution of sex and others.

Gynogenesis can be extremely helpful in selective studies among fishes. It can be used in constructing inbred strains for subsequent commercial hybridization. The rapid increase in homozygosity during gynogenesis makes it practical to obtain highly inbred strains as early as after two to three generations of gynogenesis instead of eight to ten generations of inbreeding. The success of this work depends on the fertility of the gynogenetic females. Two types of crosses are feasible for industrial hybridization: (1) crosses between gynogenetic females and unrelated outbred

Table 51. Mapping of carp genes with respect to the centromere at two levels of coincidence (K)[a]

K	Distance from the centromere (in % of the crossing-over) for individual loci[b]						
	S	Tf	Est-S	Est-F	D	L	N
0	2 (2–3)	3 (0.5–5)	5 (1–8)	14 (9–19)	35 (32–38)	37 (33–41)	49 (48–49)
0.2	2 (2–3)	3 (0.5–5)	5 (1–8)	15 (9–21)	43 (38–49)	47 (39–56)	–

[a] Coincidence is an index showing the extent of suppression of the crossing-over (interference) in the regions adjacent to that where crossing-over had taken place. The coefficient of coincidence equal to zero corresponds to the complete suppression of crossing-over in the studied region

[b] In parenthesis confidence limits at $p = 0.05$

males (a kind of top-cross) and (2) crosses between two inbred parents. The second variant will become possible if androgenesis, production of the fish offspring with exclusively the paternal heredity, becomes a reality. There is one communication describing spontaneous diploid androgenesis in the grass carp (Stanley 1976a), but techniques for the artificial construction of androgenetic fishes still need to be developed.

Highly inbred males can also be obtained by transforming females from gynogenetic broods into males using hormones, which lead to a functional sex trans-determination, and also by the repeated backcrossings of males with gynogenetic females (Nagy and Csányi 1978; see Chap. 8).

Gynogenesis can be employed to control fish reproduction in natural habitats by introducing unisexual female offspring (Stanley and Sneed 1974). The prospects for using gynogenesis for the production of purely female offspring appear to be more limited; the main obstacle in this case is the decreased fertility of gynogenetic females. Finally, gynogenesis can be used in attempts to obtain the reproduction of distant hybrids with male sterility.

Chapter 8

Problems and Methods of Fish Selection

The Purposes of Selection

In improving economically important fish qualities by selection, use is made of their variation in many morphological, physiological and biochemical features. A marked proportion of this variation is hereditary as we could see, and this circumstance guarantees the effectiveness of the selection work. The level of genetic variation in fish populations is very high, as can be deduced from studies on protein polymorphism.

Genetic manipulations and selection work are equally necessary in the domestication and creation of new fish breeds for ponds, as well as in the reproduction and commercial rearing of lake, river, anadromous and marine fishes. The protection of fish resources and the improvement of the quality of economically important wild fish species which are not cultivated by man also means that the data offered by modern genetics are required. The role of genetics and selection is also very important in work with aquarium fishes.

The Purposes of Selection Fully Domesticated Fish Species Grown in Ponds and Rearing Reservoirs. The main purpose in this case is that of increasing the productivity of the existing and newly developed breeds. Such an increase may be achieved primarily by raising the growth rate and the survival of the fish under consideration (Schäperclaus 1961; Kirpichnikov 1966 a, 1971 b and others).

The growth rate is a function of the quantity of the food consumed and of the degree of its assimilation. Correspondingly, we can distinguish two main trends in selection: selection for more complete consumption of the living organisms or artificial food mixtures used as a food (an increase in feeding activity) and selection for better food assimilation (a decrease in the food coefficient) (Kirpichnikov 1966 a; Steffens 1975; Merla 1979). In both cases, the success of genetic selection will markedly depend on the resistance of the fishes to various unfavourable factors of the environment or to disease.

A highly important purpose of selection of pond fishes and especially in rearing cages is their improved resistance to various environmental factors, in particular to high or low temperatures, a decreased oxygen content in the water, low pH, the presence in the water of industrial or agricultural wastes, the accumulation of metabolic end products etc. (Kirpichnikov 1971 b; Swarts et al. 1978). The selection for resistance is of particular significance in connection with the cultivation of fishes in rearing enclosures or in various artificial basins, aquariums and the cooling reservoirs of electric power plants (Steffens 1975; Zonova and Ponomarenko 1978).

An important place in selection programmes should be given to selection for improved resistance to various invasive and infectious diseases, particularly to widespread global viral or bacterial diseases, as well as to the local endemic ones that are difficult to treat or to prevent by ordinary means (Wolf 1954; Snieszko 1957; Schäperclaus 1961; Ehlinger 1964, 1977; Kirpichnikov 1966a, b, 1971b; Kirpichnikov et al. 1972a, 1976, 1979; Gjedrem and Aulstad 1974).

Improvement of the characteristics related to fish reproduction is extremely important for the further development of fish breeding in ponds and rearing enclosures. The aim of selection in this case may vary greatly, depending on the fish species and the conditions of cultivation. For example, delayed sexual maturation is advantageous for the common carp (*Cyprinus carpio*), the whitefish peled (*Coregonus peled*), and tilapias (*Tilapia* spp.) (Kirpichnikov 1966a; Chan May Tchien 1971; Andriyasheva et al. 1978).

In the selective breeding of rainbow trout *Salmo gairdneri* it is frequently advantageous to shift the maturation season to a more convenient time of the year or even to construct strains with different maturation seasons (Schäperclaus 1961; Steffens 1974a, b). The shift of the maturation time accompanied by a more effective response to pituitary hormone injections is essential for the selection of the grass carp (*Ctenopharyngodon idella*) and of the silver carp and the bighead (*Hypophthalmichtys molitrix, Aristichthys nobilis*) (Konradt 1973). The selective breeding of certain fish species might require artificial selection for better fertility and the higher variability of the embryos (Slutzky 1978; Mantelman 1980). One of the most important task in this field of fish selection involves the improvement of the nutrient properties of products derived from fishes, for example, an increase in the proportion of edible parts, decreased fat content, less bone share etc.; this problem is, however, extremely difficult to solve (Sengbusch and Meske 1967; Moav et al. 1971, 1979; Steffens 1975; Bakos 1979).

In countries where there is a developed system of amateur fishing in ponds, another important aim of selective breeding includes the ease of angling, using a fishing rod, while the fish retain their ability to put up resistance (Beukema 1969; Saunders 1977). For species that rapidly achieve maturity, the tilapia, for example, the primary task is the choice of parental combinations yielding unisexual offspring or the offspring among which there is the complete sterility of at least one of the sexes (Hickling 1968; Chaudhuri 1971; Chevassus 1979; Moav 1979). For the trout belonging to the genus *Salvelinus* (particularly for sea trout and brook trout hybrids or splake), released into deep ponds, selection should involve an increase in the ability of fishes to retain gases in their air bladder (Tait 1970). Other, even more specialized purposes can be envisaged in the selective breeding of pond fishes.

The achievement of each of the above listed goals requires vast, well-planned and frequently very long selective work. It is particularly difficult to obtain changes in the traits associated with reproduction. The heritability of such traits is usually low (see Chap. 4) since they are to a large extent determined by well-balanced stable polymorphic genetic systems established by natural selection which is operating permanently. No less complex is the selection for resistance to disease, for this requires the presence of specialized, isolated zones and the search for markers of resistance (Hutt 1970, 1974). The problems inherent in such selection are primarily associated with the nature of the relationships between the host (fish) and the parasite (pathogen). Usually, the pathogen undergoes reproduction much faster and more effectively than the host. In the course of its reproduction it could easily undergo the genetic changes resulting, for example, from natural selection; this again makes it dangerous for the host that has already been selected for resistance.

The transition to industrial methods of breeding in rearing cages, reservoirs etc. requires rapid adaptation of the corresponding fish strains to a new environment, new types of food, and new reproduction techniques. This must be reflected in all the selection programmes. Selection should begin simultaneously with the domestication of a new species of freshwater fishes; the delay with selective breeding may result in the decreased diversity of the genetic structure of the cultivated species or variety and may even lead to rapid degeneration.

The Purposes of Selection of the Lake, River, Anadromous and Marine Fishes Cultivated by Man. The goals of selection are different in the work with non-domesticated species, particularly with such valuable ones as the sturgeon, salmon and whitefish. The main goal in this case is to conserve the complex natural population structure of each species (Altukhov 1973; and others). This is particularly important for migrating salmonids possessing a strong homing instinct, that is returning after the feeding stage in the sea to just that river or lake and to the same spawning site from which they originated. Usually, in such cases one has to deal with many local reproductively isolated populations possessing genetic and ecological differences. In the sockeye salmon *Oncorhynchus nerka,* for example, the number of such populations is more than two thousand (Allendorf and Utter 1979). In order to avoid the decrease in intraspecific heterogeneity, one has to take measures to reproduce all the local and seasonal races comprising a given species or a big subspecies during the breeding work; the same is true of planning the catch. This requirement is equally important for the reproduction of many non-migrating freshwater and marine fish species.

It is no less important to provide for the general maintenance of the high heterogeneity of each cultivated population, particularly in work with such fertile species as the sturgeon, the whitefish, the common carp, the bream etc.

The following aims of improving anadromous fishes by selection can now be mentioned:

a) Selection for the faster growth of the fry during the freshwater period of life, that is, at the hatcheries and fish farms and the expansion of the "smoltification" process, i.e., the preparation of the fry to migrate to the sea (Gjedrem 1975, 1976; Refstie et al. 1977a; Gjedrem and Skjervold 1978; Saunders and Bailey 1978; Allendorf and Utter 1979).
b) Raising the fertility of the spawners (Saunders and Bailey 1978).
c) Selection for the increase in general viability and the resistance of the fry to disease during the freshwater phase of life (Gjedrem 1975; Saunders and Bailey 1978).
d) Accelerated growth and improved viability during the marine phase of life, that is, the increase of return coefficient (Donaldson and Menasveta 1961; Donaldson 1969; Ryman 1970; Bardach 1972; Gjedrem and Skjervold 1978).
e) Shorter staying period in the sea (Donaldson 1969; Saunders and Bailey 1978).

In his presentation of the programme of selection and genetic work with the Atlantic salmon adopted by the North-American Salmon Research Center, N.A.S.R.C., Saunders has emphasized that the programme includes the improvement of all the important economic and biological indices characteristic of a given population of the salmon. One of the Center's important tasks is the construction of strains for those rivers and lakes where the salmon population have been destroyed as a consequence of excessive fishing, poisoning of the water or destruction of the spawning grounds.

The industrial rearing of wild-living fish species such as the salmon, the eel and many others in basins, rivers or marine rearing cages has been developing rapidly over the last few years. "Sea farms", that is the restricted areas of sea used for fish rearing, have been set up in a number of countries; in a number of cases floating rearing cages have been used for the cultivation of marine and anadromous fishes. Intensive feeding and even the use of fertilizers is employed in the framework of such programmes. It will not be surprising if the combined area of such sea farms soon reaches millions of hectares. Such programmes for large-scale fish breeding require the appropriate adaptation of fishes which can be achieved by genetic selection. In the case of the salmon, this includes a certain retardation of sexual maturation, the acceleration of growth and an increase in fish survival in marine rearing cages (Saunders 1978 b; Møller et al. 1979; Shevtzova and Chuksin 1979). In more general form, such selection programms would include the adaptation of fishes to high densities under the conditions of limited mobility, the greater effectiveness of the use of artificial food, resistance to disease, etc.

Elements of Genetic Selection in Commercial Fishing. Fishing in general and intensified fishing in particular almost inevitably leads to the degeneration of wild-living fish species; this includes a decrease in the fish size, the more rapid maturation and the deterioration of fishes regarded as a food product. The most important factor underlying this phenomenon is the catching of the best and largest fishes from the population. The principle of the so-called "fishing measure", that is the smallest size of fishes of which fishing is permitted, only contributes to this process (Riggs and Sneed 1959; Donaldson and Menasveta 1961; Nikolsky 1966; Gwanaba 1973; Kirpichnikov 1973 c; Moav et al. 1978).

When fish reproduction cannot be controlled, genetic selection should be aimed at preventing degeneration of commercially important fishes and at maintaining or restoring the resources. Strict fishing rules and the quotas of the annual permitted catch compulsory for all countries should be worked out on the basis of genetic data.

The Selection of Decorative Fish Species. The main task in selecting decorative aquarium species and forms living in ponds includes the development of new varieties and strains with attractive pigmentation and body shape. There can be no doubt that it is beneficial to broaden the range of cultivated varieties. The construction of genetically marked strains of aquarium fishes suited to the investigation of certain theoretical problems of modern genetics and selection such as the problem of hereditary cancer, the regulatory mechanisms of gene action in development, genetic polymorphism, the heritability of selected traits and selection effectiveness, the inbred depression and heterosis, and many others is of considerable scientific interest.

The genetic selection of fishes therefore involves the solution of very important problems; they can be solved, but their solution is not always easy.

Methods of Selection: Mass Selection

The terms "mass selection" or "individual selection" in English literature are used to describe artificial selection and use in the subsequent reproduction of individuals having the best phenotypes (from the breeder's point of view). The features used in such selection can be quite different and generally depend on specific purposes. The traits used in selecting fishes may include increased weight or body size, good exterior indices, the necessary pigmentation, the desired pattern of scaling, the absence of any defects, resistance to unfavourable environmental effects and to disease, and certain improved physiological or biochemical characteristics easily measurable in living fishes. Individual selection for the rate of sexual maturation and for the rate of smoltification is possible (particularly for migrating salmon). Selection may include individual interior features such as the number of intermuscular bones, the size of the air bladder and others. The indices used on scoring several characters such as the rate of growth, body shape, fat content and others can be employed by analogy with the selection of certain domestic animals (Bakoš et al. 1978).

During mass selection the genotypes of the selected or of the discarded individual remains unknown. Mass selection only applies to the phenotype and is therefore always associated with greater risk of error.

The effectiveness of mass selection can be determined, using the formula (Falconer 1960):

$$R = i \cdot \sigma \cdot h^2 = Sh^2, \quad (1)$$

where R is the change in the selected trait per generation (response of selection), i = selection intensity, σ = standard deviation of the selected character, h^2 = heritability of the trait and S = the selectional differential that is the difference between the mean value of a trait in selected individuals and the value for the whole selection group prior to selection ($\bar{x}_S - \bar{x}$).

Selection intensity is equal to the selection differential expressed in terms of standard deviations:

$$i = S/\sigma. \quad (2)$$

If, for example, the weight of all the carp caught in a given pond is on average equal to 480 g with the standard deviation equal to 60 g, and the mean weight of the fishes selected for subsequent reproduction is equal to 630 g, the selection differential is equal to (630–480) = 150 g. In this case, i = 150/60 = 2.5. Heritability of weights (h^2) in fishes is not high (Chap. 4). Assuming that it is equal to 0,1, according to formula (1) we obtain:

$$R = 2.5 \cdot 60 \cdot 0.1 = 15.0 \text{ g}.$$

The calculation of selection response per year is important for fish breeders working with slowly maturing fishes. Formula (1) then acquires the following appearance:

$$R = Sh^2/I = i\,\sigma\,h^2/I, \qquad (3)$$

where I is the interfal between generations in years. For the common carp in European fish ponds I equals 4, and in our case, we obtain:

$$R = 15/4 = 3.75\ \mathrm{g}.$$

Calculations of this kind enable one to predict the results of selection work that may be expected after many years of selection. In our example the increase of mean weight of the carp equal to 10% or 48 g due to the selection of the positive genetic variants can be obtained after about 13 years of selective breeding. Any prediction of this kind cannot be very precise because all the indices in the formula (1) vary from one generation to another. Heritability usually undergoes particularly prominent changes, frequently it decreases as selection progresses, it depends frequently it decreases as selection processes, it depends on the extent of inbreeding, on the size of the group, the mode of reproduction, and other factors.

For fish breeders dealing with very fertile fishes the selection intensity can be conveniently expressed by the proportion of the number of individuals retained for subsequent reproduction relative to their initial number, this ratio being termed the severity of selection or v (Kirpichnikov 1966 a, 1971 b):

$$v = (n \cdot 100)/N, \qquad (4)$$

where N and n are the number of fishes prior to and after selection.

The high fertility of many fish species allows us to obtain high indices for the intensity and severity of selection; i can theoretically be as high as 4, while v may decrease to 0.1 and even to 0.01 (selection in the proportion 1:1000 and 1:10,000). The severity and intensity of selection are functionally interrelated (Fig. 70). At low values of v (0.01 or less) any further decrease in these characteristics hardly has any effect on the value of i and does not compensate for the expenditures necessary to grow a large number of fishes prior to the time of selection.

Comparison of the selection intensity used in fish, poultry and cattle breeding appears to indicate that fish breeding is in the most advantageous position (Fig. 70).

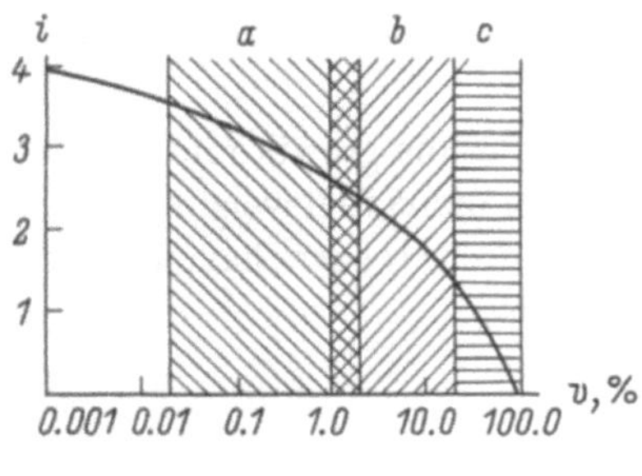

Fig. 70. The relationship between the intensity (*i*) and severity (*v*) of the selection. *a* zone characteristic of the selection of commercially important edible fishes; *b* the same for the selection of birds; *c* the same for the selection of cattle; the severity of selection is presented in the logarithmic scale

This advantage, however, should be used with caution; an excessive increase in S and i as a consequence of very strict selection may lead to harmful consequences. Extreme individuals strongly deviating in some characteristic from the mean value are frequently the carriers of the so-called "correlated response" (Falconer 1960). Frequently such changes affect viability adversely. It has been established, for example, that common carp with a high spine have defects in the structure of the vertebral column (Hofmann 1927; Moav and Wohlfarth 1967). Carp with many vertebrae show a decreased resistance to oxygen deficiency (Tzoy 1971 a). The percentage of carp with defects in gonad development is higher among the most rapidly growing individuals (Kirpichnikov 1961). The content of haemoglobin in the blood appears to be reduced in champions (Popov 1978). Increased resistance to furunculosis in salmon was found to be associated with susceptibility to gill disease (Ehlinger 1977). One could list many examples of correlated response of selection, that is, of changes in the traits which were not directly selected for. We shall only point out the drastic decrease in the survival of rainbow trout larvae after many years of selection for growth rate conducted in the USA by the Donaldson group (Herschberger et al. 1976).

The heritability of weight in fish plus-variants, just as in other animals, appears to be considerably lower than the heritability in the minus-variants (Moav and Wohlfarth 1967, 1968, 1973b; Wohlfarth and Moav 1971; Moav 1979). A special study involving selection in two directions has demonstrated that the effectiveness of selection is asymmetric: it is rather high when conducted for low weight and close to zero in cases of positive selection (Moav and Wohlfarth 1976). Selection asymmetry may be due to different factors; it has been established for the common carp, for example, that the weight variation is generally non-additive, high weight is characteristic of individuals with high heterozygosity, making mass selection ineffective. In the groups with increased heterogeneity constructed by hybridization, however, selection appears to be effective in the plus-direction as well (Kirpichnikov 1972a).

In the common carp and probably in many other fish species cultivated in ponds at high densities the largest individuals in the shoal, champions or "shoot carps", occupy the leading place due to their initial advantages in size and consequently in competition (see Chap. 4). Differences between these fishes and other individuals within the same population are, to a large extent, non-hereditary, although such individuals differ from others in some of their genetic characteristics as well (Wohlfarth 1977). The presence of shoot carp in the shoal is also a factor somewhat hampering the advance of selection in the plus-direction.

Mass selection in fish breeding should therefore be conducted at moderate intensity and severity. The intensity should not be greater than 1.5, to 2.0, in extreme cases 2.5 is the limit; optimal coefficients of selection severity are from 5% to 10% down to 1% in a few extreme cases.

A better and more precise estimate of the advantage in the growth rate can be obtained not from the weight per se, but from the ratio of individual weight gain to the mean gain of all individuals grown together. Polish fish breeders have termed this value the growth coefficient (Stegman 1965, 1967, 1969):

$$W = (V_2 - V_1)/(\bar{x}_2 - \bar{x}_1), \quad (5)$$

where V_2 and V_1 are the final and initial weight of an individual fish and $\bar{x}_2$ and $\bar{x}_1$ are the final and initial mean weights of all the individuals. Stegman has recommended that individuals should be selected with growth coefficients exceeding 1 for 2–3 years at a stretch and that they should be used for subsequent reproduction. A disadvantage of selection for better growth coefficients is the need for the individual labelling of all cultivated fishes. This limits the size of the group used for selection to several hundred individuals and therefore markedly diminishes selection opportunities. Similar limitations are inevitable in the use of other common indices of the individual growth rate, for example, those of the growth constant and the coefficient of the specific growth rate as defined by Schmalhausen (1935).

One can only partially agree with the proposals recommending that individuals deviating from the mean weight or some other mean characteristic by more than 2–3 σ should be retained for further reproduction (Wløodek 1968). We have already pointed out that excessively intense selection at $i = 2$ or more is associated with the risk of the appearance of harmful correlated responses.

Now me shall consider the means by which the response of selection can be increased. It follows from formula (3) that one can achieve this in one of three ways: (1) by increasing the selection differential, (2) by increasing heritability and (3) by decreasing the interval between generations.

The Increase in S. Higher S values will be obtained where there is greater selection severity, when v is smaller. Excessively strict selection will, however, be dangerous, and therefore the S value cannot be increased indefinitely. Since $S = i \cdot \sigma$, one can attempt to expand the variation of the selected trait, on condition that h^2 will not be decreased. It follows that either the only genetic (additive) component of the variation can be increased (that is heritability can be increased) or the genetic and paratypic variation can be expanded proportionally together.

The Increase in h^2. Heritability in the narrow sense of the word is determined, as we know, by the fraction of additive genetic variation in the total variation of the trait (σ_A^2/σ_{Ph}^2). Increased heritability can therefore be achieved by raising the relative proportion of additive genetic variance.

Generally, such an increase may be achieved by outbreeding. Inbreeding (breeding of close relatives) rapidly results in a decrease in genetic variance. The rate of homozygosity establishment has been determined for practically all modes of reproduction and found to be proportional to the degree of relatedness of the crossed individuals (Falconer 1960; Li 1976). When 50–100 parents are used for reproduction, the effects of inbreeding can be minimized, but, nevertheless, are not completely eliminated. Crosses are needed if the heritability of a selected trait is to be increased. A well-planned system of crosses allows genetic variation to be maintained at a sufficiently high level. An additional source of genetic variation can be found in the artificial mutagenesis induced by radiation or chemicals.

The decrease in the paratypic variance σ_E^2, the component of environment-related variation, plays an important part in the relative increase in additive genetic variation. Such a decrease will be effective in cases when environmental variance is not correlated with the genetic one and the decrease in the environmental component is not accompanied by any marked fall in the genetic one.

In fish selection paratypic variance can be cut down by taking a number of measures (Kirpichnikov 1966c, 1969b), the most important of which are listed below:

a) the establishment of identical conditions for all parents, particularly immediately prior to reproduction;
b) the simultaneous performance of all the crosses designed to yield the material for selection;
c) maximal standardization of environmental conditions throughout the period of fish cultivation beginning from the incubation of the eggs and up to the time of artificial selection;
d) the placing of fishes into ponds, basins or rearing cages at a moderate density only slightly exceeding the common standards in order to avoid excessive competition;
e) measures to prevent the mixing of individuals cultivated in different basins;
f) the cultivated fish should be released into the waterbody simultaneously; when this is not possible, the period during which they are released should be as short as possible;
g) selection is to be conducted at the age maximally corresponding to the marketing age.

The last condition should be discussed in somewhat greater detail. The interaction between the genotype and age in fish is not very prominent in contrast to certain important domestic animals used in farming and to cultured plants, but nevertheless such an interaction does exist. In the common carp, for example, up to 20% of the individuals having the highest weight during the first year of life show no advantage during the second year (Kirpichnikov 1966c). The early phases of life are greatly influenced by the maternal effect, that is, the impact made by the conditions under which the female was living, upon the quality of the gametes and upon the growth and viability characteristics of the fry after fertilization. The paratypic or environmental component of the total variance of weight is particularly high at the beginning, but decreases thereafter and the heritability of weight and size increases by a factor of 2–3. Such an increase has been established for the common carp (Kirpichnikov 1971b), the rainbow trout and the Atlantic salmon (Gall 1974; Refstie and Steine 1978), as well as for the tilapias (Chan May Tchien 1971). Genotypic differences are more easily detectable at later stages than during the early phases of life. At still later stages fish growth is, however, greatly affected by the differences in the rate of maturation. As a result, when selection is aimed at weight, it is optimal to work at the "middle age"; the selection conducted at the larval stages or with fingerlings or at the stage of sexually mature individuals is less effective. Recommendations advocating intense selection at the larval stage (Zonova 1978) or selection of equal severity at all stages (Brouzhinskas 1979) are, in my opinion, erroneous.

Acceleration of Generation Changes. A twofold decrease in the interval between generations implies a twofold increase in the response of selection; therefore, the shortening of the generation time appears to be an important goal in developing selection programmes. The faster maturation of fishes can generally be achieved (excluding fish living in the tropics) by keeping them in water of higher temperature for extended periods of time. The heating of water in winter in regions with a moderate or cold climate is particularly expedient. Complexes similar to the phytotrones used by botanists, that is basins and aquariums with regulated temperature, are to be planned for the large selection centres. The cultivation of fish at higher temperatures may be potentially dangerous if selection is aimed at increasing resistance to low temperatures. In such cases the fish can be transferred to heated water reservoirs immediately after the completion of artificial selection.

Another method of cutting down the generation time tested by us with the carp (Kirpichnikov and Shart 1976) is acceptable for southern regions. This method makes use of growing the fish until it has reached the marketing age and size at a drastically reduced population density. In this way, the time needed to attain maturity can be decreased or shortened by a year or two. At sufficiently high water temperatures the period up to sexual maturity is known to be associated with the fish size.

Until recently individual (mass) selection was the main technique used in fish breeding. Beginning from about the fifteenth century selection of the common carp in Europe was achieved predominantly by individual selection. Success, however, has not been very pronounced because selection techniques were primitive and many requirements of the selection work were violated. Another important factor was the high homozygosity of the common carp population on the generally small fish farms in Germany, Austria, Poland, and other European countries. The homozygosity was a direct consequence of the low numbers of spawners and of the inbreeding inevitable under such conditions.

Selection mainly included two characteristics: the growth rate and the "high spine" trait. It led to undesirable correlated changes including the deterioration of certain physiological characteristics and decreased viability (Steffens, 1964).

Selection of the common carp in China was even more primitive, the modern Chinese Big-Belly carp and other local strains retained many of the peculiar characteristics of their ancestor, the wild Far Eastern carp *Cyprinus carpio haematopterus* (Hulata et al. 1974; Moav et al. 1975 b; Wohlfarth et al. 1975 a and others).

Mass selection will continue to play a role in the future, particularly with regard to traits with a relatively high heritability. But even when selection is aimed at traits having heritability coefficients around 0.1–0.3 (involving weight and body size) mass selection can be very useful when combined with other forms of selection.

Methods of Selection: The Selection for Relatives

In contrast to mass selection, selection for relatives does, to a large extent, involve selection for genotypes; the positive characteristics of individuals chosen for subsequent reproduction are determined from an analysis of their close relatives. Two forms of such selection are used in fish breeding.

Family Selection. Several families or offspring from different pairs or small parental groups are cultivated under maximally standardized conditions. After the quality of these families in estimated the best ones are taken for further reproduction. A family is evaluated on the basis of the mean values calculated for it.

The "nests", including one female and two males, are frequently used instead of parental pairs in the common carp breeding in accordance with common techniques; less frequently the families are the progeny of several individuals (four or more).

Family selection, when correctly conducted, can be highly effective. The equation for the response of selection calculated per year is the following:

$$R_f = i_f \, \sigma_f \, h_f^2 / I \,. \qquad (6)$$

All the indices in the right part of the equation refer to the arithmetic means and not to the individual values of a trait. Selection intensity (i_f) in this case will be less than for mass selection since generally only a limited number of families can be cultivated simultaneously. The standard deviation (σ_f) will also be decreased since the variation of the means is always less than the variation of the individual values. In contrast, the heritability of the means (h_f^2) is increased. When all families are kept under standardized conditions the value of h_f^2 approaches 1; it cannot, however, be equal to 1 because of the presence of variance of common environment. This variance is the consequence of inevitable differences in the growth conditions of spawners in the stage of their maturation, and as a result, in the peculiarities of the gametes produced by them.

If the evaluation of a family in terms of the selected trait requires the dissection of the fish or damage to it, 20–30 or more individuals are examined and, if evaluation yields acceptable results, their sibs from the same family are used for reproduction. This type of family selection is called sib-selection.

The cultivation of individuals from different families can be carried out either separately when each family is released into a separate water basin or together, if the fish is labelled. Both these techniques have their merits and disadvantages (Kirpichnikov 1966 b). When the families are kept separately, the experiment needs to be repeated three or four times; therefore 30 to 40 identical isolated ponds, rearing enclosures or basins are required in evaluating ten families. Frequently this presents a serious obstacle to the fish breeder. When families are kept together in a single pool, it is not easy to label a large number of individuals belonging to different families. Considerable problems arise in this case, because of the dependence of the rate of fish growth on their initial weight. This dependence has been studied in some detail with the common carp (Moav and Wohlfarth 1968, 1974; Wohlfarth and Moav 1971, 1972). Experiments conducted in fattening ponds with carp weighing from 500 to 700 g by the end of cultivation have indicated that weight differences at the moment of release of the fishes into the pond equal 1 g (in young fishes weighing from 30 to 50 g) increased to 3–4 g by the time they are caught. This increase is a consequence of competition for food: larger individuals possess advantages in getting the necessary food in a closed waterbody. When the data obtained in experiments involving the joint cultivation of fishes from two or more families are subjected to mathematical processing, the mean weight gain of fishes from each family should be recalculated according to the equation:

$$y' = y - Kd, \qquad (7)$$

where y is the observed mean gain of all fish from a given family; y′ is the corrected mean gain; K is the linear regression coefficient of the gain normalized for the initial weight (correction coefficient) and d is the difference between the initial weight of fish from a given family and the mean initial weight of fish in all families. When two groups are compared with each other the difference in the gain between them is recalculated according to the following formula:

$$D' = D - Kd', \quad (8)$$

where D and D′ are the observed and calculated difference in gain values and d′ is the difference between the two groups of fishes in terms of the initial weight.

The differences in the initial weight at the beginning can be avoided by using the multiple-nursing method (Moav and Wohlfarth 1968, 1973 b; Moav et al. 1971). This method makes used of growing the compared groups or families at different population densities prior to the main experiment. Cultivation of a group with a lesser weight at lower population density makes it possible to equalize the mean weight (Fig. 71) since the rate of fish growth in closed water basins is inversely proportional to the fish population density in these basins.

If such equalization is impossible, one has to determine the magnitude of the correction coefficient K in the preliminary experiments; the value of this coefficient may undergo profound changes, depending on the severity of competition. Knowledge of the K value makes it easy to correct the relative weight gain of fish from different families.

Comparison of fish growth in separate and communal ponds has indicated that the genetic component of growth is manifested similarly in both cases, the correlation between fish growth in these two variants of rearing being greater than 0.9

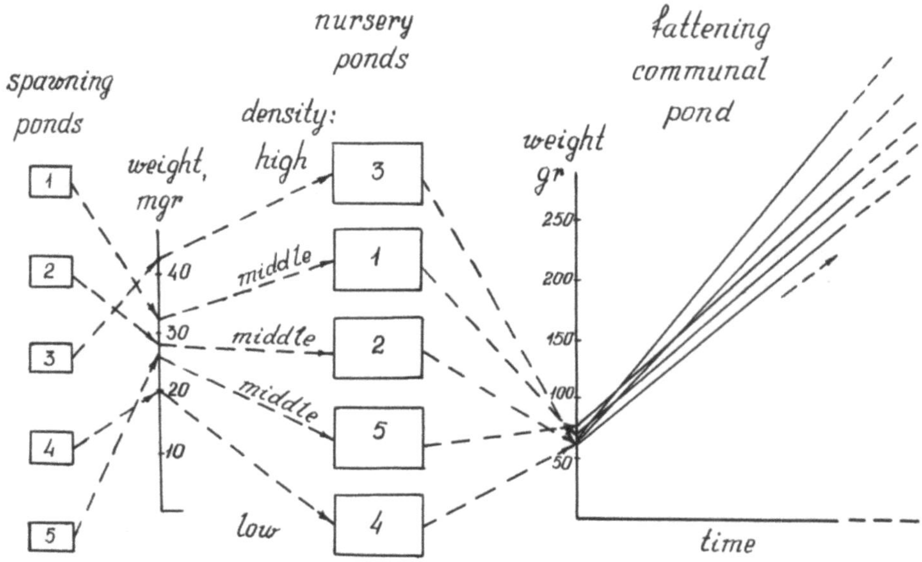

Fig. 71. The application of "multiple nursing method" for equalization of initially differing fish weight (Wohlfarth and Moav 1971, with alterations)

(Wohlfarth and Moav 1968; Moav and Wohlfarth 1973b). The value of fishes in communal pond cultivation can be expressed by the following equation (Moav and Wohlfarth 1974):

$$g_2 = (1 + \alpha) \cdot g_1 + A, \tag{9}$$

where g_2 and g_1 are the relative values of fishes in communal and separate ponds, α is a factor increasing the genetic fraction of the value due to the competition, A is the increase in the value due to fish agressiveness. Direct observations have shown that the value of A has little effect on the final value for a given group of fishes in communal ponds. The important conclusion can then be drawn that the genetic differences between families are maintained or are even more pronounced when these families are cultivated together in communal ponds. If certain requirements are fulfilled (see below), the families can be undoubtedly evaluated in communal ponds as well.

The final choice of the cultivation method requires an analysis of the interactions between the genotype and the environment, as has been conducted by Israeli authors (Moav 1979). These interactions are summarized in the following table:

Type of interaction	Magnitude
Genotype – pond	Small
Genotype – fish age	Medium
Genotype – density at release (degree of competition)	Medium
Genotype – rearing season	Marked
Genotype – rearing system	Marked

We can see that the position occupied by a given family on the quality scale hardly changes upon the transition from one pond to another and from one fish age to another; it greatly depends, however, on the season, and growing conditions (properties of the fertilizer, the feeding regimen, the presence or absence of water flow etc.). This relationship should be taken into account in developing selection programmes and in practical work on family selection.

We shall now list the more important requirements that have to be taken into account when family selection is to be practised in fish breeding (Kirpichnikov 1966b, c, 1968, 1971b; Moav et al. 1971):

a) Rearing and keeping of parents (particularly of females) prior to crosses under standardized conditions favourable for their maturation. This allows the variance of the common environment to be cut down.

b) Simultaneous breeding of crosses aimed at obtaining families for subsequent evaluation.

c) Use (in most cases) of artificial insemination of eggs in crosses.

d) Incubation of eggs in all crosses in identical apparatus. It is particularly important to standardize the temperature, oxygen content, light intensity and the rate of water current.

e) Growing of larvae, fry and older fish in waterbodies where there is a plentiful food supply in order to decrease competition for food. It should be taken into account that the conditions of cultivation should not, however, differ too much from the conditions on commercial fish farms, since the interaction between the genotype and the cultivation system in the fish is considerable.

f) The thorough standardization of a fish population density for all families at the stages of separate growth; marked differences in the population density may seriously interfere with the subsequent comparison of families.

g) No less than three replicate experiments are required for the evaluation of the fish belonging to different families, when they are grown separately; the variance of the ponds is usually much greater than the variance of families (Wohlfarth and Moav 1968) and therefore experiments conducted in the ponds require a particularly large amount of replicate determinations. The use of regions of one and the same pond isolated by nets or the use of rearing cages with similar characteristics yields lesser data scattering.

h) When fish belonging to different families are grown together, it is necessary to equalize (using the multiple-nursing method) the initial mean weight. If this equalization is impossible, it is necessary to determine the correction coefficient K and to make alterations in the observed gain values. It is very important in this case to reduce as much as possible the competition for food and guarantee the simultaneous release of fishes from different families (with serial labelling). For precise evaluation of families the use of two or three replicate ponds is generally sufficient. Combined rearing of more than 10–12 families is very difficult to accomplish. It is usually sufficient to study 30–50 specimens in each family selected at random for comparative evaluation of different families.

i) The presence in fishes of a marked "paternal" and especially of "maternal" effects, i.e., the influence of spawners' rearing conditions and their age on the quality of offspring call for carrying out the evaluation of families after complete disappearance of these influences. Maternal paratypic variance in the fish fry may be very significant at the beginning, particularly in egg size and embryo viability (Kirpichnikov 1959, 1961, 1966 b). By the end of the first year of life (in moderate climate) the maternal effect in the common carp disappears. A strong maternal effect has been discovered also in the rainbow trout (Gall 1974). Paternal effect, i.e., the influence of size and age of male spawners on the quality of offspring is much less pronounced, it can be traced in the common carp only during the first two months of its life.

j) The final evaluation and selection of the best families are to be conducted as in the case of mass selection mainly at the age when the fish attains commercial value.

Evaluation of Parents by Progeny Testing. Progeny testing can be conducted using different techniques. The simplest method involves comparison of the offspring obtained from different pairs or nests of parents (Fig. 72 a, b); in this case, the evaluation refers not to individual parents but to their combinations; selection for a common combining ability is therefore performed.

Frequently fish breeders use simplified diallele crosses (Fig. 72 c). Males or females are separately crossed with one or more individuals of the opposite sex

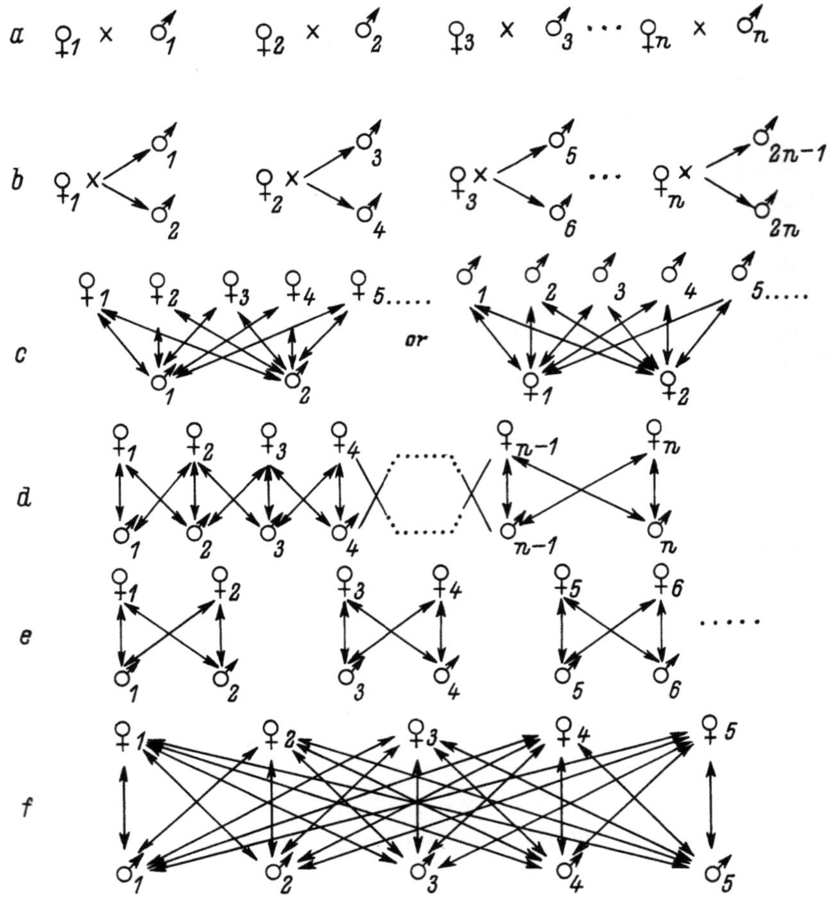

Fig. 72 a–f. Different schemes of crossings in progeny testing. **a** testing of pairs; **b** testing of three spawners ("clusters"); **c** testing of parents of one sex; **d** incomplete diallele crossing; **e, f** complete diallele crossings of 2 × 2 and 5 × 5 types

Table 52. Evaluation of male spawners of common carp in progeny testing experiments with simplified diallele crossing of 2 ♀♀×5 ♂♂ type (Kirpichnikov 1961). The place of males for growth rate of offspring (means for the data of 18 tests of combined and separate rearing)[a]

♂, number	Cross with ♀ 1		Cross with ♀ 2	
	Place	Index of evaluation[b]	Place	Index of evaluation[b]
69	1	1.71	1	1.86
74	2	2.13	3–4	3.21
65	3	3.13	2	2.71
10	4	3.38	3–4	3.25
3[c]	5	4.25	5	3.64

[a] The mechanical and isotope labelling in the tests of combined rearing was used
[b] It is the mean value from data for all the 18 tests
[c] Male N3, as well as the females 1 and 2, are heterozygous for the scale cover gene S

(Kuzema 1961, 1962; Kirpichnikov 1959, 1966b, 1971b; Polyksenov 1962). The mating of each of the males tested with the same two females provides a sufficiently reliable evaluation of the breeding value of these males (Table 52).

A so-called incomplete diallele cross has also been recommended (Fig. 72d) (Moav and Wohlfarth 1960), but has not been realized in practice.

A complete diallel cross (2 × 2, 3 × 3, 5 × 5, 10 × 10 etc.; Fig. 72e, f) also enables one to select the best individuals belonging to either sex. Since the number of offspring in this case increases in proportion to the squared number of parents of one sex subjected to examination, there are problems related to the growing of the offspring when this number is high. Breeders working with the common carp prefer crosses permitting either the males or females to be evaluated separately. In the breeding of the trout and of the Atlantic salmon multiple crosses of 2 × 2, 3 × 3 and 4 × 4 type have been conducted (Gjedrem and Skjervold 1978; Saunders 1978a, b; Møller et al. 1979). The 5 × 5 scheme has been employed in breeding the grass carp *Ctenopharyngodon idella* (Slutzky 1971a). The availability of standard incubators and rearing vessels for the incubation of eggs and larvae at fish hatcheries provides a basis for the wide use of diallel crossings in selection work with salmon. Progeny evaluation is usually limited to the embryonic and larval stages, although it should be pointed out that of late Norwegian and Canadian scientists have been practising the cultivation of a large number of offspring for one, two or even three years, including the cultivation of fish in seawater.

The main problem in evaluating the parents by progeny testing in fish breeding is not that of diallel crosses but the need for the simultaneous cultivation of many offspring under identical conditions. In the first experiments conducted with the common carp (Kuzema 1961, 1962) fishes were grown separately in different ponds, no replicates were made, and the variance between ponds was admittedly greater than the genetic variance. The technique involving the release into all ponds of easily distinguishable or labelled fishes of one and the same control group was used by Polyksenov during selection of the Byelorussian carp (1962). Unfortunately, in this case the differences in the weight at release were great and appeared to mask any hereditary differences associated with the rate of growth and survival. The absence of replicates, as well as the absence of corrections for different initial weights at release, led to meaningless results. Various other drawbacks can be mentioned in this series of experiments dealing with progeny testing.

The communal rearing of the families compared, replicate determinations and corrections for differences in the initial weight allow an objective evaluation of parents to be obtained and the optimal pairs or groups for the reproduction of a given variety to be selected. Such conclusions can be drawn in particular from the extensive experiments conducted in Israel. Eleven different offspring were evaluated and five replicate determinations made in ponds where the food supply differed drastically (Fig. 73, Table 53). The best families retained their leading position with regard to weight gain in most ponds, particularly in those where conditions were most favourable. Under more severe conditions the mutal position of families expressed in terms of the mean gain undergoes some changes, however, even in these cases the difference between plus- and minus-variants still remains. The variance analysis of numerical data has demonstrated that the variance between ponds exceeds the variance between offspring. The interaction of these two factors is not very

them with an opportunity to recover their numbers rapidly after a period of depression (Schultz 1971; Bulger and Schultz 1979). The clonal structure of the population contributes to the better and quicker occupation of the different ecological niches. All this, apparently, compensates for the decreased fertility of the unisexual forms.

Gynogenesis in fishes is of outstanding interest in connection with the problem of polyploidy in bisexual fish species. It has been demonstrated in the silkworm *Bombix mori* that triploid parthenogenetic forms can represent an intermediate stage in the process of polyploidization of bisexual species (Astaurov 1969, 1971). In crosses of triploid gynogenetic females with diploid males (♀3n×♂2n), the fusion of a triploid female gamete with the haploid sperm will result in the appearance of tetraploid gynogenetic individuals. The presence of four copies of the genome opens up the way to the restoration of meiosis and to the transition to conventional sexual reproduction. In this connection, it would be particularly interesting to continue the study of Japanese populations of *Carassius auratus gibelio,* since they were found to contain gynogenetic tetraploids. It cannot be excluded that these populations also contain bisexual tetraploid forms.

The transition from triploid gynogenesis to the tetraploid level and to bisexual reproduction is also possible in the family Poeciliidae (Moore et al. 1970; Schultz 1977; Thibault 1978).

Induced Gynogenesis

Biologists have long found attractive the possibility of induced or artificial diploid gynogenesis in fishes. The obtaining of a viable gynogenetic offspring makes it possible to solve a number of important theoretical questions of biology. Gynogenesis as a method of rapidly increasing the genetic homogeneity of a strain is also no less important in selective breeding work.

Experimental diploid gynogenesis can be achieved by solving two problems (1) the genetic inactivation of the sperm cell and (2) omission of the reduction of the female chromosomal complex. The appropriate agents have to be selected in solving these tasks.

The first evidence that gynogenesis may be induced in the trout (*Salmo trutta fario*) was presented in the paper of Opperman (1913). In 1956, positive results were reported in experiments with the loach *Misgurnus fossilis* (Neyfakh 1956). The beginning of systematic studies in this field was laid down by the work with the loach and the carp (Romashov et al. 1960; Golovinskaya et al. 1963). Gynogenesis has been observed in several species of the sturgeon (Acipenseridae), in the rainbow trout and two other species of the trout (Salmonidae), in several species of flatfishes (Pleuronectidae), in the grass carp (Cyprinidae), and in the common catfish (Siluridae) (Romashov et al. 1963; Tzoy 1972; Purdom and Lincoln 1974; Stanley and Sneed 1974; Purdom 1976; Nagy et al. 1978).

Methods of Producing Diploid Gynogenetic Offspring. Genetic inactivation of male chromosomes is observed after treating the sperm with chemical mutagens (such as dimethyl sulfate); more frequently, however, high doses of ionizing radiation e.g.,

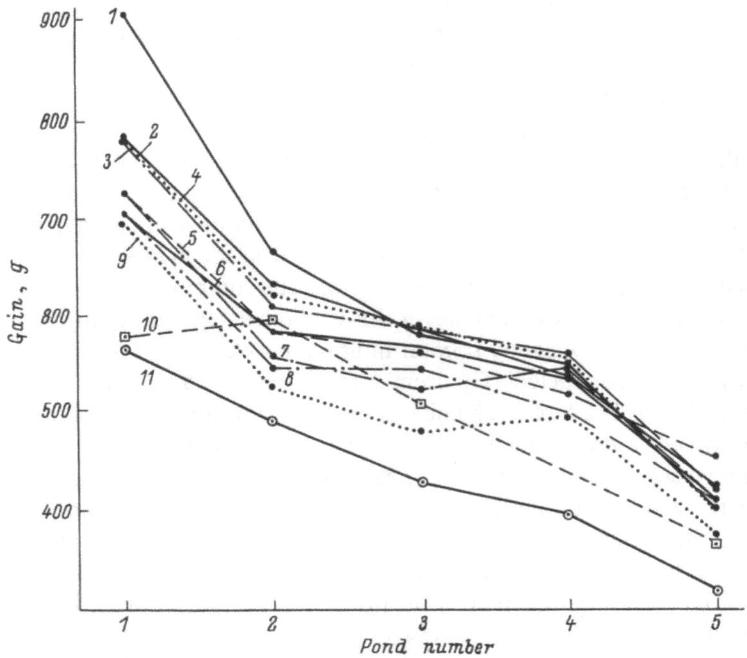

Fig. 73. Weight gain of fishes belonging to 11 families of the common carp in five ponds with different feeding status (Wohlfarth and Moav 1971)

Table 53. Results of spawners rating by progeny testing in carp fish farms at five replicates of the experiment. The normalized weight gain in grams is shown (Wohlfarth and Moav 1971)

Family No.	Replicates of the experiment					Mean values
	1st	2nd	3rd	4th	5th	
1	910	666	583	553	403	623
2	783	633	586	541	424	593
3	780	613	576	561	422	590
4	780	623	589	556	401	590
5	728	586	563	521	455	571
6	706	585	568	536	412	561
7	729	560	525	547	410	554
8	706	548	546	502	408	542
9	695	527	483	497	375	515
10	582	598	511	461	367	504
11	566	493	429	395	318	440
	Mean for five ponds					
All families	724	585	542	515	400	553

[a] Only the families cultivated in all ponds are given in the Table 53. The numbering of the families was changed. The families No. 5, 8 and 11 are full sibs; each of the families 1, 6, 9 and 10 originate from one female and several males. Other families are the product of mass crosses. Inbreeding was suspected in the family No. 11

great but it is, nevertheless, quite significant, particularly when extreme pond variants are compared.

The conditions and requirements to be observed when planning the progeny testing are similar to those used in family selection and sib selection. Both separate and communal ponds can be used. Progeny testing can be conducted separately for the most important productively significant traits or using the system of indices (Bakoš et al. 1978); this system makes use of the evaluation of several traits by a score system. One of the main drawbacks involved in the examination of parents by progeny testing results from the great dependence between the growth rate and the density of the fish populations: particular attention should be given to the equalization of the number of fish in different versions of the experiment. In our experiments conducted with the common carp, at the early stages of development when the fish were kept in rearing cages, such equalization of population density was carried out every third day.

In conclusion, we shall compare the effectiveness of family selection and progeny testing. In the case of family selection, the first from the most productive and resistant families are retained for subsequent reproduction; the approach based on progeny testing yields parents providing the best offspring. The testing of parents requires one or even two years, correspondingly the interval between generations undergoes some increase. A slowing down of the selection rate is highly undesirable and therefore family selection is to be preferred in fish breeding. This mode of selection is very easy due to the high fertility of fish.

Combined Selection

Comparison of the equations $R = Sh^2$ and $R_f = S_f \cdot h_f^2$ allows one to determine which of the selection methods is better for fish breeding. If

$$Sh^2 > S_f h_f^2 \qquad (10)$$

then mass selection is to be preferred as compared with selection for relatives and vice versa (Kirpichnikov 1968). Selection for relatives will be more effective for traits with very low heritability since, in this case, only the decrease in the selection differential by a factor of 3–4 will be compensated by a heritability increase of the familial mean values. For example, when the selection of the fry weight is performed at $h^2 = 0.2$, the transition to family selection or to the progeny testing can be advantageous if the heritability of the familial mean values will be only slightly less than 1. The variation of weight in most fish species is generally non-additive. This provides the main argument in favour of the transition to family selection, if the acceleration of the fish growth rate represents an important goal in breeding work. In the selection of the Atlantic salmon it has been recommended that mass selection should be limited solely to those traits associated with growth; the use of family selection was considered to be advantageous for such traits as the resistance of the fish to environmental factors and diseaes, and the time needed to mature and the exterior features (Gjedrem 1975). It is our opinion that mass selection is most justified for exterior and other morphological traits, the heritability of which exceeds 0.3–0.4

(see Chap. 4); the improvement of the growth indices requires the use of family selection, although it should be admitted that mass selection can be useful as well.

We have already pointed out the technical difficulties involved in the cultivation and objective comparison of a large number of families in the work with such big fish as the common carp or the salmon. The solution of the problem may be found in combined selection (Kirpichnikov 1967 a, 1968); such an approach includes consecutive application of family selection, mass selection and progeny testing in the space of one generation.

The first step in combined selection consists of crosses between heterogeneous unrelated parents. Such crosses are aimed at obtaining a small number of progeny up to 10 in carp breeding or several dozen in salmon breeding. During the cultivation of these families their productive properties are evaluated. These properties include viability, growth rate, the quality of the flesh etc., so that the best families can eventually be selected. The second stage includes mass selection in several of the best families. If each of these families contains a thousand or more individuals, the intensity and severity of selection can be quite high. At the third stage parents are examined, using progeny testing. Parents of just one sex where the onset of maturity occurs earlier are tested (males in common carp breeding). This testing is to be completed by the time of onset of maturity of individuals of the other sex.

The response of combined selection is theoretically equal to the sum total of responses of each of the selection methods used:

$$R_S = R_f + R_m + R_{pr}, \tag{11}$$

where R_f, R_m and R_{pr} refer to effectiveness of family selection, mass selection and progeny testing.

Combined selection has been employed by us in work with the selective breeding of the Ropsha carp (Kirpochnikov et al. 1972 b) and widely used in selection work with the Atlantic salmon (Gjedrem and Skjervold 1978; Gjedrem 1979); in the latter case the mass selection is combined with the evaluation of families, but progeny testing is not employed.

Inbreeding, Crosses and the Breeding System

Inbreeding and outbreeding play an important part in selection work.

Inbreeding. The extent of inbreeding can be expressed, using the inbreeding coefficient F introduced by Wright. This coefficient corresponds to the probability of the increase of the homozygosity in the offspring during one generation.[10] The value of

10 I am referring to relative homozygosity, that is to the number of loci that became homozygous during one inbreeding generation related to the initial number of homozygous loci. The true absolute homozygosity is always higher, since a marked proportion of genes is homozygous in any population, even an outbred one. True homozygosity can be determined precisely, using biochemical genetic methods (see Chaps. 5 and 6)

F depends on the degree of relatedness of crossed individuals; in the case of plant self-fertilization the inbreeding coefficient attains the maximal value equal to 0.50; in crosses between brothers and sisters and of parents with children it is equal to 0.25, in other more distant crosses it decreases to 0.125 or even lesser values. The highest values of inbreeding in fish can be obtained by gynogenesis (see Chap. 7). Theoretically all the offspring of a gynogenetic female should be completely homozygous after meiotic gynogenesis (F = 1), but the inbreeding coefficient decreases somewhat due to the crossing-over between chromatids. The actual value in the gynogenetic common carp is equal to 0.60–0.70. In a panmictic population of a limited size or in the selected strain random crosses between related individuals are inevitable. The smaller the effective population size (N_e), the greater is the value of the inbreeding coefficient for this population:

$$F = 1/2\,N_e. \tag{12}$$

In the presence of a certain degree of inbreeding in a population or a shoal the genotype frequencies in each generation are shifted towards an increase in the number of homozygotes. These shifts described by the formula (Li 1976):

$$(p^2 + Fpq)_{AA} + (2\,pq - 2\,Fpq)_{AB} + (q^2 + Fpq)_{BB} = 1. \tag{13}$$

At low F values, for example, at F = 0.01 (N_e = 50), the increase in the number of homozygotes is insignificant, but it becomes much more noticeable when the size of the population or the shoal is low.

The number of parental pairs in fish farms is fairly often small because the fertility of the fish is high; therefore when the offspring from one or two of the best pairs is left for reproduction, that is when N_e equals 2–4, the inbreeding coefficient may attain 0.1–0.25; as a consequence, the accumulation of homozygous genes occurs at a rather high rate.

A common consequence of inbreeding in most cultured plants and domestic animals, including fishes, is decreased viability and the retardation of growth which is called inbreeding depression. This depression mainly stems from an increase in the number of homozygous genes, particularly when certain harmful recessive genes become homozygous. A marked role in inbreeding depression is played by a general decrease in the heterozygosity during inbreeding. Complex polygenic heterozygous systems involving many traits related to productivity, including the growth rate and viability, developed in the course of natural selection of fishes. These systems are destroyed by inbreeding.

In the common carp one inbreeding generation after the crossing of sibs retards the growth of fish by 10% to 20%, this is accompanied by decreased viability and a marked increase in the number of malformations (Moav and Wohlfarth 1968; Wohlfarth and Moav 1971). Even moderate inbreeding leads to the inbred depression in the common carp (Kirpichnikov 1960, 1966 b, 1969 b). Deviants from the normal pattern or phenodeviants frequently appear in inbred offspring, they differ by the reduced growth rate and lowered survival (Kirpichnikov 1961; Tomilenko and Shpak 1979). The harmful consequences of inbreeding during fish reproduction have been noted by many authors (Kuzema 1953; Shaskolsky 1954; Lieder 1956;

Schäperclaus 1961; von Limbach 1970; Nagel 1970; Chan May-Tchien 1971; Ihssen 1976; Kincaid 1976; Merla 1979; Mrakovčič and Haleg 1979). According to Kincaid the slowing down of the growth rate in the rainbow trout upon tight inbreeding is 5–10%; the retardation of the growth rate has been observed in the inbred American brook trout *Salvelinus fontinalis* (Cooper 1961). Viability of the guppy *Poecilia reticulata* is decreased as a result of inbreeding (Gibson 1954).

It should be pointed out that in many aquarium fishes inbreeding does not appear to be accompanied by any marked depression. One may assume that the severe and thorough selection of the best individuals conducted by aquarium breeders overrides the harm done by inbreeding. It has recently been established in experiments with the silkworm *Bombyx mori* that the inbred depression rapidly becomes less pronounced due to the selection of compensating genes (Strunnikov 1974). A similar process should be true of fish in the case of prolonged inbreeding. It is quite possible that the low extent of inbred depression in aquarium fish is due to their relatively low fertility and to the fact that their ancestors, which had lived in nature, existed in most cases as many small isolated populations, and selection operates permanently towards the neutralization of the inbred depression.

Although inbreeding in general represents a harmful phenomenon, it can also be extremely useful in fish selection. This usefulness stems primarily from the stabilization of selective traits due to increased homozygosity and the augmented expression of several of them. Inbreeding has perhaps another, even more essential function in fish breeding just as in the selection of many other animals and plants. I am referring to the construction of commercial hybrids possessing heterosis as a consequence of crosses of individuals from different inbred strains (commercial hybridization). This problem will be discussed below.

Crossing as a Method of Increasing the Heterogeneity of the Selection Material. The selection response does, to a large extent, depend on the level of heterogeneity of the selected groups. Crosses between unrelated individuals enrich the strain, increasing the genetic component of their variation and thereby facilitating selection. Another result is the disappearance of any harmful consequences of inbreeding.

Before starting off selection with any fish species, one has to choose the optimal breeding system contributing to the establishment of the optimal structure of the breed and to the maintenance of a sifficiently high heterogeneity. Only with this approach can one guarantee success in the further improvement of the productive traits of the breed stock. One of the simplest approaches involves the separation of the selected group into two, three or an even larger number of subgroups. Each of them is reproduced separately under the conditions of moderate inbreeding with artificial selection in each generation (Kirpichnikov 1960; Golovinskaya 1962). Individuals from different subgroups are periodically crossed with each other. Meanwhile, the crosses are also used in commercial fish production. This approach has been used in the development of the Ropsha common carp breed (Kirpichnikov 1972a) and it is at present being used in selection work with the Middle Russian carp, as well as in the construction of carp strains resistent to dropsy (Golovinskaya et al. 1975; Kirpichnikov et al. 1972a, 1979).

Another technique contributing to the maintenance of heterogeneity of the selection stock makes use of the establishment of the reserve genetic pool in the form

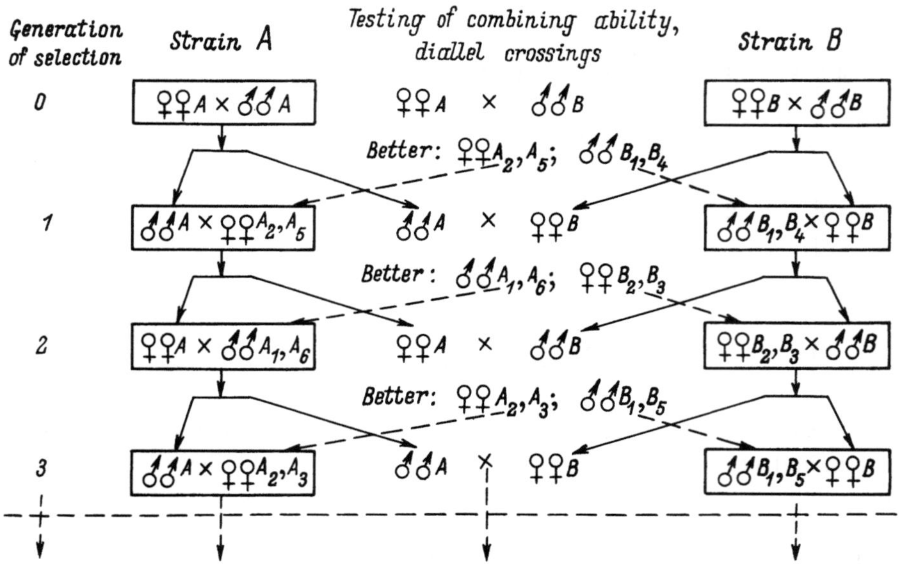

Fig. 74. The scheme of reciprocal periodic selection as applied to fish breeding. Breeding of two different strains of one variety (breed)

of a sufficiently numerous group of fishes reproduced in the absence of any inbreeding. If the genetic variation in one breed or another becomes narrower, fishes belonging to this stock are crossed with fishes from the reserve pool and the heterogeneity lost as a result of intensive selection is being restored (Moav and Wohlfarth 1967).

The "topcross" widely employed in animal breeding makes use of the establishment of several inbred and a large outbred group which also serves as a kind of reserve. The best individuals of one sex from the best inbred lines are crossed with the best individuals of the other sex from the outbred pool, and this is followed by close inbreeding in several different lines. The topcross has not been used in fish breeding so far, but can be recommended for fish that mature early such as *Tilapia* and several other species. The reciprocal periodic selection involving the determination of the combining ability of parent pairs belonging to two different strains or stocks also remains to be tested in fish selection (Fig. 74). This extremely advanced method of selection deserves testing in fish breeding.

Finally, the heterogeneity of the selection stock can be increased using a single-time or introductory cross with fish of a different breed.

However, no matter which breeding system is being employed during selection and after the establishment of the stock, it is impossible to maintain high heterogeneity of the variety for indefinite periods of time where there is intensive selection. Radical measures such as new crosses between the breeds, distant hybridization or artificial mutagenesis are necessary from time to time.

Synthetic Selection. Workers in the field of fish selective breeding frequently want to combine the useful properties present in the strains, varieties, species or even genera. Such a synthesis can be carried out in several different ways.

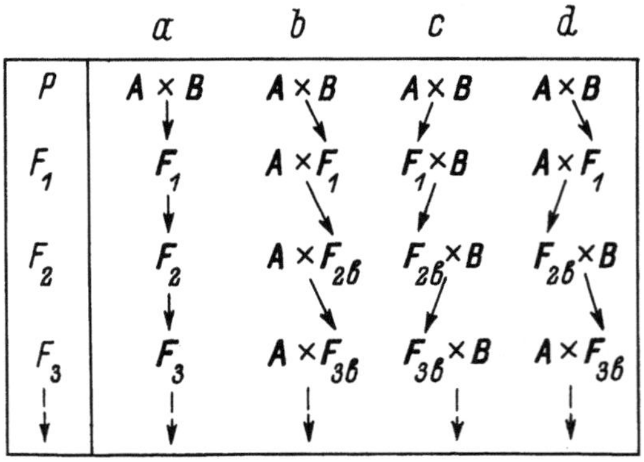

Fig. 75. Main types of crosses in the course of synthetic selection. Crosses: *a* reproducing; *b* introductory; *c* absorbing; *d* alternative

The reproducing cross (Fig. 75 a) is used when many useful traits are to be combined from both crossed breeds or species. This is easily accomplished when hybrids are completely fertile but requires very thorough selection in all hybrid generations. Reproducing crosses were employed in developing the Ukrainian and Ropsha common carp breeds (Kuzema 1953; Kirpichnikov and Golovinskaya 1966; Kirpichnikov 1972a); it has also been employes in developing the Central Russian common carp (Golovinskaya et al. 1975). It has been assumed that the Hungarian carp originated as a consequence of the combination of the traits present in two or three different breeds, mainly the local low productive carp, the German high-spine Aischgrund carp and possibly of the strain brought in from Japan (Bakoš 1974). The European rainbow trout resulted from reproducing crosses of two or three American trout (Schäperclaus, 1961). Attempts to combine the characteristic features of the two species of American trout *Salvelinus fontinalis* (brook trout) and *S. namaycush* (lake trout) appear to be very promising as well (Ihssen 1973). It would be advantageous to use the reproducing cross in many other cases of synthetic selection. By way of example, one can mention the intergeneric crosses of the beluga *Huso huso* and the sterled *Acipenser ruthenus* (Burtzev 1971; Burtzev and Serebryakova 1980) as well as of the silver carp *Hypophthalmichtys molitrix* and the bighead *Aristichthys nobilis* (Voropaev 1969, 1978; Grechkovskaja et al. 1979); interspecific crosses of the Siberian whitefishes *Coregonus nasus* and *C. peled* also seem to be quite promising (Korovina et al. 1972, 1973).

The introductory cross (Fig. 75 b) is used in those cases when the introduction of one or several valuable traits of another strain or species into the local highly productive breed is required. Hybrids obtained in the initial cross of the two forms are then back-crossed many times with the individuals of the local breed whose improvement is intended. In this process for subsequent reproduction one has to use back-cross hybrids possessing the desired traits of the donor strain used for the breed improvement. If these traits are determined by dominant, clearly manifested genes, the problem of conservation of the required characters can be solved relatively easily. When the genes involved are recessive or when inheritance is polygenic, the risk of loss of the characters whose maintenance is desired is very high.

The absorbing cross is similar to the introductory cross in its nature (Fig. 75 c). A series of back-crosses is accomplished after the initial cross of two strains but hybrids are repeatedly mated with individuals of the strain used for improvement

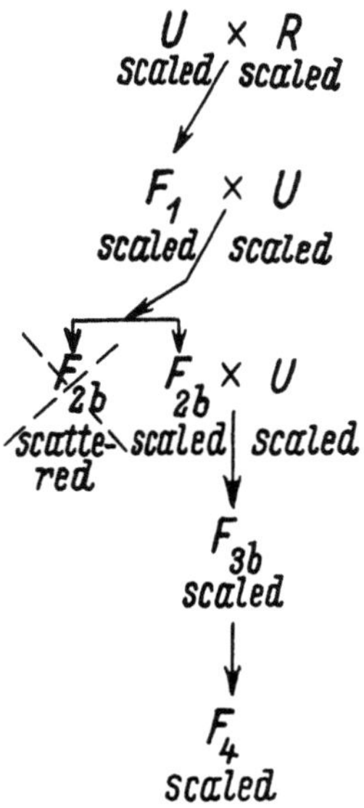

Fig. 76. The scheme of selective breeding for the creation of Nivchan common carp breed. *R* the Ropsha carp of the 3rd selected generation; *U* Ukrainian carp selected by Kuzema; F_1–F_4 1–4 generations of selection; *scaled, scattered* types of scale cover (Kuzema et al. 1968)

and not of the local strain. Here, too, one has to observe strict measures aimed at conserving the most useful traits of the absorbed strain, in this case the local one.

The alternative cross (Fig. 75 d) allows inbreeding to be avoided to a large extent when the combination of the characteristic features of two breeds by synthesis is desired. Intermittent crossing of hybrids with individuals belonging to the two initial breeds as followed by the selection of the necessary combinations of traits. After three or four generations the alternative cross is replaced by the reproducing one; otherwise it is difficult to stabilize the traits of the new hybrid breed.

Introductory or absorbing and reproducing crosses are frequently combined with each other. This was the case in particular in selection work on the Nivka common carp (Kuzema et al. 1970). Only two back-crosses were performed since there was a risk of losing the most important trait of the breed used for improvement (Ropsha carp), that is its high viability (Fig. 76).

One of the main obstacles in selection work with distant hybrids is the partial or complete infertility characteristic of many hybrid forms. It is not easy to restore the fertility of hybrids. The highly complex work on the development of fertile hybrids of the common and crucian carp (*Cyprinus carpio* and *Carassius carassius*) which was initiated in the Ukraine more than 40 years ago is of outstanding theoretical and practical significance in thes respect (Kuzema and Tomilenko 1965). A partial recovery of fertility could be achieved when the goldfish *Carassius auratus* and Ropsha carp were used as mediators. High variation with respect to fertility and viability in second-generation hybrids of the beluga and the sterlet makes one believe that selection aimed at complete recovery of fertility will be successful in this case as well (Burtzev and Serebryakova 1980).

Commercial Hybridization. The problem of the practical use of first-generation hybrids in fish breeding has been discussed in detail in the papers of Andriyasheva

(1966, 1971 a, b). It is advantageous to cultivate hybrids of the first generation commercially in the fish industry, if they possess heterosis, that is, if they are superior to their parents in a number of traits. In this book we shall not deal with the complex problem of heterosis. It has been discussed in several treatises (see Khadginov 1935; Crow 1952; Dobzhansky 1952; Hayes 1952; Haldane 1955; Mather 1955; Kirpichnikov 1967 c, 1974 a; Guzhov 1969; Strunnikov 1974).

It should merely be pointed out that heterosis appears to depend on two main complementary genetic mechanisms: the combination of useful dominant genes accumulated by both crossed forms in hybrids (the hypothesis of dominance) and the increase in hybrids of the total level of heterozygosity (the hypothesis of overdominance). An increase in biochemical versatility in hybrids occurs in both cases (Haldane 1955; Kirpichnikov 1974 a). Heterosis in natural populations is generally manifested as a rise in the fitness of hybrids, in the elevation of their adaptive value. Heterosis of this kind is typical of many intraspecific and certain interspecific fish crosses and corresponds to the concept of euheterosis advanced by Dobzhansky in 1952. It expressed predominantly in the advantage in the survival of heterozygotes. Frequently hybrids of breeds or strains created artificially by man are characteristically gigantic; this results from the acceleration of growth and the increased size of the whole organism or different parts of it (luxuriance according to Dobzhansky). Heterosis in such cases appears to be due mainly to the overall action of the dominant genes and the neutralization of the inbred depression due to homozygosity with respect to harmful recessive traits (Andriyasheva 1971 b). In these hybrids as well the viability may also be increased, particularly when completely or greatly inbred races are being crossed. In fish breeding we frequently encounter heterosis of either type.

An important prerequisite for the success of commercial fish hybridization is maximal care in carrying it out. First-generation hybrids must be completely removed from waterbodies; if they left for breeding intentionally or by chance, brood stocks of the initial forms become "contaminated", and this frequently leads to grave consequences. By way of example, we can cite the case of hybridization of the domestic common carp with its wild-living ancestor, the wild carp; such hybridization has been conducted in the USSR and in several countries of Eastern Europe. Frequently hybrids were kept for further reproduction and this resulted in contamination of the wild carp and domestic carp brood stocks and in the deterioration of their economically important traits. Another sad example is the hybridization of the beluga and the sterlet. First-generation hybrids, which were called besters, possess useful properties: from the beluga they inherit the ability to grow rapidly and from the sterlet the ability to live and to feed in freshwater basins (Burtzev 1971; Nikoljukin 1971). Unfortunately, the hybrids were released in large quantities into rivers, lakes, estuaries, and artificial reservoirs; this resulted in considerable contamination of the sturgeon stocks. The use of beluga eggs for crosses with sterlet is detrimental for the reproduction of the beluga, a valuable sturgeon species, which is now in danger of complete extinction.

Better control of commercial hybridization can be considerably facilitated by the use of genetic markers: both parents and hybrids may differ in the alleles of the genes responsible for the colour pattern, scale pattern and certain biochemical loci (Brody et al. 1976; Moav et al. 1976 a, b; Kirpichnikov and Katasonov 1978).

We shall now list the more promising hybrid combinations:

1. Interbreed hybrids of the carp *Cyprinus carpio.* Good results are obtained in crosses between the Ropsha and the Ukrainian carps (Kuzema and Tomilenko 1962; Kuzema et al. 1968; Kirpichnikov et al. 1971; Tomilenko et al. 1978; Alexeenko 1979), of the scaled and framed Ukrainian carps (Kuzema 1953), two stocks of the Para carp (Bobrova 1978), three stocks of the Krasnodar carp (Kirpichnikov et al. 1979). To increase the productivity of the Chinese carp it is recommended that European carp should be crossed with the Chinese Big-Belly carp (Moav et al. 1975 b; Wohlfarth et al. 1975 a; Moav 1979). Heterosis manifested in better survival and productivity has been noted after crosses of Hungarian and Polish carps (Rychlicki 1973; Wlødek and Matlak 1978). A similar phenomenon has been reported after crosses of the Japanese Yamato carp with the European mirror carp (Suzuki 1979).

2. Intraspecies hybrids of salmonids. Crosses between individuals of the Atlantic salmon from different populations are accompanied by heterosis with respect to the growth rate (Saunders 1978 b). An accelerated growth rate is also observed in hybrids between strains of the rainbow trout having an early and a late maturation season (Reichle 1974). Increased viability has been reported in hybrids of common rainbow trout and its albino variety (Subla Rao and Chandrasekaran 1978) as well as in hybrids between subspecies of *Oncorhynchus* – *O. masu masu* and *O. masu rhodurus* (Chevassus 1979).

3. Hybrids of the silver carp of Amur and Chinese origin (Poljarush 1979). Heterosis manifested in the survival and growth rate (around 11%) has been observed for the larvae of such hybrids.

4. Hybrids of the domestic carp and the Amur wild carp (*Cyprinus carpio haematopterus*) are now widely used in the Ukraine (Kirpichnikov 1959, 1962; Andriyasheva 1966; Karpenko 1966; Tomilenko et al. 1977).

5. Hybrids of the ludoga whitefish *Coregonus lavaretus* and the cisco *C. albula* (Nesterenko 1957; Lemanova 1960, 1965; Gorbunova 1962).

6. Splake – hybrids of the brook and lake trout *Salvelinus fontinalis* and *S. namaycush* (Ayles 1974; Ihssen 1976; Chevassus 1979).

7. Hybrids of the brook and lake trout and the Alpin char *S. alpinus* (Chevassus 1979).

8. Hybrids of the Atlantic salmon and the brown trout (*Salmo salar, S. trutta*) (Refstie and Gjedrem 1976; Chevassus 1979). Such hybrids are infertile and can therefore be used directly for release into natural waterbodies.

9. Hybrids of catostomid species belonging to the genus *Ictiobus* (Giudice 1966).

10. Hybrids of the American catfish belonging to the genus *Ictalurus* (Giudice 1964; Sneed 1971; Green et al. 1979).

11. Hybrids of *Tilapia* species (Hickling 1960, 1968; Pruginin et al. 1975; Lovshin and Da Silva 1976). The offspring of most crosses consist completely or predominantly of males, which grow more quickly than females. Crosses of *T. nilotica* and *T. aurea* have been widely used in Israel, the progeny from such crosses having advantageous combinations of some features of the parental forms (Rothbard and Pruginin 1975; Moav 1979).

12. Hybrids of the pike belonging to the genus *Esox* (Armbruster 1966).

13. Hybrids of the sunfishes belonging to the genus *Lepomis* (Childers 1971).

14. Intergeneric hybrids of the beluga *Huso huso* and the sterlet *Acipenser ruthenus* (Burtzev 1971; Nikoljukin 1971).

15. Intergeneric hybrids of the Alpine char with the Atlantic salmon and the brown trout (Refstie and Gjedrem 1976).

16. Intergeneric hybrids of the common and crucian carp (Nikoljukin 1952). These infertile hybrids fortuitiously combine several useful traits of the parents and can therefore be used for waterbodies little suited to the cultivation of the common carp.

17. Intergeneric hybrids of the silver carp *Hypophthalmichthys molitrix* and the bighead *Aristichthys nobilis* (Vinogradov and Erokhina 1964; Voropaev 1969; Grechkovskaja et al. 1979).

18. Intergeneric hybrids of the Indian carps *Catla catla* and *Labeo rohita* (Chaudhuri 1971).

Many other hybrid combinations are also possible and will probably be used in future for commercial hybridization. A large number of such combinations can be proposed among salmonids (Suzuki and Fukuda 1972; Chevassus 1979) and among cyprinids (Ryabov 1979).

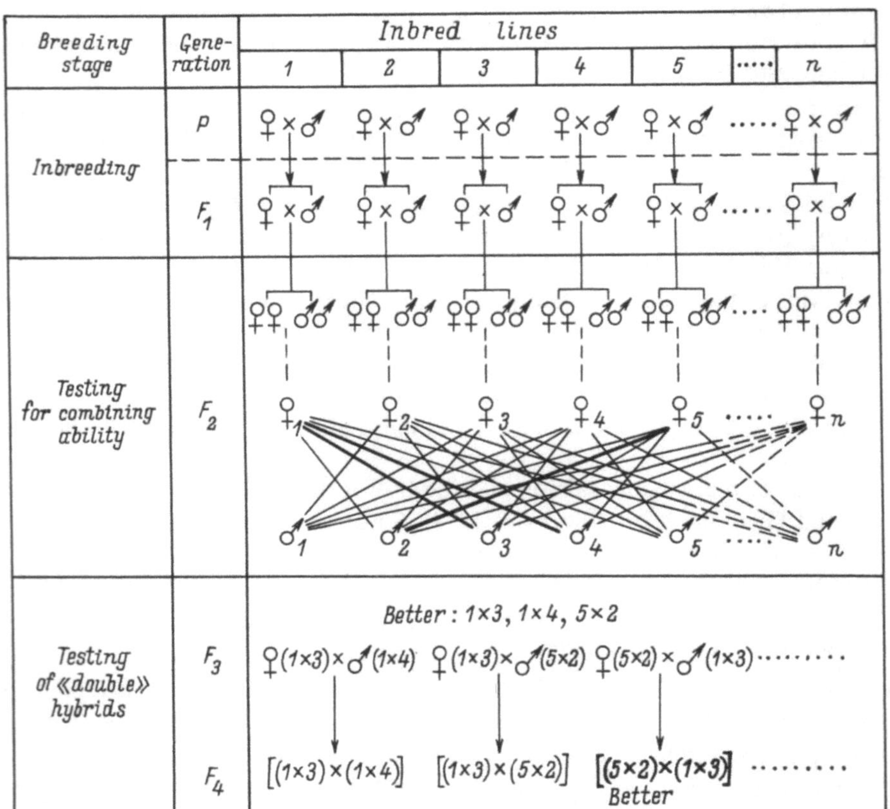

Fig. 77. The scheme of production of double commercial hybrids of the common carp (Bakoš 1974, with small simplifications)

The search for fish hybrids possessing heterosis is now being conducted in many countries; such projects appear to be most intense in the USSR, the USA, and India. Experimenters attach great importance to the development of inbred strains and of hybrids of them. As an example of an interesting and advanced attempt of this kind we can mention the work conducted in Hungary aimed at the rearing of commercial "double" (in fact quadruple) interstrain hybrids of the common carp (Bakoš 1974, 1979). In the way it is planned this work is similar to that on the production of double corn hybrids. This very difficult work has not yet been completed (Fig. 77), but several hybrid combinations have been tested and proved to be highly productive. In a moderate European climate projects of this kind need no less than 20–25 years; they can, however, be markedly accelerated if special heated rearing servoirs are built.

The development of inbred strains has been planned in selection work with the peled whitefish *Coregonus peled* (Andriyasheva 1978a), as well as in work with the Atlantic salmon and the rainbow trout (Gjedrem 1975; Gjedrem and Skjervold 1978), the Central Russian and Kazakhstan common carp (Tzoy 1978; Cherfas and Ilyasova 1980a). Gynogenesis has been used in order to accelerate the development of homozygous fish in the case of the peled and the carp.

The high fertility and external fertilization characteristic of many fish species facilitate the production of first-generation hybrids in quantities sufficient to satisfy the needs of large commercial farms. Many fish hybrids are fertile and this frequently creates a risk of contamination of brood stocks of the initial forms. Therefore a pressing problem in selection at the present time is that of developing methods to breed infertile or unisexual commercial hybrids.

New Trends in Fish Selection

Specialized genetic methods of selection have been acquiring ever-increasing importance in fish breeding over the last few years. This is due in part to the rapid development of molecular and biochemical genetics and to the accumulation of information on the special genetics of fishes. Several of these methods will be considered below.

Selection and the Special Genetics of Fishes. The information regarding the pattern of inheritance of morphological (both qualitative and quantitative), physiological and biochemical traits in fish is very helpful in organizing selection programmes. Certain genes can be used directly in selection projects. Selection can be planned in such a way as to act for or against these genes. A good example of such an application of information provided by special genetics can be seen in the purification of the brood stock of the Ropsha common carp from the recessive gene s (mirror or scattered scalling pattern); more than 400 parents were checked, using back-crosses, for homozygosity with respect to the gene S. As a result of this project, three stocks of the Ropsha carp homozygous with respect to gene S were bred, segregation for the scaling pattern was completely eliminated (Kirpichnikov 1972a).

The collection and accumulation of data on the inheritance of quantitative fish traits, particularly those related to yield capacity, is very important. This information provides a basis for planning selection; it helps in deciding whether to go in for either mass selection, or selection for relatives, and it is useful in developing a system of crosses.

Use of genes for marking the breeding stocks is another very promising trend (Brody et al. 1976, 1979; Moav et al. 1976a; Moav 1979). At present, genetic markers are successfully used for labelling different stocks of the common carp. The work on selection of the Central Russian carp includes the breeding of two stocks differing in the S (pattern of scaling) and the D (pigmentation of the spine and head) loci. The strain ssDD has a scattered pattern of scaling with a light band over the spine, the strain SSdd is scaled without any colour pattern. Commercial hybrids of these two strains having the genotype of SsDd can be easily distinguished from the carp of both parental forms by its phenotype; they have continuous scaling and a characteristic colour pattern (Katasonov 1974b; Kirpichnikov and Katasonov 1978).

Two strains of the common carp in Israel are also marked with colour recessive genes (Moav and Wohlfarth 1968). One of these stocks is homozygous in the gene bl (blue colour) and the gene gr (grey colour) while the other is homozygous in the gene g (golden colour). In hybrids all three genes are heterozygous (Bl/bl, Gr/gr, G/g) and the resulting phenotype is normal without any particular pigmentation.

In Czechoslovakia various strains of common carp are marked by alleles of the Tf locus (Valenta et al. 1976a, 1978). The gene Tf has also been used to mark carp stocks selected for resistance to dropsy (Shart and Iljasov 1979). Genetic marking has been successfully used in experiments on artificial gynogenesis with the common carp (Cherfas and Truveller 1978; Nagy et al. 1978).

Biochemical markers have been very helpful in work with other fish species as well. The selection of genotypes on the basis of the Tf alleles in the coho salmon has made it possible to establish the resistance of different genetic groups to highly infectious diseases (Suzumoto et al. 1977). Use of albumin alleles was helpful in confirming the hybrid origin of the gynogenetic fish *Poecilia formosa* (Abramoff et al. 1968). Triploids of *Poeciliopsis* belonging to the type P_{2m-1} and P_{m-21}, with similar external morphological traits could be easily distinguished from each other and from diploids by the allelic pattern of the Ldh-C locus (Vrijenhoek 1972). The fry of *Stizostedion* introduced into rivers and lakes was identified by the presence of some Mdh alleles (Schweigert et al. 1977).

The relative proportion of *Salmo salar* × *Salmo trutta* hybrids in natural populations was evaluated from the frequency of the occurrence of hybrid spectra of serum proteins (Payne et al. 1972).

Marker genes are helpful in establishing the origin of certain hybrid fish breeds. For example, they were useful in demonstrating the genetic component belonging to the Amur wild common carp in domestic carp varieties (Paaver 1980). Finally, it should be pointed out that if the correlation between marker genes and selective traits is extensive and significant, the selection involving such markers may be considerably accelerated and the selection work thereby facilitated.

Artificial Mutagenesis. The genetic variation of the fish can be markedly increased after the treatment of the gametes by nitrosoethyl- and nitrosomethyl urea (NEU and NMU), ethylenimine (EI) and by other chemical mutagens (Tzoy 1969a, b,

1978, 1980; Tzoy et al. 1973); X-ray irradiation is also highly effective in this respect (Schröder 1973). Due to high fish fertility artificial mutagenesis can be successfully used in selection work as a means of generally elevating the heterogeneity of a given variety. Attempts of this kind have been made using the common carp (Kirpichnikov et al. 1972a, b; Tzoy et al. 1974b). The work with the Kazakhstan carp has been particularly successful: the variation in productivity traits has been markedly increased by a number of mutagens predominantly by NEU and EI (Tzoy 1978).

The search for individual useful mutations in fish selective breeding is of low value. Their detection and use is impeded by slow sex maturation of fish species used for breeding.

Karyological Studies. The study of chromosomes can be useful in many cases, namely:

a) in studies of hybridization and gynogenesis,
b) in development of triploids and tetraploids for commercial breeding,
c) in investigations of sex determination mechanisms,
d) in experiments with hormonal sex-reversal, and in other studies.

Populations and stocks of the rainbow trout, Atlantic salmon, sockeye salmon, silver crucian carp and several other species differ from each other in the structure of the chromosomal complements. An analysis of karyotypes of local forms of these species can be instrumental in finding the origin of these forms and in choosing adequate methods of reproducing and selecting them.

Karyological studies are particularly important for the analysis of gonad maturation in distant fish hybrids, especially in the case of infertility, as well as in studies of the mutagenic effects of radiation in the regions where its level can be increased.

Gynogenesis in Selection. The significance of gynogenesis for raising selection response has already been considered (see Chap. 7). The main advantage of gynogenesis stems from the opportunity of constructing highly homozygous inbred lines which can then be used to develop heterotic hybrids. Hungarian authors (Nagy and Csányi 1978) recently proposed two interesting schemes for combining gynogenesis and the usual bisexual reproduction in selection work (Fig. 78a, b). In both cases, the males rapidly become highly inbred just like the females although the inbreeding coefficient will grow more slowly than under the conditions of sequential and continuous gynogenesis.

Gynogenesis has been succesfully used to accelerate homozygotization of common carp strains after mutagenic treatment (Tzoy 1980), to produce fertile hybrids of the crucian and common carp (Cherfas and Ilyasova 1980b), as well as to construct inbred lines in the selection of the peled (Mantelman 1978, 1980).

The Development of Sterile and Unisexual Commercial Hybrids. Commercial hybridization, which is becoming increasingly important in fish breeding, requires the maintenance of the purity of parental forms. If it is difficult to distinguish between hybrids and their parents, the risk of contamination of the brood stock by the hybrids may be very high. Particularly serious problems arise in crosses of *Tilapia*

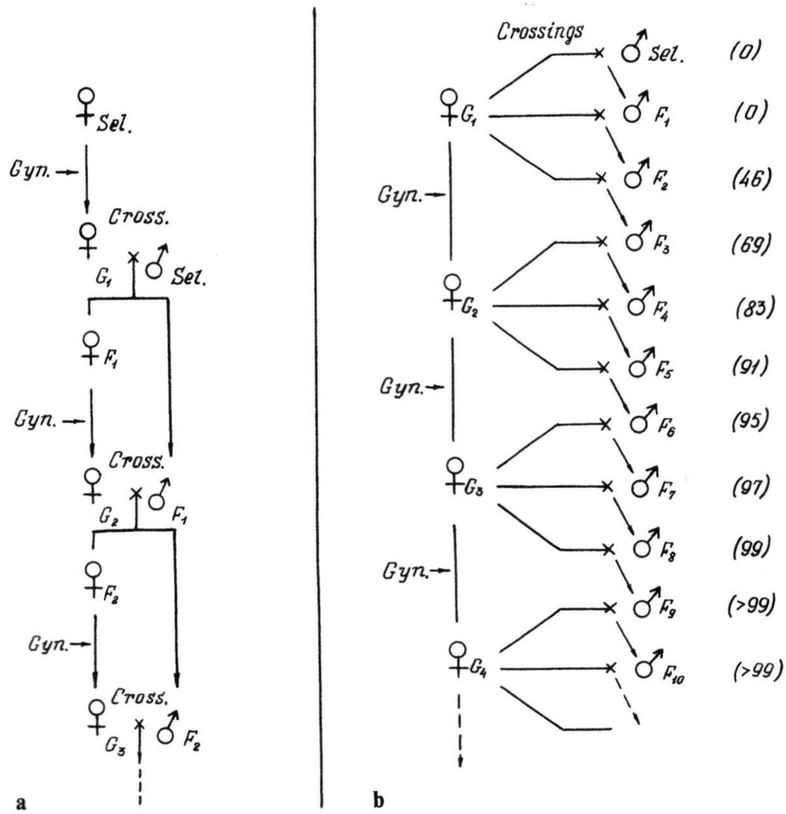

Fig. 78 a, b. Two schemes of using gynogenesis in common carp selection. **a** construction of a semi-gynogenetic strain; **b** a method for the rapid increase of male homogeneity. *Gyn.* gynogenesis; G_1, G_2, G_3, G_4 gynogenetic generations, $♀_{Sel.}$, $♂_{Sel.}$ selected spawners. *In brackets* values illustrating genotypic equivalence of the bisexual population to the gynogenetic stock. F_2, F_3, F_4, etc. one-year-old males obtained und used in successive crossings (Nagy and Csányi 1978)

species, which become mature at the age of three or four months; hybrids start multiplying in fish ponds, and this results in great overpopulation of such ponds and delayed fish growth. This can be avoided if the hybrids are infertile or if all of them are of one sex. Infertility is generally observed only in a few interspecific or more frequently intergeneric fish hybrids, for instance, in crosses of the common and the crucian carp or of the brown trout and the Atlantic salmon, and in several other cases.

The search for infertile hybrids with superior productive traits is very important for the further expansion of the work on commercial fish hybridization.

Unisexual offspring has been obtained for several interspecific fish crosses as well, for the hybridization of the *Tilapia* in particular. The following combinations of species result in exclusively male offspring (Pruginin et al. 1975; Moav 1979):

1. *T. mossambica* ♀ × *T. hornorum* ♂,
2. *T. nilotica* ♀ × *T. hornorum* ♂,
3. *T. nilotica* ♀ × *T. macrochir* ♂,
4. *T. nilotica* ♀ × *T. aurea* ♂ (not all crosses),
5. *T. nigra* ♀ × *T. hornorum* ♂.

Production and cultivation of unisexual hybrids is also advantageous because tilapia males grow much more rapidly than the females and achieve markedly larger size. Two hybrid combinations, *T. nilotica* × *T. aurea* and *T. mossambica* × *T. hornorum,* have acquired practical significance (Moav 1979). It is probable that in other fish taxons unisexual hybrid combinations promising for subsequent reproduction will be found as well.

Sex reversal resulting from the feeding of sex hormones to small fish has been successfully achieved with several species including the crucian carp (see Chap. 1). This opens up extensive prospects for growing hybrids belonging to the sex that is optimal from a commercial point of view. For example, rapidly growing males can be obtained in *Tilapia* (Jensen and Shelton 1979), or females in the common carp. Females that produce a large amount of eggs can be obtained in sturgeon hybrids. In the case of the development of inbred gynogenetic lines, the conversion of females into males after feeding the fry of fish with testosteron may solve the problem of crosses of individuals from two promising stocks, which normally, in the absence of any treatment, consist predominantly of females. Only females will again be obtained in the offspring (see Chap. 7).

The Selection of Fishes Living in Natural Waterbodies

The selection of wild fish species is possible if their reproduction is partially or completely controlled by man. It should be taken into account that a considerable part of the life of individuals of these species occurs under free living conditions where these individuals occupy certain ecological niches and a certain place in natural biocenoses. Careless or one-sided and too intensive selection may easily cause the complex relations between species in natural biocenosis to deteriorate and thus have a marked detrimental effect.

In work with anadromous fish species it is very important to select for the maximal return of the fishes obtained and reared in the hatcheries and fish farms to their native river or lake after the marine phase of life. The coefficient of commercial return should also be regarded as the main index during the selection of lake or river fish species. This coefficient appears to integrate all aspects of fish adaptation to the conditions of a natural waterbody. The selection for the increased return, that is for survival up to maturation of the gametes and up to the beginning of the spawning migration is, however, a difficult problem. To achieve this, family selection is necessary with simultaneous evaluations of dozens of families and with serial tagging of the fingerlings released by the hatchery.

Conscious and, more often than not, unconscious selection for the increased survival of eggs and fry is usually carried on at all fish culture enterprises. It should be pointed out at the same time that very strict natural selection operates in natural conditions at the early stages of fish development. This selection results in the elimination of most slow-growing deviating from the norm, deficient, unstable fishes. Under the conditions of artificial reproduction embryos and larvae are protected from enemies and unfavourable fluctuations of the environment. This leads to the greater survival of defective individuals, therefore the average viability of the fry released into natural waterbodies decreases and the genetic pool of the population appears to diminish. This dangerous side effect resulting from the improvement in processes of incubation and rearing can be eliminated by introducing specialized methods of testing the viability of larvae in specially provoked unfavourable environmental conditions, including selection at the early stages of development. In some cases, it may be advantageous, as proposed by Swedish authors, to incubate the eggs using specially constructed artificial spawning grounds directly in a natural waterbody. This may be regarded as the establishment of a kind of "natural hatcheries".

In any case, working with fish species living in nature, one has to conduct selection at lower intensity than when working with pond fish species. In the former case selection must play a predominantly guarding role and primarily involve the culling of individuals with a poor growth rate or deviating from the norm or finally showing abnormalities in the structure and functions of the gonads, etc.

It is advantageous to practice such selection at all fish farms engaged in the reproduction of a given species. In order to avoid inbreeding, a decrease in heterozygosity and the onset of inbred depression, the parental pairs used as sources of eggs and sperm should not be closely related. Their number at each reproduction site should not be less than 50–100. If it is possible, all the main populations within a species should be used for reproduction; in salmonids isolated local subpopulations should also be used. Observance of all these conditions will help to avoid the deterioration of the species when natural reproduction is replaced by artificial.

One of the most complex problems of fish breeding today is the development of the measures required to prevent a decrease in size and the deterioration of species that are subjected to an excessively intensive catch. Examples of such a decrease in size are very numerous, and the need to take special measures to protect the genetic material of commercial fish species that are not cultured artificially is becoming more and more evident (Riggs and Sneed 1959; Donaldson and Menasveta 1961; Nikolsky 1966; Kirpichnikov 1973c; Moav et al. 1978).

A possible solution to this problem is the abandonment of the rules regulating the lowest size of fish permitted to be caught. The need to protect non-mature young fishes from depredatory elimination is self-evident, but the existence of the size criterion during intensive fishing also has very negative consequences. Individuals with the lowest growth rate in each age group (with the exception of the youngest) remain in the waterbody and the best and largest individuals are eliminated completely or almost completely. This means that there is intensive negative selection in each generation. We have already seen above that selection in a negative direction is generally very effective. The harm caused by such selection exceeds the benefits obtained by the conservation of young fish that have not yet grown to the permitted limit: indeed, most fish species that are now subject to intensive fishing have become markedly smaller in size over the last few years.

We believe that the size criterion must probably be replaced of the introduction of very strict control of the total amount of fishes captured. The sites of the fry concentration must be localized and all types of fishing must be prohibited in such areas. It cannot be excluded that the protection and regeneration of the genetic pool of the most important commercial fish species will require the administration of the maximal permitted size, that is the upper size limit of individuals caught.

Another very important genetic measure involves the establishment of strictly differentiated fishing quotas for each reproductively isolated population of a given species. The average values for the species as a whole are unacceptable because frequently their application leads to the gradual destruction of all populations, one after another.

In order to enrich the genetic pool of the deteriorated species of commercially important wild fish species, a proposal was recently advanced that the large-scale hybridization of such wild species with their domesticated highly productive relatives should be conducted (Moav et al. 1978). This proposal may at present be realized only for a few fish species that possess both wild populations and good domesticated breeds common carp, rainbow trout, chars).

It should, however, be emphasized that such hybridization might have undesirable consequences as well; this is particularly true of the possible impairments of maturation and the decrease in the effectiveness of fish reproduction. In each specific case many-sided and detailed discussion of all the possible positive and negative consequences of such a hybridization must be undertaken before such work is started.

The efforts of geneticists at the present time must be aimed at the study of intraspecific and the intrapopulational structure of the most important commercial species subject to fishing in rivers, lakes, seas and oceans, particular attention being paid to the analysis of the genetic consequences of excessive fishing. This will provide an opportunity to determine real ways of regenerating resources of the species that have already suffered some deterioration, and this will prevent the destruction of populations of those species that have not yet become so far the victim of uncontrolled excessive, irresponsible fishing.

The Most Important Fish Breeds Created by Man

We do not intend to describe in detail the fish breeds developed up to the present time by geneticists and fish-breeders throughout the world. The number of breeds of commercial fishes satisfying the strict requirements made upon the breed remains very low; so far they are practically limited to three or four species cultured in ponds (this is predominantly the common carp). The description of many breeds is unfortunately lacking. Below we shall list a number of breeds, which are well described and most important from a commercial point of view.

The Common Carp (Cyprinus carpio)

Only three sufficiently distinct breeds of the carp and commercial hybrids are widely used (Kirpichnikov and Golovinskaya 1966) in the USSR.

Fig. 79 a, b. Ukrainian common carp. **a** scaled, **b** framed. (Photography of V. G. Tomilenko)

1. The Ukrainian carp (Fig. 79) created as a result of many years of synthetic selection work with several local stocks of the Ukrainian carp (Kuzema 1953). There are two varieties of Ukrainian carp with different scaling patterns, a scaled one and a framed one, are found. Both are distinguished by a high spine, rapid growth rate and great fertility. They are not, however, resistant to unfavourable environmental conditions and various diseases.

2. The Ropsha carp bred by selecting hybrids of the European Galician carp and the Amur wild carp (Figs. 80 and 81). The Ropsha carp possesses increased resistance to low temperatures and high general viability. It grows rapidly during the first and second year of life, however, thereafter its growth is retarded. This breed is well adapted to cultivation in the northern and northwestern regions of the USSR (Kirpichnikov 1967 a, 1972 a; Zonova and Kirpichnikov 1971; Zonova 1976).

3. Hybrids of the first generation of the domesticated carp and the Amur wild carp; they are characterized by heterosis with respect to survival and the growth rate (Kirpichnikov 1962; Andriyasheva 1966; Karpenko 1966).

In addition to these breeds and hybrids, which have already become quite common, several other strains may be mentioned, selection work with which is not yet completed.

Fig. 80 a, b. Ropsha northern common carp (5th–7th generation of selection). **a** age of 5+, **b** age of 2+

The Nivchan carp developed as a result of selection of the hybrids of the Ropsha and the scaled Ukrainian carp (Fig. 76 and 81). The fourth selected generation has been obtained, and the work with the Nivchan carp appears to be nearing completion (Kuzema et al. 1970; Tomilenko and Kucherenko 1975; Kucherenko 1978, 1979). The Nivchan carp possesses a fortuitous combination of the rapid growth rate of the Ukrainian carp and the increased viability of the Ropsha carp.

The Central Russian carp is reared by synthetic selection, that is by combining the features of several breeds of the Russian carp (Golovinskaya 1969; Golovinskaya et al. 1975; Kirpichnikov and Katasonov 1978).

The Kazakhstan breed of carp has been developed using artificial mutagenesis and gynogenesis; the work in this direction is proceeding very successfully (Tzoy 1980).

The Krasnodar carp has already been selected for four to six generations now; these carp are selected for increased resistance to highly dangerous disease such as

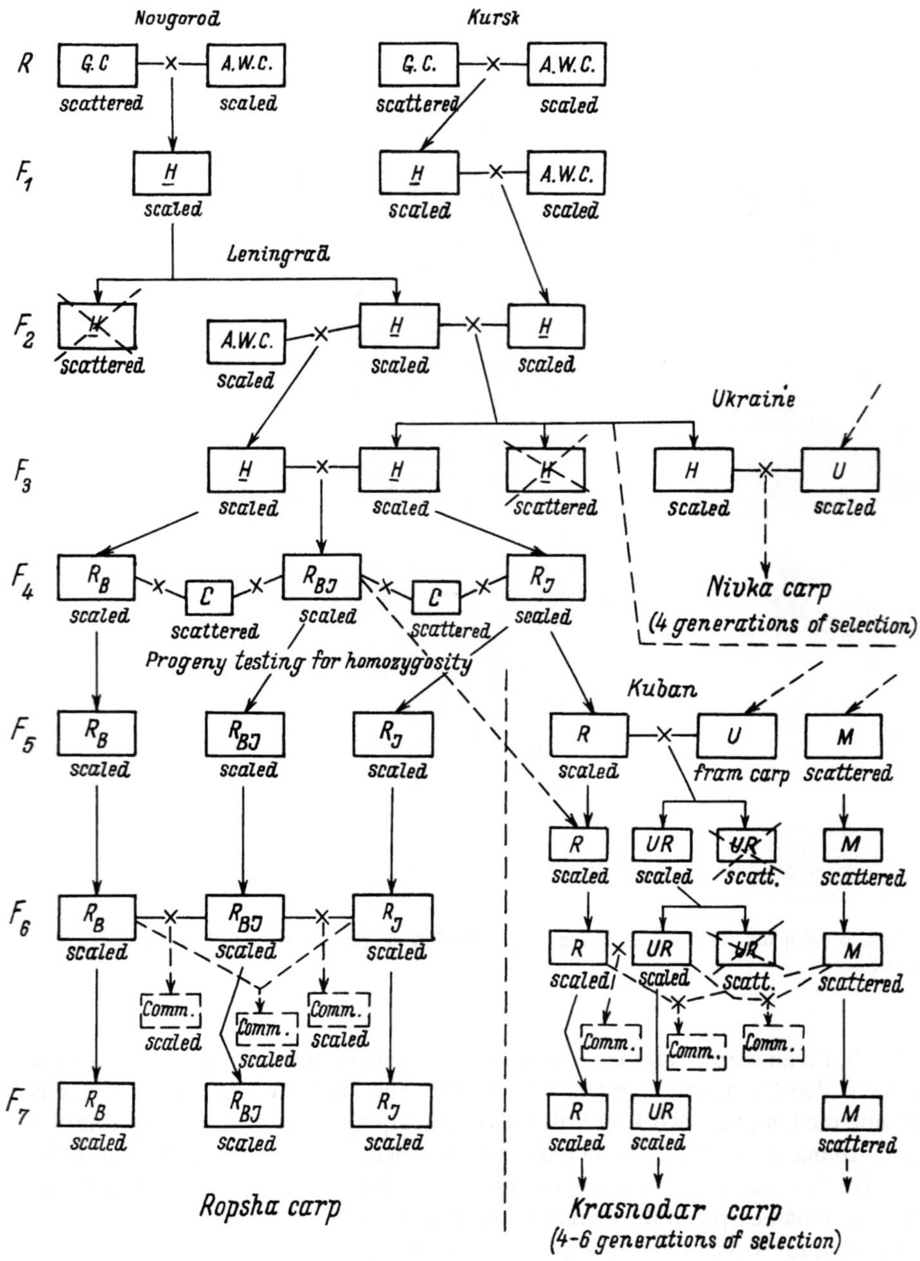

Fig. 81. The scheme of construction of three new strains of the common carp, Ropsha, Nivka und Krasnodar. Abbreviations: *G.C.* Galician carp; *A.W.C.* Amur wild carp; *H* hybrids; *U* Ukrainian carp; *C* non-selected (low selected) carp; *R*, R_B, R_{BI} and R_I different stocks of Ropsha carp; *M* local low-selected Krasnodar carp; *UR* hybrids of Ukrainian and Ropsha carps; *Comm.* commercial hybrids

infectious dropsy (Kirpichnikov et al. 1971, 1972a, 1976, 1979; Kirpichnikov and Factorovich 1972).

The Byelorussian breed of carp has also passed through four generations of selection, the rapid growth rate being the main criterion during the selection work (Polyksenov 1962; Chutaeva et al. 1975a).

The Para breed of carp is characterized by features of commercial significance; two stocks of this carp have been obtained as a result of synthetic crosses followed by selection (Bobrova 1978).

Carp selection is being practised in other regions of the USSR as well, in particular in West Ukraine, Moldavia, Lithuania, Estonia, West Siberia, Uzbekistan and Georgia (Grechkovskaja 1971, 1977; Brouzhinskas 1979; Lobchenko et al. 1979; and others).

No established breeds of common carp appear to exist at present in Germany, Poland, Romania, Hungary, Yugoslavia, and France. The best carp strains in the German Democratic Republic appear to originate from the Königswartha, Deutschbaselitzer and Koselitzer fish farms (Nagel 1970; Müller 1975; Steffens 1975); the differences between these strains are, however, limited and are, to a large extent, determined by the degree of previous inbreeding.

Intensive carp selection is being carried out in Hungary at the Szarvas fish farm. The breeding value of the hybrids of the inbred lines appears to be particularly high (Bakoš 1978, 1979).

Selection work with the common carp has been carried out since 1940 at the Dumbrava-Sibiu fish farm, Romania (Pojoga 1972); the initial selection stock was based on the German carp from Lausitz and Galicia and on local varieties of wild carp.

A number of selected carp strains has been obtained in Poland but they are not described in detail.

Intensive and diverse selection of the common carp, involving the use of modern genetic methods, is being conducted on a wide scale in Israel. The commercial characteristics of the selected strains are high (Wohlfarth et al. 1975a).

Of the Asian countries we would like to mention Indonesia, Japan and China. In Indonesia several local breeds of the common carp are being cultivated, the best-known being the following: Ikan mas, Puntener, Sinjonia, Kumpai and Tjiko (Steffens 1975). The rate of growth of the Indonesian carp is not high, but the European carp is unable to compete with its Indonesian counterpart in local conditions (Buschkiel 1938). Several breeds are known in Japan, those with the greatest productivity being Yamato, Shinshu and Asagi, all of which varieties are scaly. Decorative stocks such as the Higoi (golden carp) and the Irigoi (multicoloured or coloured or chromista carp) have also been constructed. Among the Irigoi carp all the known variants of the scaling pattern are known (Steffens 1975; Suzuki et al. 1976). The history of the development of these carp breeds unfortunately remains unknown.

Several breeds of common carp appear to exist in China as well. The most important of them is the Big-Belly; Big-Belly carp have a comparatively moderate rate of growth, they reach maturity early and possess high fertility as well as increased viability, the ability to avoid fishing nets and other fishing tackle (Wohlfarth et al. 1975b).

Carp belonging to the European, Chinese and Japanese breeds are easily crossed with each other and have high fertility.

The Rainbow Trout (Salmo gairdneri = S. irideus)

No established breeds of rainbow trout in the strict sense of the word exist at the present time, however, many strains differ greatly in their productivity (Reinitz et al. 1978, 1979). Many years of rainbow trout selection in the state of Washington (Seattle) have resulted in the marked acceleration of growth to maturation, but this was also accompanied by a decrease in viability (Donaldson 1969; Kato 1974; Herschberger et al. 1976). Useful properties have also been described for the Danish trout (Savostyanova 1971, 1976). In Virginia, USA, trout selection has been carried out since 1930, in the state of New York this work began in 1967; in the latter case three selected generations have already been obtained (Bruhn and Bowen 1973; Kincaid et al. 1977). Selection work has been started in Norway, Canada, Sweden, and other countries.

The Brook Trout (Salvelinus fontinalis)

Selection work with the brook trout aimed at accelerating its growth rate and in particular at increased resistance to disease has been conducted in the United States for more than 20 years now (Wolf 1954; Snieszko 1957; Flick and Webster 1976). This work has not yet yielded any great success.

The Goldfish (Carassius auratus)

Many varieties or breeds of goldfish have been developed in China and Japan. Domestication of this species in China began in 1163. By 1643, six breeds had already been described, and their number had reached eight by 1925. After the acclimatization of the goldfish in Japan in the sixteenth century many new breeds were reared there; we have already described the scheme of development of these breeds (see Chap. 2).

Selection work has recently been conducted with other fish species as well. These include the catfish *Ictalurus punctatus* (Green et al. 1979), the brown trout *Salmo trutta* (Ehlinger 1964), and several other species. In particular, I would like to mention the selection work with the Atlantic salmon (*Salmo salar*) initiated on a wide scale in Norway (since 1971) and Canada (since 1976). Selection work with the salmon aimed predominantly at the creation of productive and resistant breeds suitable for commercial cultivation is developing in accordance with specially elaborated, detailed programmes (Naevdal et al. 1975; Saunders 1978a, b; Saunders and Bailey 1978; Gjedrem 1979); this work is being carried on side by side with extensive genetic studies.

Considerable success has been achieved in the selection of aquarium fishes (see Chap. 3); detailed information about the breeds and varieties of the main species used in aquarium breeding can be found in more specialized reviews, for instance in the book of Schröder (1974).

One can be sure that in the very near future the rapid development of fish breeding and the domestication of many new fish species will result in the appearance of many different breeds designed for cultivation under extremely diverse conditions. Fish selection, which has been somewhat retarded in its development, will undoubtedly develop and be extended, approaching in its achievements selection work in the field of animal husbandry and plant breeding.

Chapter 9

Conclusion

Selective breeding is aimed at improving existing and developing new breeds and hybrid forms of animals, plants and microorganisms. Selection appears to be essentially a creative process and, as in any creative work, the success of selection depends primarily on the personal capabilities of the person conducting it and on his intuition involving the capacity to "foresee" the good and the bad, to select the correct approaches in his work in order to avoid serious errors. Talent and intuition, however, are not enough: nowadays rapid, effective selection can only be attained when the breeder has sufficient knowledge of genetics and the related biological disciplines such as embryology, physiology, comparative anatomy, biochemistry, mathematical statistics, etc. Those practising selection must also be well versed in the theory of selection that has now been developed in great detail.

The review of several aspects on fish genetics conducted by us has indicated that over the last few years data on the heredity and variation of fish have accumulated particularly rapidly. A lot of new experimental data have been obtained, and these must be taken into account in improving selection methods and putting them into practice.

As regards the material basis of heredity, the extensive accumulation of data concerning the karyotypes of cyclostomes and fishes has been very intensive and by the beginning of 1980 the number of fishes with known karyotypes was more than 1300. Practically all fish orders have been examined, in particular many species were studied in the orders of Clupeaformes, Cypriniformes, and Cyprinodontiformes. A most important result of all these studies was the finding of great variability in the numbers of chromosomes in fish species, the diploid numbers varying from 12 to 250. A certain tendency was noted for a moderate decrease to occur in the number of chromosomes from 48–80 to 40–46. This appears to be associated with the transition to more advanced, relatively more highly specialized fish taxons. The same tendency has been noted with regard to the amount of DNA in the chromosomal complement.

One of the most interesting results obtained in the studies of fish karyotypes was the discovery of polyploid species and of whole families of polyploids. Certain sturgeon and carp, as well as the whole family of catostomids, certain species of cobitids and characids, all the salmonids including the whitefish and the grayling have duplicated chromosomal complements and a double amount of DNA. The duplication of chromosomal complements appears to have taken place approximately 20 to 100 million years ago, while in certain groups such as sturgeon and loach the polyploids may have appeared relatively more recently. All the genes in such polyploid forms were duplicated in their evolution. Not infrequently one of the two homologous loci later became silenced or was even lost completely. Only a few genes or 30% to 60% of all the loci of polyploid forms studied exist as duplicate copies at present.

In addition to the species with a constant number of chromosomes, several polymorphic species possessing variable chromosomal complements have been found. These include in particular many salmonids such as the rainbow trout, the Atlantic salmon, the sockeye salmon and others, certain cyprinid (the crucian carp, for example) and the tooth carp (a number of species in the genus *Aphyosemion*). It has been established that the chromosomal polymorphism existing in different fish groups may depend on at least three mechanisms such as Robertsonian translocations (fusions and fissions of acro- and metacentric chromosomes), non-disjunction of chromosomes apparently accompanied by elimination of smaller elements and perhaps by inversions and translocations, as well as on ploidy changes due to the formation of diploid gametes.

Marked achievements in the study of the mechanism of sex determination and its control are worthy of mention. The sex chromosomes have been found in more than 45 fish species; both male and female heterogamety has been found (the first being more frequent); species have been described when both types of sex determination have been combined. Using artificial selection and more recently sex hormones, one is able to modify the sex even at the fry stage. This leads to the formation of a unisexual offspring.

Success appears to be more modest in the field of special fish genetics. The genetics of morphological or other qualitative traits has been developed in some detail only with the common carp and with several aquarium species. The genes studied in the case of the common carp show a distinct Mendelian pattern of inheritance and are widely used as genetic markers in selective breeding. The genetics of aquarium fishes (primarily of the platyfish *Xiphophorus maculatus*, the guppy *Poecilia reticulata* and the medaka *Oryzias latipes*) is instrumental in solving a number of important problems of modern biology, including that of the polymorphism of natural populations, the problem of hereditary factors in cancer, the problem of the genetic mechanisms involved in the determination and reversal of sex, and others. It should be pointed out that the study of tumour genetics in the platyfish (using such tumours as melanomas, erythroblastomas and others) has indicated that the development of malignant tumours is associated with the impairment of the balance of the genes controlling the development of the pigment cells; this balance is impaired by interspecific hybridization, but can also be disturbed by selection. The analysis of the mechanisms involved in the sex determination provided extremely important results: it indicated that all the possible types of sex determination are observed in fishes varying from purely environmental nongenetic in hermaphrodites (such as the family Serranidae and several others) to distinctly chromosomal determination in the vast majority of taxons. The intermediate position is occupied by species where sex determination is of a polygenic character (*Xiphophorus helleri* in poeciliids and others). In contrast to birds and mammals, as well as to many invertebrates, in most fish species where sex is determined by the chromosomes the sex chromosomes (X and Y) differ only in the presence of one gene or the little region responsible for sex determination. In the past this provided the possibility of fish polyploidization and at present this facilitates the development of methods of artificial reversal of sex.

Certain achievements in the study of quantitative fish traits should be mentioned. It is a general rule that such traits are inherited in a polygenic fashion and greatly depend on the effect of the environment. The heritability of many quantitative traits, particularly of traits linked with the adaptive value or fitness and including weight, body size and viability at the early stages of development, is not high, frequently it does not exceed 0.1–0.2. Recently the methods of heritability determination have been subjected to serious criticism (see, e.g., Nikoro 1976). In spite of this, calculations of the heritability of fish characters from regression coefficients and the correlation between relatives, on the one hand, and on the basis of a variance analysis in diallele and hierarchical complexes, on the other, generally provide similar results and an opportunity to determine with a degree of approximation the genuine fraction of genetic variation in the total variation of the trait. The precision of such determinations is sufficient to plan the use of different selection methods in work with different fish species on the basis of h^2 value.

One of the most rapidly developing chapters in fish genetics nowadays is the biochemical genetics widely used in embryological, populational and evolutionary studies. The level of biochemical variation in fish is comparable to that in other classes of vertebrates, mammals in particular. A large number of protein loci, including both enzyme loci and the loci of non-enzymatic proteins appears to be polymorphic in fish populations, the mean level of heterozygosity (mainly for enzyme loci) being equal to 4%–5%.

In addition to the genetic polymorphism, that is the presence of two or several alleles of one locus in a population, two or more independently inherited genes have been described for many proteins. In a few cases the number of a homologous loci is five or six (lactate dehydrogenase in the polyploid sturgeon, common carp and salmon). The evolution of several biochemical loci has been traced in particular with loci coding for lactate dehydrogenase and creatine kinase; the increase in the number of homologous loci is associated with polyploidization and tandem duplications and is accompanied by the divergence of the functions of the daughter genes and sometimes by the appearance of "silenced" genes. The number of isozymes of one and the same protein may attain 15–20, up to 25–30, taking into account genetic polymorphism (in heterozygotes with respect to one or two loci).

Biochemical genetic studies, including studies of blood group variability, are now being conducted with an ever greater range of species. The accumulation of the data on the inheritance of genetic differences for dozens of different biochemical systems has allowed many aspects of fish systematics to be studied in detail, and evolutionary trees and dendrograms to be constructed, and was helpful in elucidating the genesis of various species and sometimes of closely related fish genera.

Biochemical genetics has made a prominent contribution to the systematics and the philogeny of fishes (Ferguson 1980). Genetic biochemical studies have been of outstanding theoretical and practical significance for the analysis of the intraspecific populational structure of fish, for the delineation and characterization of isolated populations. The results of the genetic work are to be used for the development of recommendations regarding the conservation of fish resources, and the rational exploitation and reproduction of the most important commercial fish species.

A large section of fish biochemical genetics deals with the analysis of the regulation of the expression of specific loci during development, with the temporal and spatial differentiation of isoenzymes. The main types of homologous loci differing in the pattern of expression have been described: housekeeping isozymes, which appear to be active at all stages of development and in all tissues, as well as the more specialized isozymes, the activity of which is limited to certain tissues and to certain periods.

Many observations dealing with allele frequency in populations have been made. These apply to the geographical variation both of the clinal and the network type. Information has been collected indicating that natural selection operates on the alleles of the biochemical loci. Successful experiments with artificial and natural selection have been performed, taking into account various stress factors. Definite functional and frequently adaptive differences between the isozymes and between the allelic forms of proteins have been established; in many cases monogenic heterosis has been recorded which is expressed in the increased survival of the heterozygotes or the better functioning of the proteins in heterozygous individuals. All these facts considered together provide evidence in favour of the selective or balanced hypothesis of the origin and maintenance of biochemical polymorphism in fish populations. These data can be extrapolated to other organisms and provide evidence against the neutralist explanation of protein polymorphism and evolution.

I would also like to point out the marked progress made in investigating two rare patterns of animal reproduction, gynogenesis and hybridogenesis, found in certain amphibia, reptilia and fishes. It has been established that gynogenesis in fishes may be meiotic, that is occurring without loss of the meiotic divisions, and ameiotic when meiosis is changed in favour of a specialized mechanism providing for the transition of the entire maternal chromosomal complement without any reduction in the gamete. In the case of meiotic gynogenesis the female gamete receives the haploid chromosomal complement, and the male gamete does not participate in development. Therefore the embryos are non-viable haploids. Use of a cold shock after insemination allows gynogenetic diploids to be obtained as a result of fusion of the maternal pronucleus with the nucleus of the second polar body. Such induced diploid gynogenesis is a very important instrument in the hands of a fish breeder. This instrument provides new opportunities to accelerate the development of inbred fish strains, as well as serving a number of other purposes. Permanent high heterozygosity is maintained in fish populations in the case of natural hybridogenesis; in such cases only the genome of the female parent passes to the gamete during meiosis, while the genome of the male parent is eliminated together with the polar body; one maternal genome and the paternal genome from male belonging to a different subspecies or species are present in the embryo after fertilization. Hybridogenesis in fish species appears to be a peculiar mechanism leading to the stable fixation of heterosis.

We have summarized the most important results of studies dealing with the various aspects of fish genetics conducted over the past few decades. Their purposeful application in selection work will speed up the tempo of selection and improve its effectiveness. Still, much remains to be done: detailed studies of the genetics of the most important commercial fishes are necessary, including research into the genetics of morphological (marker and quantitative) traits. Rapid techniques for the examination of fish chromosomes have to be developed, as well as techniques for the exact differentiation of individual pairs of chromosomes including sex chromosomes, and also for the elucidation of the inheritance of biochemical loci. It is very important to identify linkage groups in the various fish species cultivated by man. The list of problems facing investigators studying fish genetics is very extensive, but even at present fish genetics appears to be a large autonomous section of present-day animal genetics; it does, to a very significant extent, affect both the organization of the selection work with fish, the protection of the natural fish resources and the reproduction of commercially significant fish species in rivers, lakes, seas, and oceans. Genetic studies of fish will also be helpful in gaining a better understanding of many theoretical questions of present-day biology.

References[1]

Abdullaev MA, Khakberdiev B (1972) Dwarf wild common carp *Cyprinus carpio* L. in lakes of Khoresm district. Vopr Ikhtiol 12 [6(77)]: 1114–1117 (R)[2]

* Abe S (1975) Karyotypes of 9 species of anabantoid fishes. CIS 19:5–7

* Abe S (1976) A cyto-taxonomical study in some freshwater cottoid fishes (Cottidae, Pisces). Cytologia 41 (2):323–329

Abe S, Muramoto J-I (1974) Differential staining of chromosomes of two salmonoid species, *Salvelinus leucomaenis* (Pallas) and *Salvelinus malma* (Walbaum). Proc Jpn Acad 50 (7):507–511

Abe T (1972) A partially albinistic specimen of *Limanda yokohamae* (Gunther). UO (Japan) 14:1–2

Abramoff P, Darnell RM, Balsano JS (1968) Electrophoretic demonstration of the hybrid of the gynogenetic teleost *Poecilia formosa*. Am Nat 102 (928):555–558

Adhasadeh H, Ritter H (1971) Polyploidisierung in der Fischfamilie Cyprinidae Ordnung Cypriniformes. Duplikation der Loci für Nad-abhängige Malatdehydrogenasen. Humangenetik 11 (2):91–94

* Ahmed M (1974) A chromosome study of two species of *Gobiosoma* from Venezuela (Gobiidae: Teleostei). Bol Inst Oceanogr Univ Oriente 13 (1–2):11–16

Ahuja MR, Anders F (1976) A genetic concept of the origin of cancer, based in part upon studies of neoplasms in fishes. In: Progress in Experimental Tumor Research, vol 20. Karger S, Basel, 380–397

Ahuja MR, Schwab M, Anders F (1977) Tissue-specific esterases in the xiphophorine fish *Platypoecilus maculatus, Xiphophorus helleri* and their hybrid. Biochem Genet 15 (7–8):601–610

Ahuja MR, Lepper K, Anders F (1979) Sex chromosome aberrations involving loss and translocation of tumor-inducing loci in *Xiphophorus*. Experientia 35 (1):28–30

Aida T (1921) On the inheritance of color in a fresh-water fish *Aplocheilus latipes* with special reference to the sex-linked inheritance. Genetics 6 (3):554–573

Aida T (1930) Further genetical studies of *Aplocheilus latipes*. Genetics 15 (1):1–16

Aida T (1936) Sex reversal in *Aplocheilus latipes* and a new explanation of sex differentiation. Genetics 21 (1):136–153

Aitken WW (1937) Albinism in *Ictalurus punctatus*. Copeia 1:64

Aksiray F (1952) Genetical contribution to the systematical relationship of Anatolian cyprinodont fishes. Hydrobiology (Istanbul) (B) 1:33–81

Akulin VN, Svetashev VI, Salmenkova EA (1975) Intraspecific genetic variability of serum phospholipids in sockeye salmon *Oncorhynchus nerka* and in chum salmon *O. keta*. J Evol Bioch Physiol 11 (3):306–308 (RE)

Alekseev AE, Alekseeva EI, Titova NV (1979) Polymorphous system of muscle esterases and ecological and population structure of species of rock grenadier (*Macrurus rupestris*). In: Biochem Popul Genet Fish. Inst Cytol Acad Sci USSR, Leningrad, pp 58–63 (RE)

1 * = articles used in Table 2; R = articles or books in Russian; RE = in Russian with English summary

Alexandrov VYa (1977) Cell, molecules and temperature. Springer, Heidelberg Berlin New York
Alexeenko AA (1979) Physiological traits of Ropsha-Ukrainian hybrid carps and its parental forms. In: Selection of pond fishes. Kolos, Moscow, pp 61–66 (RE)
Alferova NM, Nefyodov GN (1973) Electrophoretic study of muscle esterases of some fish species from the Eastern Atlantic. In: Biochem Genet Fish, Leningrad, pp 195–200 (RE)
Ali MY, Lindsey CC (1974) Heritable and temperature-induced meristic variation in the medaka *Oryzias latipes*. Can J Zool 52 (8):959–976
Allendorf F, Utter FM (1973) Gene duplication within the family Salmonidae: disomic inheritance of two loci reported to be tetrasomic in rainbow trout. Genetics 74 (4):647–654
Allendorf F, Utter FM (1975) Genetic variation of steelhead. In: Annu Rep Coll Fisher, Univ Wash 415:44
Allendorf F, Utter FM (1976) Gene duplication in the family Salmonidae. III. Linkage between two duplicated loci coding for aspartate amino transferase in the cutthroat trout (*Salmo clarkii*). Hereditas 82 (1):19–24
Allendorf FW, Utter FM (1979) Population genetics of fish. In: Fish physiology, vol 8. Academic Press, London New York, pp 136–187
Allendorf F, Utter FM, May BP (1975) Gene duplication within the family Salmonidae. II. Detection and determination of the genetic control of duplicate loci through inheritance studies and the examination of populations. In: Isozymes, vol IV. Academic Press, London New York, pp 415–432
Allendorf F, Ryman N, Stennek A, Stahl G (1976) Genetic variation in Scandinavian brown trout (*Salmo trutta* L.): evidence of distinct sympatric populations. Hereditas 83 (1):73–82
Allendorf FM, Mitchell N, Ryman N, Stahl G (1977) Isozyme loci in brown trout (*Salmo trutta* L.): detection and interpretation from population data. Hereditas 86 (2):179–190
Altukhov YuP (1962) The study of thermostability of the isolated muscle and the serological analysis of "large" and "small" horse mackerel of Black Sea. In: Tr Karadag Biol Stn Akad Nauk Ukr. SSR 18; 3–16 (R)
Altukhov YuP (1967) The study of thermostability of isolated muscle tissue in two horse mackerel species of Black and North Seas. Dokl Akad Nauk SSSR (DAN) 175 (2):467–469 (R)
Altukhov YuP (1969a) On immunogenetical approach to problem of intraspecific differentiation in fishes. In: Advances of modern genetics, vol 11. Nauka, Moscow, pp 161–195 (R)
Altukhov YuP (1969b) Interrelations of hemoglobin monomorphism and polymorphism in fish microevolution. Dokl Akad Nauk SSSR (DAN) 189 (5):1115–1117 (R)
Altukhov YuP (1973) Local stocks of fishes as genetically stable populations systems. In: Biochem Genet Fish. Inst Cytol Acad Sci USSR, Leningrad, pp 43–53 (RE)
Altukhov YuP (1974) Population genetics of fishes. "Pischchevaya Promyshlennost", Moscow (R)
Altukhov YuP, Apekin VS (1963) Serological analysis of interrelations between "large" and "small" horse mackerel from Black Sea. Vopr Ikhthiol 3 [1(26)]:39–50 (R)
Altukhov YuP, Matvejeva AI, Rusanova BM (1966) Thermostability of isolated muscle tissue and antigenic properties of erythrocytes of some cultured varieties of the carp. Tzitologiya 8 (1):100–104 (RE)
Altukhov YuP, Nefjodov GN, Pajusova AN (1967) A cytophysiological analysis of divergence between *Sebastes marinus* Linné and *S. mentella* Travin from North-West Atlantic. In: Variation in animal cell thermostability in onto- and phylogenesis. Nauka, Leningrad, pp 82–98 (RE)
Altukhov YuP, Nefyodov GN (1968) A study of blood serum protein composition by agar-gel electrophoresis in types of redfish (genus *Sebastes*). Res Bull Int Commis North-West Atl Fish 5:86–90
Altukhov YuP, Truveller KA, Zenkin VS, Gladkova NS (1968) The A-system of blood groups in the Atlantic herring (*Clupea harengus* L.). Genetika (Moscow) 4 (2):155–167 (RE)
Altukhov YuP, Limansky VV, Payusova AN, Truveller KA (1969) Immunogenetic analysis of the intraspecific differentiation in the European anchovy (*Engraulis encrasicholus*) inhabiting the Black Sea and the Sea of Azov. I. The blood groups and the hypothetical mechanism of the genic control. The heterogeneity of the Azov race. II. Elementary populations of anchovy and their connection with the genetic-populational structure of the species. Genetika (Moscow) 5 (4):50–64; 5 (5):81–94 (RE)

Altukhov YuP, Salmenkova EA, Sachko GD (1970) Duplication and polymorphism of the lactate dehydrogenase genes in Pacific salmon. Dokl Akad Nauk SSSR (DAN) 195 (2):711–714 (R)

Altukhov YuP, Rychkov YuG (1970) Population systems and their structural components. Genetical stability and variability. J Obtsch Biol 31 (5):507–526 (RE)

Altukhov YuP, Rychkov YuG (1972) Genetical monomorphism of species and its possible biological significance. J Obtsch Biol 33 (3):281–300 (RE)

Altukhov YuP, Salmenkova EA, Omelchenko VT, Sachko GD, Slynko VI (1972) Number of mono- and polymorphous loci in the population of tetraploid salmon species *Oncorhynchus keta* Walb. Genetika (Moscow) 8 (2):67–78 (RE)

Altukhov YuP, Salmenkova EA, Konovalov SM, Pudovkin AI (1975a) Stationary distribution of frequencies of lactate dehydrogenase and phosphoglucomutase genes in population system of local fish stock, *Oncorhynchus nerka* Walb. I. Stability of the stock in generations under simultaneous variability of subpopulations making up their structure. Genetika (Moscow) 11 (4):44–53 (RE)

Altukhov YuP, Pudovkin AI, Salmenkova EA, Konovalov SM (1975b) Stationary distribution of frequencies of lactate dehydrogenase and phosphoglucomutase genes in population system of local fish stock, *Oncorhynchus nerka* Walb. II. Random genetic drift, migration and selection as factors of stability, Genetika (Moscow) 11 (4):54–62 (RE)

Amend D (1976) Prevention and control of viral diseases of salmonids. J Fish Res B Can 33 (4, part 2): 1056–1066

Anders A, Anders F (1963) Genetisch bedingte XX- und XY-♀♀ und YY-♂♂ beim wilden *Platypoecilus maculatus* aus Mexiko. Z Vererbungsl 94 (1): 1–18

Anders A, Anders F (1978) Etiology of cancer as studied in the platyfish-swordtail system. Biochem Biophys Acta 516 (1):61–95

Anders A, Anders F, Förster W, Klinke R, Rase S (1970) XX-, XY-, YY-Weibchen und XX-, XY-, YY-Männchen bei *Platypoecilus maculatus* (Poeciliidae). Zool Anz Suppl B 33:333–339

Anders A, Anders F, Pursglove DL (1971) X-ray induced mutations of the genetically determined melanoma system of xiphophorin fish. Experientia 27:931–932

Anders A, Anders F, Klinke K (1973) Regulation on gene expression in the Gordon-Kosswig melanoma system, 1, 11. In: Genetics and mutagenesis of fish. Springer, Berlin Heidelberg New York, pp 33–52, pp 53–63

Anders F (1967) Tumour formation in platyfish-swordtail hybrids as a problem of gene regulation. Experientia 23 (1): 1–10

Anders F (1968) Genetische Faktoren bei der Entstehung von Neoplasmen. Zentralbl Vet Med 15 (1):29–46

Anders F, Vester F, Klinke K, Schumacher H (1962) Genetische und biochemische Untersuchungen über die Bedeutung der freien Aminosäuren für die Tumorgenese bei Artbeziehungsweise Gattungsbastarden lebendgebärender Zahnkarpfen (Poeciliidae). Biol Zentralbl 91 (1–2):45–65

Anders F, Klinke K, Vielkind U (1972) Genregulation und Differenzierung im Melanom-System der Zahnkarpflinge. Biol Unserer Zeit, Verlag Chemie GmbH, Weinheim 35–66

Anderson D, Woods DE (1979) Evaluation of intensive inbreeding for selection of trout brood stock. In: Rep Sect Fish Invest, Minnesota Dep Nat Resour 364: 1–31

Andrejeva AP (1971) The skin collagen thermostability in several species and subspecies of the gadold fish. Tzytologiya 13 (8): 1004–1008 (RE)

Andriyasheva MA (1966) Heterosis under intraspecific crossings of carp. Izvestija Gosud. nauchno-issled. Inst Ozern Rechn Rybn Khos (GosNIORKh) 61:62–79 (RE)

Andriyasheva MA (1971a) Commerical hybridization and heterosis in fish culture. In: Rep FAO/UNDP (TA) (2926), Rome, pp 248–262

Andriyasheva MA (1971b) Manifestigation of heterosis in fishes and its utilization in fish breeding. Izvestija Gosud. nauchno-issled. Inst Ozern Rechn Rybn Khos (GosNIORKh) 75: 100–113 (RE)

Andriyasheva MA (1976) Fish-cultural and biological characteristics of Endyr peled spawners. In: Izvestija Gosud. nauchno-issled. Inst Ozern Rechn Rybn Khos (GosNIORKh) 107:64–75 (RE)

Andriyasheva MA (1978a) Selection-genetical analysis of the Endyr peled population with regard to the time spawning. In: Izvestija Gosud. nauchno-issled. Inst Ozern Rechn Rybn Khos (GosNIORKh) 130:5–14 (RE)

Andriyasheva MA (1978b) Selection-genetical analysis of the Endyr peled spawning stock with regard to fertility. In: Izvestija Gosud. nauchno-issled. Inst Ozern Rechn Rybn Khos (GosNIORKh) 130:15–24 (RE)

Andriyasheva MA (1980) Genetical and selection characteristic of the peled spawners stocks of different origin. In: Problems of genetics and selection of fishes. State Res Inst Lake River Fish, Leningrad, pp 3–15 (RE)

Andriyasheva MA, Chernyaeva EV (1978) The level of phenotypic and genetic variability in diameter of the Endyr peled ovulated eggs. In: Izvestija Gosud. nauchno-issled. Inst Ozern Rechn Rybn Khos (GosNIORKh) 130:25–34 (RE)

Andriyasheva MA, Mantelman II, Kaidanova TI (1978) First results of selection and genetic investigations of peled. In: Genet Select Fishes 20:112–123 (RE)

Angus RA, Schultz RJ (1979) Clonal diversity in the unisexual fish *Poeciliopsis monacha-lucida:* a tissue graft analysis. Evolution 33 (1, p 1):27–40

Aps RA, Tanner RH (1979) Polymorphism of muscle esterases in Baltic sprat (*Sprattus sprattus*). In: Biochem Popul Genet Fishes. Inst Cytol Acad Sci USSR, Leningrad, pp 90–93 (RE)

* Arai R, Fujiki A (1978) Chromosomes of two species of atherinoid fishes. Bull Natl Sci Mus (Tokyo) Ser A 4 (2):147–150

* Arai R, Inoue M (1975) Chromosomes of nine species of Chaetodontidae and one species of Scorpidae from Japan. Bull Natl Sci Mus (Tokyo) Ser A 1 (4):217–224

* Arai R, Inoue M (1976) Chromosomes of seven species of Pomacentridae and two species of Acanthuridae from Japan. Bull Natl Sci Mus (Tokyo) Ser A 2 (2):73–78

* Arai R, Katsuyama I (1973) Notes on the chromosomes of three species of shore fishes. Bull Natl Sci Mus (Tokyo) 16 (6):405–408

* Arai R, Katsuyama I (1974) A chromosome study of four species of Japanese catfishes (Pisces, Siluriformes). Bull Natl Sci Mus (Tokyo) 17 (3):187–191

* Arai R, Kobayashi H (1973) A chromosome study on thirteen species of Japanese gobiid fishes. Jpn J Ikhthiol 20 (1):1–6

* Arai R, Koike A (1980) Chromosomes of labroid fishes from Japan. Bull Natl Sci Mus (Tokyo) Ser A 6 (2):119–135

* Arai R, Sawada Y (1974) Chromosomes of Japanese gobioid fish, I, II. Bull Natl Sci Mus (Tokyo) 17 (2):97–102; 17 (4):269–274

* Arai R, Sawada Y (1975) Chromosomes of Japanese gobioid fish, III. Bull Natl Sci Mus (Tokyo) Ser A 1 (4):225–232

* Arai R, Shiotsuki K (1973) A chromosome study of three species of the tribe *Salarini* from Japan (Pisces, Blennidae). Bull Natl Sci Mus (Tokyo) 16 (4):581–584

* Arai R, Shiotsuki K (1974) Chromosomes of six species of Japanese blennioid fishes. Bull Natl Sci Mus (Tokyo) 17 (4):261–268

* Arai R, Katsuyama I, Sawada Y (1974) Chromosomes of Japanese gobioid fishes. Bull Natl Sci Mus (Tokyo) 17 (4):269–274

* Arai R, Nagaiwa K, Sawada Y (1976) Chromosomes of *Chanos chanos* (Gonorynchiformes, Chanidae). Jpn J Ikhthiol 22 (4):241–242

Armburster D (1966) Hybridization of the chain pickerel and northern pike. Prog Fish Cult 28 (2):76–78

Aronstam AA, Borkin LYa, Pudovkin AI (1977) Isoenzymes in population and evolutionary genetics. In: Genetics of Isoenzymes. Nauka, Moscow, pp 199–249 (R)

Asano H, Kubo Y (1972) Variations of spinal curvature and vertebral number in goldfish. Jpn J Ikhthiol 19 (4):223–231

Aspinwall N (1973) Inheritance of alpha-glycerophosphate dehydrogenase in the pink salmon, *Oncorhynchus gorbuscha* (Walb.). Genetics 73 (4):639–643

Aspinwall N (1974a) Genetic analysis of duplicate malate dehydrogenase loci in the pink salmon, *Oncorhynchus gorbuscha* (Walb.). Genetics 76 (1):65–72

Aspinwall N (1974b) Genetic analysis of North American populations of the pink salmon (*Oncorhynchus gorbuscha*), possible evidence for the neutral mutation-random drift hypothesis. Evolution 28 (2):295–305

Astaurov BL (1969) Experimental polyploidy in animals with special reference to the hypothesis in an indirect origin of natural polyploidy in bisexual animals. Genetika (Moscow) 5 (7): 129–149 (RE)

Astaurov BL (1971) Parthenogenesis and polyploidy in animal evolution. Priroda (Moscow) 6:20–28 (R)

Atz JW (1964) Intersexuality in fishes. In: Intersexuality in vertebrates including man. Academic Press, London New York, pp 145–232

Aulstad D, Gjedrem T (1973) The egg size of salmon (*Salmo salar*) in Norwegian rivers. Aquaculture 2 (3):337–341

Aulstad D, Gjedrem T, Skjervold H (1972) Genetic and environmental sources of variation in length and weight of rainbow trout (*Salmo gairdneri*). J Fish Res B Can 29 (3):237–241

Austreng E, Refstie T (1979) Effect of varying dietary protein level in different families of rainbow trout. Aquaculture 18: 145–156

Avise JC (1976a) Genetic differentiation during speciation. In: Molecular evolution. Sinauer Assoc, Sunderland, pp 106–122

Avise JC (1976b) Genetics of plate morphology in an unusual population of three spine sticklebacks (*Gasterosteus aculeatus*). Genet Res 27 (1):33–46

Avise JC, Ayala FJ (1976) Genetic differentiation in speciose versus depauperate phylads: evidence from the California minnows. Evolution 30 (1):46–58

Avise JC, Felley, J (1979) Population structure of freshwater fishes. I. Genetic variation of bluegill (*Lepomis macrochirus*) populations in man-made reservoirs. Evolution 33 (1, p 1): 15–26

* Avise JC, Gold JR (1977) Chromosomal divergence and speciation in two families of North American fishes. Evolution 31 (1): 1–13

Avise JC, Kitto GB (1973) Phosphoglucose isomerase gene duplication in the bony fishes: an evolutionary history. Biochem Genet 8 (2): 113–132

Avise JC, Selander RK (1972) Evolutionary genetics of cave-dwelling fishes of the genus *Astyanax*. Evolution 26 (1): 1–19

Avise JC, Smith MH (1974) Biochemical genetic of sunfish. II. Genetic similarity between hybridizing species. Am Nat 108 (962):458–472

Avise JC, Smith J, Ayala F (1975) Adaptive differentiation with little genic change between two native California minnows. Evolution 29 (3):411–426

Avise JC, Straney DO, Smith MH (1977) Biochemical genetics of sunfish. IV. Relationships of centrarchid genera. Copeia 2:250–258

Avtalion RR, Hammerman IS (1978) Sex determination in *Sarotherodon* (*Tilapia*). I. Introduction to a theory of autosomal influence. Bamidgeh 30 (4): 110–115

Avtalion RR, Duczyminer M, Wojdani A, Pruginin Y (1976) Determination of allogenic and xenogenic markers in the genus of *Tilapia*. II. Identification of *Tilapia aurea, T. vulcani* a *T. nilotica* by electrophoretic analysis of their serum proteins. Aquaculture 7 (5):255–265

Ayala FJ (1975) Genetic differentiation during the speciation process. In: Evolutionary biology, vol 8. Plenum Press, New York London, pp 1–78

Ayala FJ (1976) Molecular genetics and evolution. In: Molecular evolution. Sinauer Assoc, Sunderland, pp 1–20

Ayala FJ, McDonald JF (1980) Continuous variation: possible role of regulatory genes. In: Animal genetics and evolution. Dr W Junk BV Publ, The Hague, pp 1–15

Ayles GB (1974) Relative importance of additive genetic and maternal sources of variation in early survival of young splake hybrids (*Salvelinus fontinalis* x *S. namaycush*). J Fish Res B Can 31 (9): 1499–1502

Ayles GB (1975) Influence of the genotype and the environment on growth and survival of rainbow trout (*Salmo gairdneri*) in central Canadian aquaculture lakes. Aquaculture 6 (2): 181–188

Ayles GB, Barnard D, Hendzei M (1979) Genetic differences in lipid and dry matter content between strains of rainbow trout (*Salmo gairdneri*) and their hybrids. Aquaculture 18 (3):253–262

Bachmann K (1972) The nuclear DNA of *Polypterus palmas*. Copeia 2:313–365

Bachmann K, Goin CB, Goin CJ (1974) Nuclear DNA amounts in vertebrates. In: Gordon, Breach (eds) Evolution of genetic systems. Brookhaven Symp Biol, vol 23. pp 419–450

Bailey GS, Lim ST (1975) Gene duplication in salmonid fish: evolution of A lactate dehydrogenase with an altered functions. In: Isozymes, vol IV. Plenum Press, New York London, pp 401–414

Bailey GS, Cocks GT, Wilson AC (1969) Gene duplication in fishes: malate dehydrogenases of salmon and trout. Biochem Biophys Res Commun 34 (5): 605–612

Bailey GS, Wilson AC, Halver J, Johnson C (1970) Multiple forms of supernatant malate dehydrogenase in salmonid fishes: biochemical, immunological and genetic studies. J Biol Chem 245 (22): 5927–5940

Bailey GS, Poulter RTM, Stockwell PA (1978) Gene duplication in tetraploid fish: model for gene silencing at unlinked duplicated loci. Proc Natl Acad Sci USA 75 (11): 5575–5579

Baker-Cohen KF (1961) Visceral and vascular transposition in fishes, and a comparison with similar anomalies in man. Am J Anat 109 (1): 37–55

Bakoš J (1974) Production of carps, having a better productive capacity, by means of crossbreeding different regional breeds. Külon Kiserl Közlem 67B: 113–125

Bakoš J (1978) The present state and prospective results of research on carp selection. In: Increasing productivity of fishes by selection and hybridization. Muller F, Szarvas, pp 1–7

Bakoš J (1979) Crossbreeding Hungarian races of common carp to develop more productive hybrids. In: Advances in aquaculture. Fish News Books Ltd, Farnham, pp 635–642

Bakoš J, Krasnai Z, Terez M (1978) Results of selection and genetic investigations of fish in Hungary. In: Genetics and selection of fishes. All-Union Res Inst Pond Fish, Moscow, 125–138 (RE)

Balachnin IA, Galagan NP (1972a) Segregation and surviving of specimens with different transferrin types in common carp progenies from different spawners. Hydrobiol Zh 8 (3): 56–61 (R)

Balachnin IA, Galagan NP (1972b) The transferrin types in wild common carp from Danube river and in other representatives of *Cyprinus carpio* from USSR waterbodies. Hydrobiol Zh 8 (6): 108–110 (R)

Balachnin IA, Romanov LM (1971) Distribution and gene frequencies of transferrin types in non-selected cultured common carp and in Amur wild carp. Hydrobiol Zh 7 (3): 84–86 (R)

Balachnin IA, Zrazhevskaya IV (1969) The time and character of the appearance of blood groups and their inheritance in Dnieper roach (taran). Tzytol Genet 3 (2): 124–127 (R)

Balachnin IA, Galagan NP, Lukjanenko VN, Popov AV (1972) Genetic polymorphism for some components of fish blood in sturgeon and common carp. Dokl Akad Nauk SSSR (DAN) 204 (5): 1250–1252 (R)

Balachnin IA, Bogdanov LV, Lasovsky AA (1973) The hemoglobin, transferrin and prealbumin types and the potassium content in common carp brood from fish farm "Volma" (Byelor. SSR). Vest Zool (Minsk) 7 (2): 26–29 (R)

Balsano JS (1969) Systematic relations of fishes of the genus *Poecilia* in eastern Mexico based upon plasma protein electrophoresis. Ph D Thesis, Marq Univ Diss Abst 29 (9): 3533–B

Balsano JS, Darnell RM, Abramoff P (1972) Electrophoretic evidence of triploidy associated with populations of the gynogenetic teleost *Poecilia formosa*. Copeia 2: 292–297

Bams RA (1976) Survival and propensity for homing as affected by presence or absence of locally adapted paternal genes in two transplanted populations of pink salmon (*Oncorhynchus gorbuscha*). J Fish Res B Can 33 (12): 2716–2725

Barash DP (1975) Behavioral individuality in the cichlid fish *Tilapia mossambica*. Behav Biol 13 (2): 197–202

Bardach JE (ed) (1972) Aquaculture: the farming and husbandry of fresh water and marine organisms. Wiley and Sons, New York

Barlow GW (1961) Causes and significance of morphological variation in fishes. System Zool 10 (3): 105–117

Baron JC (1972) Preliminary studies on the blood of *Sardinella* from the West African coast. Proc 12th Eur Conf Anim Bl Gr Bioch Polym, Budapest, pp 593–595

Baron JC (1973) Les estérases du sérum de *Sardinella aurita* Val. Application a l'étude des populations. Oceanogr Fr 11 (4): 389–418

Barrett I, Tsujuki H (1967) Serum transferrin polymorphism in some scombroid fishes. Copeia 3: 551–557

Barrett I, Williams A (1967) Soluble lens proteins of some scombroid fishes. Copeia 2: 468–471

Barshiene JV (1977a) Variability of chromosome sets in cells of various organs and tissues of the Atlantic salmon *Salmo salar* L. Tzytologiya (Leningrad) 19 (7): 791–797 (RE)

Barshiene JV (1977b) The karyotype of dwarf males of *Salmo salar* L. from the Neva population. Tzytologiya 19 (8): 906–913 (RE)

Barshiene JV (1978) Ontogenetical variability of the chromosome number in Atlantic salmon (*Salmo salar* L.). Genetika (Moscow) 14 (11): 2029–2036 (RE)

Barshiene JV (1980) Mechanism of ontogenic variability of chromosomal sets in Atlantic salmon (*Salmo salar* L.). In: Karyological variability, mutagenesis and gynogenesis of fishes. Inst Cytol Acad Sci USSR, Leningrad, pp 3–9

Basaglia F, Callegarini C (1977) Controllo delle frequenze alleliche a genotipiche nelle popolazioni Italiane di *Trachurus trachurus* (Suro) polimorfe per gli isoenzimi della lattico deidrogenasi (LDH). Rendiconti 111B: 12–16

Beall H (1963) The West Virginia centennial golden trout. W Vancouver Conserv Mag 27: 20–22

* Beamish RJ, Miller RR (1977) Cytotaxonomic study of gila trout, *Salmo gilae.* J Fish Res B Can 34 (7): 1041–1045

Beamish RJ, Tsuyuki H (1971) A biochemical and cytological study of the longnose sucker (*Catostomus catostomus*) and large and dwarf forms of the white sucker (*C. commersoni*). J Fish Res B Can 28 (11): 1745–1748

* Beamish RJ, Uyeno T (1978) Karyotype of *Hiodon tergisus* and DNA values of *Hiodon tergisus* and *H. alosoides.* CIS 24: 5–8

* Beamish RJ, Merrilees MJ, Grossman EJ (1971) Karyotypes and DNA values for members of the suborder *Esocoidei* (Osteichthyes: Salmoniformes). Chromosoma 34 (3): 436–447

Beardmore JA, Shami SA (1976) Parental age, genetic variation and selection. In: Population genetics and ecology. Academic Press, London New York, pp 3–22

Beardmore JA, Shami SA (1979) Heterozygosity and the optimum phenotype under stabilizing selection. Aquilo Ser Zool 20: 100–110

Beardmore JA, Ward RD (1977) Polymorphism, selection, and multilocus heterozygosity in the plaice, *Pleuronectes platessa* L. In: Measuring selection in natural populations. Springer, Berlin Heidelberg New York, pp 207–222

Beçak W, Beçak ML, Ohno S (1966) Intraindividual chromosomal polymorphism in green sunfish (*Lepomis cyanellus*) as evidence of somatic segregation. Cytogenetics 5 (5): 313–320

Bergot P, Chevassus B, Blanc J-M (1976) Déterminisme génétique du nombre de caeca pyloriques chez la truite fario (*Salmo trutta* L) et la truite arc-en-ciel (*Salmo gairdneri* Rich). Ann Hydrobiol 7 (2): 105–114

Bell MA (1976) Evolution of phenotypic diversity in *Gasterosteus aculeatus* superspecies on the Pacific coast of North America. System Zool 25 (2): 211–227

Bellamy AW (1923) Sex-linked inheritance in the teleost *Platypoecilus maculatus.* Anat Rec 24: 419–420

Bellamy AW (1928) Bionomic studies on certain teleosts (Poeciliinae). II. Color pattern inheritance and sex in *Platypoecilus maculatus.* Genetics 13 (3): 226–232

Bellamy AW (1933a) Bionomic studies on certain teleosts (Poeciliinae). III. Hereditary behavior of the color character gold. Genetics 18 (6): 522–530

Bellamy AW (1933b) Bionomic studies on certain teleosts (Poeciliinae). IV. Crossing over and non-disjunction in *Platypoecilus maculatus* Günth. Genetics 18 (6): 531–534

Bellamy AW (1936) Interspecific hybrids in *Platypoecilus:* one species ZZ-WZ, the other XY-XX. Proc Natl Acad Sci USA 22 (9): 531–536

Bender K, Ohno S (1968) Duplication of the autosomally inherited 6-phosphogluconate dehydrogenase gene locus in tetraploid species of cyprinid fish. Biochem Genet 2 (1): 101–107

* Berberovič Lj, Sofradžija A (1972) Pregled podataka o kromosomskim garniturama slatkovodnih riba Jugoslavije. Ichthyologia (Beograd) 4 (1): 1–21

* Berberovič Lj, Curič M, Hadžiselimovič R, Sofradžija A (1970) Hromosomka garnitura neretvanske mekausne *Salmothymus obtusirostris oxyrhynchus* (Steindachner). Acta Biol Jugosl F 2 (1): 55–63

* Berberovič Lj, Hadžiselimovič R, Pavlovič B (1973) Chromosome set of the species *Aulopige hügeli* Heckel. Bull Sci 18 (1–3): 10–11

* Berberovič Lj, Sofradžija A, Obradovič S (1975) Chromosome complement of *Ictalurus nebulosus* (Le Sueur), *Ictaluridae, Pisces.* Bull Sci Cons Acad Sci Arta RSFY A20 (5–6): 149–150

Berg O, Gordon M (1953) Relationship of atypical pigment cell growth to gonadal development in hybrid fishes. In: Pigment cell growth. Academic Press, London New York, pp 43–72

Berndt W (1924) Vererbungsstudien an Goldfischrassen. Z Indukt Abstammungs-Vererbungsl 36 (3–4): 161–349

Bernstein F (1925) Zusammenfassende Betrachtungen über die erblichen Blutstrukturen des Menschen. Z Intukt Abstammungs-Vererbungsl 37 (1): 237–270

Bernstein SC, Throckmorton LH, Hubby JL (1973) Still more genetic variability in natural populations. Proc Natl Acad Sci USA 70 (12): 3928–3931

* Bertollo LAC, Takahashi CS, Filho OM (1978) Cytotaxonomic considerations on *Hoplias lacerdae* (Pisces, Erythrinidae). Rev Bras Genet 1 (2): 103–120

* Bertollo LAC, Takahashi CS, Filho OM (1979) Karyotypic studies of two allopatric populations of the genus *Hoplias* (Pisces, Erythrinidae). Rev Bras Genet 2 (1): 17–37

Beukema JJ (1969) Angling experiments with carp (*Cyprinus carpio* L.). I. Differences between wild, domesticated and hybrid strains. Neth J Zool 19 (4): 596–609

Blacher LYa (1927) Materials for the genetics of *Lebistes reticulatus* Peters. Tr Lab Exp Biol Moscow Zoo P 3: 139–152 (R)

Blacher LYa (1928) Materials for the genetics of *Lebistes reticulatus* Peters. Tr Lab Exp Biol Moscow Zoo P 4: 245–253 (R)

Black DA, Howell WM (1979) The North American mosquitofish, *Gambusia affinis,* a unique case in sex chromosome evolution. Copeia 3: 509–513

Blake B (1976) Polymorphic forms of eye lens protein in the ray, *Raja clavata* (L). Comp Biochem Phys 54B (4): 441–442

Blanc JM, Chevassus B, Bergot P (1979) Déterminisme génétique du nombre de caeca pyloriques chez la truite fario (*Salmo trutta* L.) et la truite arc-en-ciel (*S. gairdneri* Rich). III. Effect du génotype et de la taille des oeufs sur la réalisation du caractère chez la truite fario. Ann Genet Select Anim 11 (1): 93–103

Bobrova JuP (1978) Organization and basic results of breeding work with common carp in Para fish farm. In: Genetics and selection of fishes. All-Union Res Inst Pond Fish, Moscow, pp 99–111 (RE)

Boffa GA, Fine JM, Drilhon A, Amouch T (1967) Immunoglobulins and transferrin in marine lamprey sera. Nature (London) 214 (5089): 700–702

Bogdanov LV, Flusova GD, Bylim LA, Shelobod LM (1979) A genetic and population study of the Pacific herring, *Clupea harengus pallasi.* In: Biochem Popul Genet Fish. Inst Cytol Acad Sci USSR, Leningrad, pp 74–82 (RE)

Bogyo TP, Becker WA (1965) Estimates of heritability from transformed percentage sib data with unequal subclass numbers. Biometrics 21: 1001–1007

Bolaffi JL, Booke HE (1974) Temperature effects on lactate dehydrogenase isozyme distribution in sceletal muscle of *Fundulus heteroclitus.* Comp Biochem Phys 48B (4): 557–564

* Booke HE (1968) Cytotaxonomic studies of the coregonine fishes of the Great Lakes, USA: DNA and karyotype analysis. J Fish Res B Can 25 (8): 1667–1687

* Booke HE (1974) A cytotaxonomic study of the round white fishes, genus *Prosopium.* Copeia 1: 115–119

* Booke HE (1975) Cytotaxonomy of the salmonid fish *Stenodus leucichthys.* J Fish Res B Can 32 (2): 291–296

* Boothroyd ER (1959) Chromosome studies on three Canadian populations of Atlantic salmon, *Salmo salar* L. Can J Genet Cytol 1 (2): 161–172

Borkin LJa, Darevsky IS (1980) Net (hybridogenic) speciation in vertebrates. Zh Obtsch Biol 41 (4): 485–505 (RE)

Borowsky R (1973) Melanomas in *Xiphophorus variatus* (Pisces, Poeciliidae) in the absence of hybridization. Experientia 29 (11): 1431–1433

Borowsky R, Kallman KD (1976) Patterns of mating in natural populations of *Xiphophorus* (Pisces, Poeciliidae). I. *Xiphophorus maculatus* from Belize and Mexico. Evolution 30 (4): 693–706

Borowsky R, Khouri J (1976) Patterns of mating in natural populations of *Xiphophorus*. II. *X. variatus* from Tamanlipas, Mexico, Copeia 4:727–734

Bouck GR, Ball RC (1968) Comparative electrophoretic patterns of lactate dehydrogenase in three species of trout. J Fish Res B Can 25 (7): 1323–1331

Bowler B (1975) Factors influencing genetic control in lakeward migrations of cut-throat trout fry. Trans Am Fish Soc 104 (3):474–482

Brander K (1979) The relationship between vertebral number and water temperature in cod. J Conseil 38 (3):286–292

Brannon EL (1967) Genetic control of migrating behavior of salmon. Prog Rep Int Pacific Salmon Fish Comm 16: 1–31

Braunitzer G (1966) Phylogenetic variation in the primary structure of hemoglobin. J Cell Phys 67 (Suppl 1): 1–19

Breider H (1935) Geschlechtsbestimmung und Differenzierung bei *Limia nigrofasciata, caudofasciata, vittata* und deren Artbastarden. Z Indukt Abstammungs-Vererbungsl 68 (2):265–299

Breider H (1936) Eine Allelenserie von Genen verschiedener Arten. Z Indukt Abstammungs-Vererbungsl 72 (1): 80–87

Breider H (1938) Die genetischen, histologischen und cytologischen Grundlagen der Geschwülstbildung nach Kreuzung verschiedener Rassen und Arten lebendgebärenden Zahnkarpfen. Z Zellforsch Microsk Anat 28 (5): 784–828

Breider H (1942) ZW-Männchen und WW-Weibchen bei *Platypoecilus maculatus*. Biol Zentralbl 62 (1): 187–195

Breider H (1956) Farbgene und Melanosarkomhäufigkeit (Ein Beitrag zur Physiologie der Color-Serie der *Poeciliinae)*. Zool Anz 156 (5–6): 129–140

Brett BLH, Turner BJ, Miller RR (1980) Allozymic divergences among the shortfin mollies of the *Poecilia sphenops* species complex. Isoz Bull 13: 104

Brewer CJ, Sing CP (1970) An introduction to isozyme techniques. Academic Press, London New York

Bridges RA, Freier EF (1966) Genetic and conformational heterogeneity of lactate dehydrogenase enzymes. Tex Rep Biol Med 24 (Suppl): 375–385

Bridges WR, Limbach B von (1972) Inheritance of albinism in rainbow trout. J Hered 63 (3): 152–153

Brody T, Moav R, Abramson ZV, Hulata G, Wohlfarth G (1976) Application of electrophoretic genetic markers to fish breeding. II. Genetic variation within maternal halfsibs in carp. Aquaculture 9 (4): 351–366

Brody T, Kirsht D, Parag G, Wohlfarth G, Hulata G, Moav R (1979) Biochemical genetic comparison of the Chinese and European races of the common carp. Anim Blood Gr Biochem Genet 10 (3): 141–149

Brouzhinskas IV (1979) Prospective plan of breeding work with common carp in Lithuania. In: Selection of pond fishes. Kolos, Moscow, pp 106–111 (RE)

Bruhn DS, Bowen JT (1973) Selection of rainbow trout broodstock, 1968. Prog Fish Cult 35 (2): 119

Buhler DR, Shanks WE (1959) Multiple hemoglobins in fishes. Science 129 (3353): 899–900

Bulger AJ, Schultz RJ (1979) Heterosis and interclonal variation in thermal tolerance in unisexual fishes. Evolution 33 (3): 848–859

Burmakin EV (1956) On morphological changes in wild common carp, acclimatized in Lake Balkhash basin. Zool Zh 35 (12): 1887–1895 (RE)

Burtzev IA (1971) Goals and methods of breeding and selection of the hybrids between white sturgeon (beluga) and sterlet. In: Actual problems of sturgeon fishculture. Astrakhan, pp 11–17 (R)

Burtzev IA, Serebryakova EV (1980) Progeny testing of "besters" (hybrids between the white sturgeon, *Huho huho* and the sterlet, *Acipenser ruthenus*) by cytological indices and viability of embryos. In: Karyological variability, mutagenesis and gynogenesis in fishes. Inst Cytol Acad Sci USSR, Leningrad, 63–69 (RE)

Busack CA, Halliburton R, Gall GAE (1979) Electrophoretic variation and differentiation in four strains of domesticated rainbow trout (*Salmo gairdneri*). Can J Genet Cytol 21 (1): 81–94

Buschkiel AL (1933) Teichwirtschaftliche Erfahrungen mit Karpfen in den Tropen. Z Fischer Hilfswissensch 31 (4):619–644
Buschkiel AL (1938) Grenzen der Vererblichkeit von Karpfeneigenschaften. Z Fischer Hilfswissensch 36 (1):1–22
Bushuev VP (1973) The presence of two-component haemoglobins in salmonids as a consequence of their tetraploid origin. In: Biochem Genet Fish. Inst Cytol Acad Sci USSR, Leningrad, pp 62–66 (RE)
Bushuev VP, Omelchenko VT, Salmenkova EA (1975) Species-specificity and intraspecific constancy of electrophoretic peculiarities and thermostability of hemoglobins in several fishes from order Clupeiformes. Zh Obtsch Biol 36 (3):569–578 (RE)
Bushuev VP, Shitikova OYu, Bogdanov LV (1980) Biochemical differentiation of Pacific redfishes of the genus *Tribolodon* (Cyprinidae) from the Kievka River. Vopr Ikhthiol 20 [3(122)]:445–451 (R)
Buth DG (1977a) Alcohol dehydrogenase variability in *Hypentelium nigricans*. Biochem Syst Ecol 5 (1):61–63
Buth DG (1977b) Biochemical identification of *Moxostoma rhothoecum* and *M. hamiltoni*. Biochem Syst Ecol 5 (1):57–60
Buth DG (1979a) Creatine kinase variability in *Moxostoma macrolepidotum* (Cypriniformes, Catostomidae). Copeia 1:152–154
Buth DG (1979b) Genetic relationships among the torrent suckers, genus *Thoburnia*. Biochem Syst Ecol 7 (4):311–316
Buth DG (1979c) Biochemical systematics of the cyprinid genus *Notropis*. I. The subgenus *Luxilus*. Biochem Syst Ecol 7 (1):69–79
Buth DG (1979d) Duplicate gene expression in tetraploid fishes of the tribe *Moxostomatini* (Cypriniformes, Catostomidae), Comp Biochem Phys 63 B (1):7–12
Buth DG, Burr BM (1978) Isozyme variability in the cyprinid genus *Campostoma*. Copeia 2:298–311
Butler L (1968) The potential use of selective breeding in the face of changing environment. In: A symposium on introduction of exotic species, Montreal, pp 54–72
Calaprice JR, Cushing JE (1967) A serological analysis of three populations of golden trout, *Salmo aguabonita* Jordan. Calif Fish Game 53 (4):273–281
Callegarini C (1966) Le emoglobine di alcune popolazioni di *Ictaluridae* (Teleostei) dell'Italia septentrionale. Rendiconti 100 (1):31–35
Callegarini C, Cucchi C (1968) Intraspecific polymorphism of hemoglobin in *Tinca tinca*. Biochem Biophys Acta 160 (2):264–266
Callegarini C, Cucchi C (1969) Polimorphismo intraspecifico delle emoglobino di *Cottus gobio* (Teleostei, Cottidae). Rendiconti 103 (11):269–275
* Calton MS, Denton TE (1974) Chromosomes of the chocolate gourami: a cytogenetic anomaly. Science 185 (4151):618–619
* Campos HH (1972) Karyology of three galaxiid fishes, *Galaxias maculatus, G. platei* and *Brachigalaxias vullocki*. Copeia 2:368–370
* Campos HH, Hubbs Cl (1971) Cytomorphology of six species of gambusiine fishes. Copeia 3:566–569
* Campos HH, Hubbs Cl (1973) Taxonomic implications of the karyotype of *Opsopoedus emiliae*. Copeia 1:161–163
* Capanna E, Cataudella S (1973) The chromosomes of *Calamoichthys calabaricus* (Pisces, Polypteriformes). Experientia 29 (4):491–492
Carlin B (1969) Salmon tagging experiments. In: Laxforskingsinstitutet meddelande, Stockholm, pp 2–4
* Cataudella S, Capanna E (1973) Chromosome complement of three species of *Mugilidae* (Pisces, Perciformes). Experientia 29 (4):489–490
* Cataudella S, Civitelli MV (1975) Cytotaxonomical consideration of the genus *Blennius* (Pisces Perciformes). Experientia 31 (2):167–169
* Cataudella S, Civitelli MV, Capanna E (1973) The chromosomes of some Mediterranean teleosts: Scorpaenidae, Serranidae, Labridae, Blenniidae, Gobiidae (Pisces; Scorpaeniformes, Perciformes). Bol Zool 40 (3–4):385–389
* Cataudella S, Civitelli MV, Capanna E (1974) Chromosome complements of the mediterranean mullets (Pisces, Perciformes). Caryologia 27 (1):93–105

* Cataudella S, Sola L, Accame Muratori R, Capanna E (1977) The chromosomes of 11 species of Cyprinidae and one Cobitidae from Italy, with some remarks on the problem of polyploidy in the Cypriniformes. Genetica (Hague) 43 (3): 161–171

* Cataudella S, Sola L, Capanna E (1978) Remarks on the karyotype of the Polypteriformes. The chromosomes of *Polypterus delhezi, P. endlicheri congicus* and *P. palmas*. Experientia 34 (8): 999–1000

Cederbaum SD, Yoshida A (1972) Tetrasolium oxidase polymorphism in rainbow trout. Genetics 72 (2): 363–367

Cederbaum SD, Yoshida A (1976) Glucose 6-phosphate dehydrogenase in rainbow trout. Biochem Genet 14 (3–4): 245–258

Champion MJ, Whitt GS (1976 a) Differential gene expression in multilocus isozyme system of the developing green sunfish. J Exp Zool 196 (3): 263–282

Champion MJ, Whitt GS (1976 b) Synchronous allelic expression at the glucosephosphate isomerase A and B loci in interspecific sunfish hybrids. Biochem Genet 14 (9–10): 723–738

Champion MJ, Shaklee JB, Whitt GS (1975) Developmental genetics of teleost isozymes. In: Isozymes, vol III. Academic Press, London New York, pp 417–437

Chan May Tchien (1969) Variability of some physiological traits in common carps of different genotype. In: Genetics, selection and hybridization of fishes. Nauka, Moscow, pp 117–123 (R)

Chan May Tchien (1971) Experience with the determination of realized weight heritability in the tilapia (*Tilapia mossambica* Peters). Genetika (Moscow) 7 (1): 53–59 (RE)

Chaudhuri H (1971) Fish hybridization in Asia with special reference to India. Rep FAO/UNDP(TA), (2926) Rome, pp 151–159

Chavia W, Gordon M (1951) Sex determination in *Platypoecilus maculatus*. I. Differentiation of the gonads in members of all male broods. Zoologica 36 (2): 135–146

Chen SC (1928) Transparency and mottling, a case of mendelian inheritance in the goldfish, *Carassius auratus*. Genetics 13 (2): 434–452

Chen SC (1934) The inheritance of the blue and brown colours in the goldfish, *Carassius auratus*. J. Genet 29 (1): 61–74

Chen SC (1956) A history of the domestication and the factors of the varietal formation of the common goldfish, *Carassius auratus*. Sci Sinica 5 (2): 287–321

* Chen TR (1967) Comparative karyology of selected deep-sea and shallow-water teleost fishes. Ph D Thesis, Yale Univ, New Haven

* Chen TR (1969) Karyological heterogamety of deep-sea fishes. Postilla 130 (1): 1–29

* Chen TR (1971) A comparative chromosome study of twenty killifish species of the genus *Fundulus* (Teleostei: Cyprinodontidae). Chromosoma 32 (4): 436–453

* Chen TR, Ebeling AW (1968) Karyological evidence of female heterogamety in the mosquitofish, *Cambusia affinis*. Copeia 1: 70–75

* Chen TR, Ebeling AW (1971) Chromosomes of the goby fishes in the genus *Gillichthys*. Copeia 1: 171–174

* Chen TR, Ebeling AW (1974) Cytotaxonomy of Californian myctophoid fishes. Copeia 4: 839–848

* Chen TR, Reisman HM (1970) A comparative study of the North American species of stickleback (Teleostei: Gasterosteidae). Cytogenetics 9 (5): 321–332

* Chen TR, Ruddle FH (1970) A chromosome study of four species and a hybrid of the killifish genus *Fundulus* (Cyprinodontidae). Chromosoma 29 (3): 255–267

Cherfas NB (1966 a) Analysis of meiosis in unisexual and bisexual forms of the silver crucian carp. Tr VNIIPRKh (Moscow) 14: 63–82 (R)

Cherfas NB (1966 b) Natural triploidy in the females of the unisexual variety of the silver crucian carp (*C. auratus gibelio* Bloch). Genetika (Moscow) 2 (5): 16–24 (RE)

Cherfas NB (1975) Studies on diploid gynogenesis in the carp. I. Experiments on the mass production of the diploid gynogenetic offspring. Genetika (Moscow) 11 (7): 78–86 (RE)

Cherfas NB (1977) Studies of the diploid radiation-induced gynogenesis in the carp. II. Segregation with respect to certain morphological traits in gynogenetic offsprings. Genetika (Moscow) 13 (5): 811–820 (RE)

Cherfas NB (1978) Induced gynogenesis in carp and basic trends of its utilization in selective and genetic works. In: Genetics and selection of fish, Tz VNIIPRKh, 20, Moscow, pp 149–173 (RE)

Cherfas NB, Ilyasova VA (1980a) Some results of investigations on diploid radiation gynogenesis in common carp. In: Karyological variability, mutagenesis and gynogenesis in fishes. Inst Cytol Acad Sci USSR, Leningrad, pp 74–81 (RE)

Cherfas NB, Ilyasova VA (1980b) Induced gynogenesis in silver crucian carp and carp hybrids. Genetika (Moscow) 16 (7): 1260–1269 (RE)

Cherfas NB, Shart LA (1970) On the existence of triploidy in Moldavian populations of the silver crucian carp (short communication). Collect Sci Pap VNIIPRKh 5: 276–283 (R)

Cherfas NB, Truveller KA (1978) Studies on radiation induced diploid gynogenesis in the common carp. III. The analysis of gynogenetic offsprings using biochemical markers. Genetika (Moscow) 14 (4): 599–604 (RE)

* Chernenko EV (1968) The karyotypes of the dwarf and anadromous sockeye salmon (*Oncorhynchus nerka* [Walb.]) from the Lake Dalneye (Kamchatka). Vopr Ikhthiol 8 [5(52)]: 834–846 (R)

* Chernenko EV (1971) The chromosomal complement of the sockeye salmon. Sci Rep Inst Mar Biol (Vladivostok), 2: 228–231 (R)

Chernenko EV (1976a) The differentiation of population of the sockeye salmon *Oncorhynchus nerka kennerlyi* in the Lake Kronotskoye. In: Salmonoid fishes. Zool Inst, Leningrad, p 119 (R)

Chernenko EV (1976b) Genome mutations in embryos of anadromous and local sockeye salmon *Oncorhynchus nerka* from the Dalneye Lake (Kamchatka). Vopr Ikhthiol 16 [3(98)]: 416–423 (R)

* Chernenko EV (1977) Karyotype variability in the sockeye salmon (*Oncorhynchus nerka*) from the Lake Dalneje. Ph D Thesis, Moscow, Univ (R)

* Chernenko EV (1980) Chromosomal polymorphism in sockeye salmon (*Oncorhynchus nerka* Walb). In: Karyological variability, mutagenesis and gynogenesis of fishes. Inst Cytol Acad Sci USSR, Leningrad, 24–28 (RE)

* Chernenko EV, Viktorovsky RM (1971) Chromosomal complements of the cherry salmon, the East Siberian char and the dolly warden char. Sci Rep Inst of Mar Biol (Vladivostok) 2: 232–235 (R)

Chernyshev VA (1980) On the results of DNA molecular hybridization in Baical cottids and comephorids. Dokl Akad Nauk SSSR (DAN) 252 (4): 1012–1014 (R)

Chervinski J (1967) Polymorphic characters of *Tilapia zillii* (Gerv.). Hydrobiologia 30 (1): 138–144

Chevassus B (1976) Variabilité et héritabilité des performances de croissance chez la truite arc-en-ciel (*Salmo gairdnerii* Rich). Ann Genet Select Anim 8 (2): 273–283

Chevassus B (1979) Hybridization in salmonids: results and perspectives. Aquaculture 17 (2): 113–128

Chevassus B, Blanc JM, Bergot P (1979) Genetic analysis of the number of pyloric caeca in brown trout (*Salmo trutta* L.) and rainbow trout (*Salmo gairdneri* Rich). II. Effet du genotype du milieu l'élevage et de l'alimentation sur la réalization du caractère chez la truit arc-en-ciel. Ann Genet Select Anim II (1): 79–92

* Chiarelly AB, Capanna E (1973) Checklist of fish chromosomes. In: Cytotaxonomy and vertebrate evolution. Academic Press, London New York, pp 205–252

* Chiarelly B, Ferrantelli O, Cucchi C (1969) The karyotype of some teleostean fish obtained by tissue culture in vitro. Experientia 25 (4): 426–427

Chikhachev AS, Tsvetnenko JuB (1979) Polymorphism of albumin and transferrin in Azov population of the stellate sturgeon *Acipenser stellatus*. In: Biochem Popul Genet Fish. Inst Cytol Acad Sci USSR, Leningrad, pp 111–115 (RE)

Child AR, Solomon DJ (1977) Observations on morphological and biochemical features of some cyprinid hybrids. J Fish Biol II (2): 125–131

Child AR, Burnell AM, Wilkins NP (1976) The existence of two races of Atlantic salmon in the British Isles. J Fish Biol 8 (1): 35–43

Childers WF (1971) Hybridization of fishes in North America (family Centrarchidae). In: Rep. FAO/UNDP (TA), (2926), Rome, pp 133–142

Childers WF, Whitt GS (1976) The clinal distribution and environmental selection of two MDH "B" alleles in *Micropterus salmoides*. Isoz Bull 9: 54

* Choudhuri RC, Prasad R, Das CC (1979) Chromosomes of six species of marine fishes. Caryologia 32 (1): 15–21

Christiansen FB (1977) Population genetics of *Zoarces viviparus* (L), a review. In: Measuring selection in natural populations. Springer, Berlin Heidelberg New York, pp 21–47

Christiansen FB (1980) Studies on selection components in natural populations using population samples of mother-offspring combinations. Hereditas 92 (2): 199–203

Christiansen FB, Frydenberg O (1974) Geographical patterns of four polymorphisms in *Zoarces viviparus* as evidence of selection. Genetics 77 (3): 765–770

Christiansen FB, Frydenberg O (1976) Selection component analysis of natural polymorphisms using mother-offspring samples of successive cohorts. In: Population genetics and ecology. Academic Press, London New York, pp 277–301

Christiansen FB, Simonsen V (1978) Geographic variation in protein polymorphism in the eelpout, *Zoarces viviparus* (L). In: Marine organisms, Plenum Publ Corp, New York, pp 171–194

Christiansen FB, Frydenberg O, Simonsen V (1973) Genetics of *Zoarces* populations. IV. Selection component analysis of an esterase polymorphism using population samples, including mother-offspring combinations. Hereditas 73 (2): 291–304

Christiansen FB, Frydenberg O, Cyldenholm AO, Simonsen V (1974) Genetics of *Zoarces* populations. VI. Further evidence based on age group samples, of a heterozygote deficit in the Est III polymorphism. Hereditas 77 (2): 225–236

Christiansen FB, Frydenberg O, Hjorth JP, Simonsen V (1976) Genetics of *Zoarces* populations. IX. Geographical variation at the three phosphoglucomutase loci. Hereditas 83 (2): 245–256

Christiansen FB, Frydenberg O, Simonsen V (1977) Genetics of *Zoarces* populations. X. Selection component analysis of Est III polymorphism using samples of successive cohort. Hereditas 87 (2): 129–150

Chutaeva AI, Dombrovsky VK, Filinovich EI (1975a) Fishfarming characteristic of Buelorussian carp. In: Materials of All-Union conference on organization of selective breeding work in fish-farms of USSR. All-Union Res Inst Pond Fish, Moscow, pp 44–56 (R)

Chutaeva AI, Dombrovsky VK, Guzyuk CN, Lazovsky AA (1975b) Polymorphism of the Isobelino common carp breed with respect to certain protein systems. In: The basis of bioproductivity of inland waterbodies of the Baltic Coast. Inst Zool Parasitol Littmanian Acad Sci, Vilnius, pp 311–313 (R)

Ciechomski de J, Weiss de Vigo G (1971) The influence of the temperature on the number of vertebrae in the Argentine anchovy, *Engraulis anchoita.* J Conseil 34 (1): 37–42

Cimino MC (1972a) Egg-production, polyploidization and evolution in a diploid all-female fish of the genus *Poeciliopsis.* Evolution 26 (2): 294–306

Cimino MC (1972b) Meiosis in triploid all-female fish (*Poeciliopsis,* Poeciliidae). Science 175 (4029): 1484–1485

* Cimino MC (1973) Karyotypes and erythrocyte sizes of some diploid and triploid fishes of the genus *Poeciliopsis.* J Fish Res B Can 30 (11): 1736–1737

Cimino MC (1974) The nuclear DNA content of diploid and triploid *Poeciliopsis* and other poeciliid fishes with reference to the evolution of unisexual forms. Chromosoma 47 (3): 297–307

Cimino MC, Schultz RJ (1970) Production of a diploid male offspring by a gynogenetic triploid fish of the genus *Poeciliopsis.* Copeia 4: 760–763

Clark E (1959) Functional hermaphroditism and self-fertilization in a serranid fish. Science 129 (3343): 215–216

Clark E, Aronson LR, Gordon M (1954) Mating behavior patterns in two sympatric species of xiphophorin fishes: their inheritance and significance in sexual isolation. Bull Am Mus Nat Hist 103 (2): 139–225

Clark FH (1970) Pleiotropic effects of the gene for golden color in rainbow trout. J Hered 61 (1): 8–10

Clayton JW, Franzin WG (1970) Genetics of multiple lactate dehydrogenase isozymes in muscle tissue of lake whitefish. J Fish Res B Can 27 (6): 1115–1121

Clayton JW, Gee JH (1969) Lactate dehydrogenase isozymes in longnose and blacknose dace (*Rhinichthys cataractae* a. *Rh. atratulus*) and their hybrid. J Fish Res B Can 26 (11): 3049–3053

Clayton JW, Tretiak DN, Kooyman AH (1971) Genetics of multiple malate dehydrogenase isozymes in skeletal muscle of walleye (*Stizostedion vitreum vitreum*). J Fish Res B Can 28 (4): 1005–1008

Clayton JW, Franzin WG, Tretiak DN (1973a) Genetics of glycerol-3-phosphate dehydrogenase isozymes in white muscle of lake whitefish (*Coregonus clupeaformis*). J Fish Res B Can 30 (2): 187–193

Clayton JW, Harris REK, Tretiak DN (1973b) Identification of supernatant and mitochondrial isozymes of malate dehydrogenase on electrophoregrams applied to the taxonomic discrimination of walleye (*Stizostedion vitreum vitreum*), sauger (*S. canadense*), and suspected interspecific hybrid fishes. J Fish Res B Can 30 (7): 927–938

Clayton JM, Tretiak DN, Billeck BN, Ihssen P (1975) Genetics of multiple supernatant and mitochondrial malate dehydrogenase isozymes in rainbow trout (*Salmo gairdneri*). In: Isozymes, vol IV. Academic Press, London New York, pp 433–448

Coad BW, Power G (1974) Meristic variation in the threespine stickleback, *Gasterosteus aculeatus*, in the Matamek river sistem, Quebec. J Fish Res B Can 31 (6): 1155–1157

Comparini A, Rizzotti M, Nardella M, Rodino E (1975) Ricerche elettroforetiche sulla variabilita genetica di *Anguilla anguilla*. Bol Zool 42 (2–3): 283–288

Comparini A, Rizzotti M, Rodino E (1977) Genetic control and variability of phosphoglucose isomerase (PGI) in eels from the Atlantic ocean and Mediterranean sea. Mar Biol 43 (2): 109–116

Constantinescu GK (1928) Kreuzungsversuche mit *Rivulus urophthalmus*. Z Indukt Abstammungs-Vererbungsl 47 (2): 341

* Cook PC (1978) Karyotypic analysis of the gobiid fish genus *Quietula* Jordan and Evermann. J Fish Biol 12 (2): 173–179

Cooper EL (1961) Growth of wild and hatchery strains of brook trout (*Salvelinus fontinalis*). Trans Am Fish Soc 90 (4): 424–438

Corbel MJ (1975) The immune response in fish: a review. J Fish Biol 7 (5): 539–563

Cordon AJ, Nicola SJ (1970) Harvest of four strains of rainbow trout, *Salmo gairdnerii*, from Beardsley reservoir, California. Calif Fish Game 56 (4): 271–287

Coyne JA, Felton AA, Lewontin RC (1978) Extent of genetic variation at a highly polymorphic esterase locus in *Drosophila pseudoobscura*. Proc. Natl Acad Sci USA 75 (10): 5090–5093

Cramer SP, McIntyre JD (1975) Heritable resistance to gas bubble disease in fall chinook salmon. US Natl Mar Fish Serv Fish Bull 73 (4): 934–938

Creyssel R, Richard GB, Silberzahn P (1966) Transferrin variants in carp serum. Nature (London) 212 (5068): 1362

Cross TF (1979) Isozymes of interspecific hybrids of the fish family Cyprinidae. Proc R Ir Acad 78B (22): 323–330

Cross TF, O'Rourke FJ (1978) An electrophoretic study of the haemoglobins of some hybrid fishes. Proc R Ir Acad 78B (11): 171–178

Cross TF, Payne RH (1971) NADP-isocitrate dehydrogenase polymorphism in the Atlantic salmon, *Salmo salar*. Fish Biol II (5): 493–496

Cross TE, Payne RH (1978) Geographic variation in Atlantic cod *Gadus morhua*, of Eastern-North America: a biochemical systematics approach. J Fish Res B Can 35 (1): 117–123

Cross TF, Ward RD, Abreu-Grobois A (1979) Duplicate loci and allelic variation for mitochondrial malic enzyme in the Atlantic salmon, *Salmo salar* L. Comp Biochem Phys 62B (4): 403–406

Crow JF (1952) Dominance and overdominance. In: Heterosis. Jova State College Press, Ames, pp 282–297

* Cucchi C (1977) Il cariotipo di *Tinca tinca* (prolabile eterogametia femminile). Rendiconti 117B: 101–106

Cucchi C, Callegarini C (1969) Analisi elettroforetica delle emoglobini e delle globine di due distinte popolazioni di *Gobius fluviatilis* (Teleostei, Gobiidae). Rendiconti 103 (11): 276

* Cucchi C, Mariotti A (1976) Il cariotipo dei centrarchidi (Teleostei). Ferrara Ann Univ Ser 13 21: 233–238

Cuellar O, Uyeno T (1972) Triploidy in rainbow trout. Cytogenetics 11 (6): 508–515

Cushing JE (1956) Observations on the serology of tuna. Spec Sci Rep US Fish Wildl Serv 183: 1–14

Dando PH (1971) Lactate dehydrogenase polymorphism in the flatfish (*Heterosomata*). Rapp P-V Reun 161:133
Dannevig A (1932) Is the number of vertebrae in the cod influenced by light or high temperature during early stages. J Conseil 7:60–62
Dannevig A (1950) The influence of the environment on number of vertebrae in plaice. Rep Norw Fish Invest 9:1–6
* Danzmann RG (1979) The karyology of eight species of fish belonging to the family Percidae. Can J Zool 57 (10):2055–2060
Darnell RM, Abramoff P (1968) Distribution of the gynogenetic fish, *Poecilia formosa,* with remarks on the evolution of the species. Copeia 2:354–361
Darnell RM, Lamb E, Abramoff P (1967) Matroclinous inheritance and clonal structure of a Mexican population of gynogenetic fish, *Poecilia formosa.* Evolution 21 (1):168–173
Davis HS (1931) The influence of heredity on the spawning season of trout. Trans Am Fish Soc 61 (1):43
* Davisson MT (1972) Karyotypes of the teleost family Esocidae. J Fish Res B Can 29 (5):579–582
Davisson MT, Wright JE, Atherton LM (1972) Centric fusion and trisomy for the LDH-B locus in brook trout, *Salvelinus fontinalis.* Science 178 (4064):992–993
Davisson MT, Wright JE, Atherton LM (1973) Cytogenetic analysis of pseudolinkage of LDH loci in the teleost genus *Salvelinus.* Genetics 73 (4):645–658
Dawson DE (1964) A bibliography of anomalies of fishes. Gulf Res Rep I (6):308–399
Dayhoff MO (ed) (1976) Atlas of protein sequence and structure, vol 5, Suppl 3. Natl Biomed Res Found Wash DC
Dell'Agata M, Giovannini E, di Cola D, Pierdomenico S (1979) Studi preliminari sulla lattatodeidrogenasi in *Lampetra planeri* (Bloch). Riv Biol 72 (1–2):109–129
Dementyeva TF, Plechkova EK, Rozanova MI, Tanasiychuk VS (1932) The race composition of the Barents Sea cod. Dokl I[st] Session Okeanogr Inst 2:49–68 (R)
* Denton TE (1973) Fish chromosome methodology. Thomas, Springfield (Ill)
* Denton TE, Howell WM (1973) Chromosomes of the African polipterid fishes, *Polypterus palmas* and *Calamoichthys calabaricus.* Experientia 29 (I):122–124
Diebig E, Meyer J-N, Glodek P (1979) Biochemical polymorphisms in muscle and liver extracts and in the serum of the rainbow trout *Salmo gairdneri.* Anim Blood Gr Biochem Gen 10 (3):165–174
* Dimovska A (1959) Chromosome set of fishes from the population of ochrid salmon (*Salmo lethnica* Kar.). God Zb Priv Math Fak Univ Skopye 12 (1):117–135
* Dingerkus G, Howell WM (1976) Karyotypic analysis and evidence of tetraploidy in the North America paddlefish, *Polyodon spathula.* Science 194 (4267):842–843
Dobrovolov I (1972) The electrophoretic analysis of barbel (genera *Barbus* Cuvier) haemoglobins from Danube and Kamchia. Proc Peopl Mus Varna 8:294–297
Dobrovolov I (1976) Multiple forms of lactate dehydrogenase in anchovy (*Engraulis encrasicholus* L.) from the Black Sea, the Sea of Asov and the Atlantic Ocean. Dokl Bolgar Akad Nauk (Sofia) 29 (6):877–880
Dobrovolov I, Tschayn Kh (1978) Electrophoretic investigation of the muscle myogens in sprat (*Sprattus sprattus* L.) from the Black Sea, Baltic Sea and Atlantic Ocean. Proc Inst Fish Resour (Varna) 16:123–128
Dobrovolova Sv (1978) Polymorphism of the muscle proteins of *Clupeonella delicatula* (Nordmann). Proc Inst Fish Resour (Varna) 16:129–133
Dobzhansky Th (1952) Nature and origin of heterosis. In: Heterosis. Ames, Academic Press, London New York, pp 218–223
* Donahue WH (1974) A karyotypic study of three species of *Rajiformes* (Chondrichthyes, Pisces). Can J Genet. Cytol 16 (1):203–211
Donaldson LR (1969) Selective breeding of salmonoid fishes. In: Marine Aquaculture. Newport, Oregon State Univ Press, pp 65–74
Donaldson LR, Menasveta D (1961) Selective breeding of chinook salmon. Trans Am Fish Soc 90 (2):160–164
* Dorofeeva EA (1965) Karyological evidence clarifying the taxonomic relationship of Kaspian and Black Sea salmon (*Salmo trutta caspius* Kessl. and *S. trutta labrax* Pall.). Vopr Ikhthiol 5 (1):38–45 (R)

* Dorofeeva EA (1967) The chromosomal complements of the Sevan trout (*Salmo ischchan* Kessl.) in connection with karyosystematics of salmonids. Zool Zh 46 (2):248–253 (RE)

* Dorofeeva EA (1977) The use of karyological data for the solution of the problems of taxonomy and phylogeny of salmonoid fishes. In: The principles of the classification and phylogeny in the salmonoid fishes, Zool Inst, Leningrad, pp 86–95 (R)

Drilhon A, Fine IM (1971) Les groupes de transferrines dans le genre *Anguilla* L. Rapp Pr-V Reun 161:122–125

Dubinin NP, Romashov DD (1932) The genetic basis of the structure of species and its evolution. Biol Zh 1 (5–6):52–95 (R)

Dubinin NP, Romashov DD, Geptner MA, Demidova ZA (1937) The aberrational polymorphism in *Drosophila fasciata* Meig. Biol Zh 6 (2):311–354 (R)

Dubinin NP, Altukhov YuP, Salmenkova EA, Milashnikov AN, Novikova TA (1975) Analysis of monomorphic gene markers in populations as a method for mutagenic evaluation of the environment. Dokl Akad Nauk SSSR (DAN) 225 (3):693–696 (R)

Dufour D, Barrette D (1967) Polymorphisme des lipoproteines et des glycoproteins sériques chez la truite. Experientia 23 (5):955–966

Dutschenko VV (1979) Phenotype frequencies of fast esterases in rock grenadier (*Macrurus rupestris*) of Hatton plateau. In: Biochem Popul Genet Fish. Inst Cytol Acad Sci USSR, Leningrad, pp 54–57 (RE)

Dzwillo M (1959) Genetische Untersuchungen an domestizierten Stämmen von *Lebistes reticulatus* (Peters). Mitt Hamburg Zool Mus Inst 57:143–186

Dzwillo B (1962) Über künstliche Erzeugung funktioneller Männchen weiblichen Genotyps bei *Lebistes reticulatus.* Biol Zentralbl 81 (2):575–584

Dzwillo M (1966) Über den Einfluss von Methyltestosteron auf primäre und sekundäre Geschlechtsmerkmale während verschiedener Phasen der Embryonalentwicklung von *Lebistes reticulatus.* Verh Dtsch Zool Ges Jena 29:471–476

Dzwillo M, Zander CD (1967) Geschlechtsbestimmung bei Zahnkarpfen (Pisces). Mitt Hamburg Zool Mus Inst 64:147–162

Eastman JT, Underhill JC (1973) Intraspecific variation in the pharyngeal tooth formulae of some cyprinid fishes. Copeia 1:45–53

* Ebeling AW, Chen TR (1970) Heterogamety in teleostean fishes. Trans Am Fish Soc 99 (1):131–138

Ebeling AW, Setzer PY (1971) Cytological confirmation of female homogamety in the deep-sea fish *Bathylagus milleri.* Copeia 3:560–562

Ebeling AW, Atkin NB, Setzer PY (1971) Genome size of teleostean fishes: increases in some deep-sea species. Am Nat 105 (946):549–562

Eberhardt K (1941) Die Vererbung der Farben bei *Betta splendens* Regan. Z Indukt Abstammungs-Vererbungsl 79 (3):548–560

Eberhardt K (1943) Ein Fall von geschlechtskontrollierter Vererbung bei *Betta splendens* Regan. Z Indukt Abstammungs-Vererbungsl 81 (1):72–83

Echelle AA, Echelle A-F, Taber BA (1976) Biochemical evidence for congeneric competition as a factor restricting gene flow between populations of a darter (Percidae: Etheostoma). Syst Zool 25 (3):228–235

Eckroat LR (1971) Lens protein polymorphisms in hatchery and natural populations of brook trout, *Salvelinus fontinalis* (Mitch.). Trans Am Fish Soc 100 (3):527–536

Eckroat LR (1973) Allele frequency analysis of five soluble protein loci in brook trout, *Salvelinus fontinalis* (Mitch.). Trans Am Fish Soc 102 (2):335–340

Eckroat LR (1974) Interspecific comparisons of lens proteins of Esocidae. Copeia 4:977–978

Eckroat LR (1975) Heterozygosity at the lactate dehydrogenase A locus in grass pickerel, *Esox americanus vermiculatus.* Copeia 3:466–470

Eckroat LR, Wright JE (1969) Genetic analysis of soluble lens protein polymorphism in brook trout, *Salvelinus fontinalis.* Copeia 3:466–473

Edmunds PH, Sammons JI (1973) Similarity of genic polymorphism of tetrazolium oxidase in blue fin tuna (*Thunnus thynnus*) from the Atlantic coast of France and the western North Atlantic. J Fish Res B Can 30 (7):1031–1032

Edwards D, Gjedrem T (1979) Genetic variation in survival of brown trout eggs, fry and fingerlings in acidic water. SNSF-project, Norway FR 16/79, pp 1–28

Edwards DJ, Austreng E, Risa S, Gjedrem T (1977) Carbohydrate in rainbow trout diets. I. Growth of fish of different families fed diets containing different proportions of carbohydrate. Aquaculture 11:1, 31–38

Efroimson VP (1968) The introduction into medical genetics. Medicina, Moscow (R)

Efroimson VP (1971) The immunogenetics. Medicina, Moscow (R)

Egami N (1954) Geographical variation in the male characters of the fish, *Oryzias latipes*. Annot Zool Jp 27 (1):7–12

Egami N, Hyodo-Taguchi Y (1973) Dominant lethal mutation rates in the fish, *Oryzias latipes*, irradiated at various stages of gametogenesis. In: Genetics and mutagenesis of fish. Springer, Berlin Heidelberg New York, pp 75–81

Ege V (1942) A transplantation experiment with *Zoarces viviparus*. C R Trav Lab Carlsb Ser Physiol 23 (17):65–75

Ehlinger NF (1964) Selective breeding of trout for resistance to furunculosis. NY Fish Game J 11 (2):78–90

Ehlinger NF (1977) Selective breeding of trout for resistance to furunculosis. NY Fish Game J 24 (1):25–36

El-Ibiary HM, Joyce JA (1978) Heritability of body size traits, dressing weight and lipid content in channel catfish. J Anim Sci 47 1:82–88

El-Ibiary HM, Andrews JW, Joyce JA, Page JW (1976) Source of variations in body size traits, dress out weight, and lipid content and their correlations in channel catfish, *Ictalurus punctatus*. Trans Am Fish Soc 102 (2):267–272

Endler JA (1980) Natural selection on color patterns in *Poecilia reticulata*. Evolution 34 (1):76–91

Engel W, Op't Hof J, Wolf U (1970) Genduplikation durch polyploide Evolution: die Isoenzyme der Sorbitdehydrogenase bei herings- und lachsartigen Fischen (Isospondyli). Humangenetik 9 (2):157–163

Engel W, Faust J, Wolf U (1971a) Isoenzyme polymorphism of the sorbitol dehydrogenase and the NADP-dependent isocitrate dehydrogenases in the fish family Cyprinidae. Anim Blood Gr Biochem Genet 2 (1):127–133

Engel W, Schmidtke J, Wolf U (1971b) Genetic variation of α-glycerophosphate dehydrogenase isoenzymes in clupeoid and salmonoid fish. Experientia 27 (12):1489–1491

Engel W, Schmidtke J, Vogel W, Wolf U (1973) Genetic polymorphism of lactate dehydrogenase isoenzymes in the carp (*Cyprinus carpio* L.) apparently due to a "nullallele". Biochem Genet 8 (3):281–289

Engel W, Schmidtke J, Wolf U (1975) Diploid-tetraploid relationship in teleostean fishes. In: Isozymes, vol IV. Academic Press, London New York, pp 449–465

Engel W, Kuhl P, Schmidtke J (1977) Expression of the paternally derived phosphoglucose isomerase genes during hybrid trout development. Comp Biochem Phys 56B (2):103–108

Eppenberger HM, School A, Ursprung H (1971) Tissue-specific isoenzyme patterns of creatine kinase (2.7.3.2) in trout. FEBS Lett 14 (5):317–319

Ermin R (1954) On the ocular tumor with exophthalmia in an interspecific hybrid of *Anatolichthys*. Rev Fac Sci Univ Istanbul Ser B 19 (3):203–212

Fahy WE (1972) Influence of temperature change on number of vertebrae and caudal fin rays in *Fundulus majalis* (Wallbaum). J Conseil 34 (2):217–231

Fahy WE (1979) The influence of temperature change on number of anal fine rays developing in *Fundulus majalis* (Walbaum). J Conseil 38 (3):280–285

Falconer DS (1960) Introduction to quantitative genetics. Oliver and Boyd, Edinburgh London

Farr JA (1976) Social facilitation of male sexual behaviour, intrasexual competition and sexual selection in the guppy, *Poecilia reticulata*. Evolution 30 (4):707–717

Farris JS (1972) Estimating phylogenetic trees from distance matrices. Am Nat 106 (951):645–668

Felley JD, Avise JC (1980) Genetic and morphological variation of bluegill populations in Florida lakes. Trans Am Fish Soc 109 (I):108–115

Felley JD, Smith MH (1978) Phenotypic and genetic trends in bluegills of a single drainage. Copeia I:175–177

Ferguson A (1974) The genetic relationships of the coregonid fishes of Britain and Ireland indicated by electrophoretic analysis of tissue proteins. J Fish Biol 6 (3):311–315

Ferguson A (1975) Myoglobin polymorphism in the pollan (Osteichthyes; Coregoninae). Anim Blood Gr Biochem Genet 6 (I):25–29

Ferguson A (1980) Biochemical systematics and evolution. Blackie, Glasgow London

Ferguson A, Himberg K-JM, Svärdson G (1978) Systematics of the Irish pollan (*Coregonus pollan* Thompson): an electrophoretic comparison with other holarctic Coregoninae. J Fish Biol 12 (3):221–233

Ferguson DE, Ludke JL, Murphy GC (1966) Dynamics of endrin uptake and release by resistant and susceptible strains of mosquitofish. Trans Am Fish Soc 95 (4):335–344

Ferno A, Sjölander S (1973) Some imprinting experiments on sexual preferences for colour variants in the platyfish (*Xiphophorus maculatus*). Z Tierpsychol 33 (3–4):417–423

Ferris SD, Whitt GS (1977a) Loss of duplicate gene expression after polyploidization. Nature (London) 265 (5591):258–260

Ferris SD, Whitt G (1977b) Duplicate gene expression in diploid and tetraploid loaches (Cypriniformes, Cobitidae). Biochem Genet 15 (11–12):1097–1112

Ferris SD, Whitt GS (1977c) The evolution of duplicate gene expression in the carp (*Cyprinus carpio*). Experientia 33 (10):1299–1301

Ferris SD, Whitt GS (1978a) Genetic and molecular analysis of nonrandom dimer assembly of the creatine kinase isozymes of fishes. Biochem Genet 16 (7–8):811–829

Ferris SD, Whitt GS (1978b) Phylogeny of tetraploid catostomid fishes based on the loss of duplicate gene expression. Syst Zool 27 (2):189–206

Ferris SD, Whitt GS (1979) Evolution of the differential regulation of duplicate genes after polyploidization. J Mol, Evol 12 (3):267–317

Ferris SD, Whitt GS (1980) Genetic variability in species with extensive gene duplication: the tetraploid catostomid fishes. Am Nat 115 (5):650–666

Ferris SD, Portnoy SL, Whitt GS (1979) The roles of speciation and divergence time in the loss of duplicate gene expression. Theoret Popul Biol 15 (1):114–139

Fineman R, Hamilton J, Chase G, Bolling D (1974) Length, weight and secondary sex character development in male and female phenotypes in three sex chromosomal genotypes (XX, XY, YY) in the killifish, *Oryzias latipes*. J Exp Zool 189 (2):227–233

Fineman R, Hamilton J, Chase G (1975) Reproductive performance of male and female phenotypes in three sex chromosomal genotypes (XX, XY, YY) in the killifish *Oryzias latipes*. J Exp Zool 192 (3):349–354

Fishelson L (1970) Protogynous sex reversal in the fish *Anthias squamipinnis* (Teleostei, Anthiidae) regulated by the presence fish. Nature (London) 227 (5253):90–91

Fisher RA (1970) Statistical methods for research workers. 14th edn. Oliver and Boyd, Edinburgh

Fisher SE, Whitt GS (1978a) Evolution of isozyme loci and their differential tissue expression. J Mol Evol 12 (I):25–55

Fisher SE, Whitt GS (1978b) Testis specific creatine kinase isozymes. Isoz Bull II:31

Fisher SE, Ferris SD, Whitt GS (1977) Multilocus isozyme systems in two species of sarcopterygian fishes. Isoz Bull 10:72–73

Fitzsimons JM (1972) A revision of two genera of goodeid fishes (Cyprinodontiformes, Osteichthyes) from the Mexican plateau. Copeia 4:728–756

Flick WA, Webster DA (1976) Production of wild, domestic, and interstrain hybrids of brook trout (*Salvelinus fontinalis*) in natural ponds. J Fish Res B Can 33 (7):1525–1539

* Foerster W, Anders F (1977) Zytogenetischer Vergleich der Karyotypen verschiedener Rassen und Arten lebendgebärender Zahnkarpfen der Gattung *Xiphophorus*. Zool Anz 198 (3–4):167–177

Fontana F (1976) Nuclear DNA content and cytometry of erythrocytes of *Huso huso* L., *Acipenser sturio* L. and *Acipenser naccarii* Bonaparte. Caryology 29 (I):127–138

* Fontana F, Colombo G (1974) The chromosomes of Italian sturgeons. Experientia 30 (6):739–742

* Fontana F, Chiarelli B, Rossi AC (1970) Il cariotipo di alcune specie di Cyprinidae, Centrarchidae, Characinidae studiate mediante colture "in vitro". Caryologia 23 (4):549–564

Ford EB (1966) Genetic polymorphism. Proc R Soc Ser B 164 (995):350–361

Fowler JA (1970) Control of vertebrae number in teleosts-an embryological problem. Quant Rev Biol 45 (2):148–167

Frankel JS (1978) Gene activation of alcohol dehydrogenase in *Danio* hybrids. J Hered 69 (I):57–58
Frankel JS (1979) Inheritance of spotting in the leopard danio J Hered 70 (4):287–288
Frankel JS, Hart NH (1977) Lactate dehydrogenase ontogeny in the genus *Brachydanio* (Cyprinidae). J Hered 68 (2):81–86
Franz R, Villwock W (1972) Beitrag zur Kenntnis der Zahnentwicklung bei oviparen Zahnkarpfen der Tribus *Aphanini* (Pisces Cyprinodontidae). Mitt Hamburg Zool Mus Inst 68:135–176
Franzin WG, Clayton JW (1977) A biochemical genetic study of zoogeography of lake whitefish (*Coregonus clupeaformis*) in western Canada. J Fish Res B Can 34 (5):617–625
Fraser AC, Gordon M (1928) Crossing-over between the W and Z chromosomes of the killifish *Platypoecilus*. Science 67 (1740):470
Frydenberg O, Simonsen V (1973) Genetics of *Zoarces* populations. V. Amount of protein polymorphism and degree of genic heterozygosity. Hereditas 75 (2):221–232
Frydenberg O, Moller D, Naevdal G, Sick K (1965) Haemoglobin polymorphism in Norwegian cod populations. Hereditas 53 (2):255–271
Frydenberg O, Nielsen JT, Simonsen V (1969) The maintenance of the hemoglobin polymorphism of the cod. Jpn J Genet 44 (Suppl I):160–165
Frydenberg O, Gyldenholm AO, Hjorth JP, Simonsen V (1973) Genetics of *Zoarces viviparus*. III. Geographic variation in the esterase polymorphism Est III. Hered 73 (2):233–238
Fryer G (1977) Evolution of species flocks of cichlid fishes in African lakes. Z Zool Syst Evol Forsh 15:141–165
Fujino K (1969) Atlantic skipjack tuna genetically distinct from Pacific specimens. Copeia 3:626–628
Fujino K (1970) Immunological and biochemical genetics of tunas. Trans Am Fish Soc 99 (1):152–178
Fujino K (1976) Subpopulation identification of skipjack tuna specimens from the south-western Pacific Ocean. Bull Jpn Soc Sci Fisher 42 (11):1229–1235
Fujino K (1978) Blood group and biochemical genetic research in fisheries biology. ABPI (Japan) 6:1–11
Fujino K, Kang T (1968) Transferrin groups of tunas. Genetics 59 (1):79–91
Fujio Y (1977) Natural hybridization between *Platichthys stellatus* and *Kareius bicoloratus*. Jpn J Genet 52 (2):117–124
* Fukuoka H (1972a) Chromosomes of the sockeye salmon (*Oncorhynchus nerka*). Jpn J Genet 47 (6):459–464
* Fukuoka H (1972b) Chromosome number variation in the rainbow trout (*Salmo gairdneri irideus* (Gibbons). Jpn J Genet 47 (6):455–458
Fyhn EH, Sullivan B (1974) Elasmobranch hemoglobins: dimerization and polymerization in various species. Comp Biochem Phys 50B (1):119–129
Gabriel ML (1944) Factors affecting the number and form of vertebrae in *Fundulus heteroclitus*. J Exp Zool 95 (1):105–143
Galagan NP (1973) Transferrins of Danube wild common carp. Hydrobiol Zh 9 (2):94–99 (R)
Gall GAE (1974) Influence of size of eggs and age of female on hatchability and growth in rainbow trout. Calif Fish Game 60 (1):26–36
Gall GAE (1975) Genetics of reproduction in domesticated rainbow trout. J Anim Sci 40 (1):19–28
Gall GAE (1978) Genetic control of reproductive function in domesticated rainbow trout. Abstr I4th Int Genet Cong Sect Sessions 1:505
Gall GAE, Gross SJ (1978) A genetic analysis of the performance of three rainbow trout broodstocks. Aquaculture 15 (2):113–127
Gall GAE, Busack CA, Smith RC, Gold JR, Kornblatt BJ (1976) Biochemical genetic variation in populations of golden trout, *Salmo aguabonita*. J Hered 67 (6):330–335
Gardner ML (1976) A review of factors which may influence the sea-age and maturation of Atlantic salmon *Salmo salar* L. J Fish Biol 9 (3):289–327
Gauldie RW, Smith PJ (1978) The adaptation of cellulose acetate electrophoresis to fish enzymes. Comp Biochem Phys 61B (2):421–425
Geraskin TP, Lukyanenko VI (1972) Species specificity of the fractional composition of haemoglobin from the blood of several sturgeons. Zh Obtsch Biol 33 (4):478–483 (RE)

Gervai J, Csányi V (1978) Induced triploidy in carp (*Cyprinus carpio* L.). In: Increasing productivity of fishes by selection and hybridization. F Müller, Szarvas, pp 31–36

Gervai J, Marian T, Karsznai Z, Nagy A, Csányi V (1980) Occurence of aneuploidy in radiation gynogenesis of carp. J Fish Biol 16 (4):435–439

Geyer F (1940) Abnorme Seitenlinien bei Fischen. Z Fischerei Hilfswissensch 38 (2):221–253

Gibson MB (1954) Upper lethal temperature relations of the guppy, *Lebistes reticulatus.* Can Zool 32:393

Giles MA, Vanstone WE (1976) Ontogenetic variation in the multiple hemoglobins of coho salmon (*Oncorhynchus kisutch*) and effect of environmental factors on their expression. J Fish Res B Can 33 (5):1144–1149

Gillespie J, Kojima K (1968) The degree of polymorphism in enzymes involved in energy production compared to that in nonspecific enzymes in two *D. ananassae* populations. Proc Natl Acad Sci USA 61 (3):582–585

Gillespie JH, Langley CH (1974) A general model to account for enzyme variation in natural populations. Genetics 76 (4):837–848

Giudice JJ (1964) Production and comparative growth of three buffalo hybrids. Proc South-East Assoc Game Comm 18:1–13

Giudice JJ (1966) Growth of a blue-channel catfish hybrid as compared to its parental species. Prog Fish Cult 28 (3):142–145

Gjedrem T (1975) Possibilities for genetic gain in salmonids. Aquaculture 6 (1):23–29

Gjedrem T (1976) Genetic variation in tolerance of brown trout to acid water. In: SNSF-Project, Norway, Oslo, FR 5/76, pp 1–11

Gjedrem T (1979) Selection for growth rate, and domestication in Atlantic salmon. Z Tierzücht Züchtungs-Biol 96 (1):56–59

Gjedrem T, Aulstad D (1974) Selection experiments with salmon. I. Differences in resistance to vibrio disease of salmon parr (*Salmo salar*). Aquaculture 3 (1):51–59

Gjedrem T, Skjervold H (1978) Improving salmon and trout farm yields through genetics. World Rev. Anim Prod 14 (3):29–38

Glushankova MA, Korobtzova NS, Kusakina AA, Neyfakh AA (1973) Expression of genes controlling FDP-aldolase in fish embryos. Thermostability as a genetic marker. In: Biochem Genet Fish. Inst Cytol Acad Sci USSR, Leningrad, pp 76–84 (RE)

Goodard CI, Tait JS (1976) Preferred temperature of F_3 to F_5 hybrids of *Salvelinus fontinalis* x *S. namaycush.* J Fish Res B Can 33 (2):197–202

Gold JR (1977) Systematics of western North American trout (*Salmo*), with notes on the redband trout of Sheep heaven Greek, California. Can J Zool 55 (11):1858–1873

* Gold JR (1979) Cytogenetics. In: Fish physiology, vol 8. Academic Press London New York, pp 353–405

* Gold JR (1980a) Chromosomal change and rectangular evolution in North American cyprinid fishes. Genet Res 35 (2):157–164

Gold JR, Avise JC (1976) Spontaneous triploidy in the California roach *Hesperoleucus symmetricus* (Pisces: Cyprinidae). Cytogenet Cell Genet 17 (3):144–149

* Gold JR, Avise JC (1977) Cytogenetic studies in North American minnows (Cyprinidae). I. Karyology of nine California genera. Copeia 3:541–549

Gold JR, Gall GAE (1975) Chromosome cytology and polymorphism in the California High Sierra golden trout (*Salmo aguabonita*) Can J Genet Cytol 17 (1):41–54

* Gold JR, Avise JC, Gall GAE (1977) Chromosome cytology in the cutthroat trout series *Salmo clarkii* (Salmonidae). Cytologia 42 (2):377–382

* Gold JR, Womac WD, Deal FH, Barlow JA (1978) Gross karyotype change and evolution in North American cyprinid fishes. Genet Res 32 (1):37–46

* Gold JR, Janak BJ, Barlow JA (1979a) Karyology of four North American percids (Perciformes: Percidae). Can J Genet Cytol 21 (2):187–191

* Gold JR, Whitlock CW, Karel WJ, Barlow JA (1979b) Cytogenetic studies in North American minnows (Cyprinidae). VI. Karyotypes of thirteen species in the genus *Notropis.* Cytologia 44 (2):457–466

* Gold JR, Karel WJ, Strand MR (1980a) Chromosome formulae of North American fishes. Prog Fish Cult 42 (1):10–23

* Gold JR, Womac WD, Deal FH, Barlow JA (1980b) Cytogenetic studies in North American minnows (Cyprinidae). VII. Karyotypes of 13 species from the southern United States. Cytologia (in press)

Goldberg E (1966) Lactate dehydrogenase of trout: hybridization in vivo and in vitro. Science 151 (3714): 1091–1093

Goldberg E, Cuerrier JP, Ward JC (1969) Lactate dehydrogenase ontogeny, paternal gene activation and tetramer assembly in embryos of brook trout, lake trout and their hybrids. Biochem Genet 2 (3): 335–350

Goldberg E, Kerekes J, Cuerrier JP (1971) Lactate dehydrogenase polymorphism in wild populations of brook trout from Newfoundland. Rapp P-V Reun 161: 97–99

Golovinskaya KA (1940) The pleiotropic action of scale genes in the common carp. Dokl Akad Nauk SSSR (DAN) 28 (6): 533–536 (R)

Golovinskaya KA (1946) On the linear form of the cultivated common carp. Dokl Akad Nauk SSSR (DAN) 54 (7): 637–640 (R)

Golovinskaya KA (1954) Reproduction and heredity of the silver crucian carp. Tr VNIIPRKh 6: 34–57 (R)

Golovinskaya KA (1960) On the males of the silver crucian carp and their crossings with the common carp. Rybovodstvo Rybolovstvo 6: 16–17 (R)

Golovinskaya KA (1962) Selective breeding in pond fish industry. Rybovodstvo Rybolovstvo 3: 7–10 (R)

Golovinskaya KA (1965) The selectional value of the air bladder variation in the common carp. Tr VNIIPRKh 13: 97–103 (R)

Golovinskaya KA (1968) Artificial gynogenesis in fishes and perspectives of its using for the development of hybrid combinations possessing heterosis. In: Heterosis in animal husbandry. Kolos, Moscow, pp 248–254 (R)

Golovinskaya KA (1969) First steps in the creation of the Middle Russian common carp breed. In: Pond fish breeding. All-Union Res Inst Pond Fish, Moscow, pp 139–148 (R)

Golovinskaya KA, Romashov DD (1966) Segregation for the scaling pattern during diploid gynogenesis in the common carp. Tr VNIIPRKh 14: 227–235 (R)

Golovinskaya KA, Romashov DD, with the participation of Musselius VA (1947) The study of gynogenesis in the silver crucian carp. Tr VNIIPRKh 4: 73–113 (R)

Golovinskaya KA, Romashov DD, Cherfas NB (1963) Radiation = induced gynogenesis in the common carp. Tr VNIIPRKh 12: 149–167 (R)

Golovinskaya KA, Romashov DD, Cherfas NB (1965) Unisexual and bisexual forms of the silver crucian carp *Carassius auratus gibelio* Bl. Vopr Ikhthiol 5 (4): 614–629 (R)

Golovinskaya KA, Cherfas NB, Tsvetkova LI (1974a) Results of evaluation of the reproductive function in gynogenetic common carp females. Tr VNIIPRKh 23: 20–26 (RE)

Golovinskaya KA, Tscherbina MA, Solovyeva LM, Bobrov AS (1974b) A relationship between the origin of carp fingerlings and the accumulation and use of nutrients during the winter period. Tr VNIIPRKh 23: 48–54 (RE)

Golovinskaya KA, Katasonov VYa, Bobrova YuP, Popova AA (1975) Works on the development of Middle Russian common carp breed. In: Materials of the All-Union Conference on the organization of selectional work and the improvement of spawners in fish farms of the country, Moscow, pp 14–30 (R)

Gomelsky BI, Ilyasova VA, Cherfas NB (1979) Investigation of radiation- induced diploid gynogenesis in carp (*Cyprinus carpio* L.). IV. Gonad state and evaluation of reproductive ability in carp of gynogenetic origin. Genetika (Moscow) 15 (9): 1643–1650 (RE)

Gonzalez DR, Padron M, Subero LE (1974) Analysis electroforetico de hemoglobina, lactato deshidrogenasa, esterasa y proteinas no enzymaticas de dos especies del genero *Anchoa* (*Pisces: Engraulidae*). Biol Inst Oceanogr Univ Oriente 13 (1–2): 47–52

Goodman M (1976) Protein sequences in phylogeny. In: Molecular evolution. Sinauer Assoc, Sunderland

Goodman M, Moore GW, Matsuda G (1975) Darwinian evolution in the genealogy of haemoglobin. Nature (London) 253 (5493): 603–608

Goodrich HB (1929) Mendelian inheritance in fish. Q Rev Biol 4 (1): 83–99

Goodrich HB, Smith MA (1937) Genetics and histology of the color pattern in the normal and albino paradise fish, *Macropodus opercularis* L. Biol Bull 73 (3): 527–534

Goodrich HB, Josephson ND, Trinkaus JP, State JM (1944) The cellular expression and genetics of two new genes in *Lebistes reticulatus.* Genetics 29 (6): 584–592

Goodrich HB, Hine RL, Lesner HM (1947) Interaction of genes in *Lebistes reticulatus.* Genetics 32 (3): 535–540

Gorbunova LA (1962) Hybridization of whitefishes as a way to improve the productivity of Karelian Lakes. In: Biology of inland waterbasins of the Baltic. Acad Sci USSR, Moscow, pp 77–79 (R)

Gordon AH (1975) Electrophoresis of proteins in polyacrylamide and starch gels. North-Holland, Amsterdam

Gordon H, Gordon M (1957) Maintenance of polymorphism by potentially injurious genes in eight natural populations of the platy fish, *Xiphophorus maculatus.* J Genet 55 (1): 1–44

Gordon M (1927) The genetics of a viviparous top-minnow *Platypoecilus;* the inheritance of two kinds of melanophores. Genetics 12 (2): 253–283

Gordon M (1931) Hereditary basis of melanosis in hybrid fishes. Am J Cancer 15: 1495–1523

Gordon M (1937) Genetics of *Platypoecilus.* III. Inheritance of sex and crossing-over of the sex chromosomes in the platyfish. Genetics 22 (2): 376–392

Gordon M (1941) Back to their ancestors. J Hered 32 (11): 385–390

Gordon M (1942) Mortality of albino embryos and aberrant Mendelian ratios in certain broods of *Xiphophorus helleri.* Zoologica 27 (1): 73–74

Gordon M (1946) Introgressive hybridization in domesticated fishes. I. The behavior of comet, a *Platypoecilus maculatus* gene in *Xiphophorus helleri.* Zoologica 3 (1): 77–88

Gordon M (1947a) Genetics of *Platypoecilus maculatus.* IV. The sex determining mechanism in two wild populations of the mexican platy fish. Genetics 32 (I): 8–17

Gordon M (1947b) Speciation in fishes. Adv Genet (1): 95–132

Gordon M (1948) Effects of five primary genes on the site of melanomas in fishes and the influence of two color genes on their pigmentation. In: The biology of melanomas, vol 4. Acad Sci New York, pp 216–268

Gordon M (1950) The origin of modifying genes that influence the normal and atypical growth of pigment cells in fishes. Zoologica 35 (1): 19–20

Gordon M (1951a) Genetic and correlated studies of normal and atypical pigment cell growth. Growth Symp 10: 153–219

Gordon M (1951b) Genetics of *Platypoecilus maculatus.* 5. Heterogametic sex-determining mechanism in females of a domesticated stock originally from British Honduras. Zoologica 36: 127–134

Gordon M (1952) Inheritance in aquarium fishes. 3. The genetics of the wagtail platy. Aquarist 17: 186–190

Gordon M (1953) Hereditary differences in seven natural populations of the platyfish, *Xiphophorus maculatus.* Proc XIV Int Congr Zool Copenhagen, pp 172–176

Gordon M (1956) An intricate genetic system that controls nine pigment cell patterns in the platyfish. Zoologica 41: 153–162

Gordon M (1957) Physiological genetics of fishes. In: The physiology of fishes, vol II. Academic Press, London New York, pp 431–501

Gordon M (1958) A genetic concept for the origin of melanomas. Ann NY Acad Sci 71 (6): 1213–1222

Gordon M (1959) The melanoma cell as an incompletely differentiated pigment cell. In: Pigment cell biology, New York, Academic Press, London New York, pp 215–236

Gordon M, Baker KF (1955) Post-natal lethal gene in the platy fish, *Xiphophorus maculatus.* Anat Rec 122 (2): 436–437

Gordon M, Rosen DE (1951) Genetics of species differences in the morphology of the male genitalia of xiphophorin fishes. Bull Am Mus Natl Hist 95 (7): 409–464

Gorman GC, Kim YJ, Rubinoff R (1976) Genetic relationships of three species of *Bathygobius* from the Atlantic and Pacific sides of Panama. Copeia 2: 361–364

Gorshkova GV (1978) Some peculiarities of karyotypes of the Pacific salmon. Tzitologiya 20 (12): 1431–1435 (RE)

* Gorshkova GV (1979) Comparative karyology of the Pacific salmons. Th D Thesis, Inst Cytol, Leningrad (R)

* Gorshkova GV (1980) Karyology and chromosomal polymorphism of Pacific salmon. In: Karyological variability, mutagenesis and gynogenesis in fishes. Inst Cytol Acad Sci USSR, Leningrad, pp 29–33 (RE)

Gorshkova GV, Gorshkov SA (1978) Chromosomal sets of seasonal races in the sockeye salmon of the Lake Azabachye (Kamchatka). Zool Zh 57 (9): 1382–1388 (RE)

Grag RW, McKenzie JA (1970) Muscle protein electrophoresis in the genus *Salmo* of Eastern Canada. J Fish Res B Can 27 (11): 2109–2112

* Grammeltvedt A-F (1975) Chromosomes of salmon (*Salmo salar*) by leukocyte culture. Aquaculture 5 (2): 205–209

Grant WS, Utter FM (1980) Biochemical genetic variation in walleye pollock, *Theragra chalcogramma:* population structure in the southeastern Bering Sea and the Gulf of Alaska. Can J Fish Aquat Sci 37 (7): 1093–1100

Grant WS, Milner GB, Krasnowski P, Utter FM (1980) Use of biochemical genetic variants for identifications of sockeye salmon *Oncorhynchus nerka* stocks in Cook Inlet, Alaska. Can J Fish Aquat Sci 37 (8): 1236–1247

Grechkovskaja AP (1971) Fisheries-biological characteristic of the common carps of new strain (UKN-52) and their hybrids in fish farms of West districts of Ukraine. Ph D Thesis, Chernovitzy (R)

Grechkovskaja AP, Turanov VF, Puljaeva VI (1979) Backcrossings of the hybrids between silver carp and bighead. In: Materials of All-Union scientific conference on development and intensification of fisheries in inland waters of North Caucasus, Rostov-Don, pp 58–61 (R)

Green, OL, Smitherman RO, Pardue GB (1979) Comparisons of growth and survival of channel catfish, *Ictalurus punctatus,* from distinct populations. In: Advances in aquaculture. Fish News Books Ltd, Farnham, pp 626–628

Greenberg SS, Kopac MJ (1968) Electrophoretic analysis of species relationships in the genus *Xiphophorus.* Comp Biochem Phys 28 (1): 37–54

Gregory PE, Howard-Peebles PN, Ellender RD, Martin BJ (1980) Analysis of a marine fish cell line from a male sheepshead. J Hered 71 (3): 209–211

Grimm H (1979) Veränderungen in der Variabilität von Populationen des Zahnkarpfens *Aphanius anatoliae* während 30 Jahren 1943–1974. Z Zool Syst Evol-Forsch 17 (4): 272–280

Gross HP (1977) Adaptive trends of environmentally sensitive traits in the three-spined stickleback *Gasterosteus aculeatus* L. Z Zool Syst Evol-Forsch 15 (4): 252–278

Grossman GD (1977) Polymorphism of plasma esterases in rainbow trout. Prog Fish-Cult 39 (1): 35–36

Guerrero RD (1979) Culture of male *Tilapia mossambica* produced through artificial sex reversal. In: Advances in aquaculture. Fish New Books Ltd, Farnham, pp 166–168

Gunnes K, Gjedrem T (1978) Selection experiments with salmon. IV. Growth of Atlantic salmon during two years in the sea. Aquaculture 15 (1): 19–33

Guse CJ, Ney JJ, Turner BJ (1980) Allozymic evidence of reproductive isolation between two components of a landlocked population of striped bass. Isoz Bull 13: 101

Gutierrez M (1969) Estudio electroforetico de las proteinas solubles de tres zones del cristalino de atun *Thunnus thynnus* L. Invest Pesq 33 (1): 149–169

Guyomard R (1978) Identification par electrophorese d'hybrides de salmonides. Ann Genet Select Anim 10 (1): 17–27

Guzhov YuL (1969) Heterosis and crops. Kolos, Moscow

Gwanaba JJ (1973) East Afric Wildl J 11: 317 (cit by Moav et al., 1978)

* Gyldenholm AO, Scheel JJ (1971) Chromosome numbers of fishes. J Fish Biol 3 (3): 479–486

Haaker PL, Lane ED (1973) Frequencies of anomalies in a bothid, *Paralichthys californicus* and a pleuronectid, *Hypsopsetta guttulata* flatfish. Copeia 1: 22–25

Haas R (1976) Sexual selection in *Nothobranchius guentheri* (Pisces: Cyprinodontidae). Evolution 30 (3): 614–622

Haberman H, Tammert M (1976) On connections of the individual productivity of the bream with the genotype in some Estonian lakes. In: Eston Contrib Int Biol Programme Tallinn 10: 100–113

Haen PJ, O'Rourke FJ (1968) Proteins and haemoglobins of salmon trout hybrids. Nature (London) 217 (5123): 65–67

Hagen DW (1973) Inheritance of number of lateral plates and gill rakers in *Gasterosteus aculeatus.* Heredity 30 (3):303–312
Hagen DW, Gilbertson L (1972) Geographic variation and environmental selection in *Gasterosteus aculeatus* L. in the Pacific Northwest, America. Evolution 26 (1):32–51
Hagen DW, Gilbertson LG (1973a) Selective predation and the intensity of selection acting upon the lateral plates of three-spine sticklebacks. Heredity 30 (3):273–287
Hagen DW, Gilbertson LG (1973b) The genetics of plate morphs in fresh water threespine sticklebacks. Heredity 31 (1):75–84
Hagen DW, Moodie GEE (1979) Polymorphism for breeding colors in *Gasterosteus aculeatus.* I. Their genetics and geographic distribution. Evolution 33 (2):641–648
* Haimoviči S, Ciuca L (1973) Observations concernant les chromosomes chez *Eudoontomyson maria* (Cyclostomata, Petromysonidae). Ann Sti Univ Jasi Sect 2 19 (2a):345–348
Haldane JBS (1955) On the biochemistry of heterosis and the stabilization of polymorphism. Proc R Soc London ser B 144 (915):217–220
Hammerman IS, Avtalion RR (1979) Sex determination in *Sarotherodon* (*Tilapia*) part 2. Theoret Appl Genet 55 (3–4):177–187
Handford PT (1971) An esterase polymorphism in the bleak *Alburnus alburnus* (Pisces). In: Ecological genetics and evolution. Blackwell Sci Publ, Oxford Edinburg, pp 289–297
Harrington RW (1963) Environmentally controlled induction of primary male gonochorists from eggs of the self-fertilizing hermaphroditic fish, *Rivulus marmoratus.* Biol Bull 132 (1):174–199
Harrington RW, Crossman RA (1976) Temperature induced meristic variation among three homozygous genotypes (clones) of the self-fertilizing fish *Rivulus marmoratus.* Can J Zool 54 (7):1143–1155
Harris H (1970) The principles of human biochemical genetics. North-Holland Publ Co, Amsterdam London
Harris H, Hopkinson DA, Edwards YH (1976) Polymorphism and the subunit structure of enzymes: controversy. Proc Natl Acad Sci USA, 74 (2):698–701
Harris JE (1974) Electrophoretic patterns of blood serum proteins of the cyprinid fish, *Leuciscus leuciscus* (L). Comp Biochem Phys. 48B (3):389–399
Hart NH, Cook M (1977) Esterase isozyme patterns in developing embryos of *Brachydanio rerio* (zebra danio), *B. albolineatus* (pearl danio) and their hybrids. J Exp Zool 199 (1):109–118
Hartman WL, Raleigh RV (1964) Tributary homing of sockeye salmon at brooks and Karluk Lakes, Alaska. J Fish Res B Can 21 (3):485–504
Haskins CP, Druzba JP (1938) Note on anomalous inheritance of sex-linked color factors in the guppy. Am Nat 72 (743):571–574
Haskins CP, Haskins EF (1948) Albinism, a semilethal autosomal mutation in *Lebistes reticulatus.* Heredity 2 (2):251–262
Haskins CP, Haskins EF (1951) The inheritance of certain color patterns in wild populations of *Lebistes reticulatus* in Trinidad. Evolution 5 (3):216–225
Haskins CP, Haskins EF (1954) Note on a "permanent" experimental alteration of genetic constitution in a natural population. Proc Natl Acad Sci USA 40 (7):627–635
Haskins CP, Haskins EF, Hewitt RE (1960) Pseudogamy as an evolutionary factor in the poeciliid fish *Mollienesia formosa.* Evolution 14 (4):473–483
Haskins CP, Haskins EF, McLaughlin JJ, Hewitt RE (1961) Polymorphism and population structure in *Lebistes reticulatus,* an ecological study. In: Vertebrate speciation. Univ Texas Press, New York, pp 320–395
Haskins CP, Young P, Hewitt RE, Haskins EF (1970) Stabilized heterozygosis of supergenes mediating certain Y-linked colour patterns in populations of *Lebistes reticulatus.* Heredity, 25 (4):575–589
Hasnain AU, Siddiqui AQ, Ali SA (1973) Hemoglobin polymorphism in the air-breathing climbing perch, *Anabas testudineus* (B). Curr Sci 42 (19):691–692
Haussler G (1928) Über die Melanombildung bei Bastarden von *Xiphophorus helleri* und *Platypoecilus maculatus* var. *rubra.* Klin Wochenschr. 7 (33):1561–1562
Hayes HK (1952) Development of the heterosis concept. In: Heterosis, Jova St Coll Press, Ames, pp 49–68

Healy JA, Mulcahy MF (1979) Polymorphic tetrameric superoxide dismutase in the pike *Esox lucius* L. (Pisces: Esocidae). Comp Biochem Physiol 62B (4): 563–565

Hedrick PW (1971) A new approach to measuring genetic similarity. Evolution 25 (2): 276–280

Hegenauer J, Saltman P (1975) Iron and susceptibility to infection disease. Science 188 (4192): 1038–1039

Heincke G (1898) Naturgeschichte des Herings. I. Die Lokalformen und die Wanderungen des Herings in den europäischen Meeren. Abh Dtsch Seefisch Ver 2: 1–136

Hempel G, Blaxter JHS (1961) The experimental modification of meristic characters in herring (*Clupea harengus* L.). Rapp P-V Reun 26 (3): 336–346

Henricson J, Nyman L (1976) The ecological and genetical segregation of two sympatric species of dwarfed char (*Salvelinus alpinus* L. species complex). Rep Inst Freshwater Res 55: 15–37

Henze M, Anders F (1975) Über einen Makropterinophoren-komplexlocus und dessen Expressionskontrolle im Genom von *Platypoecilus maculatus, Xiphophorus helleri* und *X. montezumae cortezi* (Poeciliidae). Verh Dtsch Zool Ges 69: 159–162

Herrera RJ (1979) Preferential gene expression of an amylase allele in interspecific hybrids of *Xiphophorus* (Pisces: Poeciliidae). Biochem Genet 17 (3–4): 223–227

Herschberger WK (1970) Some physicochemical properties of transferrins in brook trout. Trans Am Fish Soc 99 (1): 207–218

Herschberger WK (1978) The use of interpopulation hybridization in development of coho salmon stocks for aquaculture. Proc Annu Meet World Maricult Soc 9: 147–156

Herschberger WK, Brannon EL, Donaldson LR, Yokoyama GA, Roley SE (1976) Salmonid aquaculture studies: selective breeding. Annu Rep Coll Fish Univ Washington 444: 61

Heuts MJ (1949) Racial divergence in fin ray variation patterns in *Gasterosteus aculeatus*. Genetics 49 (1): 185–192

Hickling CF (1960) The Malacca *Tilapia* hybrids. J Genet 57 (1): 1–10

Hickling CF (1968) Fish hybridization. FAO Fish Rep (44) Rome 4: 1–11

Hilse K, Sorger U, Braunitzer G (1966) Zur Philogenie des Hämoglobin-moleküls. Über der Polymorphismus und die N-terminalen Aminosäuren des Karpfenhämoglobins. Z Physiol Chem 344 (1–3): 166–168

Hinegardner R (1968) Evolution of cellular DNA content in teleost fishes. Am Nat 102 (928): 517–523

Hinegardner R (1976a) Evolution of genome size. In: Molecular evolution. Sinauer Assoc, Sunderland, pp 179–199

Hinegardner R (1976b) The cellular DNA content of sharks, rays and some other fishes. Comp Biochem Phys 55B (3): 367–370

* Hinegardner R, Rosen DE (1972) Cellular DNA content and the evolution of teleostean fishes. Am Nat 106 (951): 621–644

Hines RS, Wohlfarth GW, Moav R, Hulata G (1974) Genetic differences in susceptibility to two diseases among strains of the common carp. Aquaculture 3 (2): 187–197

* Hitotsumachi S, Sasaki M, Ojima Y (1969) A comparative karyotype study in several species of Japanese loaches (Pisces, Cobitidae). Jpn J Genet 44 (3): 157–161

Hitzeroth H, Klose J, Ohno S, Wolf U (1968) Asynchronous activation of parental alleles at the tissue-specific gene loci observed on hybrid trout during early development. Biochem Genet 1 (3): 287–300

Hjorth JP (1971) Genetics of *Zoarces* populations. I. Three loci determining the phosphoglucomutase isoenzymes in brain tissue. Hereditas 69 (2): 233–241

Hjorth JP (1974) Genetics of *Zoarces* populations. VII. Fetal and adult hemoglobins and a polymorphism common to both. Hereditas 78 (1): 69–72

Hjorth JP (1975) Molecular and genetic structure of multiple hemoglobins in the eelpout, *Zoarces viviparus*. Biochem Genet 13 (5–6): 379–391

Hjorth JP, Simonsen V (1974) Genetics of *Zoarces* populations. VIII. Geographical variation common to the polymorphic loci HbI and Est III. Hereditas 81 (2): 173–184

Hochachka PW (1967) Organization of metabolism during temperature compensation. In: Molecular mechanisms of temperature adaptation. Am Assoc Adv Sci, Washington, pp 177–203

Hochachka PW (1968) Lactate dehydrogenase function in *Electrophorus* swimbladder and in the lungfish lung. Comp Biochem Phys 27 (4): 613–617

Hochachka PW, Somero GN (1973) Strategies of biochemical adaptation. WB Saunders Co, Philadelphia London Toronto

Hochachka PW, Gyppy M, Gurerley HE, Storey KB, Hulbert WC (1978) Metabolic biochemistry of water- vs airbreathing osteoglossids fishes. Can J Zool 56 (4): 736–750, 751–768

Hodges DH, Whitmore DH (1977) Muscle esterases of the mosquitofish, *Gambusia affinis*. Comp Biochem Phys 58B (4): 401–407

Hodgins H (1972) Serological and biochemical studies in racial identification of fishes. In: The stock concept in pacific salmon. Univ Br Columbia, Vancouver, pp 199–208

Hodgins H, Utter FM (1971) Lactate dehydrogenase polymorphism of sockeye salmon (*Oncorhynchus nerka*). Rapp P-V Reun 161: 100–101

Hodgins H, Ames WE, Utter FM (1969) Variants of lactate dehydrogenase isozymes in sera of sockeye salmon (*Oncorhynchus nerka*). J Fish Res B Can 26 (1): 15–19

Hofmann J (1927) Die Aischgründer Karpfenrasse. Z Fischerei Hilfswissensch 25 (2): 291–365

Holm M, Naevdal G (1978) Quantitative genetic variation in fish-its significance for salmonid culture. In: Marine organisms, Plenum Press, New York, pp 679–698

Holmes RS (1973) Evolution of lactate dehydrogenase genes. FEBS Lett 28: 51–55

Holmes RS, Masters CJ (1967) The developmental multiplicity and isoenzyme status of cavian esterases. Biochem Biophys Acta 132 (2): 379–399

Holmes RS, Whitt G (1970) Developmental genetics of the esterase isoenzymes of *Fundulus heteroclitus*. Biochem Genet 4 (4): 471–480

Holzberg S (1978) A field and laboratory study of the behaviour and ecology of *Pseudotropheus zebra* (Boul.), an endemic cichlid of Lake Malawi (Pisces; Cichlidae). Z Zool Syst Evol-Forsch 16 (2): 171–187

Holzberg S, Schröder JH (1972) Behavioural mutagenesis of the convict cichlid fish, *Cichlasoma nigrofasciatum* Guenther I. The reduction of male aggressiveness in the first generation. Mutat Res 16 (2): 289–296

Horn P (1972) A mindkét ivaru guppiu (*Poecilia reticulata* Pet.) matatkozó uj autoszomális dominánc mutació. Allattani Kozl 59 (1–4): 53–59

Horowitz JJ, Whitt GS (1972) Evolution of a nervous system specific lactate dehydrogenase isozyme in fish. J Exp Zool 180 (1): 13–32

* Howell WM (1972) Somatic chromosomes of the black ghost knifefish, *Apteronotus albifrons* (Pisces: Apteronotidae). Copeia 1: 191–193

* Howell WM, Denton TE (1969) Chromosomes of ammocoetes of the Ohio brook lamprey, *Lampetra aepyptera*. Copeia 2: 393–395

* Howell WM, Duckett CR (1971) Somatic chromosomes of the lamprey, *Ichthyomyzon gagei* (Agnatha; Petromysonidae). Experientia 27 (2): 222–223

* Howell WM, Villa I (1976) Chromosomal homogeneity in two sympatric cyprinid fishes of the genus *Rhinichthys*. Copeia 1, 112–116

Howlett G, Jamieson A (1971) A system of muscle esterase variants in the sprat (*Sprattus sprattus*). Rapp P-V Reun 161: 45–47

Hubbs CL, Hubbs LC (1932) Apparent parthenogenesis in nature in a form of fish of hybrid origin. Science 76 (1983): 628–630

Hubbs CL, Hubbs LC (1946 a) Breeding experiments with the invariably female, strictly matroclinous fish, *Mollienesia formosa*. Genetics 31 (2): 218

Hubbs CL, Hubbs LC (1946 b) Experimental breeding of the Amazon molly. Aquarium J 17 (8): 4–6

Hubbs CL, Drenry GE, Warburton B (1959) Occurrence and morphology of a phenotypic male of a gynogenetic fish. Science 129 (3357): 1227–1229

Hubby JL, Lewontin RC (1966) A molecular approach to the study of genic heterozygosity in natural populations I. The number of alleles at different loci in *Drosophila pseudoobscura*. Genetics 54 (2): 577–594

Hulata G, Moav R, Wohlfarth G (1974) The relationship of gonad and egg size to weight and age in the European and Chinese races of the common carp *Cyprinus carpio* L. J Fish Biol 6 (4): 745–758

Hulata G, Moav R, Wohlfarth G (1980) Genetic differences between the Chinese and the European races of the common carp. III. Gonad abnormalities in hybrids. J Fish Biol 16 (4):369–370

Humm DG, Clark EE, Humm JH (1957) Transplantation of melanomas from platyfish – swordtail hybrids into embryos of swordtails, platyfish and their hybrids. J Exp Biol 34 (4):518–528

Hunter RL, Markert GL (1957) Histochemical demonstration of enzymes separated by zone electrophoresis in starch gels. Science 125 (3261):1294–1295

Huntsman GR (1970) Disc gel electrophoresis of blood sera and muscle extracts from some catostomid fishes. Copeia 3:457–467

Hutt FB (1970) Genetic resistance to infection. In: Resistance to Infections Disease. Modern Press, Saskatoon, pp 1–11

Hutt FB (1974) Genetic indicators of resistance to disease in domestic animals. In: Proc 1st World Congress of genetics applied to livestock production, Madrid, pp 179–185

Huzyk L, Tsuyuki H (1974) Distribution of "LDH-B" gene in resident and anadromous rainbow trout (*Salmo gairdneri*) from streams in British Columbia. J Fish Res B Can 31 (1):106–108

Ihssen P (1973) Inheritance of thermal resistance in hybrids of *Salvelinus fontinalis* and *S. namaycush*. J Fish Res B Can 30 (3):401–408

Ihssen P (1976) Selective breeding and hybridization in fisheries management. J Fish Res B Can 33 (2):316–321

Iljasov JuI, Shart LA (1979) Polymorphous genetic systems of blood serum and their connection with selective traits in common carp. In: Biochem Popul Genet Fish Inst Cytol Acad Sci USSR, Leningrad, pp 152–156 (RE)

Imai HT (1978) On the origin of telocentric chromosomes in mammals. J Theoret Biol 71 (4):619–638

Ingram VM (1963) The hemoglobins in genetics and evolution. Columbia Univ Press, New York

* Itoh Y, Niijama H (1972) Comparative chromosome studies of two cyprinid fish, Ugui, *Trilodon hakonensis* and Ezo-ugi, *T. ezoe*. Bull Fak Fish Hokk Univ 23 (2):73–76

Ivanenkov VV (1976) Expression of paternal alleles of lactate dehydrogenase, glutamate dehydrogenase and acetylcholin esterase during the development of hybrids between species belonging to cobitid and cyprinid families. Ontogenez 7 (6):579–589 (RE)

Ivanenkov VV (1979) Esterase-2 in development of loach (*Misgurnus fossilis*). Heterogeneity of loach eggs for expression of esterase-2 alleles. In: Biochem Popul Genet Fish, Inst Cytol Acad Sci USSR, Leningrad, pp 29–35 (RE)

Ivanenkov VV (1980) Esterase-2 in the development of loach (*Misgurnus fossilis*). I. Period of expression of esterase-2 genes and duration of maternal enzyme resistance in the embryos. II. Differential expression of the allelic esterase-2 genes in the oocytes and eggs. III. Absence of feed-back in the regulation of Est-2 gene expression. Isoz Bull 13:76–78

* Ivanov VN (1969) The chromosomes of the Black Sea flatfish *Rhombus maeoticus* Pall. Dokl Akad Nauk SSSR (DAN) 187 (6):1397–1399 (R)

* Ivanov VN (1975) The chromosomes of Black Sea gobiid fish *Gobius melanostomus* and *G. batrachocephalus*. Tzitol Genet 9 (6):551–552 (R)

* Ivanova IM, Kirpichnikov VS, Rolle NN (1973) Variability of lactate dehydrogenase (LDH) in cultured and wild carp (*Cyprinus carpio* L.). In: Biochem Genet Fish. Inst Cytol Acad Sci USSR, Leningrad, pp 91–96 (RE)

Iwata M (1973) Genetic polymorphism of tetrazolium oxidase in walleye pollock. Jpn J Genet 48 (2):147–149

Iwata M (1975) Genetic identification of walleye pollock (*Theragra chalcogramma*) populations on the basis of tetrazolium oxidase polymorphism. Comp Biochem Phys 50B (1):197–201

Iwata M, Numachi K (1979) Pollock populations in north part of Pacific Ocean. In: Abstr 14th Pacific Sci Congr Comm F Sect F-II a, Moscow, pp 161–162

James GD (1972) Revision of the New Zealand flatfish genus *Peltorhamphus* with description of two new species. Copeia 2:345–355

Jamieson A (1967) New genotypes in cod at Greenland. Nature (London) 214 (5101):661–662

Jamieson A (1975) Enzyme types of Atlantic cod stocks on the North American banks. In: Isozymes, vol IV. Academic Press, London New York, pp 491–515

Jamieson A, Jonsson J (1971) The Greenland component of spawning cod at Iceland. Rapp P-V Reun 161:65–72

Jamieson A, Ligny W de, Naevdal G (1971) Serum esterases in mackerel (*Scomber scombrus*). Rapp P-V Reun 161:109–117

Jamieson A, Otterlind G (1971) The use of cod blood protein polymorphism in the Belt Sea, the Sound and the Baltic Sea. Rapp. P-V Reun 161:55–59

Jamieson A, Thompson D (1972) Blood proteins in North-Sea cod (*Gadus morhua* L.). Proc 12th Eur Conf Anim Blood Gr Biochem Polym, Budapest, pp 585–591

Jamisson A, Turner RJ (1978) The extended series of Tf alleles in Atlantic cod *Gadus morhua* L. In: Marine organisms. Plenum Press, New York, pp 699–729

Jensen GL, Shelton WL (1979) Effect of estrogens on *Tilapia aurea:* implications for production of monosex genetic male tilapia. Aquaculture 16 (3):233–242

* Jin Sonia M, Toledi V (1975) Citogenetica de *Astianax fasciatus* e *A. bimaculatus* (Characidae, Tetragonopterinae). Cienc. Cult 27 (10):1122–1124

Johnson AG (1977) A survey of biochemical variants found in ground fish stocks from the North Pacific and Bering Sea. – Anim Blood Gr Biochem Genet 8 (1):13–19

Johnson AG, Beardsley AJ (1975) Biochemical polymorphism of starry flounder, *Platichthys stellatus* from the north-western and north-eastern Pacific Ocean. Anim Blood Gr Biochem Genet 6 (1):9–18

Johnson AG, Utter FM (1976) Electrophoretic variation in intertidal and subtidal organisms in Puget Sound, Washington. Anim Blood Gr Biochem Genet 7 (1):3–14

Johnson AG, Utter FM, Hodgins HO (1970a) Electrophoretic variants of L-alpha-glycerophosphate dehydrogenase in Pacific Ocean perch (*Sebastodes alutus*). J Fish Res B Can 27 (5):943–945

Johnson AG, Utter FM, Hodgins HO (1970b) Interspecific variation of tetrazolium oxidase in *Sebastodes* (rockfish). Comp Biochem Phys 37 (2):281–285

Johnson AG, Utter FM, Hodgins HO (1971) Phosphoglucomutase polymorphism in Pacific Ocean perch, *Sebastodes alutus.* Comp Biochem Phys 39 (2):285–290

Johnson AG, Utter FM, Hodgins HO (1972) An electrophoretic investigation of the family Scorpaenidae. Fish Bull 70:403–413

Johnson AG, Utter FM, Hodgins HO (1973) Estimate of genetic polymorphism and heterozygosity in three species of rockfish (genus *Sebastes*). Comp Biochem Phys 44B (2):397–406

Johnson GB (1976) Genetic polymorphism and enzyme function. In: Molecular evolution. Sinauer Assoc, Sunderland, pp 46–59

Johnson JE (1968) Albinistic carp, *Cyprinus carpio,* from Roosevelt Lake, Arizona. Trans Am Fish Soc 97 (2):209–210

Johnson MS (1971) Adaptive lactate dehydrogenase variation in the crested blenny, *Anoplarchus.* Heredity 27 (2):205–226

Johnson MS (1974) Comparative geographic variation in *Menidia.* Evolution 28 (3):607–618

Johnson MS (1976) Biochemical systematics of the atherinid genus *Menidia.* Copeia 4:662–691

Johnson MS (1977) Association of allozymes and temperature in the crested blenny *Anoplarchus purpurescens.* Mar Biol 41 (2):147–152

Johnson MS, Mickevich MF (1977) Variability and evolutionary rates of characters. Evolution 31 (3):642–648

Johnston R, Simpson TH, Youngson AF (1978) Sex reversal in salmonid culture. Aquaculture 13 (2):115–134

Jurča V (1974) Electroforeza hemoglobinei unor specii si linii de ciprinide in gel de poliacrilamida. Stud Cerc Biochim 17 (3):259–264

Jurča V, Matei G (1975) Proteinele din musclii scheletici la unele specii si linii de ciprinide. Stud Cerc Biochim 18 (2):115–118

Kabai P, Csányi V (1979) Genetical analysis of tonic immobility in two subspecies of *Macropodus opercularis.* Acta Biol 29:3, 295–298

Kaidanova TI (1974) The study of chromosomal polymorphism in populations of the rainbow trout *Salmo irideus* G. and the brook trout *Salmo trutta* m. *fario* L. Izvestija. Gosud. nauchno-issled Inst Ozern Rechn Rybn Khos (GosNIORKh) 97:155–158 (RE)

Kaidanova TI (1976) Comparative study of chromosomal polymorphism in "Danish" and "Gostilitzian" populations of rainbow trout. Izvestija Gosud. nauchno-issled Inst Ozern Rechn Rybn Khos (GosNIORKh) 113:71–76 (RE)

* Kaidanova TI (1978) Study of karyotypes in two species of the whitefish. Izvestija Gosud nauchno-issled. Inst Ozern Rechn Rybn Khos (GosNIORKh) 130:50–55 (RE)

* Kaidanova TI (1980) A comparative karyological analysis of some salmon and whitefish species. In: Karyological variability, mutagenesis and gynogenesis of fishes. Inst Cytol Acad Sci USSR, Leningrad, pp 16–23 (RE)

Kaidanova TI, Efanov GV (1976) Karyotype of Chudskoy whitefish. Izvestija Gosud. nauchno-issled Inst Ozern Rechn Rybn Khos (GosNIORKh) 107:94–97 (RE)

Kajishima T (1965) Heredities of de-coloration in the goldfish. Jpn J Genet 40:397–398

Kajishima T (1975) In vitro analysis of gene depression in goldfish choroidal melanophores. J Exp Zool 191 (1):121–126

Kajishima T (1977) Genetic and developmental analysis of some new color mutants in the goldfish, *Carassius auratus.* Genetics, 86 (1):161–174

Kajishima T, Takeuchi IK (1977) Ultrastructural analysis of gene interaction and melanosome differentiation in the retinal pigment cells of the albino goldfish. J Exp Zool 200 (3):349–376

Kallman KD (1962a) Gynogenesis in the teleost *Mollienesia formosa* (Girard) with a discussion of the detection of parthenogenesis in vertebrates by tissue transplantation. J Genet 58 (1):7–21

Kallman KD (1962b) Population genetics of the gynogenetic teleost, *Mollienesia formosa.* Evolution 16 (4):497–504

Kallman KD (1964a) Genetics of tissue transplantation in isolated platyfish populations. Copeia 3:513–522

Kallman KD (1964b) An estimate of the number of histocompatibility loci in the teleost *Xiphophorus maculatus.* Genetics 50 (4):583–595

Kallman KD (1965a) Sex determination in the teleost *Xiphophorus milleri.* Am Zool 5 (2):246–247

Kallman KD (1965b) Genetics and geography of sex determination in the poeciliid fish, *Xiphophorus maculatus.* Zoologica 50 (1):151–190

Kallman KD (1968) Evidence for the existence of transformer genes for sex in the teleost *Xiphophorus maculatus.* Genetics 60 (4):811–828

Kallman KD (1970a) Different genetic basis of identical pigment patterns in two populations of platyfish, *Xiphophorus maculatus.* Copeia 3:472–475

Kallman KD (1970b) Sex determination and the restriction of sex-linked pigment patterns to the X- and Y-chromosomes in populations of a poeciliid fish, *Xiphophorus maculatus,* from the Belize and Sibun rivers of British Honduras. Zoologica 55 (1):1–16

Kallman KD (1970c) Stable changes in pigment patterns after crossing-over in the teleost *Xiphophorus maculatus* (Abstr). Genet 64 (2, [p]2):32

Kallman KD (1971) Inheritance of melanophore patterns and sex determination in the Montezuma swordtail, *X. montezumae cortezi.* Rosen. Zoologica 56 (3):77–94

Kallman KD (1973) The sex-determining mechanism of the platyfish, *Xiphophorus maculatus.* In: Genetics and mutagenesis of fish. Springer Berlin Heidelberg New York, pp 19–28

Kallman KD (1975) The platyfish, *Xiphophorus maculatus.* In: Handbook of genetics, vol IV. Plenum Press, New York, pp 81–132

Kallman KD, Atz JW (1966) Gene and chromosome homology in fishes of the genus *Xiphophorus.* Zoologica 51 (4):107–135

Kallman KD, Borkoski V (1978) A sex-linked gene controlling the onset of sexual maturity in female and male platyfish (*Xiphophorus maculatus*), fecundity in females and adult size in males. Genetics 89 (1):79–119

Kallman KD, Borowsky R (1972) The genetics of gonadal polymorphism in two species of poeciliid fish. Heredity 28 (2):297–310

Kallman KD, Harrington RW (1964) Evidence for the existence of homozygous clones in the self-fertilizing hermaphroditic teleost *Rivulus marmoratus* (Poey). Biol Bull 26 (1):101–114

Kallman KD, Schreibman MP (1971) The origin and possible genetic control of new stable pigment patterns in the poeciliid fish *Xiphophorus maculatus.* J Exp Zool 176 (2):147–168

Kallman KD, Schreibman MP (1973) A sex-linked gene controlling gonadotrop differentiation and its significance in determining the age of sex maturation and size of the platyfish *Xiphophorus maculatus.* Gen Comp Endocrinol 21 (2):287–304

* Kang YS, Park EH (1973 a) Studies on the karyotypes and comparative DNA values in several Korean cyprinid fishes. Korean J Zool 16 (1):97–108

* Kang YS, Park EH (1973 b) Somatic chromosomes of the Manchurian trout, *Brachymystax lenok* (Salmonidae). CIS 15:10–11

Kanis E, Refstie T, Cjedrem T (1976) A genetic analysis of egg, alevin and fry mortality in salmon (*Salmo salar*), sea trout (*S. trutta*) and rainbow trout (*S. gairdneri*). Aquaculture, 8 (3):259–268

Kao Y-HJ, Farley TM (1978 a) Thermal modulation of pyruvate substrate inhibition in the B_4^{2}'and B_4^{2}'' liver lactate dehydrogenases of rainbow trout, *Salmo gairdneri.* Comp Biochem Phys 60B (2):153–155

Kao Y-HJ, Farley TM (1978 b) Purification and properties of allelic lactate dehydrogenase isozymes at the B^2 locus in rainbow trout, *Salmo gairdneri.* Comp Biochem Phys 61B (4):507–512

Karpenko IM (1966) Hybrids between the wild and domestical common carp. Kamenjar, Lvov (R)

Kartavtsev YuF (1975) Comparative electrophoretic analysis of haemoglobins, water-soluble muscle proteins and eye lenses in five species of cottids (Cottidae). Biol Morya 2:31–38 (R)

Katasonov VYa (1971) Results of study of the Japanese decorative common carps and their hybrids. In: Development of pond fish breeding and the rational use water-basins and artificial lakes. All-Union Res Inst Pond Fish, Moscow, pp 223–225 (R)

Katasonov VYa (1973) A study of pigmentation in hybrids between the common and the decorative Japanese carp. I. A study of the dominant pigmentation types. Genetika (Moscow) 9 (8):59–69 (RE)

Katasonov VYa (1974 a) A study of pigmentation in hybrids between the common and decorative Japanese carp. II. The pleiotropic action of dominant pigmentation genes. Genetika (Moscow) 10 (12):56–66 (RE)

Katasonov VYa (1974 b) The use of Japanese decorative carps for the development of genetically labelled carp stocks. Tr VNIIPRKh (Moscow) 23:10–19 (RE)

Katasonov VYa (1976) The lethal action of the gene of light pigmentation in the common carp *Cyprinus carpio* L. Genetika (Moscow) 12 (4):152–161 (RE)

Katasonov VYa (1978) A study of pigmentation in hybrids between the common and the decorative Japanese carp. III. The inheritance of blue and orange patterns of pigmentation. Genetika (Moscow) 14 (12):2184–2192 (RE)

Kato T (1974) On domestication of salmonid fishes. Fish Cult 11 (10):56–59

Kaushik NK (1960) On the absence of pelvic fins in *Cirrhina mrigala* (Ham.) and anal fin in *Catla catla* (Ham.) Curr Sci 29 (8):316–317

Keese A, Langholz HJ (1974) Electrophoretische Studies zur Populationsanalyse bei der Regenbogenforelle. Z Tierzücht Zucht Biol 91 (1–2):109–124

Kempf CT, Underhill DK (1974) A serum esterase polymorphism in *Fundulus heteroclitus.* Copeia 3:792–794

Kepes KL, Whitt GS (1972) Specific lactate dehydrogenase gene function in the differenciated liver of cyprinid fish. Genetics 71 (Suppl):29

Kerrigan AM (1934) The inheritance of the crescent and twin spot marking in *Xiphophorus helleri.* Genetics 19 (6):581–599

Keyvanfar A (1962) Serologie et immunologie de deux especes de thonides (*Germo alalunga* Gm et *Th. thynnus* L.) de l'Atlantique et de la Mediterranée. Rev Trav Inst Pesh Mar 26 (4):407–450

Khadginov MI (1935) Heterosis. In: Theoretical bases of plant selection, vol I, Selkhosgis, Moscow Leningrad, pp 435–490 (R)

Khanna ND, Juneja RK, Larsson B (1975 a) Electrophoretic studies on esterases in the Atlantic salmon, *Salmo salar* L. Swed J Agric Res 5 (4):193–197

Khanna ND, Juneja RK, Larsson B (1975 b) Electrophoretic studies on proteines and enzymes in the Atlantic salmon, *Salmo salar* L. Swed J Agric Res 5 (4):185–192

* Khuda-Bukhsh AR (1979a) Chromosomes in three species of fishes, *Aplocheilus panchax* (Cyprinodontidae), *Lates calcerifer* (Percidae) and *Gadusia charpa* (Clupeidae). Caryologia 32 (2): 161–169

* Khuda-Bukhsh AR (1979b) Karyology of 2 species of hillstream fishes, *Barilius bendelisis* and *Rasbora daniconius* (Cyprinidae). Curr Sci 48 (17): 793–795

* Khuda-Bukhsh AR, Manna GK (1974) Somatic chromosomes in seven species of teleostean fishes. CIS 17: 5–6

Kijima A, Fujio Y (1978) Geographic distribution of IDH and LDH isozymes in chum salmon populations. Bull Jpn Soc Sci Fish 45 (3): 287–296

Kiknadze II (1972) The functional organization of the chromosomes. Nauka Leningrad (R)

Kimura Masao (1976) Hemoglobin electrophoretic patterns of the loach, *Misgurnus anguillicaudatus*. Jpn J Genet 51 (2): 143–145

Kimura Masao (1978a) Protein polymorphism and geographic variation in the loach *Misgurnus anguillicaudatus*. Anim Blood Gr Biochem Genet 9 (1): 13–20

Kimura Masao (1978b) Protein polymorphism and genic variation in a population of the loach *Cobitis delicata*. Anim Blood Gr Biochem Genet 9 (3): 183–186

Kimura Masao (1978c) Phosphoglucomutase electrophoretic patterns of the loach *Cobitis biwae*. Anim Blood Gr Biochem Gen 9 (3): 187–190

Kimura Motoo (1968a) Evolutionary rate at the molecular level. Nature (London) 217 (5129): 624–626

Kimura Motoo (1968b) Genetic variability maintained in a finite population due to mutational production of neutral and nearly neutral isoalleles. Genet Res 11 (3): 247–269

Kimura Motoo (1977) The neutral theory of molecular evolution and polymorphism. Scientia 112 (9–12): 687–707

Kimura Motoo, Ohta T (1969) The average number of generation until fixation of a mutant gene in a finite population. Genetics 61 (3): 763–771

Kimura Motoo, Ohta T (1974) On some principles governing molecular evolution. Proc Natl Acad Sci USA 71 (7): 2848–2852

Kincaid HL (1972) A preliminary report of the genetic aspects of 150-day family weight in hatchery rainbow trout. In: Proc 52nd Annu Conf West Assoc State Game Fish Comm Portland, pp 562–565

Kincaid HL (1975) Iridescent metallic blue color variant in rainbow trout. J Hered 66 (2): 100–101

Kincaid HL (1976) Inbreeding in rainbow trout (*Salmo gairdneri*). J Fish Res B Can 33 (11): 2420–2426

Kincaid HL, Bridges WR, Limbach B von (1977) Three generations of selection for growth rate in fall-spawning rainbow trout. Trans Am Fish Soc 106 (6): 621–628

Kingsburg N, Masters GI (1972) Heterogeneity, molecular weight interrelationships and developmental genetics of the esterase isoenzymes of the rainbow trout. Biochem Biophys Acta 258 (2): 455–465

* Kirby RF, Thompson UW, Hubbs Cl (1977) Karyotypic similarities between the mexican and blind tetras. Copeia 3: 578–580

Kirkpatrick M, Selander RK (1979) Genetics of speciation in lake whitefishes in the allegash basin. Evolution 33 (1, p 2): 478–485

Kirpichnikov VS (1935a) The biological and systematic essay in the smelt of the White Sea, Cheshskaya bay and Pechora river. Tr VNIRO 2: 103–194 (R)

Kirpichnikov VS (1935b) The autosomal genes of *Lebistes reticulatus* and the problem of genetic sex determination. Biol Zh 4 (2): 343–354 (RE)

Kirpichnikov VS (1937) A major genes for scale cover in the common carp. Biol Zh 6 (3): 601–632 (RE)

Kirpichnikov VS (1943) Experimental taxonomy of the wild carp *Cyprinus carpio* L. I. Growth and morphological characteristics of the Taparavan, Volga-Caspian and Amur wild carps under the conditions of pond rearing. Izv Akad Nauk SSSR 4: 189–220 (R)

Kirpichnikov VS (1945) The effect of rearing conditions on viability, growth of rate and morphology of the carps with different genotypes. Dokl Akad Nauk SSSR (DAN) 47 (7): 521–524 (RE)

Kirpichnikov VS (1948) A comparative characteristics of four major varieties of the cultured common carp cultivated in the North of the USSR. Izvestija Vseross. nauchno-ussled. Inst Ozern Rechn Rybn Khos (VNIORKh) 26: 145–170 (R)
Kirpichnikov VS (1949) The Amur wild carp in the North of the USSR. Rybnoe Khoziaystvo 8:39–44 (R)
Kirpichnikov VS (1958a) Genetic methods of selection for relatives in carp breeding. Dokl Akad Nauk SSSR (DAN) 121 (4):682–685 (RE)
Kirpichnikov VS (1958b) The extent of heterogeneity in populations of the wild common carp and hybrids between the wild and domesticated carp. Dokl Akad Nauk SSSR (DAN) 122 (4):716–719 (RE)
Kirpichnikov VS (1959) Genetic methods of fish selection. Bull Mosk O-va Ispyt Prir 64 (1):121–137 (R)
Kirpichnikov VS (1960) Organization of carp selective breeding. Nauchno-Tech Bull VNIORKh 11:38–40 (R)
Kirpichnikov VS (1961) Die genetischen Methoden der Selektion in der Karpfenzucht. Z Fisch 10 (1–3):137–163
Kirpichnikov VS (1962) Hybridization of the domesticated and wild common carp. In: Proc 2nd Plenary Meet Commitee Invest West Pacific, Moscow, pp 162–169 (R)
Kirpichnikov VS (1966a) Goals and methods of common carp selection. Izvestija Gosud. nauchno-issled. Inst Ozern Rechn Rybn Khos (GosNIORKh) 61:7–28 (RE)
Kirpichnikov VS (1966b) Methods of progeny testing in common carp farms. Izvestija Gosud. nauchno-issled Inst Ozern Rechn Rybn Khos (GosNIORKh) 61:40–61 (RE)
Kirpichnikov VS (1967a) The hybridization of the European domesticated common carp with the Amur wild carp and selection of hybrids. Dissertation, Zool Inst, Leningrad (R)
Kirpichnikov VS (1967b) Homologous hereditary variation and evolution of the wild carp (*Cyprinus carpio* L.). Genetika (Moscow) 3 (2):34–47 (RE)
Kirpichnikov VS (1967c) A general theory of heterosis. I. Genetic mechanisms of heterosis. Genetika (Moscow) 3 (10):167–180 (RE)
Kirpichnikov VS (1968) Efficiency of mass selection and selection for relatives in fish culture. FAO Fish Rep (44) Rome, 4:179–194
Kirpichnikov VS (1969a) The present state of fish genetics. In: Genetics, selection and hybridization of fishes, Nauka, Moscow, pp 9–29 (R)
Kirpichnikov VS (1969b) A theory of fish selection. In: Genetics, selection and hybridization of fishes. Nauka, Moscow, pp 44–58 (R)
Kirpichnikov VS (1971a) Genetics of the common carp (*Cyprinus carpio* L.) and other edible fishes. In: Rep FAO/UNDP(TA), (2926), Rome, pp 186–201
Kirpichnikov VS (1971b) Methods of fish selection. I. Aims of selection and methods of artificial selection. 2. Crossing, modern genetic methods of selection. In: Rep FAO/UNDP-(TA), (2926), Rome, pp 202–216, pp 217–227
Kirpichnikov VS (1972a) Methods and efficiency of selection of the Ropsha common carp breed. I. Goals of the selection, initial forms and the system of crosses. Genetika 8 (8):65–72 (RE)
Kirpichnikov VS (1972b) Biochemical polymorphism and the problem of so-called non-Darwinian evolution. Uspechi Sovrem Biol 74 [2(5)]:231–246 (R)
Kirpichnikov VS (1973a) Biochemical polymorphism and microevolution processes in fish. In: Genetics and mutagenesis of fish. Springer Berlin Heidelberg New York, pp 223–241
Kirpichnikov VS (1973b) On karyotype evolution in Cyclostomata and Pisces. Ikhthiologia 5 (1):55–77
Kirpichnikov VS (1973c) Use of genetic selection in commercial fish breeding in USSR and Eastern European countries (the present status and the prospectives). Tr VNIIPRKh 21:94–108 (R)
Kirpichnikov VS (1974a) Genetic mechanisms and evolution of heterosis. Genetika (Moscow) 10 (4):165–179 (RE)
Kirpichnikov VS (1974b) Adaptive nature of fish biochemical polymorphism. In: Functional morphology, genetics and biochemie of cell. Inst Cytol, Leningrad, pp 320–322 (R)
Kirpichnikov VS (1977) Selective character of biochemical polymorphism in Kamchatka sockeye salmon *Oncorhynchus nerka* (Walb.). In: The principles of the classification and phylogeny in the salmonoid fishes. Zool Inst, Leningrad, pp 53–60 (R)

Kirpichnikov VS (1979) Functional differences between isozymes (isoformes) and between allelic variants of proteins in fishes. In: Biochem Popul Genet Fishes. Inst Cytol Acad Sci USSR, Leningrad, pp 5–9 (RE)

Kirpichnikov VS, Balkhashina EI (1935) Materials on genetics and selection of the common carp. I. Zool Zh 14 (1):45–78 (RE)

Kirpichnikov VS, Balkhashina EI (1936) Materials on genetics and selection of the common carp, II. Biol Zh 5 (2):327–376 (RE)

Kirpichnikov VS, Faktorovitsch KA (1969) Genetische Methoden der Fischkrankheitsbekämpfung. Z Fisch 17 (1–4):227–236

Kirpichnikov VS, Faktorovitsch KA (1972) The increase of the common carp resistance to dropsy by selection. II. The course of selection and the evaluation of the selected stocks. Genetika (Moscow) 8 (5):44–54 (RE)

Kirpichnikov VS, Golovinskaya KA (1966) Characteristic of spawners of the main carp bred stocks, reared in USSR. Izvestija Gosud. nauchno-issled. Inst Ozern Rechn Rybn Khos (GosNIORKh) 61:28–39 (RE)

Kirpichnikov VS, Ivanova IM (1977) Space, temporal and agedependent variation in Pacific sockeye salmon for Ldh-BI and Pgm-I loci. Genetika (Moscow) 13 (7):1183–1193 (RE)

Kirpichnikov VS, Katasonov VJa (1978) Genetical-selection investigations and present state of breeding work in pond fisheries of USSR. In: Genetics and selection of fishes, vol 20. All-Union Res Inst Pond Fish Moscow, pp 3–51 (RE)

Kirpichnikov VS, Muske GA (1979) Functional differences between allozymes in Pacific sockeye salmon (*Oncorhynchus nerka* Walb.). In: Materials XVI Int Conf Anim Blood Gr Biochem Polym, vol IV. Leningrad, pp 228–234

Kirpichnikov VS, Muske GA (1980) The adaptive value of biochemical polymorphisms in animal and plant populations. In: Animal genetics and evolution. V Junk B V Publ, Hague, pp 183–193

Kirpichnikov VS, Shart LA (1976) Acceleration in the succession of common carp generations in the course of selective breeding of carp in southern regions of the USSR. Tr VNIIPRKh 23:55–63 (RE)

Kirpichnikov VS, Golovinskaya KA, Mikhailov FN (1937) Major types of the scaling pattern in the carp and their relation to the economically important traits. Rybn Khoz 10–11:51–59 (R)

Kirpichnikov VS, Factorovich KA, Babushkin JuP, Ninburg EA (1971) Selection of common carp for resistance to dropsy. Izvestija Gocud. nauchno-issled. Inst Ozern Rechn Rybn Khos. (GosNIORKh) 74:140–153 (RE)

Kirpichnikov VS, Faktorovich KA, Suleymanyan VS (1972a) The increase of the common carp resistance to dropsy by selection. I. Methods of selection for higher resistance. Genetika (Moscow) 8 (3):34–41 (RE)

Kirpichnikov VS, Ponomarenko KV, Tolmacheva NV, Tzoy RM (1972b) Methods and effectiveness of the Ropsha common carp selection. II. Methods of artificial selection. Genetika (Moscow), 8 (9):42–53 (RE)

Kirpichnikov VS, Faktorovich KA, Shart LA (1976) Selection of common carp for resistance to dropsy. Izvestija Gosud. nauchno-issled. Inst Ozern Rechn Rybn Khos (GosNIORKh) 106:16–28 (RE)

Kirpichnikov VS, Ilijasov JuI, Shart LA, Faktorovich KA (1979) Selection of common carp (*Cyprinus carpio*) for resistance to dropsy. In: Advances in aquaculture. Fish News Books Ltd, Farnham, pp 628–633

Kirschbaum F (1977) Zur Genetik einiger Farbmusternmutanten der Zebrabarbe *Brachydanio rerio* (Cyprinidae, Teleostei) und zum Phänotyp von Artbastarden der Gattung *Brachydanio.* Biol Zentralbl 96 (2):211–222

Kirsipuu A, Tammert M, Haberman H, Laugaste K (1972) Connection between electrophoretic fractions of blood serum proteins and some indices of productivity in bream. In: Proc 12th Eur Conf Anim Blood Gr Biochem Polym, Budapest, pp 597–600

* Kitada J-I, Tagawa M (1972) On the chromosomes of the racefield eel (*Fluta alba = Monopterus albus*). Kromosomo 88–89:2804–2807

* Kitada J-I, Tagawa M (1975) Somatic chromosomes of three species of *Cyclostomata.* CIS 18:10–12

Klar GT, Stalnaker GB (1979) Electrophoretic variation in muscle lactate dehydrogenase in Snake Valley cutthroat trout, *Salmo clarkii* subsp. Comp Biochem Phys 64B (2):391–394

Klar GT, Stalnaker GB, Farley TM (1979) Comparative blood lactate response to low oxygen concentrations in rainbow trout, *Salmo gairdneri,* LDH B2 phenotypes. Comp Biochem Phys 63A (2):237–240

Klose J, Wolf U (1970) Transitional hemizygosity of the maternally derived allele at the 6 PGD locus during early development of the cyprinid fish *Rutilus rutilus.* Biochem Genet 4 (1):87–92

Klose J, Wolf U, Hitzeroth H, Ritter H, Atkin NB, Ohno S (1968) Duplication of LDH gene loci by polyploidization in the fish order Clupeiformes. Humangenetics 5 (3):190–196

Klose J, Hitzeroth H, Ritter H, Schmidt E, Wolf U (1969 a) Persistance of maternal isoenzyme patterns of the lactate dehydrogenase and phosphoglucomutase system during early development of the hybrid trout. Biochem Genet 3 (1):91–97

Klose J, Wolf U, Hitzeroth H, Ritter H, Ohno S (1969 b) Polyploidization in the fish family Cyprinidae, order Cypriniformes. II. Duplication of the gene loci coding for LDH (E.C. 1.1.1.27) and 6 PGDH (E.C. 1.1.1.44) in various species of Cyprinidae. Humangenetics 7 (3):245–250

* Kobayashi H (1971) A cytological study on gynogenesis of the triploid ginbuna (*Carassius auratus langsdorfii*). Zool Mag 80 (9):316–322

Kobayashi H (1976) Comparative study of karyotypes in the small and large races of spinous loaches (*Cobitis biwae*). Zool Mag 85 (1):81–87

Kobayashi H, Ochi H (1972) Chromosome studies of the hybrids, ginbuna (*Carassius auratus langsdorfii*) x kinbuna (*C. auratus* subsp.) and ginbuna x loach (*Misgurnus anguillicaudatus*). Zool Mag 81 (2):67–71

* Kobayashi H, Kawashima J, Takeuchi N (1970) Comparative chromosome studies in the genus *Carassius* especially with a finding of polyploidy in the ginbuna (*C. auratus langsdorfii*). Jpn J Ikhthiol 17 (4):153–160

Kobayashi H, Ochi H, Takeuchi N (1973) Chromosome studies in the genus *Carassius:* comparison of *C. auratus grandoculis, C. a. buergeri* and *C. a. langsdorfii.* Jpn J Ikhthiol 20 (1):7–12

Kobayashi H, Nakano K, Nakamura M (1977) On the hybrids, 4 n ginbuna (*Carassius auratus langsdorfii*) x kinbuna (*C. auratus* supsp.) and their chromosomes. Bull Jpn Soc Sci Fish 43 (1):31–37

Koch HJA, Wilkins NP, Bergström E, Evans JC (1967) Further studies of the multiple components of the haemoglobins of *Salmo salar* L. Meded Vlaam Acad 29 (7):1–16

Koehn RK (1968) The component of selection in the maintenance of a serum esterase polymorphism. Proc 12th Int Congr Genet Tokyo 1:1227

Koehn RK (1969 a) Esterase heterogeneity: dynamic of a polymorphism. Science 163 (3870):943–944

Koehn RK (1969 b) Hemoglobins of fishes of the genus *Catostomus* in western North America. Copeia 1:21–30

Koehn RK (1970) Functional and evolutionary dynamics of polymorphic esterases in catostomid fishes. Trans Am Fish Soc 99 (1):219–228

Koehn RK (1971) Biochemical polymorphism: a population strategy. Rapp P-V Reun 161:147–153

Koehn RK, Johnson DW (1967) Serum transferrin and serum esterase polymorphisms in an introduced population of the bigmouth buffalofish, *Ictiobus cyprinellus.* Copeia 4:805–809

Koehn RK, Rasmussen DI (1967) Polymorphic and monomorphic serum esterase heterogeneity in catostomid fish populations. Biochem Genet 1 (2):131–144

Koehn RK, Williams GG (1978) Genetic differentiation without isolation in the American eel, *Anguilla rostrata.* II. Temporal stability of geographic pattern. Evolution 32 (3):624–637

Koehn RK, Peretz JE, Merritt RB (1971) Esterase enzyme function and genetical structure of population of the freshwater fish, *Notropis stramineus.* Am Nat 105 (941):51–68

Kok Leng Tay, Garside ET (1972) Meristic comparisons of populations of mummichog *Fundulus heteroclitus* (L) from Sable Island and mainland Nova Scotia. Can J Zool 50 (1):13–17

Konishi Y, Taniguchi N (1975) Polymorphism in the liver esterase pattern of the sparid fish *Dentex tunifrons.* Jpn J Ikhthiol 21 (4):220–222

Konovalov SM (1979) Population biology of Pacific salmon. Dissertation, Inst Zool, Leningrad, (R)

Konradt AG (1973) Problems of phytophagous fish selective breeding. Izvestija Gosud. nauchno-issled. Inst Ozern Rechn Rybn Khos (GosNIORKh) 85:2–9 (RE)

Kornfield IL (1978) Evidence for rapid speciation in African cichlid fishes. Experientia 34 (3):335–336

Kornfield IL, Koehn RK (1975) Genetic variation and speciation in New World cichlids. Evolution 29 (3):427–437

Kornfield IL, Nevo E (1976) Likely pre-Suez occurrence of a Red Sea fish *Aphanius dispar* in the Mediterranean. Nature (London) 264 (5583):289–291

Kornfield IL, Ritte U, Richler C, Wahrman J (1979) Biochemical and cytological differentiation among cichlid fishes of the sea of Galilee. Evolution 33 (1, p 1):1–14

Korochkin LI (1976a) Interaction of genes in development. Nauka, Moscow (R)

Korochkin LI (1976b) Genetics of isoenzymes and development. Ontogenez 7 (1):3–17 (RE)

Korochkin LI (1977) Activity of genes controlling isozyme synthesis in animal development. In: Genetics of isoenzymes, Nauka, Moscow, pp 149–167 (R)

Korochkin LI, Serov OL, Manchenko GP (1977) Concept of isoenzymes. In: Genetics of isoenzymes. Nauka, Moscow, pp 5–17 (R)

Korovina VM, Golovkov GA, Lebedeva LI, Prirodina VP (1972) Morphological peculiarities of reciprocal hybrids between *Coregonus nasus* and *C. peled*, I. Vopr Ikhthiol 12 (3):490–503 (R)

Korovina VM, Lebedeva LI, Prirodina VP (1973) Morphological peculiarities of reciprocal hybrids between *Coregonus nasus* and *C. peled*, 2. Vopr Ikhthiol 13 (3):423–435 (R)

Kossmann H (1971) Hermaphroditismus und Autogamie beim Karpfen. Naturwissenschaften 58 (6):328–329

Kossmann H (1972) Untersuchungen über die genetische Varianz der Zwischenmuskelgräten des Karpfens. Theoret Appl Genet 42 (2):130–135

Kosswig C (1929) Zur Frage der Geschwülstbildung bei Gattungsbastarden der Zahnkarpfen *Xiphophorus* und *Platypoecilus*. Z Indukt Abstammungs-Vererbungsl 52 (1):114–120

Kosswig C (1935a) Genotypische und phänotypische Geschlechtsbestimmung bei Zahnkarpfen und ihren Bastarden. VI. Über polyfactorielle Geschlechtsbestimmung. Roux' Arch Entw-Mech 133 (1):140–195

Kosswig C (1935b) Über Albinismus bei Fischen. Zool Anz 110 (1):41–47

Kosswig C (1936) Homogametische ZZ- und WW- Weibchen entstehen nach Artkreuzung mit dem in weiblichen Geschlecht heterogametischen *Platypoecilus maculatus*. Biol Zentralbl 56 (2):409–414

Kosswig C (1937a) Über die veränderte Wirkung von Farbgenen in fremden Genotypen. Biol Gen 13:276–293

Kosswig C (1937b) Kreuzungen mit *Platypoecilus xiphidium*. Roux' Arch Entw-Mech 136 (3):491–528

Kosswig C (1954) Zur Geschlechtsbestimmungs-Analyse bei den Zahnkarpfen. Rev Fac Sci Univ Istanbul Ser B 19 (3):187–190

Kosswig C (1961) Über sogenannte homologische Gene. Zool Anz 166 (9–12):333–356

Kosswig C (1963) Genetische Analyse Konstruktiver und degenerativer Evolutionsprozesse. Z Zool Syst Evol-Forsch 1:205–239

Kosswig C (1964a) Problems of polymorphism in fishes. Copeia 1:65–75

Kosswig C (1964b) Polygenic sex determination. Experientia 20 (1):1–10

Kosswig C (1965) 40 Jahre genetische Untersuchungen an Fischen. Abh und Verh Naturwiss Ver Hamburg 10:13–39

Kosswig C (1973) The role of fish in research on genetics and evolution. In: Genetics and mutagenesis of fish. Springer Berlin Heidelberg New York, pp 3–16

Koval EZ, Bogdanov LV (1979) Biochemical polymorphism of 9 flatfish species (subfamily Pleuronectinae) of Peter the Great Bay. In: Biochem Popul Genet Fish. Inst Cytol Acad Sci USSR, Leningrad, pp 99–105 (RE)

Koval LI (1976) Intraspecies heterogeneity of the North Sea sprat. In: Ecologic physiology of fishes, vol II. Naukova Dumka Kiev, pp 51–52 (R)

Krajnović-Ozretić M, Zikić R (1975) Esterase polymorphism in the Adriatic sardine (*Sardine pilchardus* Walb.). I. Electrophoretic and biochemical properties of the serum and tissue esterases. Anim Blood Gr Biochem Genet 6 (4):201–213

* Krasznai Z, Márián T (1978) Results of karyological and serological investigations on *Silurus glanis*. In: Increasing productivity of fishes by selection and hybridization. F Müller, Sharvas, pp 112–120

Krishnaja AP, Rege MS (1977) Haemoglobin heterogeneity in two species of the Indian carp and their fertile hybrids. Ind J Exp Biol 15 (10):925–926

Krogius FV (1975) Population dynamics and fry growth in the sockeye salmon *Oncorhynchus nerka* Walb. from the Lake Dalneye (Kamchatka). Vopr Ikhthiol 15 [4(93)]:612–629 (R)

Krogius FV (1978) The significance of genetic and ecological factors in the population dynamics of the sockeye salmon *Oncorhynchus nerka* (Walb.) from the Lake Dalneye. Vopr Ikhthiol 18 [2(109)]:211–221 (R)

Krueger CC, Menzel BW (1979) Effects of stocking on genetics of wild brook trout populations. Trans Am Fish Soc 108 (3):277–287

Kryazheva KV (1966) The influence of stocking density upon growth, variation and survival of the fry of hybrid common carp. Izvestija Gosud. nauchno-issled. Inst Ozern Rechn Rybn Khos (GosNIORKh) 61:80–101 (RE)

* Krysanov OYu (1978) The variation of chromosomal numbers in the herring. In: Ecology of the White Sea fishes. Nauka, Moscow, pp 94–97 (R)

Kryzhanovsky CG (1947) The system of cyprinids. Zool Zh 26 (1):53–64 (R)

Kucherenko AP (1978) Selection of 2d and 3d generations of the Ukrainian (Nivchan) scaly common carp and its fishery-biological characteristic. Ph D Thesis, Moscow Univ (R)

Kucherenko AP (1979) Estimation of fertility and of the sperm quality in the Ukrainian (Nivchan) scaly carp. In: Materials of All-Union Scientific Conference on development and intensification of fishieries in inland waters of North Caucasus, Rostov-Don, pp 126–128 (R)

Kuhl P, Schmidtke J, Weiler C, Engel W (1976) Phosphoglucose isomerase isozymes in the characid fish *Cheirodon axelrodi:* evidence for a spontaneous gene duplication. Comp Biochem Phys 55B (2):279–281

Kühnl P, Spielmann W (1978) Investigation on the polymorphism of phosphoglucomutase (PGM, 2.7.5.1.) by isoelectric focusing on polyacrylamide gels. In: XIV Int Congr Genet Moscow, Contrib Pap Session Abstr 1:134

Kulikova NI, Salmenkova EA (1979) An electrophoretic study of muscle proteins of Amur summer chum salmon and Amur pink salmon (*Oncorhynchus keta* and *O. gorbuscha*). In: Biochem Popul Genet Fish. Inst Cytol Acad Sci USSR, Leningrad, pp 125–128 (RE)

Kusakina AA (1959) The cytophysiological investigation of the muscle tissue in certain interspecific fish hybrids showing heterosis. Tsitologiya 1 (1):111–119 (RE)

Kusakina AA (1964) Increased stability of muscle proteins in the hybrids of the ripus and the ludoga whitefishes showing heterosis. Tsitologiya 5 (4):493–495 (RE)

Kusakina AA (1967) Thermostability of aldolase and choline esterase in related species of the poikilotherm animals. In: Variation of thermostability of animal cells during onto- and phylogenesis. Nauka, Leningrad, pp 242–249 (RE)

Kusen, SJ, Stojka RS (1980) The utilization of $AgNO_3$ for identification of lactate dehydrogenase in the loach (*Misgurnus fossilis*). Isoz Bull 13:82

Kuzema AI (1953) Ukrainian breeds of the carp. In: Proc Conf Pond Fish Breeding, Moscow, pp 65–70 (R)

Kuzema AI (1961) Experiments on progeny testing in common carp breeding. Nauk Pracy Ukrain Inst Rybn Gosp 13:72–84 (Ukrainian)

Kuzema AI (1962) Evaluation of common carp spawners by progeny testing. Nauk Pracy Ukrain Inst Rybn Cosp 14:71–84 (Ukrainian)

Kuzema AI, Tomilenko VG (1962) Reserves for the increase of pond fish productivity. Nauk Pracy Ukrain Inst Rybn Gosp 14:85–88 (Ukrainian)

Kuzema AI, Tomilenko VG (1965) Development of new breeds of the common carp using the method of distant hybridization. Rybn Khoz (Kiev) 2:3–17 (R)

Kuzema AI, Kucherenko AP, Tomilenko VG (1968) Economic efficiency of rearing of Ropsha-Ukrainian hybrid common carp. Rybn Khoz 6:68–74 (R)

Kuzema AI, Kucherenko AP, Tomilenko VG (1970) The development of a new strain of the Ukrainian scaled carp breed (UNK-59). Rybn Khoz (Kiev) 10:3–11 (R)

Kwain W (1975) Embryonic development, early growth and meristic variation in rainbow trout (*Salmo gairdneri*) exposed to combinations of light intensity and temperature. J Fish Res B Can 32 (3):397–402

Kynard BE (1979) Population decline and change in frequencies of lateral plates in threespine sticklebacks (*Gasterosteus aculeatus*). Copeia 4:635

Kynard B, Curry K (1976) Meristic variation in the threespine stickleback, *Gasterosteus aculeatus* from Auke Lake, Alaska. Copeia 4:811–816

Lagler KF, Bailey RM (1947) The genetic fixity of differential characters in subspecies of the percid fish, *Boleosoma nigrum*. Copeia 1:50–59

* Laliberte MF, Lafaurie M, Lambert JC, Ayraud N (1979) Etude préliminaire du caryotype de *Mullus barbatus* Linne. Rapp P-V Reun 25–26 (10):125–130

* Law WM, Ellender RD, Wharton JH, Middlebrooks BL (1978) Fish cell culture: properties of a cell line from the sheepshead, *Archosargus probatocephalus*. J Fish Res B Can 23 (4):767–768

Lee JV (1965) Observations sur la sérologie et l'immunologie des trons rouges (*Thunnus thynnus*) L. de Mediterranee. Rapp P-V Reun 18:225–228

* Legendre P, Steven DM (1969) Dénombrement des chromosomes chez quelques cyprins. Nat Can 96:913–918

* Le Grande WH (1975) Karyology of sic species of Louisiana flatfishes (Pleuronectiformes: Osteichthyes). Copeia 3:516–522

* Le Grande WH (1978) Cytotaxonomy and chromosomal evolution in North American catfishes (Siluriformes, Ictaluridae) with emphasis on *Naturus*. Th D Thesis, Ohio St Univ, Colbumia

* Le Grande WH, Fitzsimons IM (1976) Karyology of the mulets *Mugil curema* a. *M. cephalus* (Perciformes; Mugilidae) from Louisiana. Copeia 2:388–391

Leibel WS, Markert CL (1978) Preliminary notes on the evolution of fish esterases. I. An electrophoretic survey of Carribean reef fishes. Isoz Bull 11:58

Lemanova NA (1960) Comparative and experimental analysis of interspecific hybrids of the genus *Coregonus*. In: Distant hybridization of plants and animals, Acad Sci USSR, Moscow, pp 511–519 (R)

Lemanova NA (1965) The study of diagnostic traits, biology and gametogenesis in the reciprocal hybrids *Coregonus albula* x *C. lavaretus ludoga*. Ph D Thesis, Leningrad (R)

Lerner IM (1954) Genetic homeostasis. Oliver and Boyd, Edinburgh

Leslie JF (1979) A four point linkage group in *Poeciliopsis monacha* (Pisces; Poeciliidae). Isoz Bull 12:33

Leslie JF, Vrijenhoek RC (1977) Genetic analysis of natural populations of *Poeciliopsis monacha:* allozyme inheritance and pattern of mating. J Hered 68 (5):301–306

Leslie JF, Vrijenhoek RC (1978) Genetic dissection of clonally inherited genomes of *Poeciliopsis*. I. Linkage analysis and preliminary assessment of deleterious gene loads. Genetics 90 (4):801–811

Leuken W, Kaiser V (1972) The role of melanoblasts in melanophore pattern polymorphism of *Xiphophorus* (Pisces, Poeciliidae). Experientia 28 (11):1340–1341

Levan A, Fredga K, Sandberg AA (1964) Nomenclature for centromeric position on chromosomes. Hereditas 52 (2):201–220

* Levin B, Foster NR (1972) Cytotaxonomic studies in Cyprinodontidae: multiple sex chromosomes in *Garmanella pulchra*. Notulae natur. Acad Nat Sci Philad 446:1–5

Lewis RC (1944) Selective breeding of rainbow trout at Hot Creek hatchery. Calif Fish Game 30:95–97

Lewontin RC (1974) The genetic basis of evolutionary change. Columbia Univ Press, New York London

Lewontin RC (1978) Genetic heterogeneity of electrophoretic alleles. In: XIV Int Congr Genet Moscow, Contrib Pap Session Abstr 1:467

Lewontin RC, Hubby JL (1966) A molecular approach to the study of genic heterozygosity in natural populations. II. Amount of variation and degree of heterozygosity in natural populations of *Drosophila pseudoobscura*. Genetics 54 (2):595–609

Li ChCh (1976) First course in population genetics. Boxwood Press, Pacific Grove (Calif)

Li Sh-L, Tomita S, Riggs A (1972) The hemoglobins of the pacific hagfish, *Eptatretus stoutii*. I. Isolation, characterization and oxygen equilibria. Biochem Biophys Acta 278 (2): 344–354

Lieder U (1956) Über einige genetische Probleme in der Fischzucht. Z Fischerei 5 (1–2): 133–142

Lieder U (1957) Die Bewertung der Beschuppung des Karpfen bei der Zuchtauslese. Dtsche Fisch-Z 4: 206–213

Lieder U (1961) Untersuchungsergebnisse über die Grätenzahlen bei 17 Süsswasserfischarten. Z Fischerei 10 (2): 329–350

Lieder U (1963) Über vermutliche Gonosomen bei *Perca, Acerina* und *Anguilla*. Biol Zentralbl 82 (3): 296–302

* Lieppman M, Hubbs C (1969) A karyological analysis of two cyprinid fishes, *Notemigonus chrysoleucas* and *Notropis lutrensis*. Tex Rep Biol Med 27: 427–435

Ligny W de (1966) Polymorphism of serum transferrins in plaice. Proc. 10th Euro Conf Anim Blood Gr Biochem Polym, Paris, pp 373–378

Ligny W de (1968) Polymorphism of plasma esterases in flounder and plaice. Genet Res II (2): 179–182

Ligny W de (1969) Serological and biochemical studies on fish populations. In: Oceanogr Mar Biol Annu Rev 7: 411–513

Ligny W de, Pantelouris EM (1973) Origin of the European eel. Nature (London) 246 (5434): 518–519

Lim ST, Bailey GS (1977) Gene duplication in salmonid fishes: evidence for duplicated but catalytically equivalent A_4 lactate dehydrogenases. Biochem Genet 15 (7–8): 707–721

Lim ST, Kay RM, Bailey GS (1975) Lactate dehydrogenase isozymes of salmonid fish: evidence for unique and rapid functional divergence of duplicated H_4 lactate dehydrogenases. J Biol Chem 250 (5): 1790–1800

Limansky VV (1964) Analysis of the intraspecies differentation of certain fishes in the Black and Azov sea using the precipitation reaction. Tr AzCherNIRO 22: 31–37 (R)

Limansky VV (1965) Detection of differences in the erythrocyte antigens of the large and small forms of the Black Sea horse mackerel using heteroagglutination reaction with the normal human serum. Vopr Ikhthiol 5 (4): 695–697 (R)

Limansky VV (1967) Study of serum agglutinins in the horse mackerel of the Black Sea and West African coast. In: Metabolism and biochemistry of fishes. Nauka, Moscow, pp 308–310 (R)

Limansky VV (1969) Investigation of the erythrocyte antigens in the Atlantic anchovy of the West African coast. Vopr Ikhthiol 9 (2): 366–369 (R)

Limansky VV, Payusova AN (1969) On immunogenetic differences in elementary populations of the anchovy. Genetika (Moscow) 5 (6): 109–118 (RE)

Limbach von B (1970) Fish genetic laboratory. In: Progress in sport fishery research 1969, Res Publ 88, Washington, pp 110–117

Lin CC, Schipmann G, Kittrell WA, Ohno S (1969) The predominance of heterozygotes found in wild goldfish of lake Erie at the gene locus for sorbitol dehydrogenase. Biochem Genet 3 (6): 603–607

Lindroth A (1972) Heritability estimates of growth in fish. Aquilo Ser Zool 13: 77–80

Lindsey CC, Clayton JW, Franzin WG (1970) Zoogeographic problems and protein variation in the *Coregonus clupeaformis* white fish species complex. In: Biology of coregonid fishes. Univ Manitoba Press, Winnipeg, pp 127–146

Lobchenko VV, Ecur VV, Ryngucky VV (1979) Some results of selective breeding work with common carp in Moldavia. In: Materials of All-Union Scientific Conference on development and intensification of fisheries in inland waters of North Caucasus, Rostov-Don, pp 135–136 (R)

Locascio NJ, Wright JE (1973) A study of achromatic regions in species of Salmonidae and Esocidae. Comp Biochem Phys 45B (1): 13–16

Loch IC (1974) Phenotypic variation in the lake whitefish, *Coregonus clupeaformis*, induced by introduction into a new environment. J Fish Res B Can 31 (1): 55–62

Lodi E (1967) Un nuovo mutante di *Lebistes reticulatus* Peters. Boll Zool 34: 131–132

Lodi E (1978a) Palla: a hereditary vertebral deformity in the guppy *Poecilia reticulata* Peters (Pisces, Osteichthyes). Genetica 48 (3): 197–200

* Lodi E (1978b) Chromosome complement of the guppy, *Poecilia reticulata* Peters (Pisces, Osteichthyes). Caryologia 31 (4):475–477

Lodi E (1979) Induction of atypical gonapophyses within the gonopodial suspensorium of the palla mutant male of *Poecilia reticulata* Peters (Poeciliidae, Osteichthyes). Monit Zool Ital 13 (2–3):95–104

Lokshina AB (1980) Comparative electrophoretical analysis of some proteins in whitefishes. In: Problems of genetics and selection of fishes, GosNIORKh, Leningrad, pp 40–57 (RE)

Loudenslager EJ, Gall GAE (1980) Geographic patterns of protein variation and subspeciation in cutthroat trout, *Salmo clarkii*. Syst Zool 29 (1):27–42

Loudenslager EJ, Kitchin RM (1979) Genetic similarity of two forms of cutthroat trout, *Salmo clarkii*, in Wyoming. Copeia 4:673

* Loudenslager EJ, Thorgaard GH (1979) Karyotypic and evolutionary relationships of the Yellowston (*Salmo clarkii bouvieri*) and West-Slope (*S. c. lewisi*) cutthroat trout. J Fish Res B Can 36 (6):630–635

Lovshin LD, Da Silva AB (1976) Culture of monosex and hybrid tilapias. In: FAO/CIFA Symp Aquacult Africa. CIFA Tech Pap 4, Suppl 1:548–564

Lowe TP, Larkin JR (1975) Sex reversal in *Betta splendens* Regan with emphases on the problem of sex determination. J Exp Zool 191 (1):25–30

Lucas GA (1968) Factors, affecting sex determination in *Betta splendens*. Genetics 60 (1, part 2):199–200

Lucas GA (1972) A mutation limiting the development of red pigment in *Betta splendens*, the Siamese fighting fish. Proc Iowa Acad Sci 79 (1):31–33

Ludke JL, Ferguson DE, Burke WD (1968) Some endrin relationships in resistant and susceptible populations of golden shiners, *Notemigonus chrysoleucas*. Trans Am Fish Soc 97 (3):260–263

* Lueken W (1962) Chromosomenzahlen bei *Orestias* (Pisces, Cyprinodontidae). Mitt Hamburg Zool Mus Inst 60:195–198

Lukyanenko VI (1971) The immunobiology of fishes. "Pitschevaya promyshlennost", Moscow, (R)

Lukyanenko VI, Geraskin PP (1971) Dynamics of establishment haemoglobin fractional composition during the early development of the Russian sturgeon *Acipenser güldenstädti* Brandt. Dokl Akad Nauk SSSR (DAN) 198 (5):1242–1244 (R)

Lukyanenko VI, Popov AV (1969) Protein composition of the blood serum of two allopatric populations of the Siberian sturgeon *Acipenser baeri* Br. Dokl Akad Nauk SSSR (DAN) 186 (1):233–235 (R)

Lukyanenko VI, Sukacheva GA (1975) Peculiarities of the immunological reactivity of four common carp genotypes. In: Materials of the 6th All-Union conference on fish diseases. Moscow, pp 62–76 (R)

Lukyanenko VI, Popov AV, Mishin EA (1971) Heterogeneity and polymorphism of serum albumins in fishes. Dokl Akad Nauk SSSR (DAN) 201 (3):737–740 (R)

Lukyanenko VI, Karataeva BB, Terentyev AA (1973) The immunogenetic specificity of seasonal races of the Russian sturgeon. Dokl Akad Nauk SSSR (DAN) 213 (2):458–461 (R)

Lukyanenko VI, Popov AV, Mishin EA, Surial AI (1975) The intraspecies variation of the fractional composition of serum proteins in the sevriuga sturgeon *Acipenser stellatus*. Zh Evol Biochim Physiol 11 (2):191–193 (R)

Lush IE (1969) Polymorphism of a phosphoglucomutase isoenzymes in the herring (*Clupea harengus*). Comp Biochem Phys 30 (2):391–397

Lush IE (1970) Lactate dehydrogenase isoenzymes and their genetic variation in coalfish (*Gadus virens*) and cod (*Gadus morhua*). Comp Biochem Phys 32 (1):23–32

Lush JL (1941) Methods of measuring the heritability of individual differences among farm animals. Proc 7th Int Genet Congr, Cambridge, p 199

Macek KJ, Sanders HO (1970) Biological variation in the susceptibility of fish and aquatic invertebrates to DDT. Trans Am Fish Soc 99 (1):89–90

Mackie M, Jones BW (1978) The use of electrophoresis of the water-soluble (sarcoplastic) proteins of fish muscle to differentiate the closely related species of hake (*Merluccius* sp.) Comp Biochem Physiol 59B (2):95–98

* Makino S (1937) The chromosomes of two elasmobranch fishes. Cytologia 2:867–876

* Makino S (1939) The chromosomes of the carp, *Cyprinus carpio,* including those of some related species of Cyprinidae for comparison. Cytologia 9 (4):430–437

Makino S, Ozima Y (1943) Formation of the diploid eggs nucleus due to suppression of the second maturation division, induced by refrigeration of eggs of the carp, *Cyprinus carpio.* Cytologia 13 (1):55–60

Malecha S, Ashton GC (1968) Inbreeding in *Tilapia* spp. in Hawaii. Proc 12th Int Congr Genet Tokyo 1:224

Manchenko GP, Nikiforov SM (1979) Low level of genetic variability of non-enzymatic proteins in sea stars. Biol Morja 4:86–88 (R)

Mangaly G, Jamieson A (1978) Genetic tags applied to the European hake, *Merluccius merluccius.* Anim Blood Gr Biochem Genet 9 (1):39–48

Manna GK, Khuda-Bukhsh AR (1974) Somatic chromosomes of two hybrid carps. CIS 16:26–28

* Manna GK, Khuda-Bukhsh AR (1977) Karyomorphology of cyprinid fishes and cytological evaluation of the family. Nucleus 20 (1–2): 119–127

* Manna GK, Khuda-Bukhsh AR (1978) Karyomorphological studies in three species of teleostean fishes. Cytologia 43 (1):69–73

* Manna GK, Prasad R (1973) Chromosomes in three species of fish. Nucleus 16 (3): 150–157

* Manna GK, Prasad R (1974a) Cytological evidence for two forms of *Mystus vittatus* as two species. Nucleus 17 (1):4–8

* Manna GK, Prasad R (1974b) Chromosome analysis in three species of fishes belonging to family Gobiidae. Cytologia 39 (3):609–618

* Mantelman II (1973) A cytological investigation of several phytophagous fishes and their hybrids. Izvestija Gosud. nauchno-issled. Inst Ozern Rechn Rybn Khos (GosNIORKh) 85:87–92 (RE)

Mantelman II (1978) The use of thermal "shocks" in selective breeding of peled (*Coregonus peled*). Izvestija Gosud. nauchno-issled. Inst Ozern Rechn Rybn Khoz (GosNIORKh) 130:50–55 (RE)

Mantelman II (1980) Estimation of different peled stocks by embryonal viability. In: Problems of fish genetics and selection. GosNIORKh, Leningrad, pp 20–26 (RE)

Mantelman II, Kaidanova TI (1978) Chemical inactivation of spermatozoa with a view to obtain the gynogenetic peled larvae. In: Problems of fish selection. GosNIORKh, Leningrad, pp 35–44 (RE)

Manwell C (1963) The blood proteins of cyclostomes: a study in phylogenetic and ontogenetic biochemistry. In: The biology of myxine. Universitetsforlaget, Oslo, pp 372–455

Manwell C, Baker CMA (1970) Molecular biology and the origin of species. Heterosis, protein polymorphism and animal breeding. Sidgwick and Jackson, London

Manwell C, Baker CMA, Childers W (1963) The genetics of hemoglobin in hybrids. I. A molecular basis for hybrid vigour. Comp Biochem Phys 10 (1): 103–120

Marchalonis JJ (1972) Conservatism in the evolution of immunoglobulin. Nature (London) New Biol 236 (64):84–86

Marchalonis JJ (1977) Immunity in evolution. Harvard Univ Press, Cambridge (Massachusetts)

Marckmann K (1954) Is there any correlation between metabolism and number of vertebrae (and other meristic characters) in the sea trout (*Salmo trutta trutta* L.). Medd Dan Fisk Havunders 1 (3): 1–9

Marcus TR, Gordon M (1954) Transplantation of the Sc melanoma in fishes. Zoologica 39 (3): 123–131

* Marian T, Krasznai Z (1978a) Comparative karyological investigations on chinese carps. In: Increasing productivity of fishes by selection and hybridization. F Müller, Szarvas, pp 79–97

* Marian T, Krasznai Z (1978b) Zytologische Untersuchungen bei der Familie Cyprinidae (Pisces). Biol Zentralbl 97 (2):205–214

Markert CL (1975) Isozymes. In: Isozymes, vol I. Academic Press, London New York, pp 1–9

Markert CL, Faulhaber I (1965) Lactate dehydrogenase isozyme pattern of fish. J Exp Zool 159 (2):319–332

Markert CL, Møller F (1959) Multiple forms of enzymes: tissue, ontogenetic and specific patterns. Proc Natl Acad Sci USA 45 (5):753–763

Markert CL, Ursprung H (1971) Developmental genetics. Prentice-Hall, Englewood Cliffs
Markert CL, Shaklee JB, Whitt GS (1975) Evolution of a gene. Science 189 (4197): 102–114
Marneux M (1972) Etude de l'isotypie et de l'allotypie de la transferrine chez *Ictalurus melas.* Ann. Embryol Morphol 5 (3): 227–245
Martin FD, Richmond RC (1973) An analysis of five enzyme gene loci in four etheostomid species (Percidae: Pisces) in an area of possible introgression. J Fish Biol 5 (3): 511–517
Maskell M, Parkin DT, Vespoor E (1978) Apostatic selection by sticklebacks upon a dimorphic prey. Heredity 39 (1): 83–90
Massaro EJ (1972) Isozyme patterns of coregonine fishes: evidence for multiple cistrons for lactate and malate dehydrogenases and achromatic bands in the tissues of *Prosopium cylindraceum* (Pallas) and *P. coulteri* (Eig. and Eig.). J Exp Zool 179: 2: 247–262
Massaro EJ, Booke HE (1971) Photoregulation of the expression of lactate dehydrogenase isozymes in *Fundulus heteroclitus* (L.). Comp Biochem Phys 38B (2): 327–332
Massaro EJ, Booke HE (1972) A mutant A-type lactate dehydrogenase subunit in *Fundulus heteroclitus.* Copeia 2: 298–302
Massaro EJ, Markert CL (1968) Isozyme patterns of salmonid fishes: evidence for multiple cistrons for lactate dehydrogenase polypeptides. J Exp Zool 168 (2): 223–238
Masseyeff R, Godet R, Gombert J (1963) Les proteines seriques de *Protopterus annectens.* Etude electrophoretique. C R Seance Soc Biol 157 (1): 167–173
Mather K (1955) The genetical basis of heterosis. Proc R Soc London Ser B 144 (915): 143–162
Matsui Y (1934) Genetical studies on gold-fish of Japan. 1. On the varieties of gold-fish and the variations in their external characteristics. 2. On the Mendelian inheritance of the telescope eyes of gold-fish. 3. On the inheritance of the scale transparency of gold-fish. 4. On the inheritance of caudal and anal fins of gold-fish. J Imper Fish Inst 30: 1, 1–96
Matsui Y (1956) Gold-fish. In: Exhibits Int Genet Symp, Tokyo, Kyoto, pp 97–105
Matsumoto J, Kajishima T, Hama T (1960) Relation between the pigmentation and pterin derivatives of chromatophores during development in the normal black and transparent scaled types of goldfish (*Carassius auratus*). Genetics 45 (9): 1178–1192
Matsuzawa T, Hamilton JB (1973) Polymorphism in lactate dehydrogenases of sceletal muscle associated with YY sex chromosomes in medaka (*Oryzias latipes*). Proc Soc Exp Biol Med 142 (1): 232–236
* Matthey R (1949) Les chromosomes des vertebrés. F Rouge, Lausanne
Maurer G (1971) Disc-electrophoresis. Mir, Moscow (R)
Mauro ML, Micheli G (1979) DNA reassociation kinetics in diploid and phylogenetically tetraploid Cyprinidae. J Exp Zool 208 (3): 407–416
May B, Utter FM, Allendorf HW (1975) Biochemical genetic variation in pink and chum salmon. Inheritance of intraspecies variation and apparent absence of interspecies introgression following massive hybridization of hatchery stocks. J Hered 66 (4): 227–232
May B, Stoneking M, Wright JE (1979 a) Joint segregation of malate dehydrogenase and diaphorase loci in brown trout (*Salmo trutta*). Trans Am Fish Soc 108 (4): 373–377
May B, Wright JE, Stoneking M (1979 b) Joint segregation of biochemical loci in Salmonidae: results from experiments with *Salvelinus* and review of the literature on other species. J Fish Res B Can 36: 1114–1128
Mayorova AA, Chugunova NI (1954) Biology, distribution and evaluation of resources of the Black sea anchovy. Tr VNIRO 28: 5–33 (R)
McCabe MM, Dean DM (1970) Esterase polymorphisms in the scipjack tuna, *Katsuwonus pelamis.* Comp Biochem Phys 34 (3): 671–681
McConkey EH, Taylor BJ, Dug Phan (1979) Human heterozygosity: a new estimate. Proc Natl Acad Sci USA 78 (12): 6500–6504
McDonald JF, Ayala FJ (1978) Gene regulation in adaptive evolution. Can J Genet Cytol 20 (2): 159–175
McIntyre P (1961) Crossing over within the macromelanophore gene in the platyfish, *Xiphophorus maculatus.* Am Nat 95 (884): 323–324
McIntyre JD, Amend DF (1978) Heritability of tolerance for infectious hematopoietic necrosis in sockeye salmon (*Oncorhynchus nerka*). Trans Am Fish Soc 107: 2, 305–308
McIntyre JD, Blanc JM (1973) A genetic analysis of hatching time in steelhead trout (*Salmo gairdneri*). J Fish Res B Can 30 (1): 137–139

McKay FE (1971) Behavioral aspects of population dynamics in unisexual-bisexual *Poeciliopsis* (Pisces: Poeciliidae). Ecology 52 (5):770–790

McKenzie JA (1973) Comparative electrophoresis of tissue from blueback herring *Alosa aestivalis* (Mitchill) and caspareau, *Alosa pseudoharengus* (Wilson). Comp Biochem Phys 44B (1):65–68

McKenzie JA, Martin Ch (1975) Transferrin polymorphism in blueback herring, *Alosa aestivalis* (Mitchill). Can J Zool 53 (11):1479–1482

McKenzie JA, Pain H (1969) Variation in the plasma proteins of Atlantic salmon (*Salmo salar* L.). Can J Zool 47 (5):759–761

McPhail JD, Jones RL (1966) A simple technique for obtaining chromosomes from teleost fishes. J Fish Res B Can 23 (4):767–768

Mednikov BM, Akhundov A (1975) Systematics of the genus *Salmo* in the light of the data on DNA molecular hybridization. Dokl Akad Nauk SSSR (DAN) 222 (3):744–746 (R)

Mednikov BM, Antonov AS, Popov LS (1973a) Genosystematics and evolution of fish genomes. In: Biochem Genet Fish. Inst Cytol Acad Sci USSR, Leningrad, pp 37–42 (RE)

Mednikov BM, Popov LS, Antonov AS (1973b) Characteristics of primary DNA structure as criteria for construction of natural system of fishes. J Obtsch Biol 34 (4):516–529 (RE)

Mednikov BM, Reshetnikov YuS, Shubina EA (1977) Investigation of the relatedness between whitefishes using the molecular hybridization technique. Zool Zh 56 (3):333–341 (RE)

Mednikov BM, Maximov VA (1979) Genetic divergence in char (*Salvelinus*) of Chukotka and problems of speciation in this group. In: Biochem Popul Genet Fish. Inst Cytol Acad Sci USSR, Leningrad, pp 45–48 (RE)

Mendel G (1865) Versuche über Pflanzenhybriden. Verh Naturforsch Ver Brünn 4:1–47

Menzel BW (1976) Biochemical systematics and evolutionary genetics of the common shiner species group. Biochem Syst Ecol 4 (4):281–293

Menzel BW, Darnell RM (1973) Morphology of naturally occuring triploid fish related to *Poecilia formosa*. Copeia 2:350–352

Menzel RW (1959) Further notes on the albino catfish. J Hered 49 (6):284–293

Merla G (1959) Ein Beitrag zur Kenntnis der Anfälligkeit von Spiegel- und Nacktkarpfen gegenüber der infektiösen Bauchwassersucht. Dtsch Fisch Z 6:58–62

Merla G (1979) Grundlagen der Fischzüchtung. In: Industriemäßige Fischproduktion. Dtsch Landwirtsch Verlag, Berlin, pp 219–234

* Merlo S (1957) Oservazioni cariologicae su *Salmo carpio*. Roll Zool 24 (2):253–258

* Merrilees MJ (1975) Karyotype of *Galaxias maculatus* from New-Zealand. Copeia 1:176–178

Merritt RB (1972) Geographical distribution and enzymatic properties of lactate dehydrogenase allozymes in the fathead minnow, *Pimephales promelas*. Am Nat 106 (949):173–184

Merritt RB, Rogers JF, Kurz BJ (1978) Genic variability in the longnose dace, *Rhinichthys cataractae*. Evolution 32 (1):116–124

Mester L, Tesio C (1975) Recherches systematiques basées sur l'éléctrophorese chez certains Blennidae (Pisces) de la mer Noire. Rev Roum Biol 20 (2):113–116

Metcalf RA, Whitt GS, Childers WT (1972) Inheritance of esterases in the white crappie (*Pomoxis annularis*), black crappie (*P. nigromaculatus*) and their F_1 and F_2 interspecific hybrids. Anim Blood Gr Biochem Genet 3 (1):19–33

Meyen VA (1940) Mechanisms underlying the variation of egg size in teleost fishes. Dokl Akad Nauk SSSR (DAN) 28 (7):654–656 (R)

Miaskowski M (1957) Variabilitätsstudien an den Flossen der Cypriniden. Arch Fischereiwiss 8 (1–2):32–53

* Michele JL, Takahashi CS (1977) Comparative cytology of *Tilapia rendalli* and *Geophagus brasilliensis* (Cichlidae, Pisces). Cytologia 42 (3–4):535–537

* Michele JL, Takahashi C, Ferrari I (1977) Karyotype study of some species of the family Loricariidae (Pisces). Cytologia 42 (3–4):539–546

Millenbach C (1973) Genetic selection of steelhead trout for management purposes. Int Salmon Found Spec Publ Ser 4 (1):253–257

Miller ET, Whitt GS (1975) Lactate dehydrogenase isozyme synthesis and cellular differentiation in the teleost retina. In: Isozymes, vol II. Academic Press, London New York, pp 359–374

Miller RR (1972) Classification of the native trouts of Arizona with the description of a new species, *Salmo apache*. Copeia 3:401–422

* Miller RR, Fitzsimons JM (1971) *Ameca splendens,* a new genus and species of goodeid fish from Western Mexico, with remarks on the classification of the Goodeidae. Copeia, 1:1–13

Miller RR, Hubbs CL (1969) Systematics of *Gasterosteus aculeatus,* with particular reference to integradation and introgression along the Pacific Coast of North America: a commentary on a recent contribution. Copeia 1:62–69

Miller RR, Schultz RJ (1959) All female strains of the teleost fishes of the genus *Poeciliopsis.* Science 130 (3389): 1956–1957

Milman LS, Jurovitzky JuG (1973) Mechanisms of enzymatic regulation of carbohydrate metabolism in yearly embryogenesis. Nauka, Moscow (R)

Mirsky AE, Ris H (1951) The desoxiribonucleic acid content of animal cells and its evolutionary significance. J Gen Phys 34:451–462

Mitton JB, Koehn RK (1975) Genetic organization and adaptive response of allozymes to ecological variables in *Fundulus heteroclitus.* Genetics 79 (1):97–111

Moav R (1979) Genetic improvement in aquaculture industry. In: Advances in aquaculture, Fish News Books Ltd, Farnham, pp 610–622

Moav R, Wohlfarth G (1960) Genetic improvement of carp. I. Theoretical background. Bamidgeh 12 (1):5–12

Moav R, Wohlfarth G (1963) Breeding schemes for the genetic improvement of edible fish. In: Progress Report 1962, Jerusalem, pp 1–40

Moav R, Wohlfarth G (1967) Breeding schemes for the genetic improvement of edible fish. In: Progress Report 1964–1965, Jerusalem, pp 1–56

Moav R, Wohlfarth G (1968) Genetic improvement of yield in carp. FAO Fish Rep (44) Rome 4: 12–29

Moav R, Wohlfarth G (1973a) Genetic correlation between seine escapability and growth capacity in carp. J Hered 61 (4): 153–157

Moav R, Wohlfarth G (1973b) Carp breeding in Israel. In: Agricultural genetics. J Wiley, New York Toronto, pp 295–318

Moav R, Wohlfarth G (1974) Magnification through competition of genetic differences in yield capacity in carp. Heredity 33 (2): 181–202

Moav R, Wohlfarth G (1976) Two-way selection for growth rate in the common carp (*Cyprinus carpio* L.). Genetics 82 (1):83–101

Moav R, Wohlfarth G, Soller M (1964) Breeding schemes for the genetic improvement of edible fish. In: Progress Report 1963, Jerusalem, pp 1–46

Moav R, Ankorion J, Wohlfarth WG (1971) Genetic investigation and breeding methods of carp in Israel. Rep FAO/UNDP (TA), (2926), Rome, pp 160–185

Moav R, Finkel A, Wohlfahrt G (1975a) Variability of intermuscular bones, vertebrae, ribs, dorsal fine rays and skeletal disorders in the common carp. Theoret Appl Genet 46 (1):33–43

Moav R, Hulata G, Wohlfarth G (1975b) Genetic differences between the Chinese and European races of the common carp. I. Analysis of genotype – environment interactions for growth rate. Heredity 34 (3): 323–330

Moav R, Brody T, Wohlfahrt G, Hulata G (1976a) Applications of electrophoretic markers to fish breeding. I. Advantages and methods. Aquaculture 9 (3): 217–218

Moav R, Soller M, Hulata G (1976b) Genetic aspects of the transition from traditional to modern fish farming. Theoret Appl Genet 47 (2): 285–290

Moav R, Brody T, Hulata G (1978) Genetic improvement of wild fish populations. Science 201 (4361): 1090–1094

Moav R, Brody T, Wohlfarth G, Hulata G (1979) A proposal for the continuous production of F_1 hybrids between the European and Chinese races of the common carp in traditional fish farms of Southeast Asia. In: Advances in aquaculture. Fish News Books Ltd, Farnham, pp 635–638

Møller D (1966) Polymorphism of serum transferrin in cod. Fiskeridir Skr Ser Havunders 14:51–60

Møller D (1967) Red blood cell antigens in cod. Sarsia 29:413–430

Møller D (1968) Genetic diversity in spawning cod along the Norwegian coast. Hereditas 60 (1): 1–32

Møller D (1969) The relationship between arctic and coastal cod in their immature stages illustrated by frequencies of genetic characters. Fiskeridir Skr Ser Havunders 15:220–233

Møller D (1970) Transferrin polymorphism in Atlantic salmon (*Salmo salar*). J Fish Res B Can 27 (6):1617–1625

Møller D (1971) Concepts used in the biochemical and serological identification of fish stocks. Rapp P-V Reun 161:7–9

Møller D, Naevdal G (1966) Transferrin polymorphism in fishes. Proc 10th Eur Conf Anim Blood Gr Biochem Polym Paris, pp 367–372

Møller D, Naevdal G (1969) Studies on hemoglobins of some gadoid fishes. Fiskeridir Skr Ser Havunders 15 (2):91–97

Møller D, Naevdal G (1974) Comparison of blood proteins of coalfish from Norwegian and Icelandic waters. Fiskeridir Skr Ser Havunders 16 (5): 177–181

Møller D, Naevdal G, Valen A (1967) Serologiske undersokelser for identifisering av fiskepopulasjoner i 1966. Fisken Havet 2:15–20

Møller D, Naevdal G, Holm M, Lerøy R (1979) Variation in growth rate and age of sexual maturity in rainbow trout. In: Advances in aquaculture, Fish News Books Ltd, Farnham, pp 622–626

Moodie GEE (1972) Predation, natural selection and adaptation in an unusual threespine stickleback. Heredity 28 (2):155–168

Moon TW, Hochachka PW (1972) Temperature and the kinetic analysis of trout isocitrate dehydrogenases. Comp Biochem Phys 42B (4):724–730

Moore WS (1974) A mutant affecting chromatophore proliferation in a poeciliid fish. J Hered 65 (6):326–330

Moore WS (1977) A histocompatibility analysis of inheritance in the unisexual fish *Poeciliopsis 2monacha-lucida.* Copeia 2:213–223

Moore WS, Bradley EA (1979) The population structure of an asexual vertebrate $P_{2m\text{-}l}$ (Pisces: Poeciliidae). Evolution 33 (2):563–578

Moore WS, McKay FE (1971) Coexistence in unisexual-bisexual species complexes of *Poeciliopsis* (Pisces: Poeciliidae). Ecology 52 (5):791–799

Moore WS, Miller R, Schultz R (1970) Distribution, adaptation and probable origin of an all-female form of *Poeciliopsis* (Pisces: Poeciliidae) in northwestern Mexico. Evolution 24 (4):789–795

Morgan RP, Ulanowicz NI (1976) The frequency of muscle protein polymorphism in *Menidia menidia* (Atherinidae) along the Atlantic coast. Copeia 2:356–360

Morgan RP, Koo TSY, Krantz GE (1973) Electrophoretic determination of populations of the striped bass *Morone saxatilis* in the Upper Chesapeake Bay. Trans Am Fish Soc 102 (1):21–32

Morizot DC, Aravinda C (1977) Linkage relationships of protein coding loci in fishes of the genus *Xiphophorus.* Genetics 86 (2, p 2):46

Morizot DC, Siciliano MJ (1979) Polymorphisms, linkage and mapping of four enzyme loci in the fish genus *Xiphophorus* (Poeciliidae). Genetics 93 (4):947–960

Morizot DC, Wright DA, Siciliano MJ (1977) Three linked enzyme loci in fishes: implications in the evolution of vertebrate chromosomes. Genetics 86 (3):645–656

Mork JA, Giskeødegård R, Sundnes G (1980) LDH gene frequencies in cod samples from two locations on the Norwegian coast. J Conseil 39 (1):110–113

Morrison WJ (1970) Nonrandom segregation of the lactate dehydrogenase subunit loci in trout. Trans Am Fish Soc 99 (1):193—206

Morrison WJ, Wright JE (1966) Genetic analysis of three lactate dehydrogenase isozyme systems in trout: evidence for linkage of genes coding subunits A and B. J Exp Zool 163 (3):259–270

Morrissy NM (1973) Comparison of strains of *Salmo gairdneri* Rich. from New South Wales, Victoria and Western Australia. Bull Aust Soc Limnol 5 (1):11–20

Moskovkin LI, Truweller KA, Maslennikova NA, Romanova NI (1973) Distribution of transferrin and esterase types in connection with the types of scale cover and myogens in carp (*Cyprinus carpio* L.). In: Biochem Genet Fish. Inst Cytol Acad Sci USSR, Leningrad, pp 120–128 (RE)

Mrakovčič M, Haleg LE (1979) Inbreeding depression in the zebra fish *Brachydanio rerio* (Ham., Buch.). J Fish Biol 15 (3):323–327

Müller W (1975) Der gegenwärtige Stand der Karpfenzüchtung in der DDR. Z Binnenfischerei DDR 5:136–141

Münzing J (1959) Biologie, Variabilität und Genetik von *Gasterosteus aculeatus* L. (Pisces). Untersuchungen im Elbgebiet. Int Rev Hydrobiol 44 (3):317–382

Münzing J (1962) Ein neuer semiarmatus Typ von *Gasterosteus aculeatus* L. (Pisces) aus dem Izniksee. Mitt Hamburg Zool Mus Inst 60:181–194

Münzing J (1963) The evolution of variation and distributional patterns in European populations of the three-spined stickleback, *Gasterosteus aculeatus*. Evolution 17 (3):320–332

Münzing J (1969) Variabilität, Verbreitung und Systematik der Arten und Unterarten in der Gattung *Pungitius* Coste, 1848 (Pisces, Gasterosteidae). Z Zool Syst Evol Forsch 7 (3):208–233

Münzing J (1972) Polymorphe Populationen von *Gasterosteus aculeatus* L. (Pisces, Gasterosteidae) in sekundären Intergradationszonen der Deutschen Bucht und benachbarter Gebiete. Die geographische Variation der Lateralbeschildung. Faun Oekol Mitt 4 (3):69–84

Muramoto J (1975) A note on triploidy of the funa (Cyprinidae, Pisces). Proc Jpn Acad 51 (7):583–587

Muramoto J-I, Ohno S, Atkin NB (1968) On the diploid state of the fish order Ostariophysi. Chromosoma 24 (1):59–66

* Muramoto J-I, Igarashi K, Itoh M, Makino S (1969) A study of the chromosomes and enzymatic patterns of sticklebacks of Japan. Proc Jpn Acad 45 (9):803–807

* Muramoto J-I, Atumi J-I, Fukuoka H (1974) Karyotypes of 9 species of the Salmonidae. CIS 17:20–22

* Murofushi M, Iosida TH (1979) Cytogenetical studies of fishes. I. Karyotypes of four filefishes. Jpn J Genet 54 (3):191–196

Nace GW, Richards CM, Asher JR (1970) Parthenogenesis and genetic variability. I. Linkage and inbreeding estimation in the frog, *Rana pipiens*. Genetics 66 (2):340–368

Naevdal G (1968) Studies on hemoglobins and serum proteins in sprat from Norwegian waters. Fiskdir Skr Ser Havunders 14 (3):160–182

Naevdal G (1969) Studies on blood proteins in herring. Fiskdir Skr Ser Havunders 15 (3):128–135

Naevdal G (1970) Distribution of multiple forms of lactate dehydrogenase, aspartate aminotransferase and serum esterase in herring from Norwegian waters. Fiskdir Skr Ser Havunders 15 (6):565–572

Naevdal G (1978) Difference between marinus and mentella types of redfish by electrophoresis of haemoglobin. Fiskdir Skr Ser Havunders 16 (10):731–736

Naevdal G, Bakken E (1974) Comparison of blood proteins from East and West Atlantic populations of *Hippoglossoides platessoides*. Fiskdir Skr Ser Havunders 16:183–188

Naevdal G, Holm M, Møller D, Osthus OD (1975) Experiments with selective breeding of salmon. Int Counc Explor Sea Comm Meet 1975, M, N 22:1–9

Naevdal G, Bjerk Ø, Holm M, Lerøy R, Møller D (1979a) Growth rate and age of sexual maturity of Atlantic salmon smoltifying aged one and two years. Fiskdir Skr Ser Havunders 17:11–17

Naevdal G, Holm M, Lerøy R, Møller D (1979b) Individual growth rate and age at sexual maturity in rainbow trout. Fiskdir Skr Ser Havunders 17:1–10

Nagel L (1970) Prüfung sächsischer Zuchtkarpfenstämme auf Leistung und Resistenz. Z Fischerei 18 (3–4):217–226

Nagy A, Csányi V (1978) Utilization of gynogenesis in genetic analysis and practical animal breeding. In: Increasing productivity of fishes by selection and hybridization. F Müller, Szarvas, pp 16–30

Nagy A, Rajki K, Horvath I, Csányi V (1978) Investigation on carp, *Cyprinus carpio* L. gynogenesis. J Fish Biol 13 (2):215–224

Nagy A, Rajki K, Bakoś J, Csányi V (1979) Genetic analysis in carp (*Cyprinus carpio*) using gynogenesis. Heredity 43 (1):35–40

Nakamura N, Kasahara S (1955) A study of the phenomenon of the tobi koi or shoot carp. I. On the earliest stage at which the shoot carp appears. Bull Jpn Soc Sci Fish 21 (2):73–76

Nakamura N, Kasahara S (1957) A study of the phenomenon of the tobi koi or shoot carp. III. On the results of culturing the modal group and the growth of carp fry reared individually. Bull Jpn Soc Sci Fish 22 (11):674–678

Nakano E, Whiteley AH (1965) Differentiation of multiple molecular forms of four dehydrogenases in the teleost, *Oryzias latipes*, studied by disc electrophoresis. J Exp Zool 159 (1): 167–180

Nakaya K (1973) An albino zebra shark *Stegostoma fasciatum* from the Indian ocean with comments on albinism in elasmobranchs. Jpn J Ikhtiol 20 (2): 120–122

* Nanda A (1973) The chromosomes of *Mystus vittatus* and *Ompok pabda* (fam. Siluridae). Nucleus 16 (1): 29–32

Natali VF, Natali AI (1931) On the problem of localization of genes in the X- and Y-chromosomes of *Lebistes reticulatus*. Zh Exp Biol 7 (1): 41–70 (RE)

* Natarajan R, Subrahmanyan K (1974) A karyotype study of some teleost from Postonovo waters. Proc Ind Acad Sci Ser B 79 (5): 173–196

Nayudu PL (1975) Contributions to the genetics of *Poecilia reticulata*. Ph D Thesis, Clayton (Australia), Monash Univ

Nayudu PL (1979) Genetic studies of melanic color patterns, and atypical sex determination in the guppy, *Poecilia reticulata*. Copeia 2: 225–231

Nayudu PL, Hunter CR (1979) Cytological aspects and differential response to melatonin of melanophore based color mutants in the guppy, *Poecilia reticulata*. Copeia 2: 232–242

* Nayyar RP (1962) Karyotype studies in two cyprinids. Cytologia 27 (2): 229–231

* Nayyar RP (1964) Karyotype studies in seven species of Cyprinidae. Genetica (Holl) 35 (1): 95–104

* Nayyar RP (1965) Karyotype studies in the genus *Notopterus* (Lac.). The occurrence and fate of univalent chromosomes in spermatocytes of *N. chitala*. Genetica (Holl) 36 (3): 398–405

* Nayyar RP (1966) Karyotype studies in thirteen species of fishes. Genetica (Holl) 37 (1): 78–92

Nefyodov GN (1969) Serum haptoglobins in the sea perches from the genus *Sebastes*. Vest Mosk Univ 1: 104–107 (R)

Nefyodov GN (1971) Serum haptoglobins in the marinus and mentella types of North Atlantic redfish. Rapp P-V Reun 161: 126–129

Nefyodov GN, Alferova NM, German SM (1973) Electrophoretic study of muscle proteins of some species of fishes from Merluccidae and Carangidae. In: Biochem Genet Fish. Inst Cytol Acad Sci USSR, Leningrad, pp 201–207 (RE)

Nefyodov GN, Truveller KA, Alferova NM, Chuksin YuV (1976) The variation in the electrophoretic spectra of muscle esterases in rock grenadier. Biol Morya 4: 62–65 (R)

Nefyodov GN, Alferova NM, Chuksin YuV (1978) Polymorphism of muscle esterases in the horse mackerel of Northeastern Atlantic. Biol Morya 2: 64–74 (R)

Nei M (1972) Genetic distance between populations. Am Nat 106 (949): 283–292

Nei M, Fuerst PA, Chakraborty R (1978) Subunit molecular weight and genetic variability of proteins in natural populations. Proc Natl Acad Sci USA 75 (7): 3359–3362

Nelson B (1958) Progress report on golden channel catfish. Proc South-East Assoc Game Comm 12: 75–77

Nelson JS (1971) Absence of the pelvic complex in ninespine sticklebacks, *Pungitius pungitius*, collected in Ireland and Wood Buffalo National Park Region, Canada, with notes on meristic variation. Copeia 3: 707–717

Nelson JS (1977) Evidence of a genetic basis for absence of the pelvic sceleton in brook stickleback *Culaea inconstans* and notes on the geographical distribution and origin of the loss. J Fish Res B Can 37 (9): 1314–1320

Nenashev GA (1966) Heritability of several morphological (diagnostic) traits in the Ropsha common carp. Izvestija Gosud. nauchno-issled. Inst Ozern Rechn Rybn Khos (GosNIORKh) 61: 125–135 (RE)

Nenashev GA (1969) Heritability of several traits important in the common carp selection. Izvestija Gosud. nauchno-issled. Inst Ozern Rechn Rybn Khos (GosNIORKh) 65: 185–195 (RE)

Nenashev GA, Rybakov FYu (1978) Genetic diversity of transferrin and its connection with fishery-valuable traits in silver carp and bighead. Izvestija Gosud. nauchno-issled. Inst Ozern Rechn Rybn Khos (GosNIORKh) 130: 112–118 (RE)

Nesterenko NV (1957) On a hybridization of the Ural cisco (Ladoga ripus) with Chudski whitefish in ponds. Izvestija Gosud. nauchno-issled. Inst Ozern Rechn Rybn Khos (GosNIORKh) 39: 41–59 (RE)

Nevo E (1978) Genetic variation in natural populations: patterns and theory. Theoret Popul Biol 13 (1): 121–177
Neyfakh AA (1956) The effect of ionizing radiation on gametes of the loach (*Misgurnus fossilis* L.) Dokl Akad Nauk SSSR (DAN) 111 (3): 585–588 (R)
Neyfakh AA, Abramova NB (1979) Regulation of enzyme activity during animal development. In: Biochem Popul Genet Fish. Inst Cytol Acad Sci USSR, Leningrad, pp 18–23 (RE)
Neyfakh AA, Timofeeva MJa (1977) Molecular biology of developmental processes. Nauka, Moscow (R)
Neyfakh AA, Glushankova MA, Korobtzova NS, Kusakina AA (1973) Expression of genes controlling FDP-aldolase, in fish embryos. Thermostability as a genetic marker. Dev Biol 34 (2): 309–320
Neyfakh AA, Glushankova MA, Kusakina AA (1976) Time of function of genes controlling aldolase activity in loach embryo development. Dev Biol 50 (2): 502–510
Nikoljukin NI (1952) Interspecies hybridization in fishes. Gosud Oblast Izdatelstvo, Saratov (R)
Nikoljukin NI (1971) Hybridization of Acipenseridae and its practical significance. Rep FAO/UNDP(TA), (2926), Rome, pp 328–334
Nikolsky GV (1950) Special ichthyology. Sovetskaja Nauka, Moscow, (R)
Nikolsky GV (1966) Participation of geneticists in solution of fishery-biological problems. Vest Mosk Sk. Univ 6: 3–17 (R)
Nikolsky GV (1973) Interrelationships of trait variation, energy metabolism and karyotype in fishes. Zh Obtsch Biol 34 (4): 503–515 (RE)
Nikolsky GV, Vasilyev VP (1973) On certain tendencies in the distribution of the number of chromosomes in fishes. Vopr Ikhthiol 13 [1(78)]: 3–22 (R)
Nikoro ZS (1976) Statistical models in selection theory. In: Modelling of biological systems, vol I. University, Novosibirsk, pp 5–78 (R)
Nikoro ZS, Rokitsky PF (1972) Application and methods of determination of the heritability coefficient. Genetika (Moscow) 8 (2): 170–178 (RE)
Nikoro ZS, Vasilyeva LA (1976) Errors made when applying genetico-statistical parameters to unbalanced populations. In: Mathematical models of genetic systems. University, Novosibirsk, pp 69–111 (R)
* Nishikawa S, Sakamoto K (1977) Comparative studies on the chromosomes in Japanese fishes. III. Somatic chromosomes of three anguilliform fishes. J Shimonos Univ Fish 25 (3): 193–196
* Nishikawa S, Amaoka K, Karasawa T (1971) On the chromosomes of two species of eels (*Anguilla*). CIS 12: 27–28
* Nishikawa S, Amaoka K, Nakanishi K (1974) A comparative study of chromosomes of twelve species of gobioid fish in Japan. Jpn J Ikhthiol 21 (2): 61–71
* Nishikawa S, Honda M, Wakatsuki A (1977) Comparative studies on the chromosomes in Japanese fishes. II. Chromosomes of eight species in scorpionfishes. J Shimonos Univ Fish 25 (3): 187–191
* Nogusa S (1955 a) Chromosome studies in Pisces. IV. The chromosomes of *Mogrunda obscura* (Gobiidae) with evidence of male heterogamety. Cytologia 20 (1): 11–18
* Nogusa S (1955 b) Chromosome studies in Pisces. V. Variation of the chromosome number in *Acheilognathus rhombea* due to multiple-chromosome formation. Annot Zool Jpn 28 (4): 249–255
* Nogusa S (1957) Chromosome studies in Pisces. VI. The X-Y chromosomes found in *Cottus pollux* Günter (Cottidae). J Fac Sci Hokkaido Univ Ser VI Zool 13 (2): 289–292
Nogusa S (1960) A comparative study of the chromosomes in fishes with particular consideration on taxonomy and evolution. Mem Hyogo Univ Agric 3 (1): 1–62
Northcote TG, Williscroft SN, Tsuyuki H (1970) Meristic and lactate dehydrogenase genotype differences in stream populations of rainbow trout below and above a waterfall. J Fish Res B Can 27 (11): 1987–1995
Numachi K (1970) Polymorfism of malate dehydrogenase and genetic structure of juvenile population in saury *Cololabis saira*. Bull Jpn Soc Sci Fish 36 (12): 1235–1241

Numachi K (1971a) Genetic polymorphism of α-glycerophosphate dehydrogenase in saury, *Cololabis saira.* Seven variant forms and genetic control. Bull Jpn Soc Sci Fish 37 (6): 755–760

Numachi K (1971b) Electrophoretic variants of catalase in the black rockfish, *Sebastes inermis.* Bull Jpn Soc Sci Fish 37 (12): 1177–1181

Numachi K (1972a) Genetic control and subunit composition of lactate dehydrogenase in *Pseudorasbora parva.* Jpn J Genet 47 (3): 193–202

Numachi K (1972b) Genetic polymorphism of tetrasolium oxidase in black rockfish. Bull Jpn Soc Sci Fish 38 (7): 789

Numachi K, Matsumiya Y, Sate R (1972) Duplicate genetic loci and variant forms of malate dehydrogenase in chum salmon and rainbow trout. Bull Jpn Soc Sci Fish 38: 699–706

Nybelin O (1947) Ett fall av X-bunden nedärvning hos *Lebistes reticulatus* (Peters). Zool Bijdrag 25: 448–454

* Nygren A, Jahnke M (1972a) Cytological studies in *Myxine glutinosa* (Cyclostomata) from the Gullmaren fjord in Sweden. Swed J Agric Res 2: 83–88

* Nygren A, Jahnke M (1972b) Microchromosomes in primitive fishes. Swed J Agric Res 2: 229–238

* Nygren A, Edlund P, Hirsh H, Ashgren L (1968a) Cytological studies in perch (*Perca fluviatilis* L.), pice (*Esox lucius* L.), pice perch (*Lucioperca lucioperca* L.) and ruff (*Acerina cernua* L.). Hereditas 59 (2–3): 518–524

* Nygren A, Nilsson B, Jahnke M (1968b) Cytological studies in Atlantic salmon. Ann Acad Regiae Sci Ups 12: 21–52

* Nygren A, Nilsson B, Jahnke M (1971a) Cytological studies in *Salmo trutta* and *S. alpinus.* Hereditas, 67 (2), 259–268

* Nygren A, Nilsson B, Jahnke M (1971b) Cytological studies in *Thymallus thymallus* and *Coregonus albula.* Hereditas 67 (2): 269–274

* Nygren A, Nilsson B, Jahnke M (1971c) Cytological studies in the smelt (*Osmerus eperlanus* L.). Hereditas 67 (2): 283–286

* Nygren A, Nilsson B, Jahnke M (1971d) Cytological studies in Hypotremata and Pleurotremata (Pisces). Hereditas 67 (2): 275–282

* Nygren A, Leijon U, Nilsson B, Jahnke M (1971e) Cytological studies in *Coregonus* from Sweden. Ann Acad Regiae Sci Ups 15: 5–20

Nygren A, Nilsson B, Jahnke M (1972) Cytological studies in Atlantic salmon from Canada, and in hybrids between Atlantic salmon and sea trout. Hereditas 70 (2): 295–306

* Nygren A, Bergkvist C, Windahl T, Jahnke G (1974) Cytological studies in Gadidae. Hereditas 76 (2): 173–178

* Nygren A, Andreasson J, Jonsson L, Jahnke G (1975a) Cytological studies in Cyprinidae (Pisces). Hereditas 81 (2): 165–172

Nygren A, Nyman L, Svensson K, Jahnke G (1975b) Cytological and biochemical studies in back-crosses between the hybrid Atlantic salmon x sea trout and its parental species. Hereditas 81 (1): 55–62

Nyman L (1965) Inter- and intraspecific variations of proteins in fishes. Ann Acad Regiae Sci Ups 9: 1–18

Nyman L (1967) Protein variations in Salmonidae. Rep Inst Freshwater Res 47: 5–38

Nyman L (1969) Polymorphic serum esterase in two species of freshwater fishes. J Fish Res B Can 26 (9): 2532–2534

Nyman L (1970) Eletrophoretic analysis of hybrids between salmon (*S. salar* L.) and trout (*S. trutta* L.). Trans Am Fish Soc 99 (1): 229–236

Nyman L (1971) Plasma esterases of some marine and anadromous teleosts and their application in biochemical systematics. Rep Inst Freshwater Res 51: 109–123

Nyman L (1972) A new approach to the taxonomy of the "*Salvelinus alpinus*" species complex. Rep Inst Freshwater Res 52: 103–131

Nyman L (1975) Allelic selection in a fish (*Gymbocephalus cernua* L.) subjected to hotwater effluents. Rep Inst Freshwater Res 54: 75–82

Nyman OL, Pippy JHC (1972) Differences in Atlantic salmon, *Salmo salar,* from North America and Europe. J Fish Res B Can 29 (2): 179–185

Nyman L, Show DH (1971) Molecular weight heterogeneity of serum esterases in four species of salmonid fish. Comp Biochem Phys 40B (2): 563–566

Nyman L, Westin L (1968) On the problem of sibling species and possible intraspecific variation in fourholm sculpin, *Myoxocephalus quadricornis* (L.). Rep Inst Freshwater Res 48:57–66

Nyman L, Westin L (1969) Blood protein systematics of Cottidae in the Baltic drainage area. Rep Inst Freshwater Res 49:264–274

Odense PH, Allen TM (1971) A biochemical comparison of some Atlantic herring populations. Rapp P-V Reun 161:26

Odense PH, Leung TC (1975) Isoelectric focusing on polyacrylamide gel and starch gel electrophoresis of some gadiform fish lactate dehydrogenase isozymes. In: Isozymes, vol III. Academic Press, London New York, pp 485–501

Odense PH, Allen TM, Leung TC (1966a) Multiple forms of lactate dehydrogenase and aspartate aminotransferase in herring (*Clupea harengus* L.). Can J Biochem 44 (4): 1319–1324

Odense PH, Leung TC, Allen TM (1966b) An electrophoretic study of tissue proteins and enzymes in four Canadian cod populations. Int Counc Explor Sea Comm Mit G-14: 1–6

Odense PH, Leung TC, Allen TM, Parker E (1969) Multiple forms of LDH in the cod, *Gadus morhua* L. Biochem Genet 3 (4):317–334

Odense PH, Leung TC, Mac Dougall YM (1971) Polymorphism of lactate dehydrogenase (LDH) in some gadoid species. Rapp P-V Reun 161:75–79

Ohno S (1967) Sex chromosomes and sex-linked genes. Springer Berlin Heidelberg New York

Ohno S (1969a) The role of gene duplication in vertebrate evolution. In: The biological basis of medicine, voll IV. Academic Press, London New York, pp 109–132

Ohno S (1969b) The preferential activation of maternally derived alleles in development of interspecific hybrids. Wistar Symp Monogr 9: 137–150

Ohno S (1970a) Evolution by gene duplication. Springer Berlin Heidelberg New York

Ohno S (1970b) The enormous diversity in genome size of fish as a reflection of nature's extensive experiments with gene duplication. Trans Am Fish Soc 99 (1): 120–130

Ohno S (1974) Protochordata, Cyclostomata and Pisces. In: Animal cytogenetics, vol IV. Borntraiger, Berlin, pp 1–91

* Ohno S, Atkin NB (1966) Comparative DNA values and chromosome complements of eight species of fishes. Chromosoma 18 (3):455–466

Ohno S, Stenius C, Faisst E, Zenzer MT (1965) Postzygotic chromosomal rearrangements in rainbow trout (*Salmo irideus* Gibbons). Cytogenetics 4 (2): 117–129

Ohno S, Muramoto J, Christian L, Atkin NB (1967a) Diploid-tetraploid relationship among old-world members of the fish family Cyprinidae. Chromosoma 23 (1): 1–9

Ohno S, Klein J, Poole J, Harris C, Destree A, Morrison M (1967b) Genetic control of lactate dehydrogenase in the hagfish (*Eptatretus stouti*). Science 156 (3771):96–98

Ohno S, Wolf H, Atkin NB (1968) Evolution from fish to mammals by gene duplication. Hereditas 59 (1): 169–187

* Ohno S, Muramoto J, Klein J, Atkin B (1969a) Diploid-tetraploid relationship in clupeoid and salmonid fish. In: Chromosomes today, vol II. Israel Univers. Press, Jerusalem, pp 139–147

Ohno S, Muramoto J, Stenius C, Christian L, Kittrell WA (1969b) Microchromosomes in holocephalian, chondrostean and holostean fishes. Chromosoma 26 (1):35–40

Ohno S, Christian L, Romero M, Dofucu R, Ivey C (1973) On the question of American eels, *Anguilla rostrata* versus European eels, *Anguilla anguilla*. Experientia 29 (7):891

Ojima Y, Asano N (1977) A cytological evidence for gynogenetic development of the ginbuna (*Carassius auratus langsdorfi*). Proc Jpn Acad 53B (4): 138–142

Ojima Y, Hitotsumachi S (1967) Cytogenetic studies in lower vertebrates IV. A note on the chromosomes of the carp (*Cyprinus carpio*) in comparison with those of the funa and the goldfish (*Carassius auratus*). Jpn J Genet 42 (3): 163–167

* Ojima Y, Kashiwagi E (1979) A karyotype study of eleven species of labrid fishes from Japan. Proc Jpn Acad 55B (6):280–285

Ojima Y, Makino S (1978) Triploidy induced by cold shock in fertilized eggs of the carp. A preliminary study. Proc Jpn Acad B 54 (7):359–362

Ojima Y, Takai A (1979) Further cytogenetical studies on the origin of the gold-fish. Proc Jpn Acad 55B (7):346–350

Ojima P, Ueda T (1978) New C-banded marker chromosomes found in carp-funa hybrids. Proc Jpn Acad 54B (1): 15–20

* Ojima Y, Hitotsumachi S, Makino S (1966) Cytogenetic studies in lower vertebrates. I. A preliminary report on the chromosomes of the funa (*Carassius auratus*) and goldfish (a revised study). Proc Jpn Acad 42 (1):62–66

* Ojima Y, Hayashi M, Ueno K (1972) Cytogenetic studies in lower vertebrates. X. Karyotype and DNA studies in 15 species of Japanese Cyprinidae. Jpn J Genet 47 (6):431–440

* Ojima Y, Ueno K, Hayashi M (1973) Karyotypes of the Acheilognathinae fishes (Cyprinidae) of Japan with a discussion of phylogenetic problems. Zool Mag 82 (3):171–177

Ojima Y, Hayashi M, Ueno K (1975) Triploidy appeared in the backcross offspring from funa-carp crossing. Proc Jpn Acad 51 (8):702–711

* Ojima Y, Ueno K, Hayashi M (1976) A review of the chromosome number in fishes. Kromosomo (Tokyo) 2 (1):19–47

Ojima Y, Ueda T, Narikawa T (1979) A cytogenetic assessment on the origin of the goldfish. Proc Jpn Acad 55B (1):58–63

Öktay M (1954) Über Besonderheiten der Vererbung des Gens "fuliginosus" bei *Platypoecilus maculatus.* Istanbul Univ Fen Fak Mecm 19 (4):303–327

Öktay M (1959) Weitere Untersuchungen über eine Ausnahme XX-Sippe des *Platypoecilus maculatus* mit polygener Geschlechtsbestimmung. Rev Fac Sci Univ Istanbul Ser B 24 (3–4):225–233

Öktay M (1964) Über genbedingte rote Farbmuster bei *Xiphophorus maculatus.* In: Mitt Hamburg Zool Mus Inst Kosswig-Festschrift, pp 133–157

Omelchenko VT (1973) Species specifity and intraspecific constancy of haemoglobin electrophoregrams in some Far East fish species. In: Biochem Genet Fish, Inst Cytol Acad Sci USSR, Leningrad, pp 67–71 (RE)

Omelchenko VT (1975a) Use of electrophoresis of proteins in the systematics of the genus *Salvelinus.* Biol Morya 4:76–79 (R)

Omelchenko VT (1975b) The electrophoretic investigation of fish proteins in connection with the problem of species identification. Ph D Thesis, Vladivostok (R)

Omelchenko VT (1975c) Polymorphism of the pollock haemoglobin. Biol Morya 5:72–73 (R)

Omelchenko VT, Volokhonskaya LT, Viktorovsky RM (1971) The similarity of electrophoretic patterns of haemoglobins from several salmonids. Sci Rep Inst Mar Biol 2:176–177 (R)

Opperman K (1913) Die Entwicklung von Vorelleneiern nach Befruchtung mit radiumbestrahlten Samenfäden. Arch Mikrosk Anat 83, Abt 11 1:141–189:4:307–323

Ord WW (1976) Viral hemorragic septicemia: comparative susceptibility of rainbow trout (*Salmo gairdneri*) and hybrids (*S. gairdneri* x *O. kisutch*) to experimental infection. J Fish Res B Can 33 (5):1205–1208

Orska I (1963) The influence of temperature on the development of meristic characters of the skeleton in Salmonidae. I. Temperature-controlled variations of the number of vertebrae in *Salmo irideus* Gibb. Zool Pol 12 (3):309–339

Orzack SH, Sohn JJ, Kallman KD, Simon AL, Johnston R (1980) Maintenance of the three sex chromosome polymorphism in the platyfish, *Xiphophorus maculatus.* Evolution 34 (4):663–672

Öztan N (1954) Cytological investigation of the sexual differentiation in the hybrids of Anatolian cyprinodontids. Istanbul Univ Fen Fac Mecm 19 (4):245–280

Paaver T (1978) The investigations on the fish biochemical genetics in the lake Võrtsjärv limnology station. In: Genetics in Soviet Estonia. Acad Sci Estonian SSR, Tallinn, pp 68–75

Paaver TK (1979) On the polymorphism of some myogenes and enzymes in carp. In: Biochem Popul Genet Fish. Inst Cytol Acad Sci USSR, Leningrad, pp 162–166 (RE)

Paaver TK (1980) Genetic polymorphism of common carp (*Cyprinus carpio* L.) proteins. Ph D Thesis, Leningrad, Inst Cytol (R)

Paaver T, Tammert M (1975) Electrophoretic investigation of tissue proteins of several freshwater fish species from Estonia. In: Colect Stud Sci Pap. Univ Tartu, pp 17–22

Page LM, Whitt GS (1973a) Lactate dehydrogenase isozymes, malate dehydrogenase isozymes and tetrazolium oxidase mobilities of darters (Etheostomatini). Comp Biochem Phys 44B (2):611–623

Page LM, Whitt GS (1973b) Lactate dehydrogenase isozymes of darters and the inclusiveness of the genus *Percina.* In: Illinois Natur. Hist Surv Biol Notes. Abh Deut Seefisch 2:1–128 Urbana, pp 1–7

Pantelouris EM (1976) Aspartate aminotransferase variation in the Atlantic eel. J Exp Mar Biol Ecol 22 (2): 123–130

Pantelouris EM, Payne RH (1968) Genetic variation in the eel. I. The detection of haemoglobin and esterase polymorphism. Genet Res 11 (3):319–325

Pantelouris EM, Arnason A, Tesch FW (1970) Genetic variation in the eel. II. Transferrins, haemoglobins and esterases in the eastern North Atlantic. Possible interrelations of phenotypic frequency differences. Genet Res 16 (2):277–284

Pantelouris EM, Arnason A, Bumpus R (1976) New observations on esterase variation in the Atlantic eel. J Exp Mar Biol Ecol 22 (2): 113–121

* Park EH, Kang YS (1976) Karyotype conservation and difference in DNA amount in anguilloid fishes. Science 193 (4247):64–66

* Park EH, Kang YS (1979) Karyological confirmation of conspicuous ZW sex chromosomes in two species of Pacific anguilloid fishes (Anguilliformes: Teleostomi). Cytogenet Cell Gen 23 (1–2):33–38

Passakas T (1979) C-banding pattern in chromosomes of the European eel (*Anguilla anguilla*). Folia Biol 26 (4):301–304

Passakas T, Klekowski RZ (1972) Chromosomes of European eel (*Anguilla anguilla*) as related to in vivo sex determination. Pol Arch Hydrobiol 20 (3):517–519

Pavlú V, Kálal L, Valenta M, Čepica S (1971) Polymorfismus transferrinu krevniho séra siha severniho marény (*Coregonus lavaretus maraena*) a sivena amerckeho (*Salvelinus fontinalis*). Živočišná Výroba 16:403–407

Payne RH (1974) Transferrin variation in North American populations of the Atlantic salmon (*Salmo salar*). J Fish Res B Can 31 (6): 1037–1041

Payne RH, Child AR, Forrest A (1971) Geographical variation in the Atlantic salmon. Nature, (London) 231 (5301):240–242

Payne RH, Child AR, Forrest A (1972) The existence of natural hybrids between the European trout and the Atlantic salmon. J Fish Biol 4 (2):233–236

Payusova AN (1979) A comparison of electrophoretic spectra of hemoglobin and serum esterase in two species of Issyk-Kul daces (*Leuciscus schmidti* and *L. bergi*). In: Biochem Popul Genet Fish. Inst Cytol Acad Sci USSR, Leningrad, pp 120–124 (RE)

Payusova AN, Koreshkova ND (1973) Cytophysiological and electrophoretic analysis of *Vimba vimba* (L) differentiation from the Namunas river. In: Biochem Genet Fish. Inst Cytol Acad Sci USSR, Leningrad, pp 178–182 (RE)

Payusova AN, Koreshkova MD, Andreeva AP (1976) Cytophysiological and biochemical analysis of the *Vimba vimba* from different parts of its distribution range. In: Vimba. Vilnius, pp 109–146 (R)

Pederson RA (1971) DNA content, ribosomal gene multiplicity and cell size in fish. J Exp Zool 177 (1):65–78

Pejić K, Marić C, Hadžiselimović R (1978) Malate dehydrogenase polymorphism in some species of the genera *Alburnus* and *Salmo* (Pisces). In: XIV Int Congr. Genet (Moscow) Contrib Pap Session Abstr 1:147

Penners R (1959) Durch Röntgenstrahlen verursachte biologische Schäden. Untersuchungen mit und an dem Schwertträger. Z Nat Tech 5:403–407

Perriard JC, Scholl A, Eppenberger HM (1972) Comparative studies on creatine kinase isozymes from skeletal muscle and stomach of trout. J Exp Zool 182 (1): 119–126

Peters N, Peters G (1968) Zur genetischen Interpretation morphologischer Gesetzmässigkeiten der degenerativen Evolution. Untersuchungen am Auge einer Höhlenform von *Poecilia sphenops.* Z Morphol Tiere 62 (3):211–244

Peters N, Peters G (1973) Genetic problems in the regressive evolution of cavernicolous fish. In: Genetics and mutagenesis of fish. Springer Berlin Heidelberg New York, pp 187–201

Peters N, Scholl A, Wilkens H (1975) Der Micos-Fisch, Höhlenfisch in statu nascendi oder Bastard? Ein Beitrag zur Evolution der Höhlentiere. Z Zool Syst Evol-Forsch 13 (2): 110–124

Pfeiffer W (1966) Über die Vererbung der Schreckreaktion bei *Astyanax* (Characidae, Pisces). Z Vererbungsl 98 (2):98–105

Pfeiffer W (1967) Die Korrelation von Augengröße und Mittelhirngröße bei Hybriden aus *Astyanax* x *Anoptichthys* (Characidae, Pisces). Roux'Arch Entw-Mech 159 (3):365–378

Philipp DP, Whitt GS (1977) Patterns of gene expression during teleost embryogenesis: lactate dehydrogenase isozyme ontogeny in the medaka (*Oryzias latipes*). Dev Biol 59 (2) 183–197

Philipp DP, Parker HR, Beaty PR, Childers WF, Whitt GS (1980) The effect of genetic distance on differential gene expression during embryogenesis of sunfish hybrids. Isoz Bull 13:87

Piront A, Gosselin-Rey C (1975) Immunological cross-reactions among Gadidae parvalbumins. Biochem Syst Ecol 3 (4):251–255

Place AR, Powers DA (1977) Liver LDH allozymes of *Fundulus heteroclitus:* a selective adaptation or genetic drift? Fed Proc 36:738

Place AR, Powers DA (1978) Genetic bases for protein polymorphism in *Fundulus heteroclitus* (L.). I. Lactate dehydrogenase (Ldh B), malate dehydrogenase (Mdh-A), glucosephosphate isomerase (Gpi-B), and phosphoglucomutase (Pgm A). Biochem Genet 16 (5–6):577–591

Place AR, Powers DA (1979) Genetic variation and relative catalytic efficiencies: lactate dehydrogenase B allozymes of *Fundulus heteroclitus.* Proc Natl Acad Sci USA 76 (5):2354–2358

Plokhinsky NA (1964) Heritability. Acad Nauk Sib Otdel, Novosibirsk (R)

Plumb HA, Green OL, Smitherman RO, Pardue GB (1975) Channel catfish virus experiments with different strains of channel catfish. Trans Am Fish Soc 104 (1):140–143

Pojoga J (1969) Observations sur l'héredité de la carpe. Bull Fr Pisc 42 (235):67–73

Pojoga J (1972) Race metis et hybrides chez la carpe. Bull Fr Pisc 44 (244):134–142

Pokhiel LI (1967) Erythrocyte antigens of the common carp *Cyprinus carpio* L., the grass carp *Ctenopharyngodon idella* Vall., and the silver crucian carp *Carassius auratus gibelio* Bloch. Tr VNIIPRKh 15:278–283 (RE)

Pokhiel LI (1969) Interspecific and intraspecific differences of erythrocytary antigenes in pond fishes. In: Genetics, selection and hybridization of fishes. Nauka, Moscow, pp 114–117 (R)

Poljarush VP (1979) Heterosis in intraspecific crossing of herbivorous fishes. In: Materials of All-Union scientific conference on development and intensification of fisheries in inland waters of North Caucasus. Azov Res Inst Fish, Rostov-Don, pp 176–177 (R)

Poljarush VP, Ovechko VJu (1979) Heritability and variation of several selective traits in carp larvae. In: Selection of pond fish. All-Union Res Inst Pond Fish, Moscow, pp 111–116 (RE)

Poluhowich JJ (1972) Adaptive significance of eel multiple hemoglobin. Phys Zool 45 (3):215–222

Polyakovsky VI, Papkovskaya AA, Bogdanov LV (1973) Biochemical polymorphism in crucian carp (*Carassius auratus gibelio*) from the Sudoble Lake (Byelorussian). In: Biochem Genet Fish. Inst Cytol Acad Sci USSR, Leningrad, pp 161–166 (RE)

Polyksenov DP (1962) Creation of highly productive and viable common carp strain for the development of a new carp breed in Byelorussia. In: Problems of fish management in Byelorussia. Min Educ Byelor SSR, Minsk, pp 5–62 (R)

Pontier PJ, Hart NH (1978) Isozyme expression in interspecific hybrids between two teleosts, *Brachydanio albolineatus* and *Brachydanio rerio.* Isoz Bull 11:55

Pontier PJ, Hart NH (1979) Creatine kinase gene expression during the development of *Brachydanio.* J Exp Zool 209 (2):283–296

Pontier PJ, Baker J, Hart NH (1978) Creatine phosphokinase ontogeny in *Brachydanio albolineatus* (Cyprinidae). Isoz Bull 11:37

Popov LS, Antonov AS, Mednikov BM, Belozersky AN (1973) Natural system of fishes: results of applying of DNA hybridization method. Dokl Akad Nauk SSSR (DAN) 211 (3):737–739 (R)

Popov OP (1978) Application of hematological analysis for characterizing common carp breeding groups. In: Genetics and selection of fish, vol 20. All-Union Res Inst Pond Fish, Moscow, pp 188–198 (RE)

Popova AA (1969) The variation of the intestines and the air bladder in scaled, scattered, linear and nude common carps. In: Collect Pap Pond Fish Breeding. All-Union Res Inst Pond Fish, Moscow, pp 149–152 (R)

Porter TR, Corey S (1974) A hermaphroditic lake whitefish, *Coregonus clupeaformis,* from lake Huron. J Fish Res B Can 31 (12):1944–1945

* Post A (1965) Vergleichende Untersuchungen der Chromosomenzahlen bei süsswasser Teleosteern. Z Zool Syst Evol-Forsch 3 (1–2):47–93

* Post A (1972) Ergebnisse der Forschungsreisen des FFS "Walter Hertwig" nach Südamerica. XXIV. Die Chromosomenzahlen einiger Atlantischer Myctophidenarten (Osteichthyes, Myctophoidei, Myctophidae). Arch Fischereiwiss 23 (2): 89–93

* Post A (1973) Chromosomes of two fish species of the genus *Diretmus* (Osteichthyes, Beriformes: Diretmidae). In: Genetics and mutagenesis of fish. Springer, Berlin Heidelberg New York, pp 103–111

* Post A (1974) Die Chromosomen von drei Arten aus der Familie Gonostomatidae (Osteichthyes, Stomiatoidei). Arch Fischereiwiss 25 (1): 51–55

Potter IC, Nicol PI (1968) Electrophoretic studies on the haemoglobins of Australian lampreys. Aust J Exp Biol Med Sci 46 (5): 639–641

* Potter IC, Robinson ES (1971) The chromosomes. In: The biology of lampreys, vol I. Academic Press, New York London, pp 279–294

* Potter IC, Rothwell B (1970) The mitotic chromosomes of the lamprey, *Petromyzon marinus* L. Experientia 26 (3): 429–430

Powell JR (1975) Protein variation in natural populations of animals. In: Evol Biol, vol 8 Plenum Press New York, pp 79–119

Powers DA (1972) Hemoglobin adaptation for fast and slow habitats in sympatric catostomid fishes. Science 177 (4046): 360–362

Powers DA (1980) Molecular ecology of teleost fish hemoglobins: strategies for adapting to changing environments. Zool 20 (1): 139–162

Powers DA, Place AR (1978) Biochemical genetics of *Fundulus heteroclitus* (L.). I. Temporal and spatial variation in gene frequencies of Ldh-B, Mdh-A, Gpi-B and Pgm-A. Biochem Genet 16 (5–6): 593–607

Powers DA, Greaney GS, Place AR (1979) Physiological correlation between lactate dehydrogenase genotype and haemoglobin function in killifish. Nature (London) 277 (5693): 240–241

* Prasad R, Manna GK (1974) Somatic and germinal chromosomes of a livefish, *Heteropneustes fossilis* (Bloch). Cytologia 27 (2): 217–223

Prather EE (1961) A comparison of production of albino and normal channel catfish. Proc Auburn Univ Agric Exp Stn, Auburn (cit by Sneed, 1971)

* Prehn LM, Rasch EM (1969) Cytogenetic studies of *Poecilia* (Pisces). I. Chromosome numbers of naturally occurring poeciliid fishes and their hybrids from eastern Mexico. Can J Genet Cytol 11 (4): 880–895

Prevosti A, Ocana J, Alonso G (1975) Distances between populations of *Drosophila subobscura,* based on chromosome arrangement frequencies. Theoret Appl Genet 45 (6): 231–241

Probst E (1949 a) Der Bläuling-Karpfen. Allg Fisch-Z 74 (13): 232–238

Probst E (1949 b) Vererbungsuntersuchungen beim Karpfen. Allg Fisch-Z 74 (21): 436–443

Probst E (1950) Der Todesfaktor bei der Vererbung des Schuppenkleides des Karpfens. Allg Fisch-Z 75 (15): 369–370

Probst E (1953) Die Beschuppung des Karpfens. Münch. Beitr Fluss- u. Abwasserbiol 1: 150–227

* Prokofieva AA (1935) Chromosomal morphology of several fish and amfibian species. Tr Inst Genet 10: 153–178 (R)

Pruginin Y, Rothbard S, Wohlfahrt G, Helevy A, Moav R, Hulata G (1975) All-male broods of *Tilapia nilotica* x *T. aurea* hybrids. Aquaculture 6 (1): 11–21

Pudovkin AI (1979) Use of allozyme data for assessment of genetic similarity. In: Biochem Popul Genet Fish. Inst Cytol Acad Sci USSR, Leningrad, pp 10–17 (RE)

Purdom CE (1969) Radiation-induced gynogenesis and androgenesis in fish. Heredity 24 (3): 431–444

Purdom CE (1972) Induced polyploidy in plaice (*Pleuronectes platessa*) and its hybrid with the flounder (*Platichthys flesus*). Heredity 29 (1): 11–24

Purdom CE (1976) Genetic techniques in flatfish culture. J Fish Res B Can 33 (4, p 2): 1088–1093

Purdom CE (1979) Genetics of growth and reproduction in teleosts. In: Fish phenology: anabolic adaptiveness in teleosts. Academic Press, London New York, pp 207–217

Purdom CE, Lincoln RF (1974) Gynogenesis in hybrids within the Pleuronectidae. In: Early life history of fish. Springer, Berlin Heidelberg New York, pp 537–544

Purdom CE, Woodhead DS (1973) Radiation damage in fish. In: Genetics and mutagenesis of fish. Springer, Berlin Heidelberg New York, pp 67–73

Purdom CE, Thompson D, Dando PR (1976) Genetic analysis of enzyme, polymorphisms in plaice (*Pleuronectes platessa*). Heredity 37 (2): 193–206

Quiroz-Gutierrez M, Ohno S (1970) The evidence of gene duplication for S-form NADP-linked isocitrate dehydrogenase in carp and goldfish. Biochem Genet 4 (1): 93–99

Raicu P, Taisescu E (1972) *Misgurnus fossilis,* a tetraploid fish species. J Hered 63 (2): 92–94

Rainboth WJ, Whitt GS (1974) Analysis of evolutionary relationships among shiners of the subgenus *Luxilus* (Teleostei, Cypriniformes, Notropis) with the lactate dehydrogenase and malate dehydrogenase isozyme systems. Comp Biochem Phys 49B (2): 241–252

Raleigh RF, Chapman DW (1971) Genetic control in lakeward migrations of cutthroat trout fry. Trans Am Fish Soc 100 (1): 33–40

Rapacz J, Sløta E, Hasler J (1971) Preliminary studies on serum antigens and their development in carp. Rapp P-V Reun 161: 170–174

Rasch EM, Balsano JS (1973) Cytogenetic studies of *Poecilia* (Pisces). III. Persistence of triploid genomes in the unisexual progeny of triploid females associated with *Poecilia formosa.* Copeia 4: 810–813

Rasch EM, Darnell RM, Kallman KD, Abramoff P (1965) Cytophotometric evidence for triploidy in hybrids of the gynogenetic fish, *Poecilia formosa.* J Exp Zool 160 (2): 155–159

Rasch EM, Prehn LM, Rasch RW (1970) Cytogenetic studies of *Poecilia* (Pisces). II. Triploidy and DNA levels in naturally occurring populations associated with the gynogenetic teleost, *Poecilia formosa* (Girard). Chromosoma 31 (1): 18–40

Rattazzi MC, Pik C (1965) Haemoglobin polymorphism in cod (*Gadus morhua*): a single peptide difference. Nature (London) 208 (5009): 489–491

Raunich L, Callegarini C, Cavicchioli G (1966) Polimorfismo emoglobinico e caratteri sistematici del genere *Ictalurus* dell'Italia settentrionale. Arch Zool Ital 51: 497–510

Raunich L, Battaglia B, Callegarini C, Mozzi C (1967) Il polimorfismo emoglobinico del genere *Gobius* del la Laguna di Venezia. Atti Ist Veneto Sci Lett Arti 125: 87–105

Raunich L, Callegarini C, Cucchi C (1972) Ecological aspects of hemoglobin polymorphism in *Gasterosteus aculeatus* (Teleostei). In: Proc 5th Eur Mar Biol Symp, Padova, pp 153–162

Reagan RE, Pardue GB, Eisen EJ (1976) Predicting selection response for growth of channel catfish. J Hered 67 (1): 49–53

* Rees H (1967) The chromosomes of *Salmo salar.* Chromosoma 21 (4): 472–474

Refstie T, Gjedrem T (1976) Hybrids between Salmonidae species. Hatchability and growth rate in the fresh water period. Aquaculture 6 (3): 333–342

Refstie T, Steine TA (1978) Selection experiments with salmon. III. Genetic and environmental sources of variation in length and weight of Atlantic salmon in the freshwater phase. Aquaculture 14: 3, 221–234

Refstie T, Steine TA, Gjedrem T (1977 a) Selection experiments with salmon. II. Proportion of Atlantic salmon smoltifying at one year of age. Aquaculture 10 (3): 231–242

Refstie T, Vassvik V, Gjedrem T (1977 b) Induction of polyploidy in salmonids by cytochalasin B. Aquaculture 10 (1): 65–74

Reichle W (1974) Kann man die Regenbogenforelle züchterisch noch verbessern. Fischer Teichwirt 25 (8): 75–76

Reinitz GL (1977 a) Inheritance of muscle and liver types of supernatant NADP-dependent isocitrate dehydrogenase in rainbow trout (*Salmo gairdneri*). Biochem Genet 15 (5–6): 445–454

Reinitz GL (1977 b) Tests for association of transferrin and lactate dehydrogenase phenotypes with weight gain in rainbow trout (*Salmo gairdneri*). J Fish Res B Can 34 (12): 2333–2337

Reinitz GL, Orme LO, Lemm CA, Hitzel FN (1978) Differential performance of four strains of rainbow trout reared under standardized conditions. Prog Fish-Cult 40 (1): 21–23

Reinitz GL, Orme LO, Hitzel FN (1979) Variation of body composition and growth among strains of rainbow trout. Trans Am Fish Soc 108 (2): 204–207

Reisenbichler RR, McIntyre JD (1977) Genetic differences in growth and survival of juvenile hatchery and wild steelhead trout, *Salmo gairdneri.* J Fish Res B Can 34 (1): 123–128

Richmond MC, Zimmerman EG (1978) Effect of temperature on activity of allozymic forms of supernatant malate dehydrogenase in the red shiner, *Notropis lutrensis.* Comp Biochem Physiol 61B (3): 415–422

Richmond RC (1970) Non-Darwinian evolution: a critique. Nature (London) 225 (5237):1025–1028

Ricker WE (1972) Hereditary and environmental factors affecting certain salmonid populations. In: Symposium on the stock concept in pacific salmon. MacMillan, New York, pp 27–160

Ridgway GJ (1958) Studies on the serology. In: Mar Fish Invest Oper Rep. US Fish Wildl Serv Bur Comm Fish, New York, pp 1–13

Ridgway GJ (1962) Demonstration of blood groups in trout and salmon by isoimmunization. Ann NY Acad Sci 97 (1):111–115

Ridgway GJ (1966) A complex blood group system in salmon and trout. Proc 10th Eur Conf Anim Blood Gr Biochem Polym, Paris, pp 361–365

Ridgway GJ (1971) Problems in the application of serological methods to population studies on fish. Rapp P-V Reun 161:10–14

Ridgway GJ, Klontz GW (1961) Blood types in Pacific salmon. Bull Int North Pacific Fish Comm 5:49–55

Ridgway GJ, Utter FM (1963) Salmon serology. In: Annu Rep Int North Pacific Fish Comm 1961, Vancouver, pp 106–108

Ridgway GJ, Utter FM (1964) Salmon serology. In: Annu Rep Int North Pacific Fish Comm 1963, Vancouver, pp 149–154

Ridgway GJ, Cushing JE, Durall GL (1958) Serological differentiation of populations of sockeye salmon, *Oncorhynchus nerka.* Bull Int North-Pacific Fish Comm 3:5–10

Ridgway GJ, Sherburne SW, Lewis RD (1970) Polymorphism in the esterase of Atlantic herring. Trans Am Fish Soc 99 (1):147–151

Riggs CD, Sneed KE (1959) The effects of controlled spawning and genetic selection on the fish culture of the future. Trans Am Fish Soc 88 (1):53–60

* Rishi KK (1973) Somatic karyotypes of three teleosts. Genen Phaenen 16 (3):101–107

* Rishi KK (1975) Somatic and meiotic chromosomes of *Trichogaster fasciatus* (Teleostei, Perciformes; Osphronemidae). Genen Phaenen 18 (2–3):49–53

* Rishi KK (1976a) Karyotypic studies on four species of fishes. Nucleus 19 (2):95–98

* Rishi KK (1976b) Mitotic and meiotic chromosomes of a teleost *Callichromus bimaculatus* (Bloch) with indications of male heterogamety. Sci Cult 28 (10):1171–1173

* Rishi KK (1979) Somatic G-banded chromosomes of *Colisa fasciatus* (Perciformes; Belontidae) and confirmation of female heterogamety. Copeia 1:146–149

Rishi KK, Gaur P (1976) Cytological female heterogamety in jet-black molly, *Mollienesia sphenops.* Curr Sci 45 (18):669–670

* Roberts FL (1964) A chromosome study of twenty species of Centrarchidae. J Morphol 115 (3):401–417

* Roberts FL (1967) Chromosome cytology of the Osteichthyes. Prog Fish Cult 29 (2):75–83

* Roberts FL (1970) Atlantic salmon (*S. salar*) chromosomes and speciation. Trans Am Fish Soc 99 (1):105–111

Roberts FL, Wohnus JF, Ohno S (1969) Phosphoglucomutase polymorphism in the rainbow trout, *Salmo gairdneri.* Experientia 25 (10):1109–1110

* Robinson ES, Potter IC (1969) Meiotic chromosomes of *Mordacia praecox* and a discussion of chromosome numbers in lampreys. Copeia 4:824–828

* Robinson ES, Potter IC, Webb CJ (1974) Homogeneity of holarctic lamprey karyotypes. Caryologia 27 (4):443–454

Robinson ES, Potter IC, Atkin NB (1975) The nuclear DNA content of lampreys. Experientia 31 (8):912–913

Robinson GD, Dunson WA, Wright JE, Mamolito GE (1976) Differences in low pH tolerance among strains of brook trout (*Salvelinus fontinalis*). J Fish Biol 8 (1):5–17

Rodino E, Comparini A (1978) Genetic variability in the European eel *Anguilla anguilla* L. In: Marine organisms, Plenum Press, New York, pp 389–425

Rogers JS (1972) Measures of genetic similarity and genetic distance. In: Univ Tex Publ, N 7213, pp 145–153

Rokitsky PF (1974) The introduction into the statistical genetics. Vysshaja Shkola, Minsk, (R)

Rolle NN (1979) Comparative study of heat resistance and of isoenzyme LDH spectra in freshwater fishes. In: Physiology and biochemistry of sea- and freshwater animals, University, Leningrad, pp 181–186 (R)

Romashov DD, Belyaeva VN (1964) Cytology of the radiation induced gynogeneses and androgenesis in the loach (*Misgurnus fossilis* L.) Dokl Akad Nauk SSSR (DAN) 157 (4):964–967 (R)

Romashov DD, Belyaeva VN (1965 a) Analysis of diploidization induced by low temperature during radiation gynogenesis in loach. Tzitologiya 7 (5):607–615 (RE)

Romashov DD, Belyaeva VN (1965 b) The increase in the yield of diploid gynogenetic larvae in loach Misgurnus fossilis L. achieved by heat shocks. Bull Mosc O-va Isp Prir 60 (5):93–109 (R)

Romashov DD, Golovinskaya KA, Belyaeva VN, Bakulina ED, Pokrovskaya GL, Cherfas NB (1960) On radiation induced diploid gynogenesis in fishes. Biofizika 5 (4):461–468 (RE)

Romashov DD, Belyaeva VN, Golovinskaya KA, Prokofyeva-Belgovskaya AA (1961) Radiation damage of fishes. In: Radiation genetics. Atomizdat, Moscow, pp 247–266 (R)

Romashov DD, Nikolyukin NI, Belyaeva VN, Timofeeva NA (1963) On a possibility of the diploid radiation induced gynogenesis in the sturgeon. Radiobiologiya, (Moscow), 3 (1):104–110

Ropers HH, Engel W, Wolf U (1973) Inheritance of the S-form of NADP-dependent isocitrate dehydrogenase polymorphism in rainbow trout. In: Genetics and mutagenesis of fish, Springer, Berlin Heidelberg New York, pp 319–327

Rosenthal HL, Rosenthal RS (1950) Lordosis, a mutation in the cyprinodont, *Lebistes reticulatus.* J Hered 41 (8):217–218

* Ross MR (1973) A chromosome study of five species of Etheostominae fishes (Percidae). Copeia 1:163–165

Rothbard S, Pruginin Y (1975) Induced spawning and artificial incubation of *Tilapia.* Aquaculture 5 (3):315–321

Royal BK, Lucas GA (1972) Analysis of red and yellow pigments in two mutants of the Siamese fighting fish, *Betta splendens.* Proc Iowa Acad Sci 79 (1):34–37

Ruchkyan RG, Arakelyan GL (1980) Karyological bases of hybrid origin of Sevan whitefish *Coregonus lavaretus.* In: Karyological variability, mutagenesis and gynogenesis of fish. Inst Cytol Acad Sci USSR, Leningrad, pp 34–42 (RE)

Rudek Z (1974) Karyological investigations of two forms of *Vimba vimba* occurring in Poland. Folia Biol 22 (2):211–216

Rudloff V, Zelenik M, Braunitzer G (1966) Zur Philogenie der Hämoglobinmoleculs. Untersuchungen am Hämoglobin des Flussneunauges (*Lampetra fluviatilis*). Z Physiol Chem 344 (4–6):284–288

Rudzinski E (1928) Über Kreuzungsversuche bei Karpfen. Fisch Z 30:593–597;31:613–618; 32:636–640

Rudzinski E, Miaczynski T (1961) Zwerg-Karpfen von Pisarzowice. Acta Hydrobiol 3 (2–3):175–198

Rünger F (1934) Spiegelartige Beschuppungen bei Rotfedern und die Ursache ihrer Entstehung. Z Fischerei Hilfswissensch 32 (4):639–644

Rust W (1939) Männliche oder weibliche heterogamety bei *Platypoecilus variatus.* Z Indukt Abstammungs-Vererbungsl 77 (2):172–176

Rust W (1941) Genetische Untersuchungen über die Geschlechtsbestimmungstypen bei Zahnkarpfen unter besonderer Berücksichtigung von Artkreuzungen mit *Platypoecilus variatus.* Z Indukt Abstammungs-Vererbungsl 79 (3):336–395

Ryabov IN (1979) Hybridization of representatives of different subfamilies in family Cyprinidae. Vopr Ikhthiol 19 [6(119)]: 1025–1042 (R)

Ryabova-Sachko GD (1977 a) Isozymes of lactate dehydrogenase and some problems of ecology and evolution of *Oncorhynchus* and *Salvelinus.* In: The principles of the classification and phylogeny of the salmonoid fishes. Inst Zool Acad Sci USSR, Leningrad, pp 61–65 (R)

Ryabova (Sachko) GD (1977 b) Genetics of lactate dehydrogenase isozymes in Pacific salmonids. Ph D Thesis, Moscow, Inst Gen Genet (R)

Rychlicki Z (1973) Ocena użytkowa krzyżówki karpia wegierskiego z zatorskim. Gospod Rybna (Warshawa) 3:3–4

Ryman N (1970) A genetic analysis of recapture frequencies of released young of salmon (*S. salar* L.). Hereditas 65 (1):159–160

Ryman N (1972 a) An analysis of growth capability in full sib families of salmon (*Salmo salar* L.). Hereditas 70 (1):119–127

Ryman N (1972 b) Mortality frequencies in hatchery-reared salmon (*Salmo salar* L.). Hereditas 72 (2): 237–242

Ryman N (1973) Two-way selection for body weight in the guppy fish, *Lebistes reticulatus.* Hereditas 74 (2): 239–245

Ryman N, Ståhl G (1980) Genetic changes in hatchery stocks of brown trout (*Salmo trutta*). Can J Fish Aquat Sci 37 (1): 82–87

Ryman N, Allendorf FW, Stahl G (1979) Reproductive isolation with little genetic divergence in sympatric populations of brown trout (*Salmo trutta*). Genetics 92 (2): 247–262

Sachko GD (1971) Extent of polymorphism in the fish lactate dehydrogenase loci. In: Sci Rep Inst Mar Biol 2: 190–195 (R)

Sachko GD (1973) The genetics of lactate dehydrogenase isozymes in Pacific salmon (*Oncorhynchus*) In: Biochem Genet Fish. Inst Cytol Acad Sci USSR, Leningrad, pp 155–160 (RE)

Sadoglu P (1955) A mendelian gene for albinism in natural cave fish. Experientia 13 (2): 394

Sadoglu P (1957) Mendelian inheritance in the hybrid between the Mexican blind cave fishes and their overground ancestors. Verh Dtsch Zool Ges Graz, pp 432–439

Sadoglu P, McKee A (1969) A second gene that affects eye and color in Mexican blind cave fish. J Hered 60 (1): 10–14

Sage RD, Selander RH (1975) Trophic radiation through polymorphism in cichlid fishes. Proc Natl Acad Sci USA 72 (11): 4669–4673

Salmenkova EA (1973) Genetics of fish isozymes. Usp Sovr Biol 75 (2): 217–235 (R)

Salmenkova EA, Omelchenko VT (1978) Protein polymorphism in populations of diploid and tetraploid fish species. Biol Morja 4: 67–71

Salmenkova EA, Volokhonskaya LG (1973) Biochemical polymorphism in populations of diploid and tetraploid fish species. In: Biochem Genet Fish. Inst Cytol Acad Sci USSR, Leningrad, pp 54–61 (RE)

Salmenkova EA, Omelchenko VT, Ivaschenko II, Malanina TV, Aphanasjev KI, Zhivotovsky NA (1979) Population-genetical differences between adjacent generations in Pacific pink salmon. In: Abstr 14th Pacific Sci Congr Comm F Sect F 11: 175–176

Samokhvalova GV (1938) The effect of X-rays on fishes (*Lebistes,* swordtails and the crucian carp). Biol Zh 7 (5): 1023–1034 (RE)

Sanders BG, Wright JE (1962) Immunogenetic studies in two trout species of the genus *Salmo.* Ann NY Acad Sci 97: 116–130

Saprykin VG (1977) The effect of selection for growth rate upon the distribution of phenotypes and gene frequencies of transferrin in the one year-old fishes of the Ural common carp. In: Problems of genetics and selection in the Ural. Provincial Publ, Sverdlovsk, pp 173–174 (R)

Saprykin VG (1979) Types of transferrins and their interrelations with resistance of carp to oxygen deficiency. In: Biochem Popul Genet Fish. Inst Cytol Acad Sci USSR, Leningrad, pp 157–161 (RE)

* Sasaki M, Hitotsumachi S, Makino S, Terao T (1968) A comparative study of the chromosomes in the chum salmon, the Kokanee salmon and their hybrids. Caryologia 21 (3): 389–394

* Sasaki T, Sakamoto K (1977) Karyotype of the rockfish, *Sebastes taczanowskii* St. CIS 22: 7–8

Sassama C, Yoshiyama RM (1979) Lactate dehydrogenase – a polymorphism of *Anoplarchus purpurescens:* geographic variation in central California. J Hered 70 (5): 329–334

Sato R, Ishida R (1977) Genetic variation in malate dehydrogenase and some isozymes in white muscle of ayu (*Plecoglossus altivelis*). Bull Freshwater Fish Res Lab Tokyo 27 (2): 75–83

Saunders LH, McKenzie JA (1971) Comparative electrophoresis of arctic char. Comp Biochem Phys 38B (3): 487–491

Saunders RL (1971) Sea ranching - a promising way to enhance populations of Atlantic salmon for angling and commercial fisheries. In: IASF Spec Publ Ser 7: 17–24

Saunders RL (1978 a) Annual report to the advisory council of the North American Salmon Research Center. NASRC Rep 3: 1–7

Saunders RL (1978 b) Annual report to the advisory council of the North American Salmon Research Center. NASRC Rep 4: 1–8

Saunders RL, Bailey JK (1978) The role of genetics in Atlantic salmon management (cit from Saunders 1978 b)

Saunders RL, Sreedharan A (1977) The incidence and genetic implications of sexual maturity in male Atlantic salmon parr. In: Int Counc Explor Sea Comm Meet 1977/M, pp 1–21

Savostyanova GG (1971) Comparison of some strains of rainbow trout by their fish-cultural value. In: Izvestija Gosud. nauchno-issled. Inst Ozern Rechn Rybn Khoz (GosNIORKh), 74:87–103 (RE)

Savostyanova GG (1976) Origin, rearing and selection of the rainbow trout in the USSR and abroad. In: Biological bases of trout breeding. State Res Inst Lake River Fish, Leningrad, pp 3–13 (RE)

Schäperclaus W (1953) Bekämpfung der infektiösen Bauchwassersucht des Karpfens durch Züchtung erblich widerstandsfähiger Karpfenstämme. Z Fischerei 1 (5–6):321–353

Schäperclaus W (1954) Fischkrankheiten. Akademie Verlag, Berlin

Schäperclaus W (1961) Lehrbuch der Teichwirtschaft 2. Aufl. Paul Parey, Berlin Hamburg

* Scharma GP, Prasad R, Nayyar RP (1960) Chromosome number and meiosis in three species of fishes. Res Bull Punjab Univ 11:99–103

* Scheel JJ (1966) Taxonomic studies of African and Asian toothcarps (Rivulinae) based on chromosome numbers, haemoglobin patterns, some morphological traits and crossing experiments. Vidensk Medd Dan Naturhist Foren Khobenha 129:123–148

* Scheel JJ (1972a) Rivuline karyotypes and their evolution (Rivulinae, Cyprinodontidae, Pisces). Z Zool Syst Evol-Forsch 10 (3):180–209

* Scheel JJ (1972b) Die chromosomen der drei Neon-Tetras. Aquare-Terrar 9:307–309

* Scheel JJ (1974) Eine Übersicht zu *Aphyosemion cameronense.* Aquar-Terrar 9:306–307

* Scheel JJ, Simonsen V, Gyldenholm AO (1972) The karyotypes and some electrophoretic patterns of fourteen species of the genus *Corydorus.* Z Zool Syst Evol-Forsch 10 (2):144–152

Schemmel C (1974) Genetische Untersuchungen zur Evolution des Geschmacks Apparates bei cavernicolen Fischen. Z Zool Syst Evol-Forsch 12 (3):196–215

Schlotfeldt HJ (1968) Electrophoretische Eigenschaften des Hämoglobins von zwei in Chile wirtschaftlich wichtigen Gadiden, *Merluccius gayi gayi* und *M. polylepis,* als Versuch zu einer Populations struktur-Analyse. Arch Fischereiwissensch 19 (2–3):236–245

Schmalhausen II (1935) Determination of basic concepts and the methods of growth investigation. In: Growth of animals, Acad Nauk SSSR, Moscow-Leningrad, pp 8–60 (RE)

Schmidt J (1917) Racial investigations. I. *Zoarces viviparus* L. and local races of the same. C R Trav Lab Carlsberg 13 (3):279–396

Schmidt J (1919a) Racial investigations. III. Experiments with *Lebistes reticulatus* (Peters). C R Trav Lab Carlsberg 14 (5):1–8

Schmidt J (1919b) La valeur de l'individu a titre de générateur apprecies suivant la méthode du croisement diallele. C R Trav Lab Carlsberg 14 (6):1–34

Schmidt J (1920) Racial investigations. V. Experimental investigations with *Zoarces viviparus* L. C R Trav Lav Carlsberg 14 (9):1–14

Schmidt J (1921) Racial investigations. VII. Annual fluctuations of racial characters in *Zoarces viviparus* L. C R Trav Lab Carlsberg 14 (15):1–24

Schmidt J (1930) Racial investigations. X. The Atlantic cod (*Gadus callarias* L.) and local races of the same. C R Trav Lab Carlsberg 18 (6):1–71

Schmidt PYu (1935) Heredity and origin of the goldfish varieties. Priroda 24 (5):29–34 (R)

Schmidtke J, Engel W (1972) Duplication of the gene loci coding for the supernatant aspartate aminotransferase by polyploidization in the fish family Cyprinidae. Experientia 28 (8):976–978

Schmidtke J, Engel W (1974) On the problem of regional gene duplication in diploid fish of the orders Ostariophysi and Isospondyli. Humangenetik 21 (1):39–45

Schmidtke J, Engel W (1975) Gene action in fish of tetraploid origin. I. Cellular and biochemical parameters in cyprinid fish. Biochem Genet 13 (1–2):45–51

Schmidtke J, Engel W (1976) Gene action in fish of tetraploid origin. III. Ribosomal DNA amount in cyprinid fish. Biochem Genet 14 (1–2):19–26

Schmidtke J, Atkin NB, Engel W (1975a) Gene action in fish of tetraploid origin. II. Cellular and biochemical parameters in clupeoid and salmonoid fish. Biochem Genet 13 (5–6):301–309

Schmidtke J, Dunkhase G, Engel W (1975b) Genetic variation of phosphoglucose isomerase isoenzymes in fish of the orders Ostariophysi and Isospondyli. Comp Biochem Phys 50B (3):395–398

Schmidtke J, Kuhl P, Engel W (1976a) Transistory hemizygosity of paternally derived alleles in hybrid trout embryos. Nature (London) 260 (5549): 319–320

Schmidtke J, Schultze B, Kuhl P, Engel W (1976b) Gene action in fish of tetraploid origin. V. Cellular RNA and protein content and enzyme activities in cyprinid, clupeoid and salmonoid species. Biochem Genet 14 (11–12): 975–980

Schmidtke J, Budiman R, Engel W (1977) GPDH allele activation in brown trout x brook trout hybrids. Isoz Bull 10: 50–51

Schmidtke J, Schmidt E, Matzke E, Engel W (1979) Nonrepetitive DNA sequence divergence in phylogenetically diploid and tetraploid teleostean species of the family Cyprinidae and the order Isospondyli. Chromosoma 75 (2): 185–198

Schnakenbeck W (1927) Rassenuntersuchungen am Hering. Ber Dtsch Wiss Komm Meeresforsch 3 (2, p 2): 1–205

Schnakenbeck W (1931) Zum Rassenproblem bei den Fischen. Z Morphol Oekol Tiere 21 (2): 409–556

Scholl A (1973) Biochemical evolution in the genus *Xiphophorus*. In: Genetics and mutagenesis of fish. Springer, Berlin Heidelberg New York, pp 277–299

Scholl A, Anders F (1973a) Tissue specific preferential expression of the *Xiphophorus xiphidium* allele for 6-phosphogluconate dehydrogenase in interspecific hybrids of platyfish (Poeciliidae, Teleostei). In: Genetics and mutagenesis of fish. Springer, Berlin Heidelberg New York, pp 301–313

Scholl A, Anders F (1973b) Electrophoretic variation of enzyme proteins in platyfish and swordtails (Poeciliidae, Teleostei). Arch Genet 46 (2): 121–129

Scholl A, Holzberg S (1972) Zone electrophoretic studies on lactate dehydrogenase isoenzymes in South American cichlids (Teleostei, Percomorphi). Experientia 28 (4): 489–491

Scholl A, Schröder JH (1974) Biochemische Untersuchungen über die genetische Differenzierung mittelamerikanischer Zahnkarpfenarten (Cyprinodontiformes, Poeciliidae). Rev Suisse Zool 81 (3): 690–696

Scholl A, Corzillius B, Villwock W (1978) Contribution to the genetic relationship of Old World Aphanini (Pisces, Cyprinodontidae) with special reference to electrophoretic techniques. Z Zool Syst Evol-Forsch 16 (2): 116–132

Schreibman MP, Kallman KD (1977) The genetic control of the pituitary gonadal axis in the platyfish, *Xiphophorus maculatus*. J Exp Zool 200 (2): 277–293

Schröder JH (1964) Genetische Untersuchungen an domestizierten Stämmen der Gattung *Mollienesia* (Poeciliidae). Zool Beitr 10: 369–463

Schröder JH (1965) Zur Vererbung der Dorsalflossenstrahlenzahl bei *Mollienesia*-Bastarden. Z Zool Syst Evol-Forsch 3 (2): 330–348

Schröder JH (1966) Über Besonderheiten der Vererbung des Simpsonfactors bei *Xiphophorus helleri* Heckel (Poeciliidae, Pisces). Zool Beitr 12 (1): 27–42

Schröder JH (1969a) Die Variabilität quantitativer Merkmale bei *Lebistes reticulatus* Peters nach ancestraler Röntgenbestrahlung. Zool Beitr 15 (2–3): 237–265

Schröder JH (1969b) Erblicher Pigment-verlust bei Fischen. Aquar Terrar 16 (8): 272–274

Schröder JH (1969c) Die Vererbung von Beflossungsmerkmalen beim Berliner Guppy (*Lebistes reticulatus* Peters). Theoret Appl Genet 39 (1): 73–78

Schröder JH (1969d) Inheritance of radiation-induced spinal curvatures in the guppy *L. reticulatus*. Can J Genet Cytol 11 (4): 937–947

Schröder JH (1969e) X-ray-induced mutations in the poeciliid fish, *Lebistes reticulatus* Peters. Mutat Res 7 (1): 75–90

Schröder JH (1970) Das Züchten von Aquarienfischen – einmal wissenschaftlich betrachtet. III. Zusammenwirken mehrerer Gene – Crossing-over, drittes Mendelsches Gesetz. Aquar Mag 1: 8–17

Schröder JH (1973) Teleosts as a tool in mutation research. In: Genetics and mutagenesis of fish. Springer, Berlin Heidelberg New York, pp 91–99

Schröder JH (1974) Vererbungslehre für Aquarianer. W Keller and Co, Stuttgart

Schröder JH (1976) Genetics for aquarists. T F H Publications Inc, Neptune (New Jersey)

Schröder JH (1979) Methods of screening radiation induced mutations in fish. In: Methodology for assessing impacts of radioactivity on aquatic ecosystems. Vienna, IAEA Tech Rep Ser 190: 381–402

Schröder JH (1980) Morphological and behavioural differences between the BB/OB and B/W colour morphs of *Pseudotropheus zebra* Boul. (Pisces; Cichlidae). Z Zool Syst Evol-Forsch 18 (1):69–76

Schröder JH, Yegin MM (1968) Qualitative und quantitative Bestimmung der freien Aminosäuren bei *Mollienesia sphenops* (Poeciliidae; Pisces). Biol Zentralbl 87 (1):163–172

Schultz RJ (1961) Reproductive mechanism of unisexual and bisexual strains of the viviparous fish *Poeciliopsis*. Evolution 15 (2):302–325

Schultz RJ (1963) Stubby, a hereditary vertebral deformity in the viviparous fish *Poeciliopsis prolifica*. Copeia 2:325–330

Schultz RJ (1966) Hybridization experiments with an all-female fish of the genus *Poeciliopsis*. Biol Bull 130 (3):415–429

Schultz RJ (1967) Gynogenesis and tryploidy in the viviparous fish *Poeciliopsis*. Science 157 (3796):1564–1567

Schultz RJ (1969) Hybridization, unisexuality and polyploidy in the teleost *Poeciliopsis* (Poeciliidae) and other vertebrates. Am Nat 103 (934):605–619

Schultz RJ (1971) Special adaptive problems associated with unisexual fishes. Am Zool 11 (2):351–360

Schultz RJ (1973) Unisexual fish: laboratory synthesis of a "species". Science 179 (4069):180–181

Schultz RJ (1977) Evolution and ecology of unisexual fishes. In: Evol Biol 10:277–331

Schultz RJ, Kallman KD (1968) Triploid hybrids between the all-female teleost *Poecilia formosa* and *Poecilia sphenops*. Nature (London) 219 (5150):280–282

Schwab M, Haas J, Abdo S, Ahuja MR, Kollinger G, Anders A, Anders F (1978) Genetic basis of susceptibility for development of neoplasms following treatment with N-methyl-N-nitrosourea (MNU) or X-rays in the platyfish/swordtail system. Experientia 34 (6):780–782

Schweigert JF, Ward FJ, Clayton JW (1977) Effects of fry and fingerling introductions on walleye (*Stizostedion vitreum vitreum*) production in West Blue Lake, Manitoba. J Fish Res B Can 34 (11):2142–2150

Schwier H (1939) Geschlechtsbestimmung und -differenzierung bei *Macropodus opercularis, concolor, chinensis* und deren Artbastarden. Z Indukt Abstammungs-Vererbungsl 77 (2):291–335

Scopes RH, Gosselin-Rey C (1968) Polymorphism in carp muscle creatin kinase. J Fish Res B Can 25 (12):2715–2716

Sedov SI, Krivasova SB (1973) Comparative analysis of intraspecific heterogeneity of Caspian fishes in reference to blood protein polymorphism. In: Biochem Genet Fish. Inst Cytol Acad Sci USSR, Leningrad, pp 183–187 (RE)

Sedov SI, Krivasova SB, Komarova GV (1976) Genetic and ecophysiological characteristic of the roach of the Caspian basin. In: Ecol Physiol Fish. Naukova Dumka, Kiev, 2:57–58 (R)

Selander RH (1976) Genic variation in natural populations. In: Molecular evolution. Sinauer Assoc, Sunderland, pp 21–45

Sengbusch R von (1967) Eine Schnellbestimmungsmethode der Zwischenmuskelgräten bei Karpfen zur Auslese von "grätenfreien" Mutanten (mit Röntgen-Fernsehkamera und Bildschirmgerät). Züchter 37 (6):275–276

Sengbusch R von, Meske Ch (1967) Auf dem Wege zum grätenlosen Karpfen. Züchter 37 (2):271–274

Sengün A (1950) Beiträge zur Kenntnis der erblichen Bedingtheit von Formunterschieden der Gonopodien lebendgebärender Zahnkarpfen. Rev Fac Sci Univ Instanbul Ser B 15:110–133

Sensabaugh GF, Kaplane NO (1972) A lactate dehydrogenase specific to the liver of gadoid fish. J Biol Chem 247 (2):585–593

Sepovaara O (1962) Zur Systematik und Ökologie des Lachses und der Forellen in den Binnengewässern Finlands. Ann Zool Soc Zool Bot Fenn "Vaname" 24:1–86

* Serebriakova EV (1969) Some data on chromosomal complements of acipenserid fishes. In: Genetics, selection and hybridization of fishes. Nauka, Moscow, pp 105–113 (R)

Serebriakova EV (1970) Chromosomal complements of hybrids obtained by crosses of sturgeons possessing different karyotypes. In: Distant hybridization of plants and animals, vol II. Kolos, Moscow, pp 185–192 (R)

Serene P (1971) Esterase of the North-East Atlantic albacor stock. Rapp P-V Reun 161:118–121

Serov OL, Korochkin LI, Manchenko GP (1977) Electrophoretic methods of isozyme examination. In: Genetics of isozymes. Nauka, Moscow, pp 18–64 (R)

Sezaki K, Kobayasi H (1978) Comparison of erythrocytic size between diploid and tetraploid in spinous loach, *Cobitis biwae.* Bull Jpn Soc Sci Fish 44 (8):851–854

Sgano T, Abe T (1973) Xanthochronous examples of one or two species of sea chubs of the genus *Kyphosus* from the Bonin Island. UO (Japan) 16:1–2

Shaklee JB, Whitt GS (1977) Patterns of enzyme ontogeny in developing sunfish. Differentiation 9 (1):85–95

Shaklee JB, Kepes KL, Whitt GS (1973) Specialized lactate dehydrogenase isozymes: the molecular and genetic basis for the unique eye and liver LDH-s of teleost fishes. J Exp Zool 185 (2):217–240

Shaklee JB, Champion MJ, Whitt GS (1974) Developmental genetics of teleosts: a biochemical analysis of lake chubsucker ontogeny. Dev Biol 38 (2):356–382

Shaklee HB, Christiansen JH, Sidell BD, Prosser CL, Whitt GS (1977) Molecular aspects of temperature changes in enzyme activities and isozyme patterns to metabolic reorganization in the green sunfish. J Exp Zool 200 (1):1–20

Shami SA, Beardmore JA (1978a) Genetic studies of enzyme variation in the guppy, *Poecilia reticulata.* Genetica 48 (1):67–73

Shami SA, Beardmore JA (1978b) Stabilizing selection and parental age effects on lateral line scale number in the guppy *Poecilia reticulata* (Peters). Pak J Zool 10 (1):1–15

Sharp GD (1969) Electrophoretic study of tuna hemoglobins. Comp Biochem Phys 31B (5):749–755

Sharp GD (1973) An electrophoretic study of hemoglobins of some scombroid fishes and related forms. Comp Biochem Phys 44B (2):381–388

Shart LA, Iljasov JuI (1979) Types of transferrins and esterases in spawners of common carp selected for resistance to dropsy. In: Biochem Popul Genet Fish. Inst Cytol Acad Sci USSR, Leningrad, pp 147–151 (RE)

Shaskolsky DV (1954) On inbreeding in common carp fish farms. Tr VNIIPRKh 7:22–33 (R)

Shaw CR (1965) Electrophoretic variation in enzymes. Science 149 (3687):936–943

Shaw CR (1970) How many genes evolve? Biochem Genet 4 (2):275–283

Shaw CR, Prasad R (1970) Starch gel electrophoresis of enzymes – a compilation of recipes. Biochem Genet 4 (2):297–320

Shearer KD, Mulley JC (1978) The introduction and distribution of the carp, *Cyprinus carpio* L., in Australia. Aust J Mar Freshwater Res 29 (5):551–563

Shevtzova EE, Chuksin VS (1979) Commercial sea-farming of salmon in foreign countries. In: Sea fisheries. Pitschevaya promyshlenost, Moscow, pp 36–41 (R)

Shoemaker HH (1943) Pigment deficiency in the carp and the carp-sucker. Copeia 1:54

Shulyak TS (1961) Cases of abnormal structure of intestines in the common carp. Dokl Akad Nauk Ukrainsk SSR 3:384–386 (R)

Siciliano MJ, Wright DA (1973) Evidence for multiple unlinked genetic loci for isocitrate dehydrogenase in fish of the genus *Xiphophorus.* Copeia 1:158–161

Siciliano MJ, Wright DA (1976) Biochemical genetics of the platyfish-swordtail hybrid melanoma system. In: Prog Exp Tumor Res 20:398–411

Siciliano MJ, Wright DA, George SL, Shaw CR (1973) Inter- and intraspecific genetic distances among teleosts. In: Abstr 17th Congr Int Zool, Theme 5 (cit from Siciliano, Wright, 1976)

Sick H (1961) Haemoglobin polymorphism in fishes. Nature (London) 192 (4805):894–896

Sick K (1965a) Haemoglobin polymorphism of cod in the Baltic and the Danish Belt sea. Hereditas 54 (1):19–48

Sick K (1965b) Hemoglobin polymorphism of cod in the North Atlantic Ocean. Hereditas 54 (1):49–73

Sick K, Bahn E, Frydenberg O, Nielsen JT, Wettstein D von (1967) Haemoglobin polymorphism of the American freshwater *Anguilla.* Nature (London) 214 (5093):1141–1142

Sick K, Frydenberg O, Nielsen JT (1973) Haemoglobin patterns of second-generation hybrids between plaice and flounder. Heredity 30 (2):244–245

Sidell BD, Otto RG, Powers DA (1978) A biochemical method for distinction of striped bass and white perch larvae. Copeia 2:340–343

Siebenaller JF (1978) Genetic variability in deep sea fishes of the genus *Sebastolobus* (Scorpaenidae). In: Marine organisms, Plenum Press, New York, pp 95–122

Siebenaller J, Somero GN (1978) Pressure-adaptive differences in lactate dehydrogenase of congeneric fishes living at different depths. Science 201 (4352):255–259

* Simon RC (1963) Chromosome morphology and species evolution in the five North American species of Pacific salmon (*Oncorhynchus*). J Morphol 112 (1):77–94

* Simon RC, Dollar AM (1963) Cytological aspects of speciation in two North American teleosts *Salmo gairdneri* and *S. clarkii lewisi*. Can J Genet Cytol 5 (1):43–49

Simonarson B, Watts DC (1971) Muscle esterase and protein variation in stocks of herring from Blackwater, Dunmore and Ballantrae. Rapp P-V Reun 161:27–31

Simonsen V, Frydenberg O (1972) Genetics of *Zoarces* populations. II. Three loci determining isozymes in eye and brain tissue. Hereditas 70 (2):235–242

Simpson JC, Schlotfeldt S (1966) Algunas observaciones sobre las caracteristicas electroforeticas de la hemoglobina de anchoveta *Engraulis ringens* J. en Chile. Invest Zool Chil 13:21–45

Sindermann CJ (1962) Serology of Atlantic clupeoid fishes. Am Nat 96 (889):225–231

Sindermann CJ (1963) Use of plant hemagglutinins in serological studies of clupeoid fishes. US Fish Wildl Serv Fish Bull 63 (1):137–141

Sindermann CJ, Mairs DF (1959) A major blood group system in Atlantic sea herring. Copeia 3:228–232

Singh RC, Hubby JL, Throckmorton LH (1975) The study of genic variation by electrophoretic and heat denaturation techniques at the octanol dehydrogenase locus in members of the *Drosophila virilis* group. Genetics 80 (3):637–650

Skow LC (1976) Serum esterase variation on channel catfish: genetics and population analysis. Proc Annu Conf Southwest Assoc Fish Game Comm 29:57–62

Slota E (1973) Studies on serum antigens in carp. Anim Blood Gr Biochem Genet 4 (3):175–179

Slota E, Papaez J, Stefan L (1970) Wstepne badania ned grupanti krwi u karpia (*Cyprinus carpio*). Zesz Probl Postepów Nauk Rol 4:71–78

Slutzky ES (1971a) Variability of ovulated egg size in the grass carp. Izvestija Gosud. nauchno- issled. Inst Ozern Rechn Rybn Khos (GosNIORKh) 74:128–139 (RE)

Slutzky ES (1971b) Variability of grass carp *Ctenopharyngodon idella* (Val.) under the conditions of artificial reproduction. Ph D Thesis, Leningrad (R)

Slutzky ES (1971c) Variability of eggs and larvae in the wild common carp of the Tsimlyanskoye Artificial Lake. Izvestija Gosud. nauchno-issled. Inst Ozern Rechn Rybn Khos (GosNIORKh) 74:62–86 (RE)

Slutzky ES (1976) Variability of Ropsha carp in structure and number of intermuscular bones. Izvestija Gosud. nauchno-issled. Inst Ozern Rechn Rybn Khos (GosNIORKh) 107:41–47 (RE)

Slutzky ES (1978) Phenotypic variability in fish. In: Variation in fishes. State Res Inst Lake River Fish, Leningrad, pp 3–132 (RE)

Slynko VI (1971a) Polymorphism of malate dehydrogenase isozymes in the Pacific salmons. In: Sci Rep Inst Mar Biol 2:207–211 (R)

Slynko VI (1971b) Frequency analysis of genes coding for malate dehydrogenase isozymes in the populations of pink and chum salmon of Sakhalin rivers. In: Sci Rep Inst Mar Biol 2:212–214 (R)

Slynko VI (1976a) Multiple molecular forms of malate – and lactate dehydrogenase in the Russian sturgeon *Acipenser güldenstädti* Br. and white sturgeon (beluga) *Huso huso* L. Dokl Akad Nauk SSSR (DAN) 228 (2):470–472 (R)

Slynko VI (1976b) Electrophoretic analysis of malate dehydrogenase isozymes in fishes of fam. Salmonidae. Dokl Akad Nauk SSSR (DAN) 226 (2):448–451 (R)

Slynko VI, Kasakov RV, Semyonova SK (1980) Polymorphism of NADP-dependent isocitrate dehydrogenase and malic-enzyme in salmon. Isoz Bull 13:99

Smišek J (1978) Genetic investigation of carp in Czechoslovakia. In: Genetics and selection of fishes, vol 20. All-Union Res Inst Pond Fish, Moscow, pp 140–147 (RE)

Smišek J, Vavruska A (1975) Vysledky korelaci transferinovych fenotypu k exterieru a k biochemickym hodnotam u kapra lyseho (ssnn). Bull VURH 11 (2):3–10

Smith AC (1965) Intraspecific eye lens protein differences in yellow fin tuna, *Thunnus albacares.* Calif Fish Game 51 (3):163–167

Smith AC (1968) Electrophoretic studies of eye lens protein from marine fishes. Diss Abstr 28 (12):68

Smith AC (1969) Protein variation in the eye lens nucleus of the mackerel scad (*Decapterus pinnulatus*). Comp Biochem Phys 28 (3):1161–1168

Smith AC (1971 a) Genetic and evolutionary analysis of protein variation in eye lens nuclei of rainbow trout (*Salmo gairdneri*). Int J Biochem 2 (2):384–388

Smith AC (1971 b) Protein differences in the eye lens cortex and nucleus of individual channel rockfish, *Sebastolobus alascanus.* Calif Fish Game 57 (3):177–181

Smith AC (1978) Pathology and biochemical genetic variation in the milkfish, *Chanos chanos,* J Fish Biol 13 (2):173–177

Smith AC, Clemens HB (1973) A population study by proteins from the nucleus of bluefin tuna eye lens. Trans Am Fish Soc 102 (3):578–583

Smith AC, Goldstein RA (1967) Variation in protein composition of the eye lens nucleus in ocean whitefish, *Caulolatilus princeps.* Comp Biochem Phys 23 (2):533–539

Smith CL (1975) The evolution of hermaphroditism in fishes. In: Intersexuality in animal kingdom. Springer, Berlin Heidelberg New York, pp 295–310

Smith K (1921) Racial investigations. VI. Statistical investigations on inheritance in *Zoarces viviparus* L. C R Trav Lab Carlsberg 14 (11):1–60

Smith K (1922) Racial investigations. IX. Continued statistical investigations with *Zoarces viviparus* L. C Rend Trav Lab Carlsberg 14 (19):1–42

Smith MH, Scott SL (1975) Thermal tolerance and biochemical polymorphism of immature largemouth bass *Micropterus salmoides* Lacepede. Bull GA Acad Sci 34:180–184

Smith PJ (1979 a) Esterase gene frequencies and temperature relationships in the New Zealand snapper *Chrysophrys auratus.* Mar Biol 53 (4):305–310

Smith PJ (1979 b) Glucosephosphate isomerase and phosphoglucomutase polymorphisms in the New Zealand ling *Genypterus blacodes.* Comp Biochem Phys 62B (3):573–577

Smith PJ, Jamieson A (1978) Enzyme polymorphisms in the Atlantic mackerel, *Scomber scombrus* L. Comp Biochem Physiol 60B (3):487–489

Smith PJ, Francis RICC, Paul LJ (1978) Genetic variation and population structure in the New Zealand snapper. N Z J Mar Freshwater Res 12 (4):343–350

Smithies O (1955) Zone electrophoresis in starch gel: group variations in the serum proteins of normal human adults. Biochem J 61 (4):629–641

Sneed KE (1971) Some current North American work in hybridization and selection of cultured fishes. Rep FAO/UNDP(TA), (2926), Rome, pp 143–150

Snieszko SF (1957) Disease resistant and susceptible populations of brook trout (*Salvelinus fontinalis*). Spec Sci Rep Fisher 208:126–128

Snieszko SF, Dunbar C, Bullock G (1959) Resistance to ulcer disease and furunculosis in eastern brook trout, *Salvelinus fontinalis.* Prog Fish-Cult 21 (3):110–116

* Sofradzija A, Berberoviĉ Lj (1972) Comparative karyological investigation of *Paraphoxinus alepidotus, P. adspersus, P. pstrossi, P. metohiensis* and *P. croaticus.* Godisn Biol Inst Univ Sarajevu 25:135–173

* Sofradzija A, Berberoviĉ Lj (1973) The chromosome number of *Barbus meridionalis petenyi* Heckel (Cyprinidae, Pisces). Bull Sci (Beograd) 18 (4–6):77–78

Sohn JJ (1977) Socially induced inhibition of genetically determined maturation in the platyfish, *Xiphophorus maculatus.* Science 195 (4274):199–201

* Sola L, Cataudella S, Stefanelli S (1978) I cromosomi di quattro specie di Scorpaenidae mediterranei (Pisces, Scorpaeniformes). Atti Accad Naz Lincei Re Cl Sci Fis Mat Nat Rend 64 (4):393–396

Solomon DJ, Child AR (1978) Identification of juvenile natural hybrids between Atlantic salmon (*S. salar*) and trout (*S. trutta*). J Fish Biol 12 (5):499–501

Somero GN, Hochachka PW (1969) Isoenzymes and short term temperature compensation in poikilotherms: activation of lactate dehydrogenase isoenzymes by temperature decreases. Nature (London) 223 (5202):194–195

Somero GN, Soule M (1974) Genetic variation in marine fishes as a test of the niche-variation hypothesis. Nature (London) 249 (5458):670–672

Sorensen EP (1980) Malate dehydrogenase polymorphism in *Notropis venustus* (Cyprinidae). Isoz Bull 13:98

Spinella DG, Vrijenhoek RC (1980) Immunochemical identification of a silent allele gene product in heterozygous fish of the genus *Poeciliopsis* (Poeciliidae). Isoz Bull 13:50

Sprague LM (1967) Multiple molecular forms of serum esterase in three tuna species from the Pacific Ocean. Hereditas 57 (1–2):198–204

Sprague LM (1970) The electrophoretic patterns of skipjack tuna tissue esterases. Hereditas 65 (2):187–190

Sprague LM, Holloway JR (1962) Studies of the erythrocyte antigenes of the skipjack tuna (*Katsuwonus pelamis*). Am Nat 96 (889):233–238

Sprague LM, Vrooman AM (1962) A racial analysis of the Pacific sardine (*Sardinops caerulea*) based on studies of erythrocyte antigens. Ann N Y Acad Sci 97:131–138

Sprague LM, Holloway JR, Nakashima LI (1963) Studies of the erythrocyte antigenes of albacore, bigeye, skipjack, and yellow fin tunas and their use in subpopulation identification. Proc World Sci Meet Biol Tunas Rel Species Exp Paper 22:1–15

Spurway H (1957) Hermaphroditism with selffertilization, and the monthly extrusion of unfertilized eggs, in the viviparous fish *Lebistes reticulatus*. Nature (London) 180 (4597):1248–1251

* Srivastava MDL, Bhagwan D (1968) Somatic chromosomes of *Clarias batrachus* (Clariidae: Teleostomi). Caryologia 21 (4):349–352

* Srivastava MDL, Kaur D (1964) The structure and behaviour of chromosomes in six fresh-water teleosts. Cellule 65 (1):93–107

Stallknecht H (1975) Albinotische Rautenflecksalmler. Aquar-Terrar 3:79–81

Stanley JG (1976a) Production of hybrid androgenetic and gynogenetic grass carp and common carp. Trans Am Fish Soc 105 (1):10–16

Stanley JG (1976b) Female homogamety in grass carp (*Ctenopharyngodon idella*) determined by gynogenesis. J Fish Res B Can 33 (6):1373–1374

Stanley JG, Jones JB (1976) Morphology of androgenetic and gynogenetic grass carp, *Ctenopharyngodon idella* (Valenciennes). J Fish Biol 9 (4):523–528

Stanley JG, Sneed KE (1974) Artificial gynogenesis and its application in genetics and selective breeding of fishes. In: Early life history of fish. Springer, Berlin Heidelberg New York, pp 527–536

Steffens W (1964) Vergleichende anatomisch-physiologische Untersuchungen an Wild- und Teichkarpfen (*Cyprinus carpio* L.). Z Fischerei 12:725–800

Steffens W (1966) Die Beziehungen zwischen der Beschuppung und dem Wachstum sowie einigen meristischen Merkmalen beim Karpfen. Biol Zentralbl 85 (3):273–287

Steffens W (1974a) Aufgaben und Ziele der Forellenzüchtung. Z Binnenfischerei 21 (8):218–223

Steffens W (1974b) Methoden und Ergebnisse der Forellenzüchtung. Z Binnenfischerei 21 (8):224–232

Steffens W (1975) Der Karpfen *Cyprinus carpio*, 4. Aufl. A Ziemsen Verlag, Wittenberg Lutherstadt

Stegeman JJ, Goldberg E (1971) Distribution and characterization of hexose-6-phosphate dehydrogenase in trout. Biochem Genet 5 (6):579–589

Stegeman JJ, Goldberg E (1972) Inheritance of hexose-6-PDH polymorphism in brook trout. Biochem Genet 7 (3–4):279–288

Stegman K (1965) Variability of weight-gains in carp in the first two years of life. In: Zootechnika (Rybactwo, 2) Zesz. Nauk SGGW, Warszawa, pp 67–92

Stegman K (1967) Pedigry of Osieku carp-line. Gospod Rybna 19 (9):3–5

Stegman K (1969) Analyse der Karpfengeschlechtsregister. Z Fischerei 17 (5–7):409–421

Sterba G (1959) Über eine Mutation bei *Pterophyllum eimekei*. I. Anamnese und Beschreibung. Biol Zentralbl 78 (2):323–333

* Stevenson MM (1975) A comparative chromosome study in the pupfish genus *Cyprinodon* (Teleostei: Cyprinodontidae). Ph D Thesis, Univ Oklahoma, Norman

* Stingo V (1979) The chromosomes of cartilaginous fishes. Genetika (Holl) 50 (3):227–239

Stingo V, de Buit M-H, Odierna G (1980) The genome size of some selachian fishes (in press)

Stoneking M, May B, Wright IE (1979) Genetic variation and inheritance of quaternary structured malic enzyme in brook trout *Salvelinus fontinalis.* Biochem Genet 17 (7–8):599–619

Strommen CA, Rasch EM, Balsano JS (1975) Cytogenetic studies of *Poecilia* (Pisces): V. Cytophotometric evidence for the production of fertile offspring by triploids related to *Poecilia formosa.* J Fish Biol 7 (5):667–676

Strunnikov VA (1974) The emergence of the compensatory gene complex as a mechanism of heterosis. Zh Obtsch Biol 35 (5):666–677 (RE)

Subla Rao B, Chandrase-Karan G (1978) Preliminary report on hybridization experiments in trout – growth and survival of F_1 hybrids. Aquaculture 15 (3):297–300

Subrahmanyam K (1969) A karyotypic study of the estuarine fish *Boleophthalmus boddaerti* (Pallas) with calcium treatment. Curr Sci 38 (18):437–439

Subrahmanyam K, Natarajan R (1970) A study of the somatic chromosomes of *Therapon* Cuvier (Teleostei: Perciformes). Proc Ind Acad Sci Sect B 72 (6):288–294

Subrahmanyam̃ K, Ramamoorthi K (1971) A karyotype study in the estuarine worm eel, *Moringua linearis* (Gray). Sci Cult 37 (4):201–202

Suzuki A (1962) On the blood types of yellow fin and bigeye tuna. Am Nat 96 (889):239–246

Suzuki A (1967) Blood type of fish. Bull Jpn Soc Sci Fish 33 (4):372–381

Suzuki A, Morio T (1960) Serological studies of the races of tuna. IV. The blood groups of the bigeye tuna. Rep Nankai Reg Fish Res Lab 12:1–13

Suzuki A, Schimizy J, Morio T (1958) Serological studies of the races of tuna. I. The fundamental investigations and the blood groups of albacore. Rep Nankai Reg Fish Res Lab 8:104–116

Suzuki A, Morio T, Mimoto K (1959) Serological studies of the races of tuna. II. Blood groups frequencies of the albacore in Tg system. Rep Nankai Reg Fish Res Lab 11:17–23

Suzuki R (1977) Cross-breeding experiments on the salmonid fish in Japan. Proc 5th Japan-Soviet Joint Symp Aquacult, Tokyo-Sapporo, pp 175–188

Suzuki R (1979) The culture of common carp in Japan. In: Advances in Aquaculture, Fish News Books Ltd, Farnham, pp 161–166

Suzuki R, Fukuda Y (1972) Growth and survival of F_1 hybrids among salmonid fishes. Bull Freshwater Fish Res Lab 21 (3):117–138

Suzuki R, Yamaguchi M, Ito T, Toi J (1976) Differences in growth and survival in various races of common carp. Bull Freshwater Fish Res Lab 26 (2):59–69

Suzuki R, Yamaguchi M, Ito T, Toi J (1978) Catchability and pulling strength of various races of the common carp caught by angling. Bull Jpn Soc Sci Fish 44 (7):715–718

Suzumoto BK, Schreck CB, McIntyre JD (1977) Relative resistances of three transferrin genotypes of coho salmon (*Oncorhynchus kisutch*) and their hematological responses to bacterial kidney disease. J. Fish Res B Can 34 (1):1–8

* Svärdson G (1945 a) Chromosome studies on Salmonidae. Annu Rep Inst Freshwater Res 13:1–151

Svärdson G (1945 b) Polygenic inheritance in *Lebistes.* Ark Zool 36A (6):1–9

Svärdson G (1950) The coregonid problem. II. Morphology of two coregonid species in different environments. Annu Rep Inst Freshwater Res 31:151–162

Svärdson G (1952) The coregonid problem. IV. The significance of scales and gillrakers. Annu Rep Inst Freshwater Res 33:204–332

Svärdson G (1957) The coregonid problem. VI. The palearctic species and their intergrades. Annu Rep Inst Freshwater Res 38:267–356

Svärdson G (1958) Interspecific hybrid populations in *Coregonus.* In: Systematics of today. Univ Årsskr, 6, Uppsala, pp 231–239

Svärdson G (1970) Significance of introgression in coregonid evolution. In: Biology of coregonid fishes. Univ Manitoba Press, Winnipeg, pp 33–59

Svetovidov AN (1933) Über den europäischen und ostasiatischen Karpfen (*Cyprinus carpio* L.). Zool Anz 104 (9–10):257–268

Swarts FA, Dunson WA, Wright JE (1978) Genetic and environmental factors involved in increased resistance of brook trout to sulfuric acid solutions and mine acid polluted waters. Trans Am Fish Soc 107 (3):651–677

Swarup H (1959) Production of triploidy in *Gasterosteus aculeatus* (L.). J Genet 56 (2):129–142

Swofford DL, Branson BA, Sievert GA (1980) Genetic differentiation of cavefish populations (Amblyopsidae). Isoz Bull 13: 109

Szarski H (1970) Changes in the amount of DNA in cell nuclei during vertebrate evolution. Nature (London) 226 (5246): 651–652

Tait JS (1970) A method of selecting trout hybrids (*Salvelinus fontinalis* x *S. namaycush*) for ability to retain swimbladder gas. J Fish Res B Can 27 (1): 39–45

Takeuchi I, Kajishima T (1973) Fine structure of gold-fish melanophages appearing in the depigmentation process. Annot Zool Jpn 46: 77–84

* Taki Y, Urushido T, Suzuki A, Serizawa C (1977) A comparative chromosome study of *Puntius* (Cyprinidae, Pisces). I. Southeast Asian species. Proc Jpn Acad Sci Ser B 53 (6): 232–235

Taliev DN (1941) Serological analysis of races of the Baikal whitefish *Coregonus autumnalis migratorius* (Georgi). Tr Zool Inst 6 (4): 68–92 (RE)

Taliev DN (1946) Serological analysis of several wild and domesticated forms of the common carp (*Cyprinus carpio* L.), Tr Zool Inst 8 (1): 43–88 (RE)

Tammert M (1974) Investigation of bream blood proteins by starch gel electrophoresis. In: Hydrobiol Invest 6: 207–214

Taniguchi N, Ichiwatari T (1972) Inter- and intraspecific variation of muscle proteins in the Japanese crucian carp. I. Cellulose-acetate electrophoretic pattern. Jpn J Ikhthiol 19 (4): 217–222

Taniguchi N, Morita T (1979) Identification of European, American and Japanese eels by lactate dehydrogenase and malate dehydrogenase isozyme patterns. Bull Jpn Soc Sci Fish 45 (1): 37–41

Taniguchi N, Nakamura J (1970) Comparative electrophoregrams of two species of frigate mackerel. Bull Jpn Soc Sci Fish 36 (2): 173–176

Taniguchi N, Sakata K (1977) Interspecific and intraspecific variations of muscle protein in the Japanese crucian carp. 2. Starch gel electrophoretic pattern. Jpn J Ikhthiol 24 (1): 1–11

Taniguchi N, Tashima K (1978) Genetic variation of liver esterase in red sea bream. Bull Jpn Soc Sci Fish 44 (6): 619–622

Taniguchi N, Ochiai A, Miyazaki T (1972) Comparative studies of the Japanese platycephalid fishes by electropherograms of muscle protein, LDH and MDH. Jpn J Ikhthiol 19 (2): 89–96

Tåning ÅV (1952) Experimental study of meristic characters in fishes. Biol Rev 27 (2): 169–193

Tatarko KI (1961) The abnormalities in the structure of the gill cover and fins in the common carp. Vopr Ikhthiol 1 [3(20)]: 412–420 (R)

Tatarko KI (1963) Morphological studies of abnormal ventral fins in the common carp. Zool Zh 42 (11): 1666–1678 (RE)

Tatarko KL (1966) Abnormalities in the common carp and factors responsible for them. Zool Zh 45 (12): 1826–1834 (RE)

* Taylor KM (1967) The chromosomes of some lower chordates. Chromosoma 21 (2): 181–188

Tcherfas (Cherfas) NB (1971) Natural and artificial gynogenesis of fish. In: Rep FAO/UNDP-(TA), 2926, Rome, pp 274–291

Tegelström H (1975) Interspecific hybridization in vitro of superoxide dismutase from various species. Hereditas 81 (2): 185–198

Thibault RE (1974) Genetics of cannibalism in a viviparous fish and its relationship to population density. Nature (London) 251 (5471): 138–140

Thibault RE (1978) Ecological and evolutionary relationships among diploid and triploid unisexual fishes, associated with the bisexual species, *Poeciliopsis lucida* (Cyprinodontiformes, Poeciliidae). Evolution 32 (3): 613–623

Thompson DH (1930) Variation in fishes as a function of distance. Trans Ill State Acad Sci 23 (1): 276

Thompson DH, Adams LA (1936) A rare wild carp lacking pelvic fins. Copeia 1: 210

Thompson D, Mastert S (1974) Muscle esterase genotypes in the pilchard, *Sardinops ocellata*. J Conseil 36 (1): 50–53

* Thompson KW (1976) Some aspects of chromosomal evolution of the Cichlidae (Teleostei: Perciformes) with emphasis on neotropical forms. Ph D Thesis, Univ Texas, Austin

* Thompson KW, Hubbs Cl, Edwards RJ (1978) Comparative chromosome morphology of the blackbasses. Copeia 1:172–175

Thomson KS (1972) An attempt to reconstruct evolutionary changes in the cellular DNA content of lungfish. J Exp Zool 180 (3):363–372

Thorgaard GH (1976) Robertsonian polymorphism and constitutive heterochromatin distribution in chromosomes of the rainbow trout (*Salmo gairdneri*). Cytogenet Cell Genet 17 (4):174–184

Thorgaard GH (1977) Heteromorphic sex chromosomes in male rainbow trout. Science 196 (4292):900–902

Thorgaard GH (1978) Sex chromosomes in the sockeye salmon: a Y-autosome fusion. Can J Genet Cytol 20 (2):349–354

Thorgaard GH, Gall GAE (1979) Adult triploids in a rainbow trout family. Genetics 93 (4):961–973

Tikhonov VN (1967) The use of blood groups in animal selection. Kolos, Moscow (R)

Tills D, Mourant AE, Jamieson A (1971) Red-cell enzyme variant of Islandic and North Sea cod (*Gadus morhua*). Rapp P-V Reun 161,73–74

Timofeev AV (1979) Time of gene expression of glucoso-6-phosphate dehydrogenase in loach (*Misgurnus fossilis*). In: Biochem Popul Genet Fish. Inst Cytol Acad Sci USSR, Leningrad, pp 24–28 (RE)

Timoffeef-Ressovsky NV, Svirezhev YuM (1966) Adaptive polymorphism in the populations of *Adalia bipunctata* L. Probl Kibernet 16:137–146 (R)

Tomilenko VG, Kucherenko AP (1975) Selection of the third generation of the Ukrainian scaled Nivka common carp. In: Fish Ind, Krozhay, Kiev 20, pp 27–35 (R)

Tomilenko VG, Shpack PN (1979) Peculiarities of common carp growth rate connected with deviations in development of several traits. In: Fish Ind, Krozhay, Kiev 28, pp 25–28 (R)

Tomilenko VG, Alexeenko AA, Panchenko SM, Olenetz NI, Drok VM (1977) Winterresistance of the different common carp hybrids. Rybn Khos 2:17–19 (R)

Tomilenko VG, Panchenko SM, Zheltov YuO (1978) Common carp breeding. Urozhay, Kiev (Ukrainian)

Tran dinh-Trong (1967) Data on the intraspecific variability, biology and distribution of the common carps in North Vietnam. Genetika (Moscow) 3 (2):48–60 (RE)

Truveller KA (1978) Differentiation of herring (*Clupea harengus*) populations in North Sea for erythrocytar antigens and for electrophoretic protein spectra. Ph D Thesis, Moscow Univ

Truveller KA, Nefedov GN (1974) Multipurpose apparatus for vertical electrophoresis in parallel slabs of polyacrylamide gel. Biol Nauki 9:137–140 (R)

Truveller KA, Zenkin VS (1977a) Distribution of erythrocyte antigens in herring *Clupea harengus harengus* L. from the North Atlantic waters. I. Detection of erythrocyte antigens. Genetika (Moscow) 13 (2):238–248 (RE)

Truveller KA, Zenkin VS (1977b) Distribution of erythrocyte antigens in herring *Clupea harengus harengus* L. from the North Atlantic waters. II. Differentiation by frequencies of erythrocyte antigens. Genetika (Moscow) 13 (2):249–263 (RE)

Truveller KA, Alferova NM, Maslennikova NA (1973a) Electrophoretic studies of proteins in herring (*Clupea harengus* s.l.). In: Biochem Genet Fish. Inst Cytol Acad Sci USSR, Leningrad, pp 188–194 (RE)

Truveller KA, Maslennikova NA, Moscovkin LI, Romanova NI (1973b) Variability in disc-electrophoretic patterns of muscle myogens in carp (*Cyprinus carpio* L.). In: Biochem Genet Fish. Inst Cytol Acad Sci USSR, Leningrad, pp 113–119 (RE)

Truveller KA, Moskovkin LI, Maslennikova NA (1974) Hybridological analysis of electrophoretic spectra of esterases in the common carp *Cyprinus carpio* L. Tr VNIIPRKh 23:3–9 (RE)

Tscheglova NV, Iljasov YuI (1979) On esterases in carp (*Cyprinus carpio* L.). In: Biochem Popul Genet Fish. Inst. Cytol. Acad. Sci USSR, Leningrad, pp 176–180 (RE)

Tscherbenok YuI (1973) Correlation between esterase and transferrin types and productive traits in common carp. In: Biochem Genet Fish. Inst Cytol Acad Sci USSR, Leningrad, pp 129–137 (RE)

Tscherbenok YuI (1976) Hybridological analysis of serum esterase's and transferrin's inheritance in Ropsha common carp. Izvestija Gosud. nauchno-issled. Inst Ozern Rechn Rybn Khos (GosNIORKh) 107:48–53 (RE)

Tscherbenok YuI (1978) Selection for the resistance to oxygen deficiency among common carp larvae with different transferrin and esterase genotypes. Izvestija Gosud. nauchno-issled. Inst Ozern Rechn Rybn Khos (GosNIORKh) 130: 107–111 (RE)

Tscherbina MA, Tsvetkova LI (1974) Comparative study of the one-year-old carps of the four genotypes. III. Effectiveness of use of nutrients and energy from the food mixture. Tr VNIIPRKh 23:42–47 (RE)

* Tsitsugina VG (1970) Chromosomal complements of several Black Sea fishes. In: The problems of use and sanitary-biological regimen of the Ukrainian waterbodies. Naukova Dumka, Kiev, pp 75–76 (R)

Tsuyuki H, Roberts E (1966) Interspecies relationships within the genus *Oncorhynchus* based on biochemical systematics. J Fish Res B Can 23 (1): 101–107

Tsuyuki H, Roberts E (1969) Muscle protein polymorphism of sablefish from the eastern Pacific Ocean. J Fish Res B Can 26 (10): 2633–2641

Tsuyuki H, Ronald AP (1970) Existence in salmonid hemoglobins of molecular species with three and four different polypeptides. J Fish Res B Can 27 (6): 1325–1328

Tsuyuki H, Ronald AP (1971) Molecular basis for multiplicity of Pacific salmon hemoglobins: evidence for in vivo existence of molecular species with up to four different polypeptides. Comp Biochem Phys 39B (3): 503–522

Tsuyuki H, Williscroft SN (1973) The pH activity relations of two LDH homotetramers from trout liver and their physiological significance. J Fish Res B Can 30 (5): 1023–1026

Tsuyuki H, Williscroft SN (1977) Swimming stamina differences between genotypically distinct forms of rainbow (*Salmo gairdneri*) and steelhead trout. J Fish Res B Can 34 (7): 996–1003

Tsuyuki H, Roberts E, Vanstone WE (1965) Comparative zone electropherograms of muscle myogens and blood hemoglobins of marine and freshwater vertebrates and their application to biochemical systematics. J Fish Res B Can 22 (1): 203–213

Tsuyuki H, Roberts E, Kerr RH (1967) Comparative electropherograms of the family Catostomidae. J Fish Res B Can 24 (2): 299–304

Tsuyuki H, Roberts E, Best EA (1971) Serum transferrin systems of the Pacific halibut (*Hippoglossus stenolepis*). Rapp P-V Reun 161: 134

Tsvetkova LI (1969) Certain peculiar features of lipid metabolism in common carp of four different genotypes. In: Pond fish breeding. All-Union Res Inst Pond Fish, Moscow, pp 190–202 (RE)

Tsvetkova LI (1974) Comparative studies of one-year-old common carps of four different genotypes. I. Growth characteristics of the one-year old fishes of different genotypes under the conditions of individual and communal rearing. Tr VNIIPRKh 23:36–41 (RE)

Tsvetnenko JB (1980) Characteristic of blood proteins in Sea of Azov sturgeons. Ph D Thesis, Rostov-on Don (R)

Turner BJ (1973a) Genetic divergence of Death Valley pupfish populations: species – specific esterases. Comp Biochem Phys 46B (1): 57–70

Turner BJ (1973b) Genetic variation of mitochondrial aspartate aminotransferase in the teleost *Cyprinodon nevadensis.* Comp Biochem Phys 44B (1): 89–92

Turner BJ (1974) Genetic divergence of Death Valley pupfish species: biochemical versus morphological evidence. Evolution 28 (2): 281–294

Turner BJ, Grosse DL (1980) Trophic diversity in a genus of Mexican stream fishes: ecological polymorphism or speciation? Isoz Bull 13: 105

Turner BJ, Brett BLH, Rasch EM, Balsano JS (1980a) Evolutionary genetics of the Amazon molly, *Poecilia formosa,* a gynogenetic unisexual fish. Isoz Bull 13: 103

Turner BJ, Miller RR, Rasch EM (1980b) Significant differential gene duplication without ancestral tetraploidy in a genus of mexican fish. Experientia 36 (8): 927–930

Tzoy RM (1969a) Action of dimethylsulfate and nitrosomethylurea on the developing eggs of the rainbow trout and the peled. Tzitologiya 11 (11): 1440–1448 (RE)

Tzoy RM (1969b) Action of nitrosomethylurea and dimethylsulfate on the sperm cells of the rainbow trout and the peled. Dokl Akad Nauk SSSR (DAN) 189 (1): 411–414 (R)

Tzoy RM (1971a) Correlation between certain morphological and physiological traits in the Ropsha common carp. Izvestija Gosud. nauchno-issled. Inst Ozern Rechn Rybn Khos (GosNIORKh) 74: 39–44 (RE)

Tzoy RM (1971 b) Action of dimethylsulfate on the mutation frequency of the genes S and N in the carp (*Cyprinus carpio* L.). Dokl Akad Nauk SSSR (DAN) 197 (3):701–704 (R)

Tzoy RM (1972) Chemical gynogenesis in the rainbow trout and the peled. Genetika (Moscow) 8 (2):185–188 (RE)

Tzoy RM (1976) Problems of artificial mutagenesis in fish culture. Izvestija Gosud. nauchno-issled. Inst Ozern Rechn Rybn Khos (GosNIORKh) 107:109–118 (RE)

Tzoy RM (1978) Artificial mutagenesis in practical selection of pond fishes. In: Increasing productivity of fishes by selection and hybridization. F Müller, Szarvas, pp 121–141

Tzoy RM (1980) Chemical mutagenesis in selection of East-Kazakhstan carp (*Cyprinus carpio* L.) In: Karyological variability, mutagenesis and gynogenesis in fishes. Inst Cytol Acad Sci USSR, Leningrad, pp 55–62 (RE)

Tzoy RM, Golodov YuF, Menshova AI (1973) Influence of chemical mutagens on variability of morphological and physiological traits in carp (*Cyprinus carpio* L.). In: Biochem Genet Fish. Inst Cytol Acad Sci USSR, Leningrad, pp 97–103 (RE)

Tzoy RM, Menshova AI, Golodov YuF (1974 a) Specificity of chemical mutagen action on the sperm cells of *Cyprinus carpio* L. Genetika (Moscow) 10 (2):68–72 (RE)

Tzoy RM, Menshova AI, Golodov YuF (1974 b) Frequency of spontaneous and induced mutations in the genes coding scale pattern in common carp. Genetika (Moscow) 10 (11):60–62 (RE)

Ueda T, Ojima Y (1978) Differential chromosomal characteristics in the funa subspecies (*Carassius*). Proc Jpn Acad B 54 (6):283–288

* Ueno K, Ojima Y (1977) Chromosome studies of two species of the genus *Coreoperca* (Pisces, Perciformes), with reference to the karyotypic differentiation and evolution. Proc Jpn Acad Sci B 53 (6):221–225

Umrath K (1939) Über die Vererbung der Farben und des Geschlechts beim Schleierkampffisch, *Betta splendens.* Z Indukt Abstammungs-Vererbungsl 77 (2):450–454

* Urushido T, Takahashi E, Taki Y, Kondo N (1977) A karyotype study of polypterid fishes, with notes on their phyletic relationship. Proc Jpn Acad Sci B 53 (3):95–98

Ushakov BP, Vinogradova AI, Kusakina AA (1962) Cytophysiological analysis of intraspecies differentiation in the whitefish and grayling in the Lake Baikal. Zh Obtsch Biol 23 (1):56–63 (RE)

Uthe JF, Roberts E, Clarke LW, Tsuyuki H (1966) Comparative electropherograms of representatives of the families Petromyzontidae, Esocidae, Centrarchidae and Percidae. J Fish Res B Can 23 (11):1663–1671

Utter FM (1971) Tetrazolium oxidase phenotypes of rainbow trout (*Salmo gairdneri*) and Pacific salmon (*Oncorhynchus* spp.). Comp Biochem Phys 39B (4):891–895

Utter FM, Allendorf FW (1978) Determination of the breeding structure of steel-head populations through gene frequency analysis. In: Calif Coop Fish Res Unit Spec Rep 77 (1):44–54

Utter FM, Hodgins HO (1969) Lactate dehydrogenase isozymes of Pacific hake (*Merluccius productus*). J Exp Zool 172 (1):59–67

Utter FM, Hodgins HO (1970) Phosphoglucomutase polymorphism in sockeye salmon. Comp Biochem Phys 36 (1):195–199

Utter FM, Hodgins HO (1971) Biochemical polymorphism in the Pacific hake (*Merluccius productus*). Rapp P-V Reun 161:87–89

Utter FM, Hodgins HO (1972) Biochemical genetic variation at six loci in four stocks of rainbow trout. Trans Am Fish Soc 101 (3):494–502

Utter FM, Ames WE, Hodgins HO (1970 a) Transferrin polymorphism in coho salmon (*Oncorhynchus kisutch*). J Fish Res B Can 27 (12):2371–2373

Utter FM, Stormont CJ, Hodgins HO (1970 b) Esterase polymorphism in vitreous fluid of Pacific hake, *Merluccius productus.* Anim Blood Gr Biochem Genet 1 (2):69–82

Utter FM, Hodgins HO, Allendorf FW, Johnson AG, Mighell JL (1973 a) Biochemical variants in Pacific salmon and rainbow trout: their inheritance and application in population studies. In: Genetics and mutagenesis of fish, Springer, Berlin Heidelberg New York, pp 329–339

Utter FM, Allendorf FW, Hodgins HO (1973 b) Genetic variability and relationships in Pacific salmon and related trout based on protein variations. Syst Zool 22 (3):257–270

Utter FM, Mighell JL, Hodgins HO (1973c) Inheritance of biochemical variants in three species of Pacific salmon and rainbow trout. Annu Rep Int North Pacif Fish Comm 1971, Vancouver, pp 97–100

Utter FM, Hodgins HO, Allendorf FW (1975) Biochemical genetic studies of fishes: potentialities and limitations. In: Biochemical and biophysical perspectives in marine biology, vol I. Academic Press, London New York, pp 213–234

Utter FM, Allendorf FW, May B (1980) Genetic basis of creatine kinase (E.C. 2.7.3.2) isozymes in sceletal muscle of salmonid fishes. Biochem Genet 17 (11–12): 1079–1092

* Uyeno T (1973) A comparative study of chromosomes in the teleostean fish order Osteoglossiformes. Jpn J Ikhthiol 20 (4): 211–217

* Uyeno T, Miller RR (1971) Multiple sex chromosomes in a mexican cyprinodontid fish. Nature (London) 231 (5303): 452–453

* Uyeno T, Miller RR (1972) Second discovery of multiple sex chromosomes among fishes. Experientia 28 (2): 223–225

* Uyeno T, Miller RR (1973) Chromosomes and the evolution of the plagopterin fishes (Cyprinidae) of the Colorado river system. Copeia 4: 776–782

* Uyeno T, Smith GR (1972) Tetraploid origin of the karyotype of catostomid fishes. Science 175 (4022): 644–646

Uzzell T, Leszek B, Rainer G (1975) Diploid and triploid progeny from a diploid female of *Rana esculenta* (*Amphibia salienta*). Proc Acad Nat Sci Philad 127 (11): 81–91

Vadasz C, Kiss B, Czanyi V (1978) Defensive behaviour and its inheritance in the anabantoid fish *Macropodus opercularis* and *M. o. concolor*. Behav Process 3 (2): 107–124

Valenta M (1977) Polymorfismus a izoenzymave vzory malátdehydrogenázy u nékterych ryb celedi Cyprinidae. Živočišna Výroba, 22 (50): 801–812

Valenta M (1978a) Polymorphism of A, B and C loci of lactate dehydrogenase in European fish species of the *Cyprinidae* family. Anim Blood Gr Biochem Gen 9 (3): 139–149

Valenta M (1978b) Protein polymorphism in European fish species of the Cyprinidae family and utilization of polymorphic proteins for breeding in fish. In: Increasing productivity of fishes by selection and hybridization. F Müller, Szarvas, pp 37–78

Valenta M, Kálal L (1968) Polymorfismus sérových transferinů u kapra (*Cyprinus carpio* L.) a lina (*Tinca tinca* L.). In: Sb VŽZ Fak Agron Rada B, Praha, pp 93–103

Valenta M, Stratil A, Šlechtová V, Kálal L (1976a) Polymorphism of transferrin in carp (*Cyprinus carpio* L.): genetic determination, isolation and partial characterization. Biochem Genet 14 (1–2): 27–45

Valenta M, Šlechtová V, Šlechta V, Kálal L (1976b) Isoenzymy laktatdehydrogenazy v tkanich nekterych ryb céledi Cyprinidae. Živočišna Výroba 21 (12): 901–916

Valenta M, Šlechta V, Šlechtova V, Kálal L (1977a) Genetic polymorphism and isoenzyme patterns of lactate dehydrogenase in tench (*Tinca tinca*), crucian carp (*Carassius carassius*) and carp (*Cyprinus carpio*). Anim Blood Gr Biochem Genet 8 (3): 217–230

Valenta M, Stratil A, Kálal L (1977b) Polymorphism and heterogeneity of transferrins in some species of the fish family Cyprinidae. Anim Blood Gr Biochem Genet 8 (1): 93–109

Valenta M, Šlechtová V, Kálal L, Stratil A, Janatkova J, Šlechta V, Rab P, Pokorny J (1978) Polymorphní bilkoving kapra obecného (*Cyprinus carpio* L.), a lina obecného (*Tinca tinca* L.) a možnosti jejich využiti v plemenářské práci. Živočišna Výroba 23 (11): 797–809

Valenti RJ (1975) Induced polyploidy in *Tilapia aurea* (Steindachner) by means of temperature shock treatment. J Fish Biol 7 (4): 519–528

Valenti RJ, Kallman KD (1973) Effects of gene dosage and hormones on the expression in Dr in the platyfish, *Xiphophorus maculatus* (Poeciliidae). Genet Res 22 (1): 79–89

Vasetzky SG (1967) Change in the ploidy of the Russian sturgeon larvae induced by the heat treatment of eggs at different development stages. Dokl Akad Nauk SSSR (DAN) 172 (5): 1234–1237 (R)

* Vasilyev VP (1975a) Karyotypes of certain varieties of the arctic char *Salvelinus alpinus* (L.) from Kamchatka. Vopr Ikhthi 15 (3): 417–430 (R)

* Vasilyev VP (1975b) Karyotypes of different varieties of the mikizha *Salmo mykiss* Walb. and steelhead trout *S. gairdneri* Rich. Vopr Ikhthiol 15 (6): 998–1010 (R)

* Vasilyev VP (1977) Polyploidy in fishes and certain problems of karyotype evolution in salmonids. Zh Obtsch Biol 38 (3): 380–392 (RE)

Vasilyev VP (1980) Chromosome numbers in Cyclostomata and Pisces. Vopr Ikhthiol 20 [2(122)]: 387–422 (R)
Vasilyev VP, Makeeva AP, Ryabov IN (1975) Triploidy of distant hybrids between the common carp and species of other cyprinid subfamilies. Genetika (Moscow) 11 (8): 49–56 (RE)
Vasilyev VP, Ivanov VN, Polykarpova LK (1980) Frequencies of chromosomal morphs in various size groups of Black Sea bass (*Spicara flexuosa* Raf., Centracanthidae). In: Karyological variability, mutagenesis and gynogenesis of fishes. Inst Cytol Acad Sci USSR, Leningrad, pp 43–49 (RE)
Vasudevon P, Rao SGA, Rao SRV (1973) Somatic and meiotic chromosomes of *Heteropneustes fossilis*. Curr Sci 42 (12): 427–428
Vavilov NI (1920) The low of homologous series in hereditary variation. In: Proc 3rd All-Russian Select Congr, Saratov, pp 41–57 (R)
Veldre LA, Veldre IR (1979) On some biochemical indices of blood in the Baltic sprat, *Sprattus sprattus*. In: Biochem Popul Genet Fish. Inst Cytol Acad Sci USSR, Leningrad, pp 83–89 (RE)
Verma GK (1968) Studies on the structure and behaviour of chromosomes of certain freshwater and marine gobiid fishes. Proc Natl Acad Sci India 38: 178
Vervoort A (1979) Karyotype and DNA content of *Phractolaemus ansorgei* Bigr. (Teleostei: Gonorynchiformes). Experientia 35 (4): 479–480
Vervoort A (1980) Karyotypes and nuclear DNA contents of *Polypteridae* (Osteichthyes). Experientia 36 (6): 646–647
Vialli M (1957) Volume et contenu en ADN par noyau. In: Cytochemical methods with quantitative aims. Academic Press, London New York, pp 284–293
Vielkind J, Vielkind U, Götting KJ, Anders F (1970) Über melanotische und albinotisch-amelanotische Melanome bei lebendgebärenden Zahnkarpfen. Zool Anz 33 (Suppl): 339–341
Vielkind J, Haas-Andela H, Anders F (1976) DNA-mediated transformation in the platyfish-swordtail melanoma system. Experientia 32 (8): 1043–1045
Vielkind U (1976) Genetic control of cell differentiation in platyfish-swordtail melanomas. J Exp Zool 196 (2): 197–204
Viktorovsky RM (1969) Possible polyploidy in fish evolution. In: Genetics, selection and hybridization of fishes, Nauka, Moscow, pp 98–104 (R)
* Viktorovsky RM (1975a) Chromosomal complements of the East Siberian char *Salvelinus leucomaenis* and salmon trout *S. malma* (Salmoniformes, Salmonidae). Zool Zh 54 (5): 787–789 (RE)
* Viktorovsky RM (1975b) Mechanisms of speciation in the chars of the Lake Kronotskoye. Ph D. Thesis, Vladivostok (R)
* Viktorovsky RM (1975c) Chromosome sets of the endemic chars in Lake Kronotskoye. Tzitologiya 17 (4): 464–466 (RE)
* Viktorovsky RM (1978a) Karyotype evolution of the char (Salvelinus). Tzitologiya 20 (7): 833–838 (RE)
* Viktorovsky RM (1978b) Mechanisms of speciation in the char. Nauka, Moscow, (R)
* Viktorovsky RM (1978c) Karyotypes of East-Asian chars. In: Biolog. Investigations of Far-East Seas. Inst Sea Biol DVNC Acad Sci USSR 3, Vladivostok, pp 21–23 (R)
Viktorovsky RM, Glubokovsky MK (1977) Mechanisms and tempo of speciation in the chars from the genus *Salvelinus*. Dokl Akad Nauk SSSR (DAN) 235 (4): 946–949 (R)
* Viktorovsky RM, Maximova RA (1978) Karyotype of Amur whitefish and some problems of whitefish karyotype evolution. Tzitologiya 20 (8): 967–970 (RE)
Villwock W (1963) Genetische Analyse des Merkmals "Beschuppung" bei anatolischen Zahnkarpfen (Pisces: Cyprinodontidae) im Auflöserversuch. Zool Anz 170 (1–2): 23–45
Vinogradov VK, Erokhina LV (1964) Hybrids between the silver carp and bighead. Rybov Rybol 5: 11–13 (R)
Volf F (1956) Vady dedičného žalozeni u kapra. Social Zemed 6 (19): 1129–1132
Volokhonskaya LG, Viktorovsky RM (1971) Possibility of determination of a genetic component in the variation of fish egg size. In: Sci Rep Inst Mar Biol 2: 42–44 (R)
Voropaev NV (1969) Hybridization experiment with silver carp and bighead and breeding of F_2 hybrid fingerlings. In: Pond Fisheries, vol II. All-Union Res Inst Pond Fish, Moscow, pp 98–102 (R)

Voropaev NV (1978) Silver and bighead carp hybrids: their biology and importance for fish farming. In: Increasing productivity of fishes by selection and hybridization. F Müller, Szarvas, pp 98–104

Vrijenhoek RC (1972) Genetic relationships of unisexual hybrid fishes to their progenitors using lactate dehydrogenase isozymes as gene markers (*Poeciliopsis,* Poeciliidae). Am Nat 106 (952): 754–766

Vrijenhoek RC (1976) An allele affecting display coloration in the fish, *Poeciliopsis viriosa.* J Hered 67 (5): 324–325

Vrijenhoek RC (1979a) Genetics of a sexually reproducing fish in a highly fluctuating environment. Am Nat 113 (1): 17–29

Vrijenhoek RC (1979b) Factors affecting clonal diversity and coexistence. Am Zool 19 (3): 787–797

Vrijenhoek RC, Allendorf FW (1980) Protein polymorphism and inheritance in the fish *Poecilia mexicana* (Poeciliidae). Isoz Bull 13: 92

Vrijenhoek RC, Schultz RJ (1974) Evolution of a trihybrid unisexual fish (*Poeciliopsis,* Poeciliidae). Evolution 28 (2): 306–319

Vrijenhoek RC, Angus RA, Schultz RJ (1977) Variation and heterozygosity in sexually vs. clonally reproducing populations of *Poeciliopsis.* Evolution 31 (4): 767–781

Vrooman AM (1964) Serologically differentiated subpopulations of the Pacific sardine, *Sardinops caerulea.* J Fish Res B Can 21 (4): 691–701

* Wahl RW (1960) Chromosome morphology in lake trout *Salvelinus namaycush.* Copeia 1: 16–19

Wahlund S (1928) Zusammensetzung von Populationen und Korrelationserscheinungen vom Standpunkt der Vererbungslehre aus betrachtet. Hereditas 11 (1): 65–106

Wallbrünn HM (1958) Genetics of the Siamese fighting fish, *Betta splendens.* Genetics 43 (2): 289–298

Walter RO, Hamilton JB (1970) Supermales (YY sex chromosomes) and androgen-treated XY males: competition for mating with female killifish *Oryzias latipes.* Anim Behav 18 (1): 128–131

Wanstein MP, Yerger RW (1976) Protein taxonomy of the Gulf of Mexico and Atlantic Ocean sea trouts genus *Cynoscion.* Fish Bull 74 (3): 599–607

Ward RD (1977) Relationship between enzyme heterozygosity and quaternary structure. Biochem Genet 15 (1–2): 123–135

Ward RD (1978) Subunit size of enzymes and genetic heterozygosity in vertebrates. Biochem Genet 16 (7–8): 799–810

Ward RD, Beardmore JA (1977) Protein variation in the plaice, *Pleuronectes platessa* L. Genet Res 30 (1): 45–62

Ward RD, Galleguillos RA (1978) Protein variation in the plaice, dab and flounder, and their genetic relationships. In: Marine organisms: Genetics, ecology and evolution. Plenum Publ Corp, New York, pp 71–93

Ward RD, McAndrew BJ, Wallis GP (1979) Purine nucleoside phosphorylase variation in the brook lamprey *Lampetra planeri* Bloch (Petromizonae, Agnatha); evidence for a trimeric enzyme structure. Biochem Genet 17 (3–4): 251–256

Watson JD (1976) Molecular biology of the gene, 3rd edn WB Benjamin Inc, Menlo Park London Amsterdam

Watson JD, Crick FAC (1953) Genetical implications of the structure of deoxiribonucleic acid. Nature (London) 171 (4361): 964–967

Westrheim SJ, Tsuyuki H (1967) *Sebastodes reedi,* a new scorpaenid fish in the Northeast Pacific Ocean. J Fish Res B Can 24 (9): 1945–1954

Wheat TE, Childers WF, Miller ET, Whitt GS (1971) Genetic and in vitro molecular hybridization of malate dehydrogenase isozymes in interspecific bass (*Micropterus*) hybrids. Anim Blood Gr Biochem Genet 2 (1): 3–14

Wheat TE, Whitt GS, Childers WF (1972) Linkage relationships between the homologous malate dehydrogenase loci in teleosts. Genetics 70 (2): 337–340

Wheat TE, Whitt GS, Childers WF (1973) Linkage relationships of six enzyme loci in interspecific sunfish hybrids (genus *Lepomis*). Genetics 74 (2): 343–350

Wheat TE, Childers WF, Whitt GS (1974) Biochemical genetics of hybrid sunfish: differential survival of heterozygotes. Biochem Gen 11 (3): 205–219

Whitt GS (1969) Homology of lactate dehydrogenase genes: A gene function in the teleost nervous system. Science 166 (3909): 1156–1158

Whitt GS (1970a) Genetic variation of supernatant and mitochondrial malate dehydrogenase isoenzymes in the teleost *Fundulus heteroclitus*. Experientia 26 (3): 734–736

Whitt GS (1970b) Developmental genetics of the lactate dehydrogenase isozymes of fish. J Exp Zool 175 (1): 1–35

Whitt GS (1975a) A unique lactate dehydrogenase isozyme in the teleost retina. In: Vision in fishes, Plenum Publ Co, New York, pp 459–470

Whitt GS (1975b) Isozymes and developmental biology. In: Isozymes, vol III. Academic Press, London New York, pp 1–8

Whitt GS, Maeda FS (1970) Lactate dehydrogenase gene function in the blind cave fish, *Anoptichthys jordani*, and other characin. Biochem Genet 4 (6): 727–741

Whitt GS, Childers WF, Wheat TE (1971) The inheritance of tissue-specific LDH isozymes in interspecific bass (*Micropterus*) hybrids. Biochem Genet 5 (3): 257–273

Whitt GS, Cho PL, Childers WF (1972) Preferential inhibition of allelic isozyme synthesis in an interspecific sunfish hybrid. J Exp Zool 179 (2): 271–282

Whitt GS, Childers WF, Cho PL (1973a) Allelic expression at enzyme loci in an intertribal hybrid sunfish. J Hered 64 (2): 55–61

Whitt GS, Childers WF, Tranquilli J, Champion M (1973b) Extensive heterozygosity in three enzyme loci in hybrid sunfish populations. Biochem Genet 8 (1): 55–72

Whitt GS, Miller ET, Shaklee JB (1973c) Developmental and biochemical genetics of lactate dehydrogenase isozymes in fishes. In: Genetics and mutagenesis of fish. Springer, Berlin Heidelberg New York, pp 243–276

Whitt GS, Shaklee JB, Markert CL (1975) Evolution of the lactate dehydrogenase isozymes of fishes. In: Isozymes, vol IV. Academic Press, London New York, pp 381–400

Whitt GS, Childers WF, Shaklee JB, Matsumoto J (1976) Linkage analysis of the multilocus glucosephosphate isomerase isozyme system in sunfish (Centrarchidae, Teleostei). Genetics 82 (1): 35–42

Whitt GS, Philipp DP, Childers WF (1977) Aberrant gene expression during the development of hybrid sunfishes (Perciformes, Teleostei). Differentiation 9 (1): 97–109

Wilkens H (1970) Beiträge zur Degeneration des Auges bei Cavernicolen, Genzahl und Manifestationsart. Z Zool Syst Evol-Forsch 8 (1): 1–47

Wilkens H (1971) Genetic interpretation of regressive evolutionary processes; studies on hybrid eyes of the *Astyanax* cave population (Characidae, Pisces). Evolution 25 (3): 530–544

Wilkins NP (1966) Immunology, serology and blood group research in fishes. Proc 10th Eur Conf Anim Blood Gr Biochem Polym, Paris, pp 355–359

Wilkins NP (1971a) Haemoglobin polymorphism in cod, whiting and pollock in Scottish waters. Rapp P-V Reun 161: 60–63

Wilkins NP (1971b) Biochemical and serological studies on Atlantic salmon (*Salmo salar* L.). Rapp P-V Reun 161: 91–95

Wilkins NP (1972) Biochemical genetics of Atlantic salmon *Salmo salar* L. II. The significance of recent studies and their application in population identification. J Fish Biol 4 (4): 505–517

Wilkins NP, Iles TD (1966) Haemoglobin polymorphism and its ontogeny in herring (*Clupea harengus*) and sprat (*Sprattus sprattus*). Comp Biochem Phys 17 (4): 1141–1158

Williams GC, Koehn RK, Mitton JB (1973) Genetic differentiation without isolation in the American eel, *Anguilla rostrata*. Evolution 27 (2): 192–201

Williscroft SN, Tsuyuki H (1970) LDH systems of rainbow trout: evidence for polymorphism in liver and additional subunits in gills. J Fish Res B Can 27 (9): 1563–1567

Wilson JL (1975) Haemoglobin comparisons of green sunfish (*Lepomis cyanellus* Raf.) x bluegill (*L. macrochirus* Raf.) hybrids to the parental species. Trans Am Fish Soc 104 (1): 148–149

Winge O (1922) One-sided masculine and sex-linked inheritance in *Lebistes reticulatus*. C R Trav Lab Carlsberg 14 (18): 1–20

Winge O (1923) Crossing-over between the X- and the Y-chromosome in *Lebistes*. C R Trav Lab Carlsberg 14 (20): 1–19

Winge O (1927) The location of eighteen genes in *Lebistes reticulatus*. J Genet 18 (1): 1–43

Winge O (1934) The experimental alteration of sex chromosomes into autosomes and vice versa, as illustrated by *Lebistes.* C R Trav Lab Carlsberg 21 (1): 1–49
Winge O, Ditlevsen E (1938) A lethal gene in the Y-chromosome of *Lebistes.* C R Trav Lab Carlsberg 22 (11): 203–210
Winge O, Ditlevsen E (1948) Colour inheritance and sex determination in *Lebistes.* C R Trav Lab Carlsberg 24 (20): 227–248
Wiseman ED, Echelle AA, Echelle AF (1978) Electrophoretic evidence for subspecific differentiation and intergradation in *Etheostoma spectabile* (Teleostei: Percidae). Copeia 2: 320–327
Wlødek JM (1963) Der blaue Karpfen aus der Teichwirtschaft Landek. Acta Hydrobiol 5 (4): 383–401
Wlødek JM (1968) Studies on the breeding of carp (*Cyprinus carpio* L.) at the experimental pond farms of the Polish Acad. of Science in South Silesia, Poland. FAO Fish Rep (44): Rome 4: 93–116
Wlødek JM, Matlak O (1978) Comparative investigations of the growth of Polish and Hungarian carp in southern Poland. In: Increasing productivity of fishes by selection and hybridization, F Müller, Szarvas, pp 154–194
Wohlfarth G (1977) Shoot carp. Bamidgeh 29 (2): 35–40
Wohlfarth G, Moav R (1968) The relative efficiency of experiments conducted in individed ponds and in ponds divided by nets. FAO Fish Rep (44) Rome 4: 487–492
Wohlfarth G, Moav R (1971) Genetic investigation and breeding methods of carp in Israel. In: Rep FAO/UNDP(TA), 2926, Rome, pp 160–185
Wohlfarth G, Moav R (1972) The regression of weight gain on initial weight in carp I. Methods and results. Aquaculture 1 (1): 7–28
Wohlfarth G, Lahman M, Moav R (1963) Genetic improvement of carp. IV. Leather and line carp in fish ponds of Israel. Bamidgeh 15 (1): 3–8
Wohlfarth G, Moav R, Hulata G (1975a) Genetic differences between the Chinese and European races of the common carp. II. Multicharacter variation – a responce to the diverse methods of fish cultivation in Europe and China. Heredity 34 (3): 341–350
Wohlfarth G, Moav R, Hulata G, Beiles A (1975b) Genetic variation in seine escapability of the common carp. Aquaculture 5 (3): 375–387
Wolf B, Anders F (1975) *Xiphophorus.* I. Farbmuster. Gießen, Univ
Wolf LE (1954) Development of disease-resistant strains of fish. Trans Am Fish Soc 83 (2): 342–349
Wolf U, Ritter H, Atkin NB, Ohno S (1969) Polyploidization in the fish family Cyprinidae order Cypriniformes. I. DNA content and chromosome sets in various species of Cyprinidae. Humangenetik 7 (2): 240–244
Wolf U, Engel W, Faust J (1970) Zum Mechanismus der Diploidisierung in der Wirbeltiereevolution: Koexistenz von tetrasomen und disomen Genloci der Isocitrat Dehydrogenasen bei der Regenbogenforelle (*Salmo irideus*). Humangenetik 9 (1): 150–156
Wolfe GW, Brauson BA, Jones SL (1979) An electrophoretic investigation of six species of darters in the subgenus *Catonotus.* Biochem Syst Ecol 7 (1): 81–85
Wright JE (1972) The palamino rainbow trout. Pa Angler Mag 41: 8–9 (cit from Kincaid, 1975)
Wright JE, Atherton LM (1970) Polymorphism for LDH and transferrin loci in brook trout populations. Trans Am Fish Soc 99 (1): 179–192
Wright JE, Sklenaric R, James SM (1963) Immunogenetic relationships of trout. Proc 11th Int Congr Genet 11: 4
Wright JE, Atherton L, de Buhr A, Eckroat LR, Herschberger WK, Morrison WJ (1966) Polymorphism of soluble protein types in Salmonidae. Proc 11th Pacific Sci Congr Tokyo 7: 11
Wright JE, Siciliano MJ, Baptist JN (1972) Genetic evidence for the tetramer structure of glyceraldehyde-3-phosphate dehydrogenase. Experientia 28 (8): 888–889
Wright JE, Heckman JR, Atherton LM (1975) Genetic and developmental analysis of LDH isozymes in trout. In: Isozymes, vol III. Academic Press, London New York, pp 375–401
Wright S (1951) The genetical structure of populations. Ann Eugenics 15: 323–354
Wright TD, Hasler AD (1967) An electrophoretic analysis of the effects of isolation and homing behavior upon the serum proteins of the white bass (*Roccus chrysops*) in Wisconsin. Am Nat 101 (921): 401–413

Wunder W (1931) Über erbliche Fehler beim Karpfen. Z Fischerei Hilfswissensch 29 (1):97–112

Wunder W (1932) Beschädigungen bei Karpfen, ihre Ursache und Vermeidung. Z Fischerei Hilfswissensch 30 (1):127–140

Wunder W (1934) Beobachtungen über Knochenerweichung und nachfolgende Wirbelsäulenverkrümmung beim Karpfen (*Cyprinus carpio* L.) Z Fischer 32 (1):37–67

Wunder W (1949a) Shortened spine, an hereditary character of the Aischgrund carp (*Cyprinus carpio* L.). Roux' Arch Entw-Mech 144 (1):1–24

Wunder W (1949b) Fortschrittliche Karpfenteichwirtschaft. Schweizerbartsche Verlagsbuch, Stuttgart

Wunder W (1960) Erbliche Flossenfehler beim Karpfen und ihr Einfluß auf die Wachstumsleistung. Arch Fischereiwiss 11 (2):106–119

Wuntch T, Goldberg E (1970) A comparative physico-chemical characterization of lactate dehydrogenase isozymes in brook trout, lake trout and their hybrid splake trout. J Exp Zool 174 (3):233–252

Yamagishi H (1965) Comparative study on the growth of the fry of three races of Japanese crucian carp, *Carassius carassius* L., with special reference to behaviour and competition. Jpn J Ecol 15 (1):100–113

Yamagishi H (1969) Postembryonal growth and its variability of the three marine fishes with special reference to the mechanism of growth variation in fishes. Res Popul Ecol (Kyoto) 11 (1):14–33

Yamamoto T-O (1955) Progeny of artificially induced sex reversals of male genotype (XY) in the medaka (*Oryzias latipes*) with special reference to YY male. Genetics 40 (3):406–419

Yamamoto TO (1958) Artificial induction of functional sex reversal in genotypic females of the medaka (*Oryzias latipes*). J Exp Zool 137 (2):227–264

Yamamoto T-O (1959a) A further study on induction of functional sex reversal in genotypic males of the medaka (*Oryzias latipes*) and progenies of sex reversals. Genetics 44 (4, p 2):739–757

Yamamoto T-O (1959b) The effect of estrone dosage level upon the percentage of sex reversals in genetic male (XY) on the medaka (*Oryzias latipes*). J Exp Zool 141 (1):133–153

Yamamoto T-O (1961) Progenies of sex reversal females mated with sex reversal males in the medaka, *Oryzias latipes*. J Exp Zool 146 (2):163–179

Yamamoto T-O (1963) Induction of reversal in sex differentiation of YY zygotes in the medaka, *Oryzias latipes*. Genetics 48 (2):293–306

Yamamoto T-O (1964) The problem of viability of YY zygotes in the medaka, *Oryzias latipes*. Genetics 50 (1):45–58

Yamamoto T-O (1967) Estrone-induced white YY females and mass production of white YY males in the medaka, *Oryzias latipes*. Genetics 55 (2):329–336

Yamamoto T-O (1968) Matings of YY males with estron-induced YY females in the medaka *Oryzias latipes*. Proc 12th Int Congr Gen (Tokyo) 1:153

Yamamoto T-O (1969) Inheritance of albinism in the medaka, *Oryzias latipes*, with special reference to gene interaction. Genetics 62 (4):797–809

Yamamoto T-O (1973) Inheritance of albinism in the goldfish, *Carassius auratus*. Jpn J Genet 48 (1):53–64

Yamamoto T-O (1975a) YY male goldfish from mating estrogene-induced XY female and normal male. J Hered 66 (1):2–4

Yamamoto T-O (1975b) The medaka, *Oryzias latipes* and the guppy, *Lebistes reticulatus*. In: Handbook of genetics, vol IV. Plenum Publ Corp, New York, pp 133–149

Yamamoto T-O (1977) Inheritance of nacreous-like scaleness in the ginbuna, *Carassius auratus langsdorfii*. Jpn J Genet 52 (5):373–378

Yamamoto T-O, Kajishima T (1968) Sex hormone induction of sex reversal in the goldfish and evidence for male heterogamety. J Exp Zool 168 (2):215–222

Yamamoto T-O, Oikawa T (1963) Linkage between albino gene (i) and color interferer (ci) in the medaka, *Oryzias latipes*. Jpn J Genet 38 (5):361–375

Yamamoto T-O, Tomita H, Matsuda N (1963) Hereditary and nonheritable vertebral anchylosis in the medaka *Oryzias latipes*. Jpn J Genet 38 (1):36–47

Yamanaka H, Yamaguchi K, Hashimoto K, Matsuura H (1967) Starch gel electrophoresis of fish hemoglobins. III. Salmonid fishes. Bull Jpn Soc Sci Fish 33 (2):195–207

Yamauchi T, Goldberg E (1973) Glucose-6-phosphate dehydrogenase from brook, lake and splake trout: an isozymic and immunological study. Biochem Genet 10 (2): 121–134

Yamauchi T, Goldberg R (1974) Asynchronous expression of glucose-6-phosphate dehydrogenase in splake trout embryos. Dev Biol 39 (1): 63–68

* Yamazaki F (1971) A chromosomes study of the ayu, a salmonid fish. Bull Jpn Soc Sci Fish 37: 707–710

Yardley D, Avise JS, Gibbons JW, Smith MH (1974) Biochemical genetics of sunfish. III. Genetic subdivision of fish populations inhabiting heated waters. In: Therm Ecol AEC Symp Ser, New York, pp 255–263

Yndgaard CF (1972) Genetically determined electrophoretic variants of phosphoglucose isomerase and 6-phosphogluconate dehydrogenase in *Zoarces viviparus.* Hereditas 71 (1): 151–154

Zaks MG, Sokolova MM (1961) Immuno-serological differences between various shoals of the sockeye salmon. Vopr Ikhthiol 1 [4(21)]: 707–715 (R)

* Zan Ruiguang, Song Zheng (1979) Analysis and comparison of *Ctenopharyngodon idella* and *Megalobrama amblycephala* karyotypes. Acta Genet Sinica 6 (2): 205–210

* Zanandrea G, Capanna E (1964) Contributo alla cariologia del genere *Lampetra.* Boll Zool 31 (2, p 1): 669–670

Zander CD (1965) Die Geschlechtsbestimmung bei *Xiphophorus montezumae cortezi,* Rosen (Pisces). Z Vererbungsl 96 (1): 128–141

Zander CD (1968) Über die Vererbung von Y-gebundenen Farbgenen des *Xiphophorus pygmaeus nigrensis* Rosen (Pisces). Mol Gen Genet 101 (1): 29–42

Zander CD (1969) Über die Entstehung und Veränderung von Farbmustern in der Gattung *Xiphophorus.* Mitt Hamburg Zool Mus Inst 66: 241–271

Zander CD (1974) Genetische Merkmalsanalyse als Hilfsmittel bei der Taxonomie der Zahnkarpfen-Gattung *Xiphophorus.* Z Zool Syst Evol-Forsch 13 (1): 63–78

Zelenin AM (1974) Growth pattern of scaled and mirror carps at different conditions of rearing. In: Biol Resour Moldavian Waterbodies (Kishinev) 12: 182–189 (R)

Zenkin VS (1969) Immunogenetic study in populations of spring and autumn Baltic herring. In: All-Union conference young scientists, PINPO, Murmansk, pp 87–99 (R)

Zenkin VS (1971) Immunogenetical studies of Baltic populations of herring. Rapp P-V Reun 161: 40–44

Zenkin VS (1972) Identity of "C" and "A" erythrocyte antigens in the Atlantic herring and the analysis of blood group distribution in herrings of Georges Bank. Tr VNIRO (Moscow) 85: 95–102 (R)

Zenkin VS (1973) Population analysis of the Atlantic herring *Clupea harengus harengus* L. by the frequency of blood groups. Vopr Ikhthiol 13 [5(82)]: 798–804 (R)

Zenkin VS (1974) Analysis of Baltic herring (*Clupea harengus membras* L.) populations by frequency of occurence of blood groups. Rapp P-V Reun 166: 124–125

Zenkin VS (1976) Biochemical studies on polymorphism of myogens and esterases in the Atlantic and the Baltic herring (*Clupea harengus harengus* and *C. h. membras*). Tr AtlantNIRO (Kaliningrad) 60: 111–116

Zenkin VS (1978) Intraspecific structure in herring (*Clupea harengus harengus* L.) of North Atlantic for blood groups and for allozymes of muscular esterases. Ph D Thesis, Moscow Univ (R)

Zenkin VS (1979) Biochemical polymorphism and population-genetical analysis of the Atlantic herring, *Clupea harengus* L. In: Biochem Popul Genet Fish. Inst Cytol Acad Sci USSR, Leningrad, pp 64–69 (RE)

Zenzes MT, Voiculescu I (1975) C-banding patterns in *Salmo trutta,* a species of tetraploid origin. Genetica (Holl) 45 (4): 531–536

Zhukov VV (1974) Antigenic relationships of several species of the genus *Coregonus* L. Vopr Ikhthiol 14 (4): 558–565 (R)

Ziuganov VV (1978) Factors determining morphological differentiation in *Gasterosteus aculeatus* (Pisces, Gasterosteidae). Zool Zh 57 (11): 1686–1694 (RE)

Ziuganov VV, Khlebovich VV (1979) Analysis of mechanisms which determine differences in the reaction of spermatozoa to water salinity in the threespined stickleback marine and freshwater forms. Ontogenez 10 (5): 506–509 (RE)

Zonova AS (1976) Certain results and problems in further selection of the Ropsha common carp. Izvestija Gosud. nauchno-issled. Inst Ozern Rechn Rybn Khos (GosNIORKh) 107: 18–24 (RE)

Zonova AS (1978) Experiment with mass selection of Ropsha common carp breed fry for growth rate. Izvestija Gosud. nauchno-issled. Inst Ozern Rechn Rybn Khos (GosNIORKh) 130: 70–83 (RE)

Zonova AS, Kirpichnikov VS (1971) The selection of Ropsha carp. In: Rep FAO/UNDP(TA), (2926), Rome, pp 233–247

Zonova AS, Ponomarenko KV (1978) Elaboration of the biological bases of common carp selection in warm waters. In: Increasing productivity of fishes by selection and hybridization. F Müller, Szarvas, pp 142–153

Subject Index

Index of Fish Names

K.H. Chadwick, H.P. Leenhouts

The Molecular Theory of Radiation Biology

1981. 236 figures, 2 tables. XVII, 377 pages
(Monographs on Theoretical and Applied Genetics, Volume 5)
ISBN 3-540-10297-3

The book presents a unique theoretical model derived to provide a comprehensive description of the effect of radiation on living cells. Based on the association between DNA double strand breaks and their biological effects, it permits a quantitative analysis of cell killing, chromosomal aberrations, somatic mutations, malignancy and hereditary defects. The known role of the DNA in the cell, its structure and ability to repair damage are included in the model. One chapter is devoted to an analysis of the interaction of other DNA damaging agents, such as chemical mutagens, with radiation. This extension indicates ways in which an analogous theoretical approach to chemical toxicity can be developed. The final chapter discusses the implications of the model for radiologic protection, chemical toxicology, radiation therapy and plant mutation breeding. The model clearly illustrates that the physics of radiation interaction, the chemistry of radiolysis products and the biochemistry and biology of DNA in the cell all interact to define the magnitude of radiobiologic damage.

Fish Diseases

IIIrd COPRAQ-Session
Editor: W. Ahne
1980. 120 figures. X, 252 pages
(Proceedings in Life Sciences)
ISBN 3-540-10406-2

The Third COPRAQ Fish Disease session drew together more than 100 scientists from all over the world to discuss the major problems affecting the health of cultured fish. The papers presented at this session pointed out that bacterial diseases such as vibriosis, furunculosis and carp erythrodermatitis are becoming the predominant threats to piscine health. They also show the effectiveness of immunization in preventing some fish diseases. With its information on recent research findings (viral, bacterial and parasitic diseases) and on emerging problems in and approaches to fish diseases, this volume will be of considerable interest to all scientists engaged in ichthyology and in ichthyopathology.

M.H.A. Keenleyside

Diversity and Adaption in Fish Behaviour

1979. 67 figures, 15 tables. XIII, 208 pages
(Zoophysiology, Volume 11)
ISBN 3-540-09587-X

The book investigates the behaviour of fish and the influence it has on their adaptation to a wide range of marine and freshwater habitats around the world. The first comprehensive work of its kind, it will be appreciated by ethologists, physiological ecologists, and fisheries biologists.

R. Rieger, A. Michaelis, M.M. Green

Glossary of Genetics and Cytogenetics

Classical and Molecular

Springer Study Edition
4th completely revised edition
1976. 100 figures, 8 tables. 647 pages
ISBN 3-540-07668-9

"... the Glossary is a most useful publication, and should be on the shelf of any library with coverage of genetics. It is the most comprehensive publication of its type available, and the fact that it has now gone to a 4th edition, and has been translated into Russian and Polish, is testimony to the value placed on it by libraries and individuals." *Animal Breeding Abstracts*

J. Schulz-Schaeffer

Cytogenetics

Plants, Animals, Humans

1980. 219 figures, 11 tables. XIII, 446 pages
ISBN 3-540-90467-0

Chromosomes, the physical cornerstones of genetic understanding, are entities present in every organism. Their universality is the basis for this unique exposition of the principles of cytogenetics, covering all forms of life from plants to humans beings.
Cytogenetics introduces the subject with examples from classical research and recent experimental results. The author traces the history of the field and makes frequent reference to its leading personalities in the course of the book. The unified approach is developed in the emphasis on the central role and basic similarities of chromosomes in eukaryotes, and in the treatment of the normal and variant structure, function and movement of chromosomes. Extrachromosomal inheritance is also covered. This book fills a long-standing need for a scholarly, advanced text on the undergraduate and graduate level. It will also be welcomed by post-graduate researchers seeking a comprehensive reference for the field.

Springer-Verlag
Berlin Heidelberg New York

MIX
Papier aus verantwortungsvollen Quellen
Paper from responsible sources
FSC® C105338

If you have any concerns about our products,
you can contact us on
ProductSafety@springernature.com

In case Publisher is established outside the EU,
the EU authorized representative is:
Springer Nature Customer Service Center GmbH
Europaplatz 3, 69115 Heidelberg, Germany

Printed by Libri Plureos GmbH
in Hamburg, Germany